METHODS IN MOLECULAR BIOLOGY

Series Editor
John M. Walker
School of Life and Medical Sciences
University of Hertfordshire
Hatfield, Hertfordshire, UK

For further volumes:
http://www.springer.com/series/7651

For over 35 years, biological scientists have come to rely on the research protocols and methodologies in the critically acclaimed *Methods in Molecular Biology* series. The series was the first to introduce the step-by-step protocols approach that has become the standard in all biomedical protocol publishing. Each protocol is provided in readily-reproducible step-by-step fashion, opening with an introductory overview, a list of the materials and reagents needed to complete the experiment, and followed by a detailed procedure that is supported with a helpful notes section offering tips and tricks of the trade as well as troubleshooting advice. These hallmark features were introduced by series editor Dr. John Walker and constitute the key ingredient in each and every volume of the *Methods in Molecular Biology* series. Tested and trusted, comprehensive and reliable, all protocols from the series are indexed in PubMed.

Quantitative Methods in Proteomics

Second Edition

Edited by

Katrin Marcus and Martin Eisenacher

*Medizinisches Proteom-Center (MPC), Medical Faculty, Ruhr-Universität Bochum, Bochum, Germany;
Medical Proteome Analysis, Center for Proteindiagnostics (PRODI), Ruhr-University Bochum, Bochum,
Germany*

Barbara Sitek

*Medizinisches Proteom-Center (MPC), Medical Faculty, Ruhr-Universität Bochum, Bochum, Germany;
Klinik für Anästhesiologie, Intensivmedizin und Schmerztherapie, Universitätsklinikum
Knappschaftskrankenhaus Bochum, Bochum, Germany; Medical Proteome Analysis,
Center for Proteindiagnostics (PRODI), Ruhr-University Bochum, Bochum, Germany*

Humana Press

Editors
Katrin Marcus
Medizinisches Proteom-Center (MPC)
Medical Faculty
Ruhr-Universität Bochum
Bochum, Germany

Medical Proteome Analysis
Center for Proteindiagnostics (PRODI)
Ruhr-University Bochum
Bochum, Germany

Martin Eisenacher
Medizinisches Proteom-Center (MPC)
Medical Faculty
Ruhr-Universität Bochum
Bochum, Germany

Medical Proteome Analysis
Center for Proteindiagnostics (PRODI)
Ruhr-University Bochum
Bochum, Germany

Barbara Sitek
Medizinisches Proteom-Center (MPC)
Medical Faculty
Ruhr-Universität Bochum
Bochum, Germany

Klinik für Anästhesiologie
Intensivmedizin und Schmerztherapie
Universitätsklinikum Knappschaftskrankenhaus Bochum
Bochum, Germany

Medical Proteome Analysis
Center for Proteindiagnostics (PRODI)
Ruhr-University Bochum
Bochum, Germany

ISSN 1064-3745   ISSN 1940-6029   (electronic)
Methods in Molecular Biology
ISBN 978-1-0716-1023-7   ISBN 978-1-0716-1024-4   (eBook)
https://doi.org/10.1007/978-1-0716-1024-4

Preface

The main approach and structure of the book have been retained in the second edition, but some methodical updates and new aspects in data analysis have been added.

Today, protein identification is an almost routine requirement. However, reliable techniques for quantifying unmodified proteins (including those that escape detection under standard conditions, such as protein isoforms and membrane proteins) as well as the detection and quantification of post-translational protein modifications is still far from being a matter of routine. Hence, there is a need for a profound understanding of the principles underlying modern protein analysis, in order to apply and improve established and novel methods successfully.

In the eight years, since the first edition of this book was published the data analysis and digitalization became increasingly important. Even in the public and media discussions, it is nowadays commonly accepted that digitalization is part of each and every area of life. This is of course true for science and has been lived reality in the mass spectrometry community since its beginning. It is obligatory that spectra are identified and quantified with algorithmic methods from computer science and that peptides/protein abundances are assessed for significance with methods from statistics.

This new edition gives a detailed survey of *Quantitative Methods in Proteomics*, addressing topics and methods from statistical issues when planning proteomics experiments (Chap. 1), new chapters dealing with protein quantification as a basis for realization of quantitative studies (Chaps. 2 and 3), gel-based (Chaps. 4–7), and mass spectrometry-based quantification techniques as TMT, IPTL, PRM, and MALDI Imaging (Chaps. 8–12). In this edition, the main meaning has been given to application of different methods as possibility for the readers to get a wide overview on how proteomics quantification can be used in diverse questions. In these chapters (Chaps. 13–26), different techniques (e.g., TMT, SILAC, PTM analysis, DIA, cross-linking) are applied for various kinds of samples (e.g., human, yeast, bacteria, cell culture samples).

The subsequent section is addressing the up-to-date topics of software and data analysis (Chaps. 27–31). As the peptide undersampling (low coverage) is one of the most important challenges of Proteomics, a specific chapter about missing value monitoring is dealing with this. Two chapters are dedicated to give a meaning to pure numbers: one about variance-sensitive clustering that helps to find probably unknown structure in the data and the other about network meta-analysis that helps to compare own data to previously published data although having another experimental group structure.

As the guest editors of this volume *Quantitative Methods in Proteomics*, we would like to thank all of the authors and coauthors for sharing their experience, knowledge, and time to make this new edition possible. We hope that the reader will take advantage of his/her research work from this comprehensive and competent overview of the important and still growing field of quantitative proteomics. Enjoy!

Bochum, Germany

Katrin Marcus
Martin Eisenacher
Barbara Sitek

Contents

Contributors

ANGELA BACHI • *IFOM, FIRC Institute of Molecular Oncology, Milan, Italy*

MARCUS BANTSCHEFF • *Cellzome GmbH, GlaxoSmithKline, Heidelberg, Germany*

KATALIN BARKOVITS • *Medizinisches Proteom-Center (MPC), Medical Faculty, Ruhr-University Bochum, Bochum, Germany; Medical Proteome Analysis, Center for Proteindiagnostics (PRODI), Ruhr-University Bochum, Bochum, Germany*

JÜRGEN BARTEL • *Department of Microbial Proteomics, Center for Functional Genomics of Microbes, Institute of Microbiology, University of Greifswald, Greifswald, Germany*

MARIE BARTH • *Interdisciplinary Research Center HALOmem, Charles Tanford Protein Center, Institute for Biochemistry and Biotechnology, Martin Luther University Halle-Wittenberg, Halle, Germany*

NATHAN BASISTY • *Buck Institute for Research on Aging, Novato, CA, USA*

DÖRTE BECHER • *Department of Microbial Proteomics, Center for Functional Genomics of Microbes, Institute of Microbiology, University of Greifswald, Greifswald, Germany*

CHRISTOPH H. BORCHERS • *Gerald Bronfman Department of Oncology, Jewish General Hospital, McGill University, Montreal, QC, Canada; Segal Cancer Proteomics Centre, Lady Davis Institute, Jewish General Hospital, McGill University, Montreal, QC, Canada; Center for Computational and Data-Intensive Science and Engineering, Skolkovo Institute of Science and Technology, Moscow, Russia*

THILO BRACHT • *Medizinisches Proteom-Center (MPC), Medical Faculty, Ruhr-Universität Bochum, Bochum, Germany; Medical Proteome Analysis, Center for Proteindiagnostics (PRODI), Ruhr-University Bochum, Bochum, Germany; Klinik für Anästhesiologie, Intensivmedizin und Schmerztherapie, Universitätsklinikum Knappschaftskrankenhaus Bochum, Bochum, Germany*

FREDERIC BROSSERON • *Medizinisches Proteom-Center (MPC), Medical Faculty, Ruhr-University Bochum, Bochum, Germany*

VICTOR CORASOLLA CARREGARI • *Lab of Neuroproteomics, Department of Biochemistry and Tissue Biology, Institute of Biology, University of Campinas (UNICAMP), Campinas, Brazil*

PIOTR CHARTOWSKI • *Medizinisches Proteom-Center (MPC), Medical Faculty, Ruhr-University Bochum, Bochum, Germany*

WEIQIANG CHEN • *Medizinisches Proteom-Center (MPC), Ruhr-Universität Bochum, Bochum, Germany; Medical Proteome Analysis, Center for Proteindiagnostics (PRODI), Ruhr-University Bochum, Bochum, Germany*

LENZ CHRISTOF • *Bioanalytical Mass Spectrometry Group, Max Planck Institute for Biophysical Chemistry, Goettingen, Germany; Bioanalytics Group, Institute of Clinical Chemistry, University Medical Center Goettingen, Goettingen, Germany*

MARK R. CONDINA • *Future Industries Institute, University of South Australia, Adelaide, Australia*

STEFAN DANNENMAIER • *Biochemistry and Functional Proteomics, Institute of Biology II, Faculty of Biology, University of Freiburg, Freiburg, Germany*

BROOKE A. DILMETZ • *Future Industries Institute, University of South Australia, Adelaide, Australia*

Ute Distler • *Institute of Immunology, University Medical Center of the Johannes Gutenberg-University Mainz, Mainz, Germany; Research Center for Immunotherapy (FZI), University Medical Center, Johannes Gutenberg-University Mainz, Mainz, Germany*

H. Christian Eberl • *Cellzome GmbH, GlaxoSmithKline, Heidelberg, Germany*

Britta Eggers • *Medizinisches Proteom-Center (MPC), Medical Faculty, Ruhr-University Bochum, Bochum, Germany; Medical Proteome Analysis, Center for Proteindiagnostics (PRODI), Ruhr-University Bochum, Bochum, Germany*

Martin Eisenacher • *Medizinisches Proteom-Center (MPC), Medical Faculty, Ruhr-University Bochum, Bochum, Germany; Medical Proteome Analysis, Center for Proteindiagnostics (PRODI), Ruhr-University Bochum, Bochum, Germany*

Pan Fang • *Bioanalytical Mass Spectrometry Group, Max Planck Institute for Biophysical Chemistry, Goettingen, Germany*

Kristin Fuchs • *Medizinisches Proteom-Center (MPC), Medical Faculty, Ruhr-Universität Bochum, Bochum, Germany; Medical Proteome Analysis, Center for Proteindiagnostics (PRODI), Ruhr-University Bochum, Bochum, Germany; Klinik für Anästhesiologie, Intensivmedizin und Schmerztherapie, Universitätsklinikum Knappschaftskrankenhaus Bochum, Bochum, Germany*

Klaus Gerwert • *Biospectroscopy, Center for Proteindiagnostics (PRODI), Ruhr-Universität Bochum, Bochum, Germany*

Frederik Großerueschkamp • *Biospectroscopy, Center for Proteindiagnostics (PRODI), Ruhr-Universität Bochum, Bochum, Germany*

Shubham Gupta • *Department of Molecular Genetics, University of Toronto, Toronto, ON, Canada; Donnelly Centre for Cellular and Biomolecular Research, University of Toronto, Toronto, ON, Canada*

Christina E. Hagensen • *Department of Biochemistry and Molecular Biology, University of Southern Denmark, Odense, Denmark*

Peter Hoffmann • *Future Industries Institute, University of South Australia, Adelaide, Australia*

Anja Holtz • *Buck Institute for Research on Aging, Novato, CA, USA*

Yanlong Ji • *Bioanalytical Mass Spectrometry Group, Max Planck Institute for Biophysical Chemistry, Goettingen, Germany; Hematology/Oncology, Department of Medicine II, Johann Wolfgang Goethe University, Frankfurt, Germany*

Klaus Jung • *Institute for Animal Breeding and Genetics, University of Veterinary Medicine Hannover, Foundation, Hannover, Germany*

Christian J. Koehler • *Department of Biosciences, University of Oslo, Oslo, Norway*

Michael Kohl • *Medizinisches Proteom-Center (MPC), Ruhr-Universität Bochum, Bochum, Germany; Medical Proteome Analysis, Center for Proteindiagnostics (PRODI), Ruhr-University Bochum, Bochum, Germany*

Laxmikanth Kollipara • *Leibniz-Institut für Analytische Wissenschaften—ISAS—e.V., Bunsen-Kirchhoff-Straße, Dortmund, Germany*

Cécile Lelong • *Chemistry and Biology of Metals, Univ. Grenoble-Alpes, CNRS UMR5249, IRIG-DIESE-CBM, CEA Grenoble, Grenoble Cedex 9, France*

Sandra Maaß • *Department of Microbial Proteomics, Center for Functional Genomics of Microbes, Institute of Microbiology, University of Greifswald, Greifswald, Germany*

Katrin Marcus • *Medizinisches Proteom-Center (MPC), Medical Faculty, Ruhr-University Bochum, Bochum, Germany; Medical Proteome Analysis, Center for Proteindiagnostics (PRODI), Ruhr-University Bochum, Bochum, Germany*

DANIEL MARTINS-DE-SOUZA • *Lab of Neuroproteomics, Department of Biochemistry and Tissue Biology, Institute of Biology, University of Campinas (UNICAMP), Campinas, Brazil; Instituto Nacional de Biomarcadores em Neuropsiquiatria (INBION), Conselho Nacional de Desenvolvimento Científico e Tecnológico, São Paulo, Brazil; Experimental Medicine Research Cluster (EMRC), University of Campinas, Campinas, Brazil; D'Or Institute for Research and Education (IDOR), São Paulo, Brazil*

VITTORIA MATAFORA • *IFOM, FIRC Institute of Molecular Oncology, Milan, Italy*

CAROLINE MAY • *Medizinisches Proteom-Center (MPC), Medical Faculty, Ruhr-University Bochum, Bochum, Germany; Medical Proteome Analysis, Center for Proteindiagnostics (PRODI), Ruhr-University Bochum, Bochum, Germany*

HELMUT E. MEYER • *Medizinisches Proteom-Center (MPC), Medical Faculty, Ruhr-University Bochum, Bochum, Germany*

ASHUTOSH MISHRA • *Center for Proteomics and Metabolomics, St. Jude Children's Research Hospital, Memphis, TN, USA*

MIROSLAV NIKOLOV • *Bioanalytical Mass Spectrometry Group, Max Planck Institute for Biophysical Chemistry, Goettingen, Germany*

SILKE OELJEKLAUS • *Biochemistry and Functional Proteomics, Institute of Biology II, Faculty of Biology, University of Freiburg, Freiburg, Germany; Signalling Research Centres BIOSS and CIBSS, University of Freiburg, Freiburg, Germany*

OLIVER PAGEL • *Leibniz-Institut für Analytische Wissenschaften—ISAS—e.V., Bunsen-Kirchhoff-Straße, Dortmund, Germany*

KUAN-TING PAN • *Bioanalytical Mass Spectrometry Group, Max Planck Institute for Biophysical Chemistry, Goettingen, Germany; Hematology/Oncology, Department of Medicine II, Johann Wolfgang Goethe University, Frankfurt, Germany; Frankfurt Cancer Institute, Goethe University, Frankfurt am Main, Germany*

JUNMIN PENG • *Departments of Structural Biology and Developmental Neurobiology, St. Jude Children's Research Hospital, Memphis, TN, USA; Center for Proteomics and Metabolomics, St. Jude Children's Research Hospital, Memphis, TN, USA*

ANDREW J. PERCY • *Cambridge Isotope Laboratories, Inc., Tewksbury, MA, USA*

KATHY PFEIFFER • *Medizinisches Proteom-Center (MPC), Medical Faculty, Ruhr-University Bochum, Bochum, Germany; Medical Proteome Analysis, Center for Proteindiagnostics (PRODI), Ruhr-University Bochum, Bochum, Germany*

KATHARINA PODWOJSKI • *Medizinisches Proteom-Center (MPC), Medical Faculty, Ruhr-University Bochum, Bochum, Germany; Bayer Schering Pharma AG, Wuppertal, Germany*

GEREON POSCHMANN • *Institute of Molecular Medicine I, Proteome Research, University Hospital Düsseldorf, Heinrich Heine University Düsseldorf, Düsseldorf, Germany*

NINA PRESCHER • *Institute of Molecular Medicine I, Proteome Research, University Hospital Düsseldorf, Heinrich Heine University Düsseldorf, Düsseldorf, Germany*

THIERRY RABILLOUD • *Chemistry and Biology of Metals, Univ. Grenoble-Alpes, CNRS UMR5249, IRIG-DIESE-CBM, CEA Grenoble, Grenoble Cedex 9, France*

GUILHERME REIS-DE-OLIVEIRA • *Lab of Neuroproteomics, Department of Biochemistry and Tissue Biology, Institute of Biology, University of Campinas (UNICAMP), Campinas, Brazil*

HANNES RÖST • *Department of Molecular Genetics, University of Toronto, Toronto, ON, Canada; Donnelly Centre for Cellular and Biomolecular Research, University of Toronto, Toronto, ON, Canada*

SVITLANA ROZANOVA • *Medizinisches Proteom-Center (MPC), Medical Faculty, Ruhr-University Bochum, Bochum, Germany; Medical Proteome Analysis, Center for Proteindiagnostics (PRODI), Ruhr-University Bochum, Bochum, Germany*

BIRGIT SCHILLING • *Buck Institute for Research on Aging, Novato, CA, USA*

CARLA SCHMIDT • *Interdisciplinary Research Center HALOmem, Charles Tanford Protein Center, Institute for Biochemistry and Biotechnology, Martin Luther University Halle-Wittenberg, Halle, Germany*

KARIN SCHORK • *Medizinisches Proteom-Center (MPC), Medical Faculty, Ruhr-University Bochum, Bochum, Germany; Medical Proteome Analysis, Center for Proteindiagnostics (PRODI), Ruhr-University Bochum, Bochum, Germany*

VEIT SCHWÄMMLE • *Department of Biochemistry and Molecular Biology, University of Southern Denmark, Odense, Denmark*

BETTINA SERSCHNITZKI • *Medizinisches Proteom-Center (MPC), Medical Faculty, Ruhr-University Bochum, Bochum, Germany; Medical Proteome Analysis, Center for Proteindiagnostics (PRODI), Ruhr-University Bochum, Bochum, Germany*

ALBERT SICKMANN • *Leibniz-Institut für Analytische Wissenschaften—ISAS—e.V., Bunsen-Kirchhoff-Straße, Dortmund, Germany; Department of Chemistry, College of Physical Sciences, University of Aberdeen, Aberdeen, UK; Medizinisches Proteom-Center (MPC), Medical Faculty, Ruhr-University Bochum, Bochum, Germany*

MALTE SIELAFF • *Institute of Immunology, University Medical Center of the Johannes Gutenberg-University Mainz, Mainz, Germany; Research Center for Immunotherapy (FZI), University Medical Center, Johannes Gutenberg-University Mainz, Mainz, Germany*

IVAN SILBERN • *Bioanalytical Mass Spectrometry Group, Max Planck Institute for Biophysical Chemistry, Goettingen, Germany; Bioanalytics Group, Institute of Clinical Chemistry, University Medical Center Goettingen, Goettingen, Germany*

BARBARA SITEK • *Medizinisches Proteom-Center (MPC), Medical Faculty, Ruhr-University Bochum, Bochum, Germany; Medical Proteome Analysis, Center for Proteindiagnostics (PRODI), Ruhr-University Bochum, Bochum, Germany; Klinik für Anästhesiologie, Intensivmedizin und Schmerztherapie, Universitätsklinikum Knappschaftskrankenhaus Bochum, Bochum, Germany*

MICHAEL STEIDEL • *Cellzome GmbH, GlaxoSmithKline, Heidelberg, Germany*

MARKUS STEPATH • *Medizinisches Proteom-Center (MPC), Medical Faculty, Ruhr-Universität Bochum, Bochum, Germany; Medical Proteome Analysis, Center for Proteindiagnostics (PRODI), Ruhr-University Bochum, Bochum, Germany*

CHRISTIAN STEPHAN • *KAIROS GmbH, Bochum, Germany; Medical Faculty, Ruhr-Universität Bochum, Bochum, Germany*

KAI STÜHLER • *Institute of Molecular Medicine I, Proteome Research, University Hospital Düsseldorf, Heinrich Heine University Düsseldorf, Düsseldorf, Germany*

HAIYAN TAN • *Center for Proteomics and Metabolomics, St. Jude Children's Research Hospital, Memphis, TN, USA*

STEFAN TENZER • *Institute of Immunology, University Medical Center of the Johannes Gutenberg-University Mainz, Mainz, Germany; Research Center for Immunotherapy (FZI), University Medical Center, Johannes Gutenberg-University Mainz, Mainz, Germany*

BERND THIEDE • *Department of Biosciences, University of Oslo, Oslo, Norway*

ANKE TRAUTWEIN-SCHULT • *Department of Microbial Proteomics, Center for Functional Genomics of Microbes, Institute of Microbiology, University of Greifswald, Greifswald, Germany*

MICHAEL TUREWICZ • *Medizinisches Proteom-Center (MPC), Medical Faculty, Ruhr-University Bochum, Bochum, Germany; Medical Proteome Analysis, Center for Proteindiagnostics (PRODI), Ruhr-University Bochum, Bochum, Germany*

HENNING URLAUB • *Bioanalytical Mass Spectrometry Group, Max Planck Institute for Biophysical Chemistry, Goettingen, Germany; Bioanalytics Group, Institute of Clinical Chemistry, University Medical Center Goettingen, Goettingen, Germany; Hematology/Oncology, Department of Medicine II, Johann Wolfgang Goethe University, Frankfurt, Germany*

JULIAN USZKOREIT • *Medizinisches Proteom-Center (MPC), Medical Faculty, Ruhr-University Bochum, Bochum, Germany; Medical Proteome Analysis, Center for Proteindiagnostics (PRODI), Ruhr-University Bochum, Bochum, Germany*

ZHEN WANG • *Departments of Structural Biology and Developmental Neurobiology, St. Jude Children's Research Hospital, Memphis, TN, USA*

BETTINA WARSCHEID • *Biochemistry and Functional Proteomics, Institute of Biology II, Faculty of Biology, University of Freiburg, Freiburg, Germany; Signalling Research Centres BIOSS and CIBSS, University of Freiburg, Freiburg, Germany*

THILO WERNER • *Cellzome GmbH, GlaxoSmithKline, Heidelberg, Germany*

CHRISTINE WINTER • *Institute for Animal Breeding and Genetics, University of Veterinary Medicine Hannover, Foundation, Hannover, Germany*

KATHRIN E. WITZKE • *Medizinisches Proteom-Center (MPC), Medical Faculty, Ruhr-University Bochum, Bochum, Germany; Medical Proteome Analysis, Center for Proteindiagnostics (PRODI), Ruhr-University Bochum, Bochum, Germany*

ZHIPING WU • *Departments of Structural Biology and Developmental Neurobiology, St. Jude Children's Research Hospital, Memphis, TN, USA*

KAIWEN YU • *Departments of Structural Biology and Developmental Neurobiology, St. Jude Children's Research Hospital, Memphis, TN, USA*

Chapter 1

Important Issues in Planning a Proteomics Experiment: Statistical Considerations of Quantitative Proteomic Data

Karin Schork, Katharina Podwojski, Michael Turewicz, Christian Stephan, and Martin Eisenacher

Abstract

Mass spectrometry is frequently used in quantitative proteomics to detect differentially regulated proteins. A very important but unfortunately oftentimes neglected part in detecting differential proteins is the statistical analysis. Data from proteomics experiments are usually high-dimensional and hence require profound statistical methods. It is especially important to already correctly design a proteomic experiment before it is conducted in the laboratory. Only this can ensure that the statistical analysis is capable of detecting truly differential proteins afterward. This chapter thus covers aspects of both statistical planning as well as the actual analysis of quantitative proteomic experiments.

Key words Experimental design, Data preprocessing, Normalization, Sample size calculation, Statistical hypothesis test, Multiple testing, Fold change, Volcano plot

1 Introduction

The development and improvement of experimental procedures and technical equipment for protein detection and quantification have resulted in increasing numbers of conducted experiments each yielding vast amounts of data. Mass spectrometry-based quantitative proteomics is frequently used for the search for biomarkers for many different diseases, e.g., different cancer types or neurological diseases, and different types of tissue or body fluids, e.g., blood, urine or cerebrospinal fluid [1–3]. Often, the aim is to be able to understand biological processes behind the diseases and offer personalized therapies in the future [4]. However, the strength of most proteomic methods of measuring hundreds or thousands of protein abundances in complex biological samples within one experiment is also the soft spot of these methods. The derived data are high-dimensional and thus require sophisticated statistical analyses in order to draw the right conclusions. Unfortunately, in mass

Katrin Marcus et al. (eds.), *Quantitative Methods in Proteomics*, Methods in Molecular Biology, vol. 2228,
https://doi.org/10.1007/978-1-0716-1024-4_1, © The Author(s) 2021

spectrometry proteomic experiments, often only little attention is given to correctly planning and performing the statistical analysis.

Typically, differential proteomic experiments compare several sample groups and aim at finding differences in the protein abundances between the groups. Statistical methods to deal with this kind of scenario and the amount of hypothesis test conducted were often originally developed for genomics data and can in most cases also be applied to proteomics data. However, it is important to consider statistical issues already during the planning phase of a proteomic experiment. Only this can ensure an unobstructed statistical analysis that has the power to detect truly differential proteins.

Issues and pitfalls in experimental design are detailed in the next section. This is followed by a short section on preprocessing of the derived data. While preprocessing is a very important issue to derive reliable measures of protein quantification, it is far too broad to be covered in detail in this section. Instead, emphasis is put on the actual statistical analysis based on the derived quantitative measures. In Subheading 4, basic statistical principles are explained. These comprise statistical testing, adjusting for multiple testing, as well as sample size planning. Finally, the actual applications of statistical analyses to proteomic experiments are detailed in Subheading 5.

2   Planning a Proteomic Experiment

Typically, scientists take great care in planning all experimental procedures to be used in their experiments. Every step of the workflow is assessed, and the different steps combined in such a way that they are capable of answering the scientific questions. Sometimes even pilot experiments are performed to optimize individual steps of the workflow. All in all, the experimenters usually spend a lot of time in planning and of course also in conducting their experiments.

Once the experiment is done and the scientists have returned from the lab and are back at their computers, a "quick" statistical evaluation is sought. But all too often, the experimenters will find out that the questions they posed are not answerable with the data obtained from the experiment. And very often the cause for the failure of the experiment will be inadequate or even completely missing considerations of statistical issues during the planning phase of the experiment [5, 6].

2.1 Experimental Design for Proteomic Experiments

Quantitative proteomic experiments usually aim at detecting differentially regulated proteins between different sample groups. In the simplest differential experiment, biological samples from two different groups, e.g., tumor tissue samples vs. healthy control samples, are compared. In this case, there is one experimental factor with two possible categories (i.e., tumor or healthy) that is studied. Of course, this example can be arbitrarily expanded both to a factor with more than two categories (i.e., samples from several tumor stages) and/or several experimental factors (i.e., gender, treatment with different substances, etc.) of interest. In the case of several experimental factors, there exist two basic types of designs. In a cross-classification design, each category of one factor is combined with each category of the other factor. In hierarchical designs, the possible categories of one factor depend on the category of the other factor. Thus not all possible combinations can be studied.

The first step in planning a proteomic experiment is thus the specification of all factors of interest and their possible categories that are to be studied within the experiment. It is of course also possible to incorporate continuous variables (i.e., age) into the design. Once all factors of interest are specified, several samples should be obtained per category of each factor. Using multiple samples ensures that detected differences are actually attributable to the studied factors and are not due to technical variation or intragroup biological variations. The necessary number of subjects per group can be assessed by sample size planning which is further detailed in Subheading 4.2.

It is however also possible that other factors apart from the ones of interest (e.g., factors coming from the experimental procedure) have an influence on the measured protein abundance. Problems arise especially, when a factor of interest is completely overlapping with such an uninteresting factor. Consider, for example, an experiment where the difference between two sample groups is measured. The samples have been processed in the laboratory in two batches on two different days. Now imagine the first batch only consisted of samples from the first experimental group, while the other batch only consisted of samples from the second group. If a differential protein is detected, one cannot tell whether the difference is due to the different group or because the samples have been processed on different days. In this case the time of processing is called a confounding factor. Typically all factors that contribute to sample handling may confound the results of an experiment. Also variables like age, gender, or additional underlying diseases may not be the factors of interest and have not been incorporated into the experimental design but still may confound the analysis.

A way to avoid confounding factors in an experiment is to already incorporate appropriate methods during experimental design. First of all, samples should be assigned randomly to the different categories of the factors to study. This should be done for all factors the researcher can influence (i.e., treatments or measurement order). This way, systematic errors from variables like age or gender can be diminished. Second, when different centers, instruments, batches, or the like are used within an experiment, samples from each category should be allocated ideally with equal sample sizes to each of the different batches [7]. This experimental design is called block design [8]. Finally, an experimental procedure or protocol should never be changed halfway through an experiment as this is an almost certain source of error. The same is valid for the technical equipment as a whole as well as the individual parts. For example, the change of the LC column can already have an enormous effect that possibly obscures any true biological effect.

2.2 Design Considerations in Labeled Mass Spectrometry Experiments

In label-free mass spectrometry experiments [9, 10], the aforementioned considerations often are sufficient for planning the experiment. However, in mass spectrometry using isotopic labeling, additional aspects have to be considered. Through the use of isotopic labeling, it is possible to measure two or more samples within the same mass spectrometry run [11]. In general, these techniques reduce variances between the measurements of samples in the same run, as the samples are exposed to the same experimental conditions after the pooling. On the other hand, experiments with many samples still need to be distributed over multiple runs.

The natural question is which samples should be measured together in one run. Generally, two cases should be discerned. First there is the case of paired samples. Paired samples are present if each sample from one group is connectable to exactly one sample from the other group. This is, for example, the case, when two samples are drawn from the same subject, e.g., before and after a treatment. Also, pairings between more than two sample groups are possible, e.g., in time-course experiments. In these cases, each sample pair or group should naturally be measured together within the same mass spectrometry run.

In the second case when samples are independent or unpaired, there are generally two possibilities to advance. The first possibility is to randomly pair up one sample from each group. For this approach it is however necessary to have equal sample sizes in each group.

The other possibility to measure unpaired samples with isotopic labeling is to incorporate an internal standard (also called master mix), which can be established by pooling one aliquot of each sample. In each mass spectrometry run, samples are measured together with this internal standard, which can later be used to standardize the different MS runs. This reduces technical variations and thus makes all the different samples well comparable, no matter in which run they were measured. This procedure can especially also be used in the case of unequal sample sizes. On the other hand, this may require more MS runs and also more labeling reagents.

Another important aspect to consider is the possible effect of different isotopic labels on the measured protein abundances. By incorporating a label swapping strategy in the experimental design, measuring one experimental group always with the same label is avoided. This prevents bias that may be introduced by the different labels [12, 13].

3 Data Preprocessing

Suppose the proteomic experiment has been planned and finally conducted in the laboratory. The obtained raw results from mass spectrometry are available in the form of binary or text files representing the corresponding mass spectra. However, a direct measure for each peptide or protein is not readily available. To derive such measures, several steps of preprocessing have to be performed that connect the raw data to the different biological entities like peptides or proteins. This, for example, includes spectrum preprocessing, peptide identification, and protein inference and quantification. The preprocessing steps generally comprise methods from computer science or information technology but also from statistics. As there is a multitude of necessary steps and a great variety of suitable methods for each step, the reader is referred to corresponding reviews on this matter for more information [14–17]. Both commercial and free software solutions are available for each step or as complete workflows [18–21]. At this point, it is assumed that a suitable software solution is used for the derivation of quantitative values for peptides or proteins. In the following paragraphs, preprocessing steps for a quantitative proteomics data set are discussed, which are an essential preparation for the following statistical analysis.

3.1 Missing Values

In quantitative proteomics datasets, missing values (*see* Chap. 27) are very common and can make up a large fraction of the whole dataset. The way missing values are handled can have a huge impact on the results of the following analysis.

A data point can be missing for various reasons. For example, a missing value will occur if the corresponding protein is not present in the sample or was present, but its abundance was below the detection limit. Other reasons are based on the nature of mass spectrometry measurement, e.g., the fact that the selection of precursor ions for fragmentation is stochastic or that noisy MS/MS spectra may not be identified with high enough confidence. But also additional filters of the quantitative data, as requiring at least two unique peptides, or normalization steps may lead to missing values. These different missing value types may require different treatment, but unfortunately, often for a certain missing value in the data, it is not clear why exactly it is missing.

There are three different methods to handle missing values: (1) remove proteins with missing values, (2) perform analysis only on valid values, and (3) impute missing values. Removing all proteins that contain missing values can lead to a substantial information loss, while replacing missing values with a valid value (missing value imputation) has also a lot of drawbacks, especially in the case of many missing values or unknown origin of them [22]. For instance, imputing a constant value (e.g., the mean or median of the valid values) for a protein will lead to an underestimated variance that can easily lead to false-positive findings.

Often the strategy is to first remove proteins with too few valid values (e.g., <50%) and then continue either with the remaining valid values or choose an appropriate imputation strategy.

3.2 Normalization

Quantitative proteomics data contains biological variation that is interesting to investigate but also unwanted technical variation. Even if possible confounding factors where considered during experiment planning (*see* Subheading 2.1), there will still be small variations in experimental conditions and sample handling (e.g., temperature, pipetting), which lead to technical bias. Often, the exact reasons for this bias are unknown.

Normalization strategies reduce the technical bias while keeping the interesting biological differences. After normalization, the samples are in general more comparable which makes the following statistical analysis more reliable. In general, high-throughput data and the assumption that the majority of proteins do not change between the experimental groups are required (*see* Subheading 6.1).

There are many different normalization methods that were originally developed for genomics data or especially tailored at proteomics. Although there are several comparisons of these methods on proteomics data [23–25], in most cases it is not clear beforehand which method will work best for a certain dataset. Because of this, it is advisable to try different normalization methods and evaluate those using different types of plots [26].

Boxplots can give an overview over the whole dataset and are suitable to compare non-normalized with normalized data. The length of the boxes can show differences in variance between samples and may indicate outliers that stick out even after normalization.

An MA plot compares two samples with each other and helps to assess whether these samples have been normalized appropriately. On the x-axis the average of the $\log_2$-transformed intensities (A-value) and on the y-axis the difference (M-value) is shown. Low-abundant proteins will show up on the left side and high-abundant proteins on the right side. Proteins with a high difference between the two samples are shown on the top and the bottom. Often, a local linear regression (loess) curve is drawn into the MA plot. For well-normalized data, this line will be very close to the horizontal line at $M = 0$ (*see* Fig. 1). Deviations from this line might indicate unnormalized data or a normalization method that does not fit to the data (e.g., when median normalization is not able to adjust differences in variance).

Even if the experiment is carefully planned, batch effect sometimes cannot be completely avoided. A principal component analysis plot (PCA plot) can uncover possible batch effects by

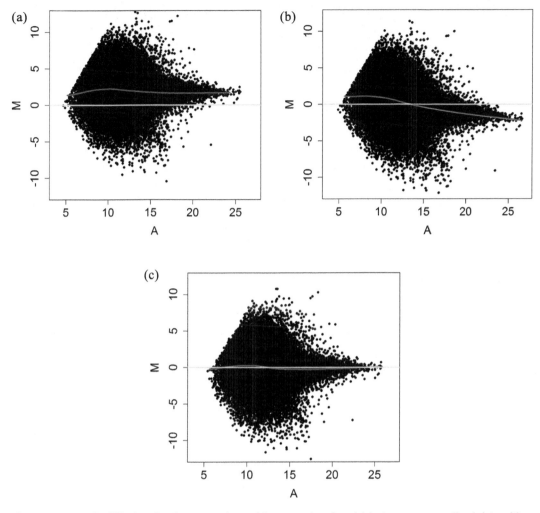

Fig. 1 Examples for MA plots for the comparison of two samples. Panel (**a**) shows unnormalized data with a high bias between the two samples. The normalization used for (**b**) is not ideal as the MA plot still shows a deviation of the local regression line from $M = 0$. Panel (**c**) shows a suitable normalization method where the regression line is almost equal to $M = 0$

representing the whole dataset in a two-dimensional plot. As batch effects are sometimes not completely removed by standard normalization methods, special batch normalization methods have been proposed [27].

4   Basic Statistical Concepts for Difference Detection

When performing quantitative proteomics experiments, one naturally wants to measure the abundances of the individual proteins within the samples. However, the calculated peptide or protein intensities are not absolute abundance measures in a standard proteomics experiment [28]. This means that one cannot directly confer the absolute amount of each protein (either molecule number or volume). This is because the different peptides do not ionize equally well (i.e., they have different ionization efficiencies) and the factor by which their intensities are reduced differs from peptide to peptide.

What is however possible is to compare the abundances in one sample to those of another sample. This is also called *relative quantification* (*see* Chaps. 9, 10, 13–21, 23, 24, 26).

Through this procedure it is especially possible to quantify differences in protein expression between different sample groups. Typically, the abundance ratio or the fold change is used to quantify the relative expression change between two samples or sample groups. Basically these measures compare the mean expression in one sample group to that of the other sample group. The exact formulae are given in Subheading 6.2.

4.1 Statistical Hypothesis Tests

The mere quantification of the magnitude in abundance change between different groups by abundance ratio or fold change is only the first step toward identifying differentially regulated proteins. Declaring a protein differentially expressed if the corresponding ratio is different from one is the solution that first comes to mind. However, this strategy neglects some of the typical characteristics of experiments in general.

First, every experimental procedure has its typical variability and imprecision resulting in measurements that slightly deviate from the true values. Second, an experiment normally only uses a small subset of subjects from the whole population of interest.

There are also always variations of the measured values (e.g., protein abundances) between different subjects of a population. In summary, there will always be deviations between the observed values of the sample subset and the true mean value of the complete population of interest. All these variations result almost certainly in observed abundance ratios differing from one for every single measured protein even though most of the proteins in truth are not differentially expressed between the different groups. To detect truly differential proteins, one will thus have to distinguish proteins with ratios truly different from one from those proteins where the measured difference is only due to experimental imprecision.

To solve this problem, a statistical test can be performed. A statistical test is based on hypotheses about the characteristics of both the populations. The *null hypothesis* usually describes the state one would like to rebut. In differential proteomics this would be "the protein *is not* differentially expressed." The *alternative hypothesis* states the opposing characteristic, i.e., "the protein *is* differentially expressed." To assess which of the two hypotheses holds true, a test statistic based upon the observed measurements is calculated. The derived score of this test statistic is then used to decide which hypothesis is to be chosen.

Table 1
Possible results from statistical tests and measures for assessing a statistical testing procedure

		Test result		
		Protein differential	Protein *not* differential	Measures
Unknown reality	Protein differential	True Positive	False Negative (type II error (β))	Sensitivity = TP/ (TP + FN)
	Protein *not* differential	False Positive (type I error (α))	True Negative	Specificity = TN/ (TN + FP)

The procedure for deciding which of the hypotheses is true is comparable to a criminal proceeding where the accused is presumed to be innocent until the contrary can be proven. In the context of statistical testing, the null hypothesis is assumed to be true unless it can be proven to be false. To this end, assumptions are made about the distribution of the test statistic under the null hypothesis. The distribution particularly takes the variability of the data into account. If the obtained value of the test statistic is too extreme in terms of the distribution under the null hypothesis, then the null hypothesis is rejected in favor of the alternative. The test result is then called *significant*. Otherwise there is not enough evidence, and the null hypothesis has to be retained.

Oftentimes the so called *p-value* is calculated to assess whether a test is significant. The *p*-value is the probability of obtaining a test statistic at least as extreme as the calculated one under the assumption that the null hypothesis is true. If the *p*-value is below or equal to the pre-specified α-*level*, the null hypothesis can be rejected.

As the test decision is based on probabilities, it is possible to decide erroneously. All possible test decisions are outlined in Table 1 together with the two types of errors that may occur depending on the decision made. The α-error occurs if the null hypothesis is rejected even though it is true. The β-error is present if the null hypothesis is retained even though it is false. Unfortunately, the two possible errors are conflictive. If one of the error types is decreased and all other experimental characteristics are kept unchanged, the other error type will automatically increase. Returning to the court example, the presumption of innocence is based on the concept that it is generally deemed worse to convict someone innocent than failing to convict someone actually guilty. In the context of statistical testing, it is worse to declare a difference where in fact there is none than to miss one true difference. In accordance with this concept, the probability for the type I error is controlled through specifying the α-level of the test to ensure that the probability for wrongly rejecting the null hypothesis is small. A typically chosen α-level in statistical testing is 5%. If one wants to be very restrictive, this can be decreased further, e.g., to 1%. In any case, the α-level has to be specified before the performance of the experiment and its statistical evaluation.

4.2 Power and Sample Size

So far, the importance of keeping the type I error small has been shown. However, a researcher will generally also want to make sure to be able to detect a true difference.

The probability of rejecting the null hypothesis in favor of the alternative hypothesis when the alternative actually is true is called the *power* of the test. It is the opposite of the probability of a type II error (power $= 1 - \beta$). The power depends on the true effect size, the variance, and the sample size. A large effect size, a small variance, and a high sample size will increase the power. Additionally, the power also depends on the significance niveau (α-level). Once the α-level has been fixed, the only remaining aspect a researcher can usually influence is the sample size of the experiment.

By increasing the sample size, it is possible to increase the power of a test. The relationship between power and sample size under several scenarios for effect size and variance is depicted in Fig. 2 for the special case of the two-sample *t*-test. It shows that the sample size will have to be increased with decreasing true effect size to obtain a certain power of the test. Also, for increasing variance the sample size will have to be increased to keep a certain power of the test.

As all the described characteristics are interdependent, it is necessary to specify all except one beforehand to be able to calculate the value of the remaining one. It is hence necessary to know and

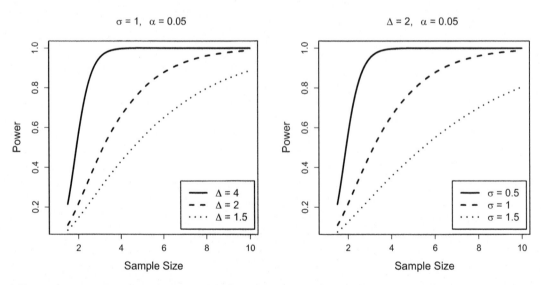

Fig. 2 Relationship of power and sample size for the two-sample *t*-test. *Left:* Fixed *α-level* and standard deviation *σ* with three different choices for the true effect size. *Right:* Fixed *α-level* and effect size Δ with three different choices for the standard deviation

specify the expected variance and the true effect size one desires to be able to detect before the experiment (*see* Subheading 6.3). The expectable true effect size depends on the samples used in the experiment and especially on how different the studied groups are. It is generally necessary to make a compromise between the possibility to detect very small expression changes and keeping the sample size reasonable.

However, one should try to specify reasonable expectable effect sizes one would like to detect, because an overestimation of expectable effect sizes can easily impede a complete experiment. Then the detectable effect size will be too large and reasonable, but smaller differences are detectable any more.

Specifying the expectable variance is even more difficult than specifying the effect size. Information about the variance are generally only be accessible through earlier experiments or from the literature. In this case it is necessary to make sure the experimental procedure to be used will be (almost) identical to the one described in the literature. Only this ensures that the variances are transferable. If no information is found in the literature, one could perform a small pilot experiment beforehand and estimate the variance from this experiment [29].

Finally, the anticipated power to detect a differential protein needs to be specified. The power is generally not set as strict as the α-level of a test. Typically, 80% are used in many statistical applications. Dropping the power much further would decrease the probability of detecting a true expression change accordingly.

4.3 Multiple Testing   Especially when measuring complex biological samples like blood or tissue, up to several hundreds or even thousands of peptides or proteins are measured simultaneously. Typically, a separate statistical test is performed for each protein to evaluate, if it is differentially expressed. For each of these tests, there is the possibility of a false test decision. The expected number of false-positive test decisions increases linearly with the number of tests performed (Table 2). And the probability for at least one false-positive test decision increases even more dramatically and converges to a 100% very fast.

Especially when considering the time and effort generally put into downstream validation of each single detected protein, one would like to keep the number of false positives to a minimum and avoid unnecessary work spent on truly unregulated proteins. But also when no further work is performed in the lab, a researcher will make sure that the published results do indeed reflect truly regulated proteins. These circumstances make it obvious that the number of false positives has to be controlled.

The typical way to control the number of false-positive test decisions in a multiple testing setting is to apply multiple testing correction methods that control certain error rates [30], the family-wise error rate (FWER) or the false discovery rate (FDR).

Table 2
Relation between false-positive test decisions and number of tests performed

Number of tests performed	Probability for at least one false-positive test decision ($\alpha = 0.05$)	Expected number of false-positive test decisions ($\alpha = 0.05$)
1	5.00%	≈ 0
2	9.75%	≈ 0
5	22.62%	≈ 0
10	40.13%	0–1
20	64.15%	1
100	99.41%	5
1000	100.00%	50

The FWER is the probability of having at least one false-positive test decision among all test decisions. The so-called Bonferroni correction adjusts the p-values so that the FWER is controlled by multiplying the original p-values with the number of performed tests. An adjusted p-value thus reflects the probability of having at least one false-positive test result. The formula for the adjusted p-values is given in Subheading 6.4. However, this method is very conservative as it basically allows *no* false-positive test decision among all tests.

The FDR is defined as the fraction of false-positive test decisions among all positive test decisions. Thus, one can allow several false-positive tests, given that there are enough true positive results. Because false positives are allowed to a certain degree, controlling the FDR generally is less strict. Several algorithms have been developed that control the FDR under different assumptions. Especially, methods for deriving adjusted p-values (called q-values) have been introduced. A q-value directly reflects the FDR, as it gives the fraction of false positives among all positives under the condition that the corresponding protein is still considered significantly regulated. A formula for deriving q-values is given in Subheading 6.4.

4.4 Statistical Significance and Biological Relevance

In proteomics literature often a *fold change* (FC, ratio of means of the two compared groups) cutoff is used in addition to the p-value. From a statistical point of view, this is not necessary. The p-value already contains the complete information if a protein is significantly regulated. This means, if the p-value is small enough (i.e., less than 0.05), then there is statistical evidence that the mean abundance of a protein between the experimental groups is different. However, from a biological point of view, even though a difference is significant, it might not be relevant, because it is too small.

Classically the terms *statistical significance* and *biological relevance* are used to characterize these different concepts. The fold change cutoff is a means of introducing this latter concept of biological relevance.

Now the question remains of how to choose such a fold change cutoff. Even though a fold change cutoff is regularly used, its derivation is rarely described in the proteomics literature. To deduce the cutoff, one might try to consider what is a biologically relevant difference in mean abundance. Opposing to the connotation of "biological," the relevance of a difference in the context of proteomic experiments is not connected to biological processes that might be controlled by it. Rather, this concept takes into consideration the technical circumstances of the experiment. Each experimental procedure has an intrinsic variability or inaccuracy in measuring outcomes. This is the so-called *technical variation* of the experiment. In proteomics experiments such variation is introduced, for example, through sample handling and through the technical equipment like LC systems or mass spectrometer. Due to this technical variation, different outcomes are observed when measuring the same sample several times. When considering this variability, it might be possible that, even though significant, a difference in mean abundance is below the typical technical variation of the experiment. Then it cannot be assessed, whether the difference found is in fact due to biological difference or just due to the experimental procedure. Biologically relevant is thus a result which lies above the technical variation of the experimental procedure, and a fold change cutoff should be chosen accordingly.

The technical variation of one's experimental procedure is dependent on the sample type, handling, and equipment used. It can, for example, be estimated through technical replicates. Details on this estimation are given in Subheading 6.5. Additionally, the fold change may be used as a second criterion to reduce the list of significant proteins for further validation. By choosing a high enough fold change cutoff, it is also ensured that abundance differences are visible, e.g., in Western blots, and that are often used for validating biomarker candidates with another experimental method.

P-values and fold changes can be simultaneously depicted in a so-called *volcano plot* (Fig. 3). On the *x*-axis the $\log_2(\text{FC})$ and on the *y*-axis the $-\log_{10}(p - \text{value})$ are shown. Cutoffs for fold changes and *p*-values can also be added. The most interesting biomarker candidates (those with a low *p*-value and a fold change away from one) will show up in the left and right upper corner of the volcano plot. It is also possible to combine fold changes and *p*-values into a single measure (*see* Subheading 6.6).

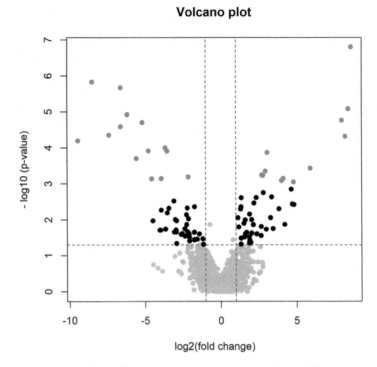

Fig. 3 Example of a volcano plot combining information of biological and statistical relevance. Cutoffs of 0.05 for the p-value and 0.5 resp. 2 for the fold change are applied. Proteins that do not reach these cutoffs are depicted in gray. Significantly regulated proteins after FDR correction using the Benjamini-Hochberg procedure are highlighted in orange

5   Statistical Tests in Proteomic Experiments

As described above, a statistical test suitable to detect the differences between different sample groups has to be applied once the proteomic experiment has been performed, data are preprocessed, and abundance measures are obtained for each peptide or protein. Standard statistical tests generally can be used for difference detection in proteomics. Which test is suitable is directly dependent on the design of the conducted experiment.

5.1 Comparing Two Sample Groups

Commonly, proteomic experiments are performed to find differences in the proteome between two different sample groups. The t-test is used most often in the literature to detect differential proteins. The t-test is the traditional parametric test for difference between two sample means. However, the t-test requires normally distributed measurements, which may not be the case with proteomic measurements. Abundance measurements based on either peak areas or heights can only take positive values, and their distribution is often skewed, i.e., they do not follow a normal distribution.

However, log-transformed abundance values will approximately be normally distributed. In any case, the normality assumption should be checked before the t-test is applied. This can be done visually through QQ plots, for example. For these plots, the theoretical quantiles of the standard normal distribution are plotted against the measured quantiles. If the points lie approximately on a straight line, then normal distribution can be assumed.

Once the conditions for the application of the t-test have been verified, there are further things to consider. Data analysis software usually offers several options for the t-test. The first option is to either use a *paired* or an *unpaired* t-test. This decision depends on the experimental design. In the case of unpaired samples, mean values are calculated separately for each group. The means are afterward compared to detect differential proteins. In the paired samples setting, the difference for each pair of samples is calculated first, and afterward the mean of these differences is calculated and compared to zero (i.e., no difference between groups).

The second selectable option is to use a *one-sided* vs. a *two-sided* alternative hypothesis. A two-sided alternative is equivalent to testing if there is any change in abundance between the groups regardless of the direction. This test is thus able to detect both up- and downregulation. If it can be assumed that the experiment will result in *only* down- or *only* upregulated proteins, the one-sided test may be chosen as it has a higher power in the specified direction. However, it is never able to find any changes in the opposite direction (that may still occur unexpectedly).

If the data is not normally distributed (even after log transformation), there are a number of nonparametric alternatives to the t-test that do not rely on normally distributed data. Especially the two-sample Kolmogorov-Smirnov test (K-S-test) or the Mann-Whitney-U test can be used for difference detection when the data is not normal.

5.2 Analysis of Multiple Sample Groups and Additional Factors

Of course, it is also possible that more than two sample groups are to be compared in a proteomic experiment. A typical example is the comparison of samples from different tumor stages. But also more advanced experimental designs where additional factors are incorporated into the analysis are possible and have been introduced in Sect. 2. In these cases analysis of variance (ANOVA) and related methods can be used. The basic idea behind ANOVA is to study the effect of each incorporated factor on the peptide or protein abundance. This is done by allocating portions of the overall variance to the different factors and performing a statistical test for each factor based on the apportioned variances. Each test evaluates if the corresponding factor has a significant influence on the abundance level. If there is a significant influence, the abundance will differ for different levels the factor takes on. In the case of comparing different tumor stages, for example, a significant influence of the factor "tumor stage" means that the abundance is different between at least one pair of tumor stages. It is also possible to assess interactions between different factors with ANOVA methods. When an interaction is present, the influence of one factor is different for the different levels of the interacting factor.

Similarly to the *t*-test, classical linear models and ANOVA methods generally assume normally distributed values. Thus the log transformation is again advisable. Aside from that, ANOVA methods are very flexible. Hence they are applicable to a multitude of different experimental designs with complicated factor combinations. However, the reader is strongly recommended to seek advice from a statistician *before* performing such complicated experimental designs. Good knowledge of the analysis method is needed to ensure the factors of interest are really assessable by ANOVA. This can only be guaranteed by careful planning of such sophisticated experiments with several factors and possibly even interactions.

Once a factor with more than two levels is found to have a significant influence, the natural question is which pairs of levels actually are differential. This can be assessed by additionally performing a test for each possible pair of factor levels. There exists special post hoc tests like Tukey's honestly significant difference (HSD) test, which is based on the standard *t*-test. Additionally, a multiple testing correction is incorporated into the post hoc test by taking into account the total number of pairwise tests performed. This way the α-level is conserved across all tests.

6   Notes

6.1   Normalization in Case of Many Changing Proteins

For some datasets the assumption for normalization that the majority of proteins do not change between experimental groups may not hold. It may come to mind to do a groupwise normalization, i.e., normalizing every group separately and combining the data for the analysis. This type of normalization is dangerous and not recommended, as it can introduce an artificial bias that can make proteins look significantly differentially expressed if there are not and increase the number of false positives. Alternatively, there exists normalization methods that can cope with a larger portion of changing proteins between groups than standard normalization methods. An example is the least trimmed squares (LTS) normalization [31].

6.2   Fold Change

Consider the expression of a protein is to be compared between two sample groups. The first group consists of n samples and the second group of m samples. Let x_i be the protein abundance measured in the ith sample in the first group and let y_j be the protein abundance in the jth sample in the second group. If the samples in one group are independent from the samples in the other group (unpaired samples), the fold change is calculated as

$$FC_{\text{unpaired}} = \frac{1}{n} \sum_{i=1}^{n} x_i \bigg/ \frac{1}{m} \sum_{j=1}^{m} y_j = \frac{\bar{x}}{\bar{y}}$$

In the case of paired samples, e.g., x_i and y_i originate from the same sample, or they have been measured in the same LC–MS run in a labeled experiment, the fold change is defined as

$$FC_{\text{paired}} = \frac{1}{n} \sum_{i=1}^{n} \frac{x_i}{y_i} = \overline{\left(x/y \right)}$$

Note that with paired samples, the sample size has to be the same in both groups, i.e., $n = m$.

6.3 Expected Effect Size and Variance for Choosing the Sample Size

To specify a correct effect size and variance for the determination of the optimal sample size, several things have to be considered. In the classical scenario where protein abundances from two different sample groups are compared, usually the two-sample t-test is used. The true effect size in this case is a difference based on (the logarithm of) the true abundances of the two groups (*see* also Subheading 5.1):

$$\Delta = \mu_{\log(x)} - \mu_{\log(y)}$$

This true effect size Δ can be estimated through the arithmetic means of the measured log-transformed intensities:

$$d = \overline{\log(x)} - \overline{\log(y)}$$

This difference is *not* identical with a fold change. Thus, while using the fold change as a measure for the size of abundance change between the two groups, the specifiable effect size for the determination of the correct sample size needs to be the above difference. Keep in mind that the variance also has to be specified based on the logarithmized scale.

Another problem to consider is that there are usually hundreds of proteins to be tested within one proteomic experiment. Each protein will have its own effect and variance. However, only one value can be used in the power analysis to derive the correct sample size. When a pilot experiment is used for the derivation, one can, for example, take the mean, median, or a specific quantile of all the observed variances. Generally one should keep in mind that it is reasonable to be conservative (i.e., overestimate the true variance). This will result in slightly higher sample sizes but at the same time ensure a true effect of the specified size is actually detectable with the conducted experiment.

6.4 Adjusted p-Values

Assume a differential proteomic experiment has been performed where n proteins have been quantified and tested for differential expression. For each test i, a p-value p_i has been derived. Adjusted p-values that control the FWER are derived by applying the following procedure of Bonferroni to each p-value:

$$p_i^* = \min\left(n \cdot p_i, 1\right)$$

A protein is then assumed to be differentially expressed, if the corresponding adjusted p-value p_i^* is below a pre-specified α. This procedure ensures that the probability for at least one false-positive test decision among all n test decisions is less or equal to α.

An alternative and less strict multiple testing correction is the control of the FDR. In the case of independent test statistics, the procedure by Benjamini and Hochberg [32] uses the n ordered p-values, $p_{(1)} \leq p_{(2)} \leq \ldots \leq p_{(n)}$ to derive adjusted q-values through

$$q_{(i)} = \min_{k=i,\ldots,n}\left(\min\left(\frac{n}{k}\, p_{(k)}, 1\right)\right)$$

Again, a protein is assumed to be differentially expressed, if the corresponding q-value is below a pre-specified α. The expected fraction of false-positive test decisions among all positive test decisions then is less or equal to α.

6.5 Fold Change Cutoff

To estimate the technical variation of a procedure, several technical replicates should be measured, for example, within a pilot experiment. The easiest way to assess the variation is to use the coefficient of variation. The coefficient of variation is a measure of the percentage variation and is defined as

$$CV = \frac{\text{standard deviation}}{\text{mean}}$$

To derive a fold change cutoff for a proteomic experiment, compute the coefficient of variation for each protein. Then take the maximum or, for example, the 95% quantile of all coefficients. Finally set the fold change cutoff for positive fold changes to

$$cut = \frac{1 + \widetilde{CV}}{1}$$

and to $-cut$ for negative fold changes, respectively. $\widetilde{CV}$ is the chosen maximum or quantile of the measured coefficients from all proteins.

6.6 Euclidean Distance Measure in Volcano Plots

Fold changes and p-values can be combined into a single measure using the volcano plot. The Euclidean distance measures the distance of a point in the volcano plot to the origin of the coordinate system [33]. The higher the Euclidean distance, the more interesting is the corresponding protein. However, it has to be taken care that the x- and y-axis of the volcano plot are comparable and the plot is close to quadratic. If not, either fold changes or p-values will gain a higher weight. Adjustment of the volcano plot can be achieved by changing the base for the logarithm of either p-values or fold changes.

Acknowledgments

This work was supported by de.NBI, a project of the German Federal Ministry of Education and Research (BMBF) [grant number FKZ 031 A534A], as well as P.U.R.E. (Protein Research Unit Ruhr within Europe) and ProDi (Center for protein diagnostics), Ministry of Innovation, Science and Research of North-Rhine Westphalia, Germany.

References

1. Crutchfield CA, Thomas SN, Sokoll LJ et al (2016) Advances in mass spectrometry-based clinical biomarker discovery. Clin Proteomics 13:1

2. Thomas S, Hao L, Ricke WA et al (2016) Biomarker discovery in mass spectrometry-based urinary proteomics. Proteomics Clin Appl 10:358–370

3. Bharucha T, Gangadharan B, Kumar A et al (2019) Mass spectrometry-based proteomic techniques to identify cerebrospinal fluid biomarkers for diagnosing suspected central nervous system infections. A systematic review. J Infect 79:407–418

4. Kowalczyk T, Ciborowski M, Kisluk J et al (2020) Mass spectrometry based proteomics and metabolomics in personalized oncology. Biochim Biophys Acta Mol Basis Dis 1866:165,690

5. Hu J, Coombes KR, Morris JS et al (2005) The importance of experimental design in proteomic mass spectrometry experiments: some cautionary tales. Br Funct Genomic Proteomic 3:322–331

6. Karp NA, Lilley KS (2007) Design and analysis issues in quantitative proteomics studies. Proteomics 7(Suppl 1):42–50

7. Cairns DA (2011) Statistical issues in quality control of proteomic analyses: good experimental design and planning. Proteomics 11:1037–1048

8. Tocher KD (1952) The design and analysis of block experiments. J R Stat Soc Ser B 14:45–91

9. Neilson KA, Ali NA, Muralidharan S et al (2011) Less label, more free: approaches in label-free quantitative mass spectrometry. Proteomics 11:535–553

10. Megger DA, Bracht T, Meyer HE et al (2013) Label-free quantification in clinical proteomics. Biochim Biophys Acta 1834:1581–1590

11. Chahrour O, Cobice D, Malone J (2015) Stable isotope labelling methods in mass spectrometry-based quantitative proteomics. J Pharm Biomed Anal 113:2–20

12. Park S-S, Wu WW, Zhou Y et al (2012) Effective correction of experimental errors in quantitative proteomics using stable isotope labeling by amino acids in cell culture (SILAC). J Proteome 75:3720–3732

13. Maes E, Valkenborg D, Baggerman G et al (2015) Determination of variation parameters as a crucial step in designing TMT-based clinical proteomics experiments. PLoS One 10: e0120115

14. Shteynberg D, Nesvizhskii AI, Moritz RL et al (2013) Combining results of multiple search engines in proteomics. Mol Cell Proteomics 12:2383–2393

15. Griss J (2016) Spectral library searching in proteomics. Proteomics 16:729–740

16. Audain E, Uszkoreit J, Sachsenberg T et al (2017) In-depth analysis of protein inference algorithms using multiple search engines and well-defined metrics. J Proteome 150:170–182

17. Blein-Nicolas M, Zivy M (2016) Thousand and one ways to quantify and compare protein abundances in label-free bottom-up proteomics. Biochim Biophys Acta 1864:883–895

18. Sinitcyn P, Rudolph JD, Cox J (2018) Computational methods for understanding mass spectrometry–based shotgun proteomics data. Annu Rev Biomed Data Sci 1:207–234

19. Navarro P, Kuharev J, Gillet LC et al (2016) A multicenter study benchmarks software tools

for label-free proteome quantification. Nat Biotechnol 34:1130–1136

20. Misra BB (2018) Updates on resources, software tools, and databases for plant proteomics in 2016-2017. Electrophoresis 39:1543–1557

21. Turewicz M, Kohl M, Ahrens M et al (2017) BioInfra.Prot: a comprehensive proteomics workflow including data standardization, protein inference, expression analysis and data publication. J Biotechnol 261:116–125

22. Webb-Robertson B-JM, Wiberg HK, Matzke MM et al (2015) Review, evaluation, and discussion of the challenges of missing value imputation for mass spectrometry-based label-free global proteomics. J Proteome Res 14:1993–2001

23. Callister SJ, Barry RC, Adkins JN et al (2006) Normalization approaches for removing systematic biases associated with mass spectrometry and label-free proteomics. J Proteome Res 5:277–286

24. Karpievitch YV, Dabney AR, Smith RD (2012) Normalization and missing value imputation for label-free LC-MS analysis. BMC Bioinformatics 13(Suppl 16):S5

25. Välikangas T, Suomi T, Elo LL (2018) A systematic evaluation of normalization methods in quantitative label-free proteomics. Brief Bioinform 19:1–11

26. Willforss J, Chawade A, Levander F (2019) NormalyzerDE: online tool for improved normalization of omics expression data and high-sensitivity differential expression analysis. J Proteome Res 18:732–740

27. Lazar C, Meganck S, Taminau J et al (2013) Batch effect removal methods for microarray gene expression data integration: a survey. Brief Bioinform 14:469–490

28. Unwin RD, Evans CA, Whetton AD (2006) Relative quantification in proteomics: new approaches for biochemistry. Trends Biochem Sci 31:473–484

29. Cairns DA, Barrett JH, Billingham LJ et al (2009) Sample size determination in clinical proteomic profiling experiments using mass spectrometry for class comparison. Proteomics 9:74–86

30. Ge Y, Sealfon SC, Speed TP (2009) Multiple testing and its applications to microarrays. Stat Methods Med Res 18:543–563

31. Huber W, Von Heydebreck A, Sültmann H et al (2002) Variance stabilization applied to microarray data calibration and to the quantification of differential expression. Bioinformatics 18:S96–S104

32. Benjamini Y, Hochberg Y (1995) Controlling the false discovery rate: a practical and powerful approach to multiple testing. J R Stat Soc Ser B 57:289–300

33. Großerueschkamp F, Bracht T, DIehl HC et al (2017) Spatial and molecular resolution of diffuse malignant mesothelioma heterogeneity by integrating label-free FTIR imaging, laser capture microdissection and proteomics. Sci Rep 7:44,829

Chapter 2

Good Old-Fashioned Protein Concentration Determination by Amino Acid Analysis

Caroline May, Bettina Serschnitzki, and Katrin Marcus

Abstract

Although a lot of new methods for protein concentration determination have been developed and established the last years, the amino acid analysis has still this relevance within proteomics for multiple reasons especially in the quantitative protein analysis. Amino acid analysis enables indirectly both the protein and peptide concentration determination which are essential for using the same amounts for comparative quantitative experiments. Moreover, the quantity and quality of synthetic peptides can be verified. The method itself is robust in comparison with colorimetric assays, especially when detergents or chaotropes are present in the sample buffer. Furthermore, it is highly sensitive. Nevertheless, amino acid analysis needs a certain experience to be set up and is time-consuming compared to other protein concentration determination techniques.

Key words Protein concentration determination, Quantitative proteome analysis, Amino acid analysis, Acid hydrolysis

1 Introduction

Amino acid analysis allows measuring of the amino acid concentration and therewith indirectly the concentration of proteins as well as peptides within a certain sample. The methodological approach can be subdivided into four main steps: first, unless the amino acids are not present in a free state within the sample of interest, but rather as peptides or proteins, they have to be released by, e.g., acid hydrolysis. In a second step, the amino acids are derivatized and then separated using HPLC. The peak areas in the resulting chromatogram are compared to a standard sample in a third step to calculate the concentration of the single amino acids (*see* Fig. 1). To determine the concentration of peptides or proteins within the sample of interest, the concentrations of the single amino acids are finally summed up.

Katrin Marcus et al. (eds.), *Quantitative Methods in Proteomics*, Methods in Molecular Biology, vol. 2228, https://doi.org/10.1007/978-1-0716-1024-4_2, © Springer Science+Business Media, LLC, part of Springer Nature 2021

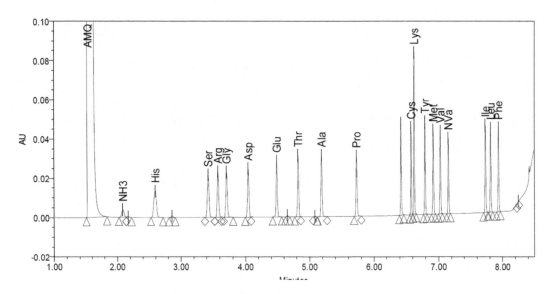

Fig. 1 *Exemplary amino acid chromatogram.* After appropriate sample preparation for amino acid analysis, including hydrolysis of peptides/proteins followed by derivatization, amino acids are separated by HPLC. The peak area of each single amino acids in the resulting chromatogram is compared to a standard sample to calculate their concentration. Finally, to determine the concentration of peptides or proteins within the present sample, the concentrations of the single amino acids are summed up

The main benefits in contrast to colorimetric assays like Lowry [1], Bradford [2], or Biuret [3] are:

- The concentration of proteins as well as peptides can be determined.

- High concentrations of chaotropic substances as well as detergents do not negatively influence the determination result.

- The derivatization using amino acids is independent from the characteristics of amino acids present in the sample in contrast to the dyes in colorimetric assays, which react unequally or preferably with defined amino acids.

- Amino acid analysis is one of the most sensitive concentration determination methods. The detection limit ranges from 40 to 800 fmol depending on the analyzed amino acid.

- Single peptides (e.g., synthetic peptides for absolute protein quantification) can be used to examine their quality and exact quantity.

The main disadvantages compared to colorimetric assays are that very expensive equipment is needed and sample preparation is time intensive. In former times, the measurement itself took more than 1 h at least. Using new devices and UPLC (ultra-performance liquid chromatography), the measurement including data analysis for one sample can be achieved within 15 min.

2 Materials

2.1 Sample Preparation

1. 5µL protein or peptide lysate (*see* **Note 1**) in duplicate (*see* **Note 2**).
2. −20 °C freezer.
3. Ice box.
4. Vacuum concentrator.
5. Glass engraving device.
6. Glass vials.
7. Muffle furnace.

2.2 Acid Hydrolysis

1. Vacuum vessel (*see* Fig. 2a).
2. Vacuum concentrator.
3. Vacuum pump.
4. Drying oven.
5. Heat-resistant gloves.
6. Beaker.
7. Tweezer.
8. Three-way valve.

A

B
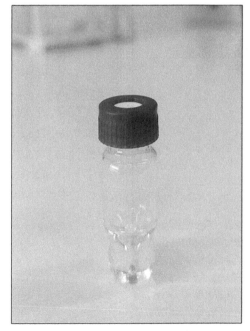

Fig. 2 (**a**) Exemplary vacuum vessel for amino acid hydrolysis. (**b**) Total recovery vial kit with screw cap

9. Argon gas flask.

10. 6 M HCl.

11. Phenol crystals.

12. Parafilm.

2.3 Derivatization of Samples

1. Vortexer.

2. 20 mM HCl.

3. AccQ-Tag™ Ultra Derivatization Kit includes borate buffer, reagent powder, and reagent diluent (Waters GmbH, Eschborn, Germany).

4. Thermomixer.

5. Parafilm.

6. Microcentrifuge with adjustable rotor speed up to $16,000 \times g$.

2.4 External Amino Acid Standard for System Calibration/ Data Analysis

1. Amino Acid Standard (AccQ-Tag, Pico-Tag, AccQ-Tag Ultra) (Waters GmbH).

2. Vacuum concentrator.

3. Vortexer.

4. 20 mM HCl.

5. AccQ-Tag™ Ultra Derivatization Kit (Waters GmbH).

6. Thermomixer.

7. Parafilm.

8. Microcentrifuge with adjustable rotor speed up to $16,000 \times g$.

2.5 HPLC Measurement

1. Acquity UPLC system for amino acid analysis including UV detector (Waters GmbH).

2. Empower Software package (Waters GmbH).

3. C18 reversed-phase separation column (2.1 mm × 100 mm in length) (Waters GmbH).

4. AccQ-Tag Ultra Eluent A (Waters GmbH).

5. AccQ-Tag Ultra Eluent B (Waters GmbH).

6. "Total Recovery Vial Kit with Screw Cap," Clear, 1 mL volume (Waters GmbH) (*see* Fig. 2b).

2.6 Generation of Calibration Curves for Sample Analysis Using the External Amino Acid Standard and Data Analysis

1. Empower Software package (Waters GmbH).

3 Methods

3.1 Sample Preparation

1. For removing possible amino acid contaminations, heat glass vials up to 400 °C for 4 h in a muffle furnace.

2. To ensure later correct sample assignments, engrave muffled glass vials with the sample name respectively the sample number.

3. Defrost frozen protein/peptide sample on ice.

4. To ensure an exact determination of the amino acid concentration, suck in and drain off the sample for in total 10 times before transferring 10µL into a muffled glass vial (*see* **Note 3**).

5. Duplicate determination is recommended for accurate quantity determination; therefore repeat **step 4**.

6. Place the glass vials in a vacuum concentrator and dry the samples completely.

7. Until further usage, dried samples can be stored at −20 °C.

3.2 Acid Hydrolysis

1. Depending on the number of samples, transfer 300–400µL of 6 M HCl and a phenol crystal to an evacuation vessel.

2. Place the open glass vials containing the dried samples with a tweezer in the evacuation vessel.

3. Connect the vacuum pump as well as the argon gas flask with the evacuation vessel using a three-way valve.

4. Evacuate and aerate with argon alternately for in total four times. This cycle should always end with the evacuation step.

5. Close the lid of the evacuation vessel and transfer it to a beaker glass.

6. Preheat the drying oven up to 150 °C for proteins and peptide mixtures and 100 °C for single peptides.

7. Incubate the beaker glass with the evacuation vessel in the drying oven. For proteins and peptide mixtures, use an incubation time of 60 min, and for single peptides, increase the incubation time up to 24 h (*see* **Note 4**).

8. Retain the beaker glass with the evacuation vessel using heat-resistant gloves.

9. Open the lid of the evacuation vessel so that excess HCl can evaporate.

10. The samples can be stored and sealed with parafilm for 1 week at −20 °C until further usage.

3.3 Derivatization of Samples

1. Prepare the AccQ-Tag™ Ultra Derivatization Kit according to manufacturers' instructions.

2. Dissolve the hydrolyzed samples in 10μL of 20 mM HCl by vortexing.

3. Add 30μL of AccQ-Fluor borate buffer with the internal standard Norvaline and 10μL of AccQ-Fluor reagent (10 mM 6-aminoquinolyl-N-hydroxysuccinimidylcarbamate in acetonitrile) to the hydrolyzed samples.

4. Seal the glass vials with parafilm.

5. Primary and secondary amines are converted to stable derivatives by incubating for 10 min at 55 °C in a thermomixer.

6. Shortly centrifuge the samples to spin down possibly condensed solution.

3.4 External Amino Acid Standard for System Calibration/Data Analysis

1. Transfer 5, 10, and 20 pmol of the amino acid standard in separate glass vials (*see* **Note 2**).

2. Completely dry the amino acid standard samples in a vacuum concentrator.

3. Prepare the AccQ-Tag™ Ultra Derivatization Kit according to manufacturers' instructions.

4. Dissolve the dried amino acid standard samples in 10μL of 20 mM HCl by vortexing.

5. Add 30μL of AccQ-Fluor borate buffer with the internal standard Norvaline and 10μL of AccQ-Fluor reagent (10 mM 6-aminoquinolyl-N-hydroxysuccinimidylcarbamate in acetonitrile) to the hydrolyzed samples.

6. Seal the glass vials with parafilm.

7. Primary and secondary amines are converted to stable derivatives by incubating for 10 min at 55 °C in a thermomixer.

8. Shortly centrifuge the samples to spin down possibly condensed solution.

3.5 HPLC Measurement of the Samples and External Amino Acid Standard

1. Transfer 45μL of the derivatized standard or sample completely into a "Total Recovery Vial" (part of the "Total Recovery Vial Kit with Screw Cap") (*see* **Note 5**).

2. Close the lid of the "Total Recovery Vial."

3. Inject 1μL of the sample into the Acquity UPLC system.

4. Separate the derivatives using a C18 reversed-phase separation column. Elute the derivatives with a flow rate of 0.7 mL/min and a column temperature of 55 °C by increasing the solvent A content in the solvent mixture (*see* Table 1).

5. The amino acid derivatives are detected at an emission wavelength of 260 nm.

6. Save data.

Table 1
Solvent gradient profile for the elution of amino acids [4]

Time (min)	% Eluent A	% Eluent B
0	99.9	0.1
0.54	99.9	0.1
5.74	90.9	0.1
7.74	78.8	21.2
8.04	40.4	59.6
8.05	10.0	90
8.64	10.0	90
8.73	99.9	0.1
9.50	99.9	0.1

3.6 Generation of Calibration Curves for Sample Analysis Using the External Amino Acid Standard

1. Assign the peaks to the different amino acids and calculate the peak areas for every dilution of the external amino acid standard.

2. Generate calibration curves for each amino acid. The deviation of each measured amino acid should be less than 5%.

3. Save calibration.

3.7 Data Analysis of Samples

1. Quantitative analyses are performed using the area under the peaks of the measured amino acids in comparison with the external amino acid standard. Therefore, integrate the data and quantify.

2. Save the results.

4   Notes

1. The original protein/peptide amount needs to be estimated roughly. The optimal range is in the range of calibration (5–20 pmol). If necessary, samples need to be diluted and measured again.

2. It is advisable to measure at least duplicates in order to exclude possible pipetting errors and to obtain the most reliable quantification results possible.

3. Proteins and peptides bind to a certain extend to the surface of plastic pipette tips. Especially, when the protein/peptide concentration of the sample of interest is low, the loss will be severe influence in the result of amino acid analysis. It means the determined concentration will be significantly lower than the actual one.

4. In case of using acid hydrolysis, amino acids like serine, threonine, and methionine can be partially destroyed. Cysteine and tryptophan can be nearly complete destroyed and asparagine as well as glutamine possibly deamidated. To what extend depends on the temperature and time that is used for the acid hydrolysis. In case the sample of interest contains these amino acids to a large extent, it is recommended to optimize the time as well as the temperature that is used for the acid hydrolysis to receive these amino acids.

5. It is important to use the specialized "Total Recovery Vials" in order to prevent sample loss. The shape of the vessels is specially designed to allow the final volume of 1 μL to be pipetted completely out of the vessel again.

Process a reference sample of known concentration with your sample set to determine a recovery.

Acknowledgments

This work was supported by the Deutsche Parkinson Gesellschaft, the Bundesministerium für Bildung und Forschung, Germany (WTZ mit Brasilien, FKZ 01DN14023), the HUPO Brain Proteome Project, and P.U.R.E. (Protein Research Unit Ruhr within Europe) and ProDi (Center for protein diagnostics), Ministry of Innovation, Science and Research of North-Rhine Westphalia, Germany, a federal German state, FoRUM (F781-13).

References

1. Lowry OH, Rosebrough NJ, Farr AL et al (1951) Protein measurement with the Folin phenol reagent. J Biol Chem 193(1):265–275

2. Bradford MM (1976) A rapid and sensitive method for the quantitation of microgram quantities of protein utilizing the principle of protein-dye binding. Anal Biochem 72:248–254

3. Gornall AG, Bardawall CJ, David MM (1949) Determination of serum proteins by means of the biuret reaction. J Biol Chem 177(2):751–766

4. Guntermann A, Steinbach S, Serschnitzki B et al (2019) Human tear fluid proteome dataset for usage as a spectral library and for protein modeling. Data Brief 23:103,742

Chapter 3

Protein Quantification Using the "Rapid Western Blot" Approach

Katalin Barkovits, Kathy Pfeiffer, Britta Eggers, and Katrin Marcus

Abstract

For the quantification of certain proteins of interest within a complex sample, Western blot analysis is the most widely used method. It enables detection of a target protein based on the use of specific antibodies. However, the whole procedure is often very time-consuming. Nevertheless, with the development of fast blotting systems and further development of immunostaining methods, a reduction of the processing time can be achieved. Major challenges for the reliable protein quantification by Western blotting are adequate data normalization and stable protein detection. Usually, normalization of the target protein signal is performed based on housekeeping proteins (e.g., glyceraldehyde 3-phosphate dehydrogenase, ß-actin) with the assumption that those proteins are expressed constitutively at the same level across experiments. However, several studies have already shown that this is not always the case making this approach suboptimal. Another strategy uses total protein normalization where the abundance of the target protein is related to the total protein amount in each lane. This approach is independent of a single loading control, and precision of quantification and reliability is increased. For Western blotting several detection methods are available, e.g., colorimetric, chemiluminescent, radioactive, fluorescent detection. Conventional colorimetric staining tends to suffer from low sensitivity, limited dynamic range, and low reproducibility. Chemiluminescence-based methods are straightforward, but the detected signal does not linearly correlate to protein abundance (from protein amounts >5µg) and have a relatively narrow dynamic range. Radioactivity is harmful to health. To overcome these limitations, stain-free methods were developed allowing the combination of fluorescent standards and a stain-free fluorescence-based visualization of total protein in gels and after transfer to the membrane. Here, we present a rapid Western blot protocol, which combines fast blotting using the iBlot system and fast immunostaining utilizing ReadyTector® all-in-one solution with the Smart Protein Layers (SPL) approach.

Key words Protein quantification, Western blot, Smart Protein Layers (SPL), Fluorescence labeling, Stain-free technology, iBlot, ReadyTector®

1 Introduction

Western blotting is a method for the detection but also for the quantification of proteins in a protein mixture (e.g., cell lysate) [1, 2]. Thereby proteins are first separated from a protein mixture by gel electrophoresis according to the protein properties. Subsequently, the separated proteins are transferred from the gel to a

Katrin Marcus et al. (eds.), *Quantitative Methods in Proteomics*, Methods in Molecular Biology, vol. 2228, https://doi.org/10.1007/978-1-0716-1024-4_3, © Springer Science+Business Media, LLC, part of Springer Nature 2021

solid carrier membrane (e.g., nitrocellulose, nylon, or PVDF), which is called blotting. The proteins are attached to the membrane surface and are therefore accessible to further methods like detection of certain proteins with specific antibodies [3]. Depending on the required criteria for the separation of the proteins, such as size, charge, or isoelectric point, different gel-based electrophoresis applications can be used. The most suitable methods are SDS-PAGE, native PAGE, or 2D-PAGE (see Chap. 5). Depending on the characteristics of the protein of interest (e.g., low or high molecular weight) different gel systems can be used (see below). Common to all gel-based approaches is that due to the separation method used, a particular protein pattern (protein band or protein spot) is present in the gel, which can be visualized by specific staining methods [4]. However, the identity of the underlying proteins cannot be deduced from the protein pattern. Western blot analysis enables the identification of certain proteins in a complex sample with specific antibodies [3]. In the blotting process, the gel with the separated proteins is placed in a specialized electrophoresis apparatus, which also contains the membrane. By applying an electrical voltage, the proteins migrate out of the gel and are transferred and bound to a membrane where they are immobilized. The protein pattern that has formed in the gel is thus transferred 1 to 1 to the membrane. During immunostaining the membrane is incubated in a solution containing the protein-specific primary antibody (mono- or polyclonal) that recognizes and binds to the target protein [5]. A secondary antibody is added, which in turn recognizes and binds the first antibody. This secondary antibody is conjugated to a dye, an enzyme, or radioactive substance allowing indirect detection of the defined targeted protein [6–9]. Alternatively, the primary antibody can itself be conjugated and visualized which saves the additional step via the second antibody. Visualization of the protein signal is performed using various imaging systems (luminescence, color reaction, autoradiography, etc.) depending on the conjugated molecule. Beside, assessing the presence or absence of a protein Western blotting is also used to compare and quantify protein levels in different conditions or in different tissues [10]. For the protein quantification within an experiment in different samples, a reliable and reproducible standardization and normalization strategy is required. Normalization is necessary to correct for variations due to changes in concentration or volume [10]. The standard methods for normalization of Western blot signals uses loading controls, such as proteins present in every experimental sample, the so-called housekeeping proteins (e.g., glyceraldehyde 3-phosphate dehydrogenase, ß-actin, detected by an independent antibody) or total protein (quantified by staining like Coomassie Brilliant Blue, Ponceau S, Congo red, or SYPRO Ruby) [11–16]. Here, for example, individual signals from the immunodetected proteins of interest are

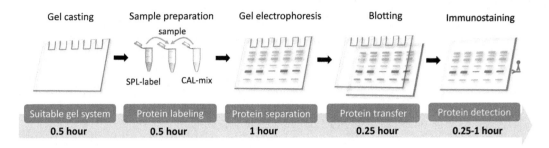

Fig. 1 Schematic overview of rapid Western blotting. Overall the whole procedure from gel casting over sample preparation, gel electrophoresis, blotting to the final protein detection is performed under 4 h. Utilization of the Smart Protein Layers (SPL) kit (NH DyeAGNOSTICS) enables accurate quantification of the target protein and is compatible with the fast protein blotting system iBlot (Thermo Fisher Scientific) and the ReadyTector® all-in-one detection solution for Western blotting (CANDOR Bioscience)

normalized to other individual signals from housekeeping proteins within the same sample. The use of housekeeping proteins as a reference should allow the quantification of the protein of interest to normalize for differences in the total protein amount. However, these proteins do not reflect the real total protein amount of a sample because, contrary to the assumption that housekeeping proteins are expressed constitutively, they are subject to changes under different conditions [17, 18]. Therefore, stain-free methods for stain-free, quantitative, and standardized analysis of proteins in the gels as well from blots offer a more reliable and accurate quantification [19–22]. Those methods employ an in-gel chemistry for subsequent imaging, detection, and quantification. This chemistry does not affect the protein transfer from gel to membrane or the subsequent antibody binding.

By combining different methods, we have established a time-efficient and for quantification precise Western blot system which is described in this chapter. The combination of the iBlot system (Thermo Fisher Scientific) for blotting, the ReadyTector® all-in-one solution (CANDOR Bioscience) for immunostaining, and the stain-free Smart Protein Layers (SPL) technology (NH DyeAGN OSTICS) allows for a time-effective and accurate protein quantification in less than 4 h (Fig. 1).

2 Materials

2.1 Gel Casting

1. Bis(2-hydroxyethyl)amino-tris(hydroxymethyl)methane (BisTris).

2. 2-Amino-2-(hydroxymethyl)-1,3-propanediol (Tris).

3. Tris hydrochloride (Tris–HCL).

4. Sodium dodecyl sulfate (SDS).

5. Hydrochloric acid (HCl).

6. Acrylamide 40% (AA/BisAA 37.5:1).

7. Tetramethylethylenediamine (TEMED).

8. Ammonium persulfate (APS).

9. 3-(N-morpholino) propanesulfonic acid (MOPS).

10. 2-(N-morpholino)ethanesulfonic acid (MES).

11. MilliQ H_2O.

12. Eppendorf tubes or glass beakers.

2.2 SDS Gel Electrophoresis

1. Favored Gel-System (precast or self-made).

2. Smart Protein Layers SPL red-IR Kit (NH DyeAGNOSTICS, Halle, Germany).

3. 60 mM DTT (*see* **Note 1**).

4. Sample (10–100µg maximal 10µL).

5. Eppendorf tubes.

2.3 Blotting

1. iBlot Western Blot system (Thermo Fisher Scientific, Inc.) or in-house Western blot system (*see* **Note 2**).

2. iBlot Dry Blotting System.

3. iBlot Gel Transfer Stack (mini, midi).

4. De-Bubbling Roller.

2.4 Immunostaining

1. ReadyTector® all-in-one solution (CANDOR Bioscience, Wangen im Allgäu, Germany, *see* **Note 3**).

2. Primary antibody.

3. Secondary antibody (e.g., IR-labelled).

4. Incubation tray.

2.5 Imaging and Data Analysis

1. Imaging system for the detection of red and infrared fluorescence with Wavelengths red 650 nm/665 nm (excitation/emission) and Wavelengths green 740 nm/770 nm (excitation/emission).

2. LabImage 1D SPL Software (Kapelan Bio-Imaging, Leipzig, Germany).

3   Methods

3.1 Gel Casting

1. Depending on the molecular weight of the target protein, select a suitable gel system either purchasing a commercial one or casting a self-made (*see* Table 1 for the respective molecular weight ranges of different gel systems). Here we provide recipes for different continuous gel systems (*see* **Note 4**).

Table 1
Gel buffer composition for different continuous gel systems

MW range	Gel buffer	Chemicals (concentration)	pH
10–200 kDa	Bis-Tris (sevenfold)	BisTris (2.5 M)	6.5
15–250 kDa	Laemmli (twofold)	Tris (570 M) Tris–HCl (180 mM) SDS (0.2%)	8.3
3–100 kDa	Schägger (threefold)	Tris (3 M) HCl (1 M) SDS (0.3%)	8.5

Table 2
Solutions and amount for one mini gel

Gel system	Acrylamide 40% (AA/BisAA 37.5:1) (mL)	Gel buffer (mL)	APS 40% (μL)	TEMED (μL)	H$_2$0 (mL)
Bis-Tris 12%	3	1.42	10	2	Ad 10
Laemmli 15%	3.75	5	10	2	Ad 10
Schägger 10%	2.5	3.3	10	2	Ad 10

2. Based on the selected gel system, gel casting requires different gel buffers and running buffers. Prepare the required gel buffers according to Table 1, and mix the required solution as listed in Table 2 to cast a mini gel (dimension of approx. 10 × 10 cm). In addition, prepare the required electrophoresis buffer as specified in Table 3.

3.2 SDS Gel Electrophoresis

1. Sample preparation: add 10–100μg of sample to 10μL buffer maximum. The normal sample buffer in which the sample was present before denaturation can be used (*see* **Note 5**).

2. The use of the SPL kit requires the preparation of different solutions (*see* **Note 6**). The Smart Label working solution (SLW) only needs to be prepared for the first use. SPL reaction and Loading Mix (RL-mix) must be prepared for each new experiment. Optionally, specific SPL calibrators (CAL A for gel-to-gel comparison of the fluorescence signal and CAL B for gel-to-gel comparison of the antibody signal) can be employed and must also be prepared for each new experiment (*see* **Note 7**).

3. For the preparation of SLW solution (only before first usage!) allows the vials containing Smart Label reagent A and Smart Label reagent B to reach room temperature. Then spin down the respective reagents briefly. To dissolve the Smart Label, add

Table 3
Electrophoresis buffer for the different continuous gel systems

Electrophoresis buffer	Chemicals (concentration)	Comment
Bis-Tris gel (MOPS)	MOPS (50 mM) Tris (50 mM) SDS (0.1%) EDTA (2.7 mM)	• One-buffer system • Do not adjust pH • Can be prepared as a 20fold stock solution
Bis-Tris gel (MES)	MES (50 mM) Tris (50 mM) SDS (0.1%) EDTA (2.7 mM)	• One-buffer system • Do not adjust pH • For better separation of smaller proteins <30 kDa • Can be prepared as a 20fold stock solution
Laemmli gel	Tris (25 mM) Glycine (192 mM) SDS (0.1%)	• One-buffer system • Do not adjust pH • Can be prepared as a tenfold stock solution
Schägger gel	Cathode buffer Tris (100 mM) Tricine (100 mM) SDS (0.1%) Anode buffer Tris (100 mM) HCl (22.5 mM)	• Two-buffer system • Do not adjust pH in cathode buffer • Can be prepared as a tenfold stock solution

15 μL from Smart Label reagent B to Smart Label reagent A, mix, and centrifuge briefly. Then transfer the complete volume from the Smart Label reagent vial to the Smart Label reagent vial and mix. The ready-to-use Smart Label Working Solution is now in Vial B and can be stored for 6 months at −20 °C to −80 °C.

4. To prepare the RL-mix, the following solutions per sample are mixed in a new Eppendorf tube (*see* **Note 8**):

   (a) 6 μL SPL buffer

   (b) 2 μL 60 mM DTT

   (c) 2 μL SMA basic S (12.5 kDa) or SMA basic L (80 kDa)

5. For the preparation of the CAL mix, 8 μL CAL A (three polypeptides labeled red- and infrared-fluorescent in the sizes 12.5, 25, 80 kDa) and 2 μL CAL B (one polypeptide nonfluorescent 50 kDa) per gel are mixed in a new Eppendorf tube. Then add 2 μL 60 mM DTT (fresh), mix, and centrifuge briefly. Incubate at 95 °C for 5 min to denature the proteins (*see* **Notes 9–11**).

6. For labelling of proteins, 10 μL sample (10–100 μg) is transferred into a new Eppendorf tube. Add 10 μL RL-mix to the sample, mix, and centrifuge briefly. Then add 1 μL SWL solution, mix, and centrifuge briefly. The preparation is incubated at 95 °C for 5 min to denature and label the proteins.

7. For gel electrophoresis, the complete sample volume (*see* **step 6**) is applied. In addition, 12μL CAL-Mix is applied per gel into one lane.

8. Carry out electrophoresis with the in-house standard parameters (*see* **Note 12**).

9. Scan the gel directly after the run. Do not fix or stain (*see* **Note 13**).

3.3 Blotting

1. Blotting within 10 min using the iBlot System (*see* **Note14**).

2. Place the bottom stack (anode) onto the iBlot Dry Blotting System (*see* **Note 15**).

3. Place your pre-run gel on top of the membrane.

4. Place the presoaked iBlot filter paper on the pre-run gel, and remove air bubbles using the roller (*see* **Note 16**).

5. Place the top stack (cathode) over the presoaked filter paper, and remove air bubbles again using the roller.

6. Place the disposable sponge with the metal contact on the upper right into the lid of the iBlot System and close the lid.

7. Choose a suitable program for your protein of interest, e.g., when you want to blot a protein with a molecular weight (>200 kDa), choose a program with a high voltage, e.g., P1. When blotting a protein, <200 kDa P3 or P4 should be sufficient (*see* **Note 17**).

8. Start the program.

9. Turn the iBlot System off when the run is finished and disassemble the system.

10. Collect cooper based materials in a special bag, because it has to be recycled separately (*see* **Note 18**).

11. Collect the membrane and proceed directly with immunostaining as described below.

3.4 Immunostaining

1. For immunostaining of the target protein, use ReadyTector® the all-in-one solution for Western blotting according to the manufactures' instruction (*see* **Note 19**).

Wash the blot three times for 5 min at RT with ReadyTector® wash buffer. Afterward incubate the blot for 15 min up to 1 h at RT under gentle agitation with the ReadyTector® all-in-one solution including your primary antibody in your normal dilutions. Remove the all-in-one solution and wash the blot again three times for 5 min at RT with ReadyTector® wash buffer (*see* **Notes 3** and **20**).

**3.5 Imaging an Data
Analysis
with LabImage1D SPL
Software**

1. Scan the blot directly, no staining is necessary (*see* **Notes 21** and **22**).

4 Notes

1. Always prepare fresh.

2. If you use in-house Western blot system, use nitrocellulose or low-fluorescence PVDF blotting membrane.

3. Depending on the final detection method ReadyTector® all-in-one solution is available in different versions with respect to the secondary antibody. It is offered with an already included secondary antibody like Anti-Mouse-HRP or Anti-Rabbit-HRP, but also as a PURE solution to which additionally to the first antibody the second antibody of choice (e.g., fluorescence labeled) is added. If the PURE solution is required, please contact the manufacturer directly (https://www.candor-bioscience.de/en/products/readytectorr/, and state the 701 as the order number). The final ReadyTector® all-in-one solution contains the blocker, primary antibody, and secondary antibody, and the entire incubation can be performed within one single step—blocking, binding of the primary and secondary antibody occur simultaneously.

4. Continuous gel electrophoresis is characterized by simplicity and speed, as both only one buffer is used for the preparation of the separating gel and the electrophoresis buffer. In contrast to discontinuous SDS gel electrophoresis, however, a lower resolution and band sharpness is achieved.

5. All common buffers are compatible, but make sure that the buffer contains less than 400 mM amines or ammonium salts.

6. Since the SPL-based detection and the blotting and immunostaining steps are exchangeable, the individual steps can be replaced/combined with alternative methods like your in-house detection method.

7. A detailed protocol including information about available SPL kits depending on the fluorescence detection system is described by Faden et al. [23].

8. RL-Mix should be prepared as a master mix for all samples. As a general guideline the best application is to prepare master mix as follow n protein samples = $n + 1$. Furthermore, select the SMA according to the gel, for gels with less than 15% AA/BisAA SMA basic L and more than 15% AA/BisAA SMA basic S. In addition, care should be taken that the selected SMA is not separated at the level of the target protein; otherwise the signal will be overlaid.

9. CAL B must be selected to match the selected secondary antibody. If the secondary antibody comes from, e.g., Rabbit, CAL B Rabbit must be used.

10. For each gel 12μL CAL mix is required. A master mix should be prepared for all gels used in an experiment.

11. The amount of CAL B should be tested. If the intensity of CAL B is significantly higher than that of the target protein, less CAL B can be used, if it is significantly lower, more CAL B can be used.

12. When using the SMA basic S, ensure that electrophoresis is stopped before the SAM band of 12.5 kDa migrates from the gel.

13. Be sure to scan both wavelengths (red and green). Make sure that no saturation is obtained. To do this, adjust the scan intensity. Save the file in the original, no adjustment of intensity, brightness, or contrast after scanning.

14. There are several other fast blotting devices available, e.g., the BioRad System. If you do not have a fast blotting system perform your in-house blotting protocol. After blotting proceed with Subheading 3.4.

15. Be careful not to touch the membrane to avoid signals in the infrared.

16. Soak the filter paper in deionized water. Do not incubate too long to avoid overall wetness in the system.

17. The iBlot system has nine default programs with defined run time and voltages, which can be looked up in the quick reference and the user manual (available at the manufacture's homepage under the following link: https://www.thermofisher.com/de/de/home/references/protocols/proteins-expression-isolation-and-analysis/western-blot-protocol/western-blotting-using-dry-blotting-system.html).

18. Please check your countries recycling guidelines for metals.

19. If the ReadyTector® kit is not used, then of course your in-house immunostaining protocol can be used since blotting, immunostaining and the SPL-based detection steps are freely interchangeable.

20. The ReadyTector® all-in-one solution including the primary and the secondary antibody can be used several times when stored at 4 °C.

21. Be sure to scan both wavelengths (red and green). Make sure that no saturation is reached. To do this, adjust the scan intensity. Save the file in the original, no adjustment of intensity, brightness, or contrast after scanning.

22. A detailed protocol for target protein detection and data analysis is described by Faden et al. [23].

Acknowledgments

A part of this study was funded by P.U.R.E. (Protein Research Unit Ruhr within Europe) and ProDi (Center for protein diagnostics), Ministry of Innovation, Science and Research of North-Rhine Westphalia, Germany, and by the H2020 project NISCI, (GA no. 681094).

References

1. Bass JJ, Wilkinson DJ, Rankin D et al (2017) An overview of technical considerations for Western blotting applications to physiological research. Scand J Med Sci Sports 27(1):4–25. https://doi.org/10.1111/sms.12702

2. Ni D, Xu P, Gallagher S (2017) Immunoblotting and Immunodetection. Curr Protoc Protein Sci 88:10.10.11–10.10.37. https://doi.org/10.1002/cpps.32

3. Towbin H, Staehelin T, Gordon J (1979) Electrophoretic transfer of proteins from polyacrylamide gels to nitrocellulose sheets: procedure and some applications. Proc Natl Acad Sci U S A 76(9):4350–4354. https://doi.org/10.1073/pnas.76.9.4350

4. Sundaram P (2018) Protein stains and applications. Methods Mol Biol 1853:1–14. https://doi.org/10.1007/978-1-4939-8745-0_1

5. MacPhee DJ (2010) Methodological considerations for improving Western blot analysis. J Pharmacol Toxicol Methods 61(2):171–177. https://doi.org/10.1016/j.vascn.2009.12.001

6. Fradelizi J, Friederich E, Beckerle MC et al (1999) Quantitative measurement of proteins by western blotting with Cy5-coupled secondary antibodies. Biotechniques 26(3):484–486; 488, 490 passim.

7. Burnette WN (1981) "Western blotting": electrophoretic transfer of proteins from sodium dodecyl sulfate—polyacrylamide gels to unmodified nitrocellulose and radiographic detection with antibody and radioiodinated protein A. Anal Biochem 112(2):195–203. https://doi.org/10.1016/0003-2697(81)90281-5

8. Kricka LJ, Thorpe GH (1986) Photographic detection of chemiluminescent and bioluminescent reactions. Methods Enzymol 133:404–420. https://doi.org/10.1016/0076-6879(86)33082-9

9. Kurien BT, Scofield RH (2006) Western blotting. Methods 38(4):283–293. https://doi.org/10.1016/j.ymeth.2005.11.007

10. Taylor SC, Posch A (2014) The design of a quantitative western blot experiment. Biomed Res Int 2014:361,590. https://doi.org/10.1155/2014/361590

11. Welinder C, Ekblad L (2011) Coomassie staining as loading control in Western blot analysis. J Proteome Res 10(3):1416–1419. https://doi.org/10.1021/pr1011476

12. Ranganathan V, De PK (1996) Western blot of proteins from Coomassie-stained polyacrylamide gels. Anal Biochem 234(1):102–104. https://doi.org/10.1006/abio.1996.0057

13. Litovchick L (2020) Staining the blot for total protein with ponceau S. Cold Spring Harb Protoc 2020(3):098459. https://doi.org/10.1101/pdb.prot098459

14. Wang JL, Zhao L, Li MQ et al (2020) A sensitive and reversible staining of proteins on blot membranes. Anal Biochem 592:113,579. https://doi.org/10.1016/j.ab.2020.113579

15. Steinberger B, Brem G, Mayrhofer C (2015) Evaluation of SYPRO ruby total protein stain for the normalization of two-dimensional Western blots. Anal Biochem 476:17–19. https://doi.org/10.1016/j.ab.2015.01.015

16. Goldman A, Harper S, Speicher DW (2016) Detection of proteins on blot membranes. Curr Protoc Protein Sci 86:10.18.11. https://doi.org/10.1002/cpps.15

17. Ferguson RE, Carroll HP, Harris A et al (2005) Housekeeping proteins: a preliminary study illustrating some limitations as useful references in protein expression studies. Proteomics 5(2):566–571. https://doi.org/10.1002/pmic.200400941

18. Thellin O, Zorzi W, Lakaye B et al (1999) Housekeeping genes as internal standards: use and limits. J Biotechnol 75(2–3):291–295. https://doi.org/10.1016/s0168-1656(99)00163-7

19. Colella AD, Chegenii N, Tea MN et al (2012) Comparison of stain-free gels with traditional immunoblot loading control methodology. Anal Biochem 430(2):108–110. https://doi.org/10.1016/j.ab.2012.08.015

20. Gürtler A, Kunz N, Gomolka M et al (2013) Stain-free technology as a normalization tool in Western blot analysis. Anal Biochem 433 (2):105–111. https://doi.org/10.1016/j.ab.2012.10.010

21. Gilda JE, Gomes AV (2013) Stain-free total protein staining is a superior loading control to β-actin for Western blots. Anal Biochem 440 (2):186–188. https://doi.org/10.1016/j.ab.2013.05.027

22. Gilda JE, Gomes AV (2015) Western blotting using in-gel protein labeling as a normalization control: stain-free technology. Methods Mol Biol 1295:381–391. https://doi.org/10.1007/978-1-4939-2550-6_27

23. Faden F, Eschen-Lippold L, Dissmeyer N (2016) Normalized quantitative Western blotting based on standardized fluorescent labeling. Methods Mol Biol 1450:247–258. https://doi.org/10.1007/978-1-4939-3759-2_20

Chapter 4

The Whereabouts of 2D Gels in Quantitative Proteomics

Thierry Rabilloud and Cécile Lelong

Abstract

Two-dimensional gel electrophoresis has been instrumental in the development of proteomics. Although it is no longer the exclusive scheme used for proteomics, its unique features make it a still highly valuable tool, especially when multiple quantitative comparisons of samples must be made, and even for large samples series. However, quantitative proteomics using two-dimensional gels is critically dependent on the performances of the protein detection methods used after the electrophoretic separations. This chapter therefore examines critically the various detection methods, (radioactivity, dyes, fluorescence, and silver) as well as the data analysis issues that must be taken into account when quantitative comparative analysis of two-dimensional gels is performed.

Key words 2D-PAGE, Fluorescence dyes, Image analysis, Organic dyes, Polyacrylamide gels, Quantification, Radioisotopes, Silver staining

1 Introduction

Since its introduction in the mid-seventies [1, 2], 2D electrophoresis has always been used as a quantitative technique of protein analysis, and it is fair to say that such quantitative analyses (e.g., in [3–8]) (*see* Chaps. 5–7) have preceded the real onset of proteomics, hallmarked by the first protein identification techniques, at that time based on Edman sequencing [9–13]. In the current proteomics landscape, completely dominated by tandem mass spectrometry [14, 15], and where 2D gel-based proteomics is now perceived for only niche applications [16], it represents an exception in the sense that this is a proteomics setup where:

1. Protein quantification is not made in a mass spectrometer (and there has been only very limited attempts to break this rule [17]).

2. The on-gel quantification step is often used as a screening process to select a limited set of protein spots of interest, that are then further characterized by mass spectrometry, thereby limiting the mass spectrometry analysis time.

Katrin Marcus et al. (eds.), *Quantitative Methods in Proteomics*, Methods in Molecular Biology, vol. 2228, https://doi.org/10.1007/978-1-0716-1024-4_4, © Springer Science+Business Media, LLC, part of Springer Nature 2021

These two cardinal features put an enormous pressure on the performances that the on-gel protein detection methods must show, as it is quite clear that what is not detected is never analyzed and thus completely ignored. Thus, the on-gel protein detection methods must be very sensitive but also linear in response (to be able to detect accurately abundance variations), homogeneous (so that all classes of proteins are detected), and of course fully compatible with mass spectrometry (to ensure easy and accurate protein characterization).

Although these constraints have been quite clear to the community for numerous years, and have led to the implementation of many protein detection schemes, an often overlooked problem is the performance of the 2D electrophoresis itself. In other words, what is the quantitative yield of 2D electrophoresis and how homogeneous this yield is for various classes of proteins. There are very few papers dealing with this issue, but one paper [18] showed that the yield of 2D electrophoresis was rather moderate (20–40%), which is an often overlooked parameter when the overall efficiency of the system is considered. Furthermore, work on membrane proteins [19, 20] has strongly suggested that such losses are not homogeneous and may be much greater for poorly soluble proteins such as membrane proteins.

Of course, the overall yield of the process will also depend on the efficiency of the protein extraction during the sample preparation process, but this process is so variable from one experimental model to another that it is really beyond the scope of this chapter. Furthermore, this chapter will deal mainly with the quantification issues in 2D gel-based proteomics. Other important issues, such as the scope of gel-based proteomics, the interest of protein electrophoresis in modern proteomics, and how electrophoresis systems can be modulated to improve their performances, have been reviewed elsewhere [21–23].

2   The Protein Detection Methods

Over the years, numerous on-gel protein detection methods have been used, each having its advantages and drawbacks in the sensitivity/linearity/homogeneity/compatibility multiple criteria selection guide. Over the numerous years of use of 2D electrophoresis, some techniques have almost disappeared, while others are now standards in 2D gel-based proteomics. Many different protein detection schemes have been devised over the several decades of use of 2D gels [24], and only the most important ones will be reviewed in this chapter.

2.1 Protein Detection Via Radioisotopes

This is an example of a technique that has subsided now, although it played a key role in the early days of proteomics, before the name was even coined.

Due to its exquisite sensitivity [25, 26] and linearity [27], protein detection via radioisotopes has been associated with almost all the early success stories of 2D electrophoresis, from the determination of protein numbers in cells [28], to the first identification of a protein from 2D gels, i.e., PCNA [29, 30], to single cell proteomics [31] or to phosphoprotein studies [32]. These exquisite sensitivity and linearity have even increased with new detection technologies such as phosphor screens [33]. However, except for special purposes where it is almost irreplaceable [18, 34], detection via radioisotopes has almost disappeared from modern proteomics. In addition to the fact that not all samples are easily accessible to this type of detection (e.g., human samples), increasing safety and regulatory issues hastened the decline of radioactivity in proteomics.

2.2 Protein Detection Via Organic Dyes

In this mode of detection, the name of the game is to bind as many dye molecules per protein molecules as possible, in order to create a light absorption signal that is detectable. Of course, this process must be as reproducible as possible, and the molar extinction coefficient of the dye also plays a major role in the signal intensity.

For all these reasons, and despite a few attempts to use other dyes [35], colloidal Coomassie Blue, as introduced in 1988 by V. Neuhoff [36], reigns supreme in this subfield. However, its rather limited sensitivity requires high protein loads to be able to detect a decent number of spots, which results in turn in increased sample consumption and decreased separation performances due to protein precipitation, leading to various artifacts such as streaks, most often in the isoelectric focusing dimension.

2.3 Protein Detection Via Silver Staining

To alleviate the sensitivity problem mentioned above, while keeping the ease of use and low costs associated with methods dealing with visible light absorbance, silver staining was introduced a few years after 2D electrophoresis [37]. However, it is fair to say that the early days of silver staining were troublesome, with erratic backgrounds and sensitivities, and this was due to the complex chemical mechanisms prevailing in silver staining [38]. However, decisive progress was made at the end of the 1980s [39], and silver staining is now a reliable technique allying high sensitivity, good reproducibility, low cost, and ease of use [40]. This is further discussed in another chapter of this book (see Chap. 6).

The main drawbacks of silver staining in modern proteomics are its limited dynamic range but also its weak compatibility with mass spectrometry [41, 42], although very high performances have been claimed [43]. It has been shown that the formaldehyde used in image development was the main culprit [44], and formaldehyde-free protocols have been developed that offer much increased compatibility with mass spectrometry [45]. In addition, the considerable increase of the sensitivity of the mass spectrometers over the years has decreased the severity of the problem.

2.4 Protein Detection Via Fluorescence

To alleviate the problems shown by silver staining and Coomassie blue, protein detection by fluorescence has been developed and has shown great expansion over the past few years. Opposite to the strict mechanisms that prevail in visible staining, either with Coomassie Blue or with silver, protein detection via fluorescence can been achieved via multiple mechanisms, thereby offering great versatility to this technique.

The first and oldest mechanism is covalent binding, quite often of probes that are not fluorescent but become so when the covalent binding takes place [46, 47]. While the performances of such probes were not very impressive, and thus of limited use, a quantum leap was achieved when probes with much higher light absorption and emission characteristics were used. Furthermore, with the development of laser scanners, use of a set of closely related probes could be designed to achieve multiplexing [48], resulting in the very popular DIGE technique [49] (*see* Chap. 7).

While this system has shown exquisite performances, it must be kept in mind that only a few fluorescent molecules are bound per protein molecule, resulting in an overall low signal intensity for many proteins. This is not a problem for pure detection, as the enormous contrast allows to use the full power of laser scanners, but this becomes a problem in some instances, e.g., spot excision for mass spectrometry, where more primitive illumination devices are used, e.g., UV tables.

Thus, another popular mode of protein detection via fluorescence uses non covalent binding, which takes place after migration and therefore does not interfere with protein migration, and which can also take place at much more numerous sites on the proteins, thereby resulting in a much higher signal, although the free fluorescent molecule remaining in the gels decreases the contrast compared to covalent binding. Although other candidates have been recently proposed [50, 51], two molecules dominate this field, namely, epicocconone [52], and ruthenium-based organometallic complexes [53–55]. These molecules offer a detection sensitivity that is very close to that obtained with silver staining, with a much better linearity and a much better compatibility with mass spectrometry. However, in this latter aspect, they do not perform as well as Coomassie Blue [42].

While these two modes of detection (covalent and non-covalent binding) are light emission counterparts of modes that have been used in visible detection (light absorption) [56], there is a third detection mode that is specific to fluorescence, which is the use of environment-sensitive probes. The molecules that are used for protein detection do not fluoresce in polar environments such as water, but do fluoresce in less polar environments such as protein-SDS complexes. Several molecules have been shown to achieve protein detection in this general scheme. Protocols using protein fixing and denaturing conditions have been devised with some probes of the styryl class [57, 58], while completely non-denaturing conditions could be used for other probes such as Nile Red [59], carbazolyl vinyl dyes [60] and more recently carbocyanines [61].

While the fixing schemes offer no real advantage over the classical non-covalent probes, the nondenaturing schemes offer distinct advantages such as speed, blotting ability [60, 61] and more importantly a very good sequence coverage in subsequent mass spectrometry [61].

Last but certainly not least, fluorescence can be used to detect specific motifs on gel-separated proteins, such as sugars [62] or phosphate groups [63], thereby offering a very wide palette of detection schemes with high versatility.

3   The Data Analysis Issues

In most instances where 2D gel-based proteomics is used, the production of the gel image by any of the protein detection methods listed above is not the end of the story, and it is very uncommon that all detected protein spots are excised for protein identification by mass spectrometry. Most often, comparative image analysis is used to determine a few spots whose expression profile within the complete experiment matches biologically relevant events. This image analysis process can be split in four majors steps. First data acquisition [4], then spot detection and quantification [5], then gel matching [6] (although gel matching can be carried out prior to spot detection in some analysis systems) and finally data analysis [7]. It must be stressed that this image analysis process has been used very early after the introduction of 2D gel electrophoresis [3, 64, 65], quite often with sophisticated data analysis tools [7, 8, 66, 67], long before the word proteomics even existed.

Although very cumbersome at these early times, image analysis has dramatically progressed over the years, greatly helped by the considerable increase in computer power. However, image analysis is very dependent on the quality of the experimental data, and especially on their reproducibility. In this respect, the generalization of immobilized pH gradient has had a major impact by bringing a level of positional reproducibility that could never be achieved with conventional isoelectric focusing with ampholytes [68–70]. However, image reproducibility is a complex process, and even with the use of immobilized pH gradients, reproducibility is maximized by parallelizing the gels in dedicated instruments [71]. Fortunately enough, such parallel electrophoresis instruments had been developed during the early days of 2D electrophoresis [72, 73], when the inter-run variability was very high.

Even though the reproducibility of 2D gel-based proteomics is much higher than for other setups, as shown by higher requirements [74] and stricter practices [75], there is an important concern that has emerged over the past few years, i.e., the problem of false positives. Although false positives can have an experimental origin [76], a certain proportion is due to the statistical processes used to determine modulated spots and thus to the problem of multiple testing [77–79]. Although purely statistical tools such as false discovery estimates have been proposed to address this concern [77–79], these tools are not completely well-adapted to the analysis of 2D gel images [80].

In this game of quantitative image analysis to determine spots that show changes in the biological process of interest and thus deserve further studies, there is another experimental parameter that plays a key role besides separation reproducibility, namely, sample variability and especially biological variability, i.e., from one biological sample to another, before any technical variability introduced by the protein extraction process. This variability grows along with two parameters. One is the plasticity of the proteome, so that variability is often greater in cultured prokaryotic cells than in mammalian ones, as a consequence of the much shorter generation times. The second factor is of course the physiological and genetic heterogeneity, so that in vitro systems are less variable than in vivo systems on inbred laboratory animals, which are in turn less variable than samples obtained in conditions where neither the precise physiological state nor the genetic background can be controlled (typically human samples). In some of the latter cases, the biological variability is so high that it becomes very difficult to find any protein spots showing a statistically significant variation in the experimental process. In such cases, it is tempting to pool samples within the same experimental group, in order to average out the biological variability and facilitate the discovery of modulated proteins. However, it must be kept in mind that interindividual variations are an important part of the problems that do exist in clinics. Thus, this factor cannot be evacuated easily, and pooling must be carefully understood and controlled to achieve correct results [81, 82].

4 The Protein Unicity Issue

In the early days of protein identification from 2D gel spots, e.g., by Edman sequencing or low sensitivity mass spectrometry, the situation was simple: at most one protein was detected per spot. With the increase of sensitivity of the mass spectrometers, several proteins are more and more often identified from one 2D gel spots. This has been perceived as a problem by some scientists [83], but not by others [84]. Data obtained under the worst conditions, i.e., with sub-milligram protein loads and a high sensitivity mass spectrometer, have shown that in most of the cases, the protein identified with the most peptides in one spot accounts for more than 75% of the mass spectrometry signal, so that a "winner-takes-all" protein assignment scheme is legitimate in most cases. In the rare cases where there is still ambiguity from the mass spectrometry signal, an additional separation of the spot(s) of interest can be carried out to settle the matter [85].

5 Conclusions

At the beginning of the twenty-first century, 2D gel-based proteomics is often depicted as an outdated technique, on the basis of its poor ability to analyze membrane proteins [86] and of its moderate analysis depth [87, 88]. However, in the landscape of proteomic techniques, it still offers unique advantages that make it stand apart (and ahead) of the other proteomic setups for many applications. For example, it is the simplest technique that is able to resolve complete proteins with their trail and combination of post-translational modifications, including protein fragmentation, and this unique ability should be highly valued with our increased knowledge on the importance of post-translational modifications on protein activity. This ability has been used in a variety of studies (reviewed in [22]). Thus, 2D gel-based proteomics has still a lot to offer to the researchers who will be able to use its strengths.

References

1. MacGillivray AJ, Rickwood D (1974) The heterogeneity of mouse-chromatin nonhistone proteins as evidenced by two-dimensional polyacrylamide-gel electrophoresis and ion-exchange chromatography. Eur J Biochem 41(1):181–190

2. O'Farrell PH (1975) High resolution two-dimensional electrophoresis of proteins. J Biol Chem 250(10):4007–4021

3. Anderson NL, Taylor J, Scandora AE, Coulter BP, Anderson NG (1981) The TYCHO system for computer analysis of two-dimensional gel electrophoresis patterns. Clin Chem 27(11):1807–1820

4. Vincens P, Paris N, Pujol JL, Gaboriaud C, Rabilloud T, Pennetier JL, Matherat P, Tarroux P (1986) Hermes—a 2nd generation approach to the automatic-analysis of two-dimensional electrophoresis gels. 1. Data acquisition. Electrophoresis 7(8):347–356

5. Vincens P (1986) Hermes—a 2nd generation approach to the automatic-analysis of two-dimensional electrophoresis gels 2. Spot detection and integration. Electrophoresis 7(8):357–367

6. Vincens P, Tarroux P (1987) Hermes—a 2nd generation approach to the automatic-analysis of two-dimensional electrophoresis gels 3. Spot list matching. Electrophoresis 8(2):100–107

7. Tarroux P, Vincens P, Rabilloud T (1987) Hermes—a 2nd generation approach to the automatic-analysis of two-dimensional electrophoresis gels 4. Data-analysis. Electrophoresis 8(4):187–199

8. Pun T, Hochstrasser DF, Appel RD, Funk M, Villars-Augsburger V, Pellegrini C (1988) Computerized classification of two-dimensional gel electrophoretograms by correspondence analysis and ascendant

hierarchical clustering. Appl Theor Electrophor 1(1):3–9

9. Matsudaira P (1987) Sequence from picomole quantities of proteins electroblotted onto polyvinylidene difluoride membranes. J Biol Chem 262(21):10,035–10,038

10. Aebersold RH, Leavitt J, Saavedra RA, Hood LE, Kent SBH (1987) Internal amino-acid sequence-analysis of proteins separated by one-dimensional or two-dimensional gel-electrophoresis after insitu protease digestion on nitrocellulose. Proc Natl Acad Sci U S A 84 (20):6970–6974

11. Aebersold RH, Pipes G, Hood LE, Kent SBH (1988) N-terminal and internal sequence determination of microgram amounts of proteins separated by isoelectric-focusing in immobilized Ph gradients. Electrophoresis 9 (9):520–530

12. Rosenfeld J, Capdevielle J, Guillemot JC, Ferrara P (1992) In-gel digestion of proteins for internal sequence analysis after one- or two-dimensional gel electrophoresis. Anal Biochem 203(1):173–179

13. Rasmussen HH, van Damme J, Puype M, Gesser B, Celis JE, Vandekerckhove J (1992) Microsequences of 145 proteins recorded in the two-dimensional gel protein database of normal human epidermal keratinocytes. Electrophoresis 13(12):960–969

14. Yates JR, Eng JK, McCormack AL, Schieltz D (1995) Method to correlate tandem massspectra of modified peptides to amino-acidsequences in the protein database. Anal Chem 67(8):1426–1436

15. Yates JR, McCormack AL, Schieltz D, Carmack E, Link A (1997) Direct analysis of protein mixtures by tandem mass spectrometry. J Protein Chem 16(5):495–497

16. Rogowska-Wrzesinska A, Le Bihan MC, Thaysen-Andersen M, Roepstorff P (2013) 2D gels still have a niche in proteomics. J Proteome 88:4–13

17. Smolka M, Zhou H, Aebersold R (2002) Quantitative protein profiling using two-dimensional gel electrophoresis, isotope-coded affinity tag labeling, and mass spectrometry. Mol Cell Proteomics 1(1):19–29

18. Zhou SB, Bailey MJ, Dunn MJ, Preedy VR, Emery PW (2005) A quantitative investigation into the losses of proteins at different stages of a two-dimensional gel electrophoresis procedure. Proteomics 5(11):2739–2747

19. Santoni V, Molloy M, Rabilloud T (2000) Membrane proteins and proteomics: un amour impossible? Electrophoresis 21 (6):1054–1070

20. Eravci M, Fuxius S, Broedel O, Weist S, Krause E, Stephanowitz H, Schluter H, Eravci S, Baumgartner A (2008) The whereabouts of transmembrane proteins from rat brain synaptosomes during two-dimensional gel electrophoresis. Proteomics 8 (9):1762–1770

21. Rabilloud T, Vaezzadeh AR, Potier N, Lelong C, Leize-Wagner E, Chevallet M (2009) Power and limitations of electrophoretic separations in proteomics strategies. Mass Spectrom Rev 28(5):816–843

22. Rabilloud T, Chevallet M, Luche S, Lelong C (2010) Two-dimensional gel electrophoresis in proteomics: past, present and future. J Proteome 73(11):2064–2077

23. Rabilloud T (2010) Variations on a theme: changes to electrophoretic separations that can make a difference. J Proteome 73 (8):1562–1572

24. Miller I, Crawford J, Gianazza E (2006) Protein stains for proteornic applications: which, when, why? Proteomics 6(20):5385–5408

25. Bonner WM, Laskey RA (1974) Film detection method for tritium-labeled proteins and nucleic-acids in polyacrylamide gels. Eur J Biochem 46(1):83–88

26. Perng GS, Rulli RD, Wilson DL, Perry GW (1988) A comparison of fluorographic methods for the detection of S-35-labeled proteins in polyacrylamide gels. Anal Biochem 173 (2):387–392

27. Laskey RA, Mills AD (1975) Quantitative film detection of H-3 and C-14 in polyacrylamide gels by fluorography. Eur J Biochem 56 (2):335–341

28. Duncan R, McConkey EH (1982) How many proteins are there in a typical mammalian cell? Clin Chem 28(4 Pt 2):749–755

29. Bravo R, Fey SJ, Bellatin J, Larsen PM, Arevalo J, Celis JE (1981) Identification of a nuclear and of a cytoplasmic polypeptide whose relative proportions are sensitive to changes in the rate of cell proliferation. Exp Cell Res 136 (2):311–319

30. Bravo R, Frank R, Blundell PA, Macdonald-Bravo H (1987) Cyclin/PCNA is the auxiliary protein of DNA polymerase-delta. Nature 326 (6112):515–517

31. Bravo R, Fey SJ, Small JV, Larsen PM, Celis JE (1981) Coexistence of 3 major isoactins in a single sarcoma-180 cell. Cell 25(1):195–202

32. Sobel A, Tashjian AH Jr (1983) Distinct patterns of cytoplasmic protein phosphorylation related to regulation of synthesis and release of prolactin by GH cells. J Biol Chem 258 (17):10,312–10,324

33. Patterson SD, Latter GI (1993) Evaluation of storage phosphor imaging for quantitative-analysis of 2-D Gels using the Quest-Ii system. Biotechniques 15(6):1076

34. Zhou SB, Mann CJ, Dunn MJ, Preedy VR, Emery PW (2006) Measurement of specific radioactivity in proteins separated by two-dimensional gel electrophoresis. Electrophoresis 27(5–6):1147–1153

35. Choi JK, Tak KH, Jin LT, Hwang SY, Kwon TI, Yoo GS (2002) Background-free, fast protein staining in sodium dodecyl sulfate polyacrylamide gel using counterion dyes, zincon and ethyl violet. Electrophoresis 23(24):4053–4059

36. Neuhoff V, Arold N, Taube D, Ehrhardt W (1988) Improved staining of proteins in polyacrylamide gels including isoelectric focusing gels with clear background at nanogram sensitivity using coomassie brilliant blue G-250 and R-250. Electrophoresis 9(6):255–262

37. Switzer RC, Merril CR, Shifrin S (1979) Highly sensitive silver stain for detecting proteins and peptides in polyacrylamide gels. Anal Biochem 98(1):231–237

38. Rabilloud T (1990) Mechanisms of protein silver staining in polyacrylamide gels—a 10-year synthesis. Electrophoresis 11(10):785–794

39. Blum H, Beier H, Gross HJ (1987) Improved silver staining of plant-proteins, RNA and DNA in polyacrylamide gels. Electrophoresis 8(2):93–99

40. Chevallet M, Luche S, Rabilloud T (2006) Silver staining of proteins in polyacrylamide gels. Nat Protoc 1(4):1852–1858

41. Scheler C, Lamer S, Pan Z, Li XP, Salnikow J, Jungblut P (1998) Peptide mass fingerprint sequence coverage from differently stained proteins on two-dimensional electrophoresis patterns by matrix assisted laser desorption/ionization-mass spectrometry (MALDI-MS). Electrophoresis 19(6):918–927

42. Chevalier F, Centeno D, Rofidal V, Tauzin M, Martin O, Sommerer N, Rossignol M (2006) Different impact of staining procedures using visible stains and fluorescent dyes for large-scale investigation of proteomes by MALDI-TOF mass spectrometry. J Proteome Res 5(3):512–520

43. Shevchenko A, Wilm M, Vorm O, Mann M (1996) Mass spectrometric sequencing of proteins from silver stained polyacrylamide gels. Anal Chem 68(5):850–858

44. Richert S, Luche S, Chevallet M, Van Dorsselaer A, Leize-Wagner E, Rabilloud T (2004) About the mechanism of interference of silver staining with peptide mass spectrometry. Proteomics 4(4):909–916

45. Chevallet M, Luche S, Diemer H, Strub JM, Van Dorsselaer A, Rabilloud T (2008) Sweet silver: a formaldehyde-free silver staining using aldoses as developing agents, with enhanced compatibility with mass spectrometry. Proteomics 8(23–24):4853–4861

46. Jackowski G, Liew CC (1980) Fluorescamine staining of non-histone chromatin proteins as revealed by two-dimensional polyacrylamide-gel electrophoresis. Anal Biochem 102(2):321–325

47. Urwin VE, Jackson P (1991) A multiple high-resolution mini 2-dimensional polyacrylamide-gel electrophoresis system—imaging 2-dimensional gels using a cooled charge-coupled device after staining with silver or labeling with fluorophore. Anal Biochem 195(1):30–37

48. Unlu M, Morgan ME, Minden JS (1997) Difference gel electrophoresis: a single gel method for detecting changes in protein extracts. Electrophoresis 18(11):2071–2077

49. Tonge R, Shaw J, Middleton B, Rowlinson R, Rayner S, Young J, Pognan F, Hawkins E, Currie I, Davison M (2001) Validation and development of fluorescence two-dimensional differential gel electrophoresis proteomics technology. Proteomics 1(3):377–396

50. Suzuki Y, Yokoyama K (2008) Design and synthesis of ICT-based fluorescent probe for high-sensitivity protein detection and application to rapid protein staining for SDS-PAGE. Proteomics 8(14):2785–2790

51. Cong WT, Jin LT, Hwang SY, Choi JK (2008) Fast fluorescent staining of protein in sodium dodecyl sulfate polyacrylamide gels by palmatine. Electrophoresis 29(2):417–423

52. Mackintosh JA, Choi HY, Bae SH, Veal DA, Bell PJ, Ferrari BC, Van Dyk DD, Verrills NM, Paik YK, Karuso P (2003) A fluorescent natural product for ultra sensitive detection of proteins in one-dimensional and two-dimensional gel electrophoresis. Proteomics 3(12):2273–2288

53. Berggren K, Chernokalskaya E, Steinberg TH, Kemper C, Lopez MF, Diwu Z, Haugland RP, Patton WF (2000) Background-free, high sensitivity staining of proteins in one- and two-dimensional sodium dodecyl sulfate-polyacrylamide gels using a luminescent ruthenium complex. Electrophoresis 21(12):2509–2521

54. Rabilloud T, Strub JM, Luche S, van Dorsselaer A, Lunardi J (2001) Comparison between sypro ruby and ruthenium II tris (bathophenanthroline disulfonate) as

fluorescent stains for protein detection in gels. Proteomics 1(5):699–704

55. Lamanda A, Zahn A, Roder D, Langen H (2004) Improved ruthenium II tris (bathophenantroline disulfonate) staining and destaining protocol for a better signal-to-background ratio and improved baseline resolution. Proteomics 4(3):599–608

56. Fazekas SDS, Webster RG, Datyner A (1963) 2 New staining procedures for quantitative estimation of proteins on electrophoretic strips. Biochim Biophys Acta 71(2):377–391

57. Steinberg TH, Jones LJ, Haugland RP, Singer VL (1996) SYPRO orange and SYPRO red protein gel stains: one-step fluorescent staining of denaturing gels for detection of nanogram levels of protein. Anal Biochem 239 (2):223–237

58. Malone JP, Radabaugh MR, Leimgruber RM, Gerstenecker GS (2001) Practical aspects of fluorescent staining for proteomic applications. Electrophoresis 22(5):919–932

59. Daban JR, Bartolome S, Samso M (1991) Use of the hydrophobic probe nile red for the fluorescent staining of protein bands in sodium dodecyl-sulfate polyacrylamide gels. Anal Biochem 199(2):169–174

60. Steinberg TH, Lauber WM, Berggren K, Kemper C, Yue S, Patton WF (2000) Fluorescence detection of proteins in sodium dodecyl sulfate-polyacrylamide gels using environmentally benign, nonfixative, saline solution. Electrophoresis 21(3):497–508

61. Luche S, Lelong C, Diemer H, Van Dorsselaer A, Rabilloud T (2007) Ultrafast coelectrophoretic fluorescent staining of proteins with carbocyanines. Proteomics 7 (18):3234–3244

62. Hart C, Schulenberg B, Steinberg TH, Leung WY, Patton WF (2003) Detection of glycoproteins in polyacrylamide gels and on electroblots using Pro-Q Emerald 488 dye, a fluorescent periodate Schiff-base stain. Electrophoresis 24 (4):588–598

63. Schulenberg B, Goodman TN, Aggeler R, Capaldi RA, Patton WF (2004) Characterization of dynamic and steady-state protein phosphorylation using a fluorescent phosphoprotein gel stain and mass spectrometry. Electrophoresis 25(15):2526–2532

64. Garrels JI (1979) Two dimensional gel electrophoresis and computer analysis of proteins synthesized by clonal cell lines. J Biol Chem 254(16):7961–7977

65. Vo KP, Miller MJ, Geiduschek EP, Nielsen C, Olson A, Xuong NH (1981) Computer analysis of two-dimensional gels. Anal Biochem 112 (2):258–271

66. Tarroux P (1983) Analysis of protein-patterns during differentiation using 2-D electrophoresis and computer multidimensional classification. Electrophoresis 4(1):63–70

67. Appel R, Hochstrasser D, Roch C, Funk M, Muller AF, Pellegrini C (1988) Automatic classification of two-dimensional gel electrophoresis pictures by heuristic clustering analysis: a step toward machine learning. Electrophoresis 9(3):136–142

68. Corbett JM, Dunn MJ, Posch A, Gorg A (1994) Positional reproducibility of protein spots in 2-dimensional polyacrylamide-gel electrophoresis using immobilized Ph gradient isoelectric-focusing in the first dimension—an interlaboratory comparison. Electrophoresis 15(8–9):1205–1211

69. Blomberg A, Blomberg L, Norbeck J, Fey SJ, Larsen PM, Larsen M, Roepstorff P, Degand H, Boutry M, Posch A, Gorg A (1995) Interlaboratory reproducibility of yeast protein-patterns analyzed by immobilized Ph gradient 2-dimensional gel-electrophoresis. Electrophoresis 16(10):1935–1945

70. Choe LH, Lee KH (2003) Quantitative and qualitative measure of intralaboratory two-dimensional protein gel reproducibility and the effects of sample preparation, sample load, and image analysis. Electrophoresis 24 (19–20):3500–3507

71. Eravci M, Fuxius S, Broedel O, Weist S, Eravci S, Mansmann U, Schluter H, Tiemann J, Baumgartner A (2007) Improved comparative proteome analysis based on two-dimensional gel electrophoresis. Proteomics 7(4):513–523

72. Anderson NG, Anderson NL (1978) Analytical techniques for cell fractions. 21. 2-Dimensional analysis of serum and tissue proteins—multiple isoelectric-focusing. Anal Biochem 85(2):331–340

73. Anderson NL, Anderson NG (1978) Analytical techniques for cell fractions. 22. 2-Dimensional analysis of serum and tissue proteins—multiple gradient-slab gel-electrophoresis. Anal Biochem 85(2):341–354

74. Celis JE (2004) Gel-based proteomics: what does MCP expect? Mol Cell Proteomics 3 (10):949

75. Hackett M (2008) Science, marketing and wishful thinking in quantitative proteomics. Proteomics 8(22):4618–4623

76. Fuxius S, Eravci M, Broedel O, Weist S, Mansmann U, Eravci S, Baumgartner A (2008) Technical strategies to reduce the amount of "false significant" results in quantitative proteomics. Proteomics 8 (9):1780–1784

77. Karp NA, Lilley KS (2007) Design and analysis issues in quantitative proteomics studies. Proteomics 7(Suppl 1):42–50

78. Karp NA, McCormick PS, Russell MR, Lilley KS (2007) Experimental and statistical considerations to avoid false conclusions in proteomics studies using differential in-gel electrophoresis. Mol Cell Proteomics 6 (8):1354–1364

79. Eravci M, Mansmann U, Broedel O, Weist S, Buetow S, Wittke J, Brunkau C, Hummel M, Eravci S, Baumgartner A (2009) Strategies for a reliable biostatistical analysis of differentially expressed spots from two-dimensional electrophoresis gels. J Proteome Res 8(5):2601–2607

80. Diz AP, Carvajal-Rodriguez A, Skibinski DO (2011) Multiple hypothesis testing in proteomics: a strategy for experimental work. Mol Cell Proteomics 10(3):M110.004374

81. Diz AP, Truebano M, Skibinski DO (2009) The consequences of sample pooling in proteomics: an empirical study. Electrophoresis 30 (17):2967–2975

82. Karp NA, Lilley KS (2009) Investigating sample pooling strategies for DIGE experiments to address biological variability. Proteomics 9 (2):388–397

83. Campostrini N, Areces LB, Rappsilber J, Pietrogrande MC, Dondi F, Pastorino F, Ponzoni M, Righetti PG (2005) Spot overlapping in two-dimensional maps: a serious problem ignored for much too long. Proteomics 5 (9):2385–2395

84. Hunsucker SW, Duncan MW (2006) Is protein overlap in two-dimensional gels a serious practical problem? Proteomics 6(5):1374–1375

85. Colignon B, Raes M, Dieu M, Delaive E, Mauro S (2013) Evaluation of three-dimensional gel electrophoresis to improve quantitative profiling of complex proteomes. Proteomics 13(14):2077–2082

86. Rabilloud T (2009) Membrane proteins and proteomics: love is possible, but so difficult. Electrophoresis 30(Suppl 1):S174–S180

87. Petrak J, Ivanek R, Toman O, Cmejla R, Cmejlova J, Vyoral D, Zivny J, Vulpe CD (2008) Deja vu in proteomics. A hit parade of repeatedly identified differentially expressed proteins. Proteomics 8(9):1744–1749

88. Wang P, Bouwman FG, Mariman EC (2009) Generally detected proteins in comparative proteomics—a matter of cellular stress response? Proteomics 9(11):2955–2966

Chapter 5

Proteome Analysis with Classical 2D-PAGE

Caroline May, Frederic Brosseron, Kathy Pfeiffer, Kristin Fuchs, Helmut E. Meyer, Barbara Sitek, and Katrin Marcus

Abstract

Two-dimensional polyacrylamide gel electrophoresis (2D-PAGE) is based on the combination of two orthogonal separation techniques. In the first dimension, proteins are separated by their isoelectric point, a technique known as isoelectric focusing (IEF). There are two important variants of IEF, which are carrier-ampholine (CA)-based IEF and immobilized pH-gradient (IPG)-based IEF. In the second dimension, proteins are further separated by their electrophoretic mobility using SDS-PAGE. Finally, proteins can be visualized and quantified by different staining procedures such as Coomassie, silver staining, or fluorescence labeling. This article gives detailed protocols for 2D-PAGE, using both CA- and IPG-based separation in the first dimension.

Key words Two-dimensional polyacrylamide gel Electrophoresis (2D-PAGE), Carrier ampholyte system (CA), Isoelectric focusing (IEF), Immobilized pH gradient (IPG), Sodium dodecyl sulfate (SDS)

1 Introduction

Capturing the content of the proteome under defined circumstances is one major goal of the proteome analysis [1]. In the past decades, several methods were developed to study the proteome. Gel-based approaches, especially two-dimensional polyacrylamide gel electrophoresis (2D-PAGE), for the separation and qualitative, as well as quantitative analyses of the proteome, are, next to mass spectrometry approaches, the most frequently used techniques in proteomics (for review, *see* Chap. 4).

2D-PAGE is a combination of two orthogonal separation techniques and has an essentially higher resolution than 1D-PAGE [2]. In the first dimension, proteins are separated by their isoelectric point (isoelectric focusing; IEF) and in the second dimension by their electrophoretic mobility with sodium dodecyl sulfate (SDS) as negatively charged detergent using conventional SDS polyacrylamide gel electrophoresis (SDS-PAGE) according to Laemmli

Katrin Marcus et al. (eds.), *Quantitative Methods in Proteomics*, Methods in Molecular Biology, vol. 2228,
https://doi.org/10.1007/978-1-0716-1024-4_5, © Springer Science+Business Media, LLC, part of Springer Nature 2021

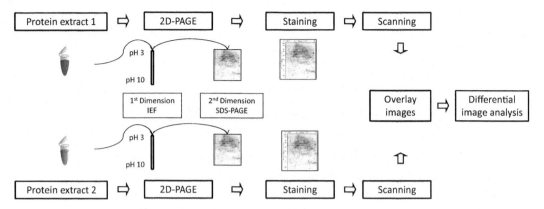

Fig. 1 Overview of 2D-PAGE workflow. Classical 2D-PAGE is a combination of two orthogonal gel-based separation approaches: IEF as first dimension and SDS-PAGE as second dimension. After gel electrophoresis, the proteins are visualized with an appropriate staining method (e.g., silver stain), and the gels are digitalized using conventional scanners. Protein patterns of two different samples can be compared manually or using image analysis software to find differentially expressed or modified proteins

[3]. By combination of those two techniques, 2D-PAGE allows the separation of up to 10,000 protein species and therefore a global quantitative proteome analysis, pinpointing relative changes of respective protein species in different samples [4–6]. After separation, proteins can be visualized (*see* Chaps. 6 and 7), cut out of the gel, and identified by mass spectrometry. 2D-PAGE in combination with image analysis and mass spectrometry is still a highly popular method for the analysis of complex protein mixtures (Fig. 1) [7].

Two different IEF techniques are generally applied for proteome analysis nowadays [8, 9]. The first is a method introduced by Klose and O'Farrell, where the pH gradient is formed during the focusing process via amphoteric, oligoaminooligocarbonic acids with high buffer capacity at their pI (carrier ampholytes, (CA-based IEF) [10, 11]. The second method, described by Bjellqvist and Görg, is based on an immobilized pH gradient (IPG-based IEF), which is directly polymerized into the gel [12, 13]. Due to its simple handling and commercialization, IPG-based IEF is typically used for 2D-PAGE-based proteome analysis allowing widespread applications. A number of different IPG strips with variations in length (7–24 cm) and pH range (narrow or broad, e.g., pH 4–5 or 3–11; linear or nonlinear) are provided by different manufacturers (Fig. 2). Both types of IEF can be combined with conventional SDS-PAGE as second dimension. Afterward, the protein pattern can be visualized by different staining methods, e.g., colloidal Coomassie, silver staining (*see* Chap. 6), Sypro Ruby, or imidazole-zinc, all differing in sensitivity and dynamic detection range [8].

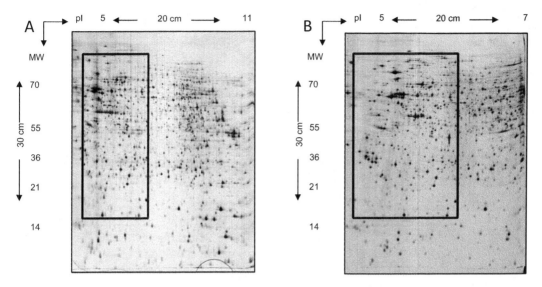

Fig. 2 2D-PAGE of SH-SY5Y cell lysate using different IPG gradients. (**a**) Wide-range IPG for focusing between pH 3 and 11. This range comprises the p*I* of the majority of cytoplasmic proteins. (**b**) Narrow-range IPG for focusing between pH 4 and 7. The resolution is enhanced for more acidic proteins, while proteins with a p*I* higher than 7 are not displayed. The basic region is compressed, whereas proteins with acidic p*I* are significantly better resolved. The black rectangle points the same p*I* and MW region in both gels out to indicate the differences in resolution

2   Materials

2.1   Technical Equipment (See Note 1)

2.1.1   IPG-Based IEF

1. Horizontal electrophoresis apparatus incl. Electrodes.
2. Power supply.
3. Thermostatic circulator.
4. Immobiline DryStrips (*see* **Note 2**).
5. Reswelling cassette.
6. Electrode papers.
7. Sample cups.
8. Sample cup holders.

2.1.2   CA-Based IEF

1. Vertical electrophoresis apparatus and 22 cm glass tubes, with three marks at 20, 20.5, and 21 cm (handmade as described by Klose [14].
2. Power supply.
3. Thermostatic circulator.
4. Gel carrier grooves (handmade as described by Klose [14].
5. Power supply.
6. Water line recirculator.

2.1.3 SDS-PAGE

1. Running chamber (Desaphor VA300,Desaga, Heidelberg, Germany).

2. Glass plates, cramps, and spacer.

3. Gel dryer.

2.2 Buffers and Solutions

2.2.1 IPG-Based IEF

1. Rehydration buffer: 8 M (w/v) urea, 2 M (w/v) thiourea, 2% (w/v), 3-[(cholamidopropyl)-dimethylammonio]-propanesulfonic acid (CHAPS), 2% (v/v) Servalyte, 100 mM (v/v) DTT (freshly added, store as 2 M stock solution at −20 °C) (*see* **Notes 3–5**).

2. Silicone oil.

3. Equilibration solution: 6 M (w/v) urea, 2% (w/v) SDS, 30% (w/v) glycerol, 3.3% (v/v) separation gel buffer, 65 mM (w/v) DTT, 280 mM (w/v) iodoacetamide (IAA).

2.2.2 CA-Based IEF

1. CA mixture (pH 2–11): 12.5% ampholine (pH 3.5–10), 12.5% Servalyte (pH 2–4), 25% Pharmalyte (pH 5–8), 37.5% Pharmalyte (pH 4–6.5), 12.5% Servalyte (pH 6.5–9) (*see* **Notes 3 and 4**).

2. Separation gel solution (degassed) (pH 2–11): 3.5% (w/v) acrylamide, 0.3% (w/v) piperazindiacrylamide, 4% (v/v) CA mixture (pH 2–11), 9 M urea, 5% glycerol, 0.06% (v/v) N, N, N′, N′-tetramethylethylenediamine (TEMED) (*see* **Note 5**).

3. Cap gel solution: 12.3% (w/v) acrylamide, 0.13% (w/v) piperazindiacrylamide, 4% (v/v) CA mixture (pH 2–11), 9 M urea, 5% glycerol, 0.06% (v/v) TEMED, 1.2% (w/v) ammonium persulfate (APS).

4. Sephadex solution: 272 mg Sephadex suspension (20 g Sephadex swollen in 500 mL aqua bidest and resuspended in 1000 mL 25% (w/v) glycerol solution and filtered), 233 mg urea (7 M final), 98 mg thiourea (2 M final), 25 μL ampholine mixture, 25 μL DTT (1.08 g/5 mL, 1.4 M).

5. Protection solution: 30% (w/v) urea, 5% (w/v) glycerol, 2% Servalyte (pH 2–4).

6. Anodic solution: 3 M (w/v) urea, 4.3% (w/v) phosphoric acid.

7. Cathodic solution: 9 M (w/v) urea, 5% (w/v) glycerol, 0.75 M (v/v) ethylenediamine.

8. Equilibration solution: 125 mM (w/v) Tris, 40% (w/v) glycerol, 65 mM (w/v) DTT, 3% (w/v) SDS.

2.2.3 SDS-PAGE

1. Running buffer: 20 mM (w/v) Tris, 0.01% (w/v) SDS, 0.19 mM (w/v) glycine (*see* **Notes 3 and 4**).

2. Separation gel buffer (15%): 375 mM (w/v) Tris, 0.1% (w/v) SDS, 0.03% (v/v) TEMED, 15% (v/v) acrylamide, 0.2% (v/v) bis-acrylamide, 40% (w/v) APS.

3. Protection solution: 375 mM (w/v) Tris/Tris–HCL 82.4:1), 0.1% SDS.

4. Agarose solution: 0.4% (w/v) agarose (in running buffer), 0.01% (w/v) bromophenol blue.

3   Methods

3.1   IPG-Based IEF

The method is modified from the IPG-based IEF originally described by Görg et al. [13]. The sample can be applied in two different ways, either by sample cup loading [13] or by in-gel rehydration [15]. The application via sample cup will be described in detail for 24 cm strips.

1. Rehydrate the IPG gel stripes upside down before starting the IEF for 6 h at RT in 550 μL rehydration solution.

2. Place the aligned tray in the horizontal electrophoresis apparatus.

3. Add the rehydrated IPG strips gel-side up with the basic end toward the cathode in the aligned tray.

4. Apply one electrode paper and soaked with water to each end of the strips.

5. Position the electrodes on the top of the electrode strips and fix gently.

6. Place and fix the sample cup on the more acid or basic end of the strip depending on your sample characteristics (*see* **Notes 6–10**).

7. Pipette sample into the sample cup. The sample volume should not exceed 150 μL.

8. Cover the sample in the sample cup and the strip with silicone oil.

9. Start isoelectric focusing by applying a stepwise voltage gradient. The optimal gradient depends on the sample and on the IPG strip. For samples with a low salt concentration, the following program seems to be optimal: gradient 300 V for 3 h, gradient 1000 V for 4 h, hold 1000 V for 4 h, gradient 8000 V 2 h, hold 8000 V for 5 h, and hold 500 V till end. The ampere should be limited to 75 μA/strip and the temperature to 20 °C (*see* **Note 7**).

10. After the IEF run, the strips can be stored at −80 °C.

11. Before applying on the SDS gels, the strips have to be equilibrated in two 15 min steps under gentle shaking in suitable

vessels, e.g., plastic tubes. The first step is reduction of the proteins by addition of. 65 mM DTT to the equilibration solution. After 15 min, the solution is replaced by equilibration buffer containing 250 mM IAA for protein alkylation. After that, the buffer will be removed.

12. Rinse the strip three times in running buffer using a plastic Pasteur pipette.

3.2　CA-Based IEF

The CA-based IEF gels described in the following have a pH gradient from 2 to 11. Perform all steps at room temperature. Isoelectric focusing of the proteins is accomplished by application of a voltage gradient. Keep in mind that several substances can have negative influence on the IEF (*see* **Notes 7–10**). Afterward, the gels are incubated in an incubation buffer for SDS charging, which is essential for the SDS-PAGE afterward. The glass tubes used (Schott AG, Wertheim, Germany) have a length of 22 cm and an inner diameter of 1.5 mm (modified according to [4]).

1. Place a syringe at that end of the glass tube.

2. Add 45 μL APS to the separation gel solution, shake it gently to avoid air bubbles, and pull the syringe until the separation gel solution reaches the 20 cm mark.

3. In the next step, add 14 μL APS to the cap gel solution, shake gently, and pull the syringe until the gel solution reaches the 20.5 cm mark.

4. In a third step, pull the syringe until the 21 cm mark in order to get an air cushion, which should avoid urea crystallization.

5. The polymerization takes 2 days at RT. Temperature variability should be avoided to prevent urea crystallization. A heating cabinet may be useful. Tubes should be stored in wet tissue during polymerization.

6. Place Sephadex solution 2 mm high on top of the gel to prevent sample precipitation at the border of the gel. Afterward, place the sample on the Sephadex. Avoid air bubbles.

7. The protection solution is applied on the sample to prevent a direct contact of the sample with the cathodic focusing buffer.

8. Air bubbles have to be prevented in every step, to avoid disturbance of the focusing.

9. The bottom of the chamber is filled with anodic and the top with cathodic buffer.

10. Isoelectric focusing takes place by applying a stepwise voltage program:

    100 V for 1 h, 200 V for 1 h, 400 V for 17.5 h, 650 V for 1 h, 1000 V for 0.5 h, 1500 V for 10 min, and 2000 V for 5 min (*see* **Notes 7, 11**).

11. After the isoelectric focusing is finished, the round gels are extruded of the glass tubes into the gel groove by using a nylon fiber.

12. Incubate the round gel for 15 min in equilibration solution.

13. Remove the equilibration solution and rinse the gel three times in the running buffer.

3.3 SDS-PAGE

1. Clean the glass plates before assembling with water, followed by 100% and 70% ethanol with a lint-free tissue.

2. A thin line of silicone is placed on both edges of the glass plates, followed by two spacers and another glass plate with silicone (clamp in place).

3. Add 144 μL APS to 77 mL gel solution, mix gently to avoid air bubbles, and cast the gel top down between the glass plates in the polymerization chamber until 0.5 cm below the top if CA-based IEF is used. If using IPG-based IEF, leave 0.5–1.0 cm below the top to apply the slightly broader IPG stripes (*see* **Note 12**).

4. Cover the separation gel with isopropanol to prevent air inclusion and for getting a straight surface.

5. Polymerization takes 45 min.

6. Remove isopropyl alcohol and wash the surface once with protection solution.

7. The gels can be stored with an overlay of protection solution at 4 °C for up to 1 week.

8. Before placing the round CA gel or the IPG strip on the top of the gel, rinse it three times with running buffer. Tilt the gel sideward and use a filter paper on one end of the gel to remove the remaining buffer. Still, the surface should not be completely dry.

9. Apply the strip between the glass plates on the top of the SDS gel using tweezers and spatulas.

10. Apply the round CA gels by placing the groove at the edge of the glass plate of the SDS gel and transfer the gel with a suitable formed wire to the gel surface.

11. Fix the strip or the round gel with molten agarose solution.

12. Fill up the electrophoresis chamber with running buffer.

13. Insert the gels into the electrophoresis chamber.

14. Fill up with running buffer and start focusing. For protein entry into the SDS gel, the electrical power should be limited to 2 W/gel for 45 min. The separation is done by 20 W/gel for 4 h. The temperature for the whole run should be limited to 20 °C.

15. For storage and documentation, the gels can be dried to avoid protein loss and to improve gel stability. Thus, it is possible to make an image analysis in a differential study and to identify the differential proteins later, out of the dried gels (*see* **Note 13**).

4   Notes

1. Suitable equipment is available from GE Healthcare Life Sciences, Munich, Germany, or Bio-Rad, Munich, Germany.

2. The optimal pH range for the first dimension depends on the sample and should be analyzed to optimize the resolution. Most cytosolic proteins seem to have a pI between 4 and 8 and nuclear proteins between 8 and 10 [16]. Nonlinear gradients in these pH ranges could also improve resolution. For the analyses of whole cell lysates, nonlinear IPG stripes from pH 3–10, with a spreading of the gradient between 4 and 8, provide a good coverage of proteins.

3. For all solutions and buffers, use high-purity water and p.a. reagents and never change reagents during one study.

4. To ensure a high reproducibility and to minimize the technical variability of 2D gels, all solutions and buffers should be produced in large batches for one proteome study. Most of the buffers and solutions can be stored in aliquots at −20 °C or −80 °C. Sample application via sample cup should be done on the basic or acidic end of the IPG stripe depending on where most of the proteins in the sample have their pI.

5. Avoid heating of urea solutions above 37 °C to reduce the risk for protein carbamylation.

6. If the majority of the proteins in the sample have an acidic pI, apply the sample cup at the acidic site of the IPG stripe and vice versa.

7. If protease inhibitors (which often contain various types of salts) or buffers, as well as solutions with high concentrations of salts, are used, elongate the sample entry phase to remove salts during focusing. High concentrations of salts may cause an inefficient resolution of the proteins during IEF. However, CA-based IEF is less sensitive to disturbances from these sources than IPG-based IEF.

8. Lipids should be removed during sample preparation, because they can disturb the sample entry into the first dimension.

9. The sample preparation protocol should be as simple as possible to avoid protein loss and artificial variations in the interesting proteome. Various types of sample preparation protocols

for tissues and cell types have already been described in literature.

10. The protein concentration should be determined, and a defined protein amount should be applied on the gels to allow the comparison of proteomes. The sample amount which is necessary to detect as many proteins as possible depends on the staining or labeling method which is used.

11. Overfocusing results in horizontal streaking should be prevented; therefore the total kVh count should be in the range between 50 and 60 kVh.

12. The percentage of the acrylamide in the SDS gels can be varied depending on the molecular weight of the proteins, which should be detected.

13. After finishing the focusing, the protein patterns can be visualized by different staining techniques, e.g., Coomassie or silver (*see* Chap. 6) staining.

Acknowledgments

This work was supported by the Bundesministerium für Bildung und Forschung (NGFN, FZ 01GS08143) as well as the European Regional Development Fond (ERDF) of the European Union and the Ministerium für Innovation, Wissenschaft und Forschung des Landes Nordrhein-Westfalen (ParkChip,FZ 280381102), and P.U.R.E. (Protein Research Unit Ruhr within Europe), Ministry of Innovation, Science and Research of North-Rhine Westphalia, Germany a federal German state.

References

1. Wilkins MR, Sanchez JC, Gooley AA et al (1996) Progress with proteome projects: why all proteins expressed by a genome should be identified and how to do it. Biotechnol Genet Eng Rev 13:19–50

2. Rabilloud T, Chevallet M, Luche S et al (2010) Two-dimensional gel electrophoresis in proteomics: past, present and future. J Proteome 73(11):2064–2077

3. Laemmli UK (1970) Cleavage of structural proteins during the assembly of the head of bacteriophage T4. Nature 227:680–685

4. Klose J, Kobalz U (1995) Two-dimensional electrophoresis of proteins: an updated protocol and implications for a functional analysis of the genome. Electrophoresis 16:1034–1059

5. Nesterenko MV, Tilley M, Upton SJ (1994) A simple modification of Blum's silver stain method allows for 30 minutes detection of proteins in polyacrylamide gels. J Biochem Biophys Methods 28:239–242

6. Neuhoff V, Stamm R, Pardowitz I et al (1990) Essential problems in quantification of proteins following colloidal staining with coomassie brilliant blue dyes in polyacrylamide gels, and their solution. Electrophoresis 11:101–117

7. May C, Brosseron F, Chartowski P et al (2011) Instruments and methods in proteomics. Methods Mol Biol 696:3–26

8. Marcus K, Joppich C, May C et al (2009) High-resolution two-dimensional gel electrophoresis (2-DE)—different methods and applications. Methods Mol Biol 519:221–240

9. Stühler K, Pfeiffer K, Joppich C et al (2006) Pilot study of the human proteome organisation brain proteome project: applying different

2-DE techniques to monitor proteomic changes during murine brain development. Proteomics 18:4899–4913

10. Klose J (1975) Protein mapping by combined isoelectric focusing and electrophoresis of mouse tissues. A novel approach to testing for induced point mutations in mammals. Humangenetik 26:231–243

11. O'Farrell PH (1975) High resolution two-dimensional electrophoresis of proteins. J Biol Chem 250:4007–4021

12. Bjellqvist B, Ek K, Righetti PG et al (1982) Isoelectric focusing in immobilized pH gradients: principle, methodology and some applications. J Biochem Biophys Methods 6:317–339

13. Görg A, Postel W, Gunther S (1988) The current state of two-dimensional electrophoresis with immobilized pH gradients. Electrophoresis 9:531–546

14. Klose J (1999) Large-gel 2-D electrophoresis. Methods Mol Biol 112:147–172

15. Rabilloud T, Valette C, Lawrence JJ (1994) Sample application by in-gel rehydration improves the resolution of two-dimensional electrophoresis with immobilized pH gradients in the first dimension. Electrophoresis 15 (12):1552–1558

16. Schwartz R, Ting CS, King J (2001) Whole proteome pI values correlate with subcellular localizations of proteins for organisms within the three domains of live. Genome Res 11:703–709

Chapter 6

Silver Staining of 2D Electrophoresis Gels

Thierry Rabilloud

Abstract

Silver staining is used to detect proteins after electrophoretic separation on polyacrylamide gels. It combines excellent sensitivity (in the low nanogram range) with the use of very simple and cheap equipment and chemicals. For its use in proteomics, two important additional features must be considered, compatibility with mass spectrometry and quantitative response. Both features are discussed in this chapter, and optimized silver staining protocols are proposed.

Key words Mass spectrometry, Quantification, Polyacrylamide gels, Protein visualization, Silver staining

1 Introduction

Silver staining of polyacrylamide gels was introduced in 1979 by Switzer et al. [1] and rapidly gained popularity owing to its high sensitivity, ca. 100 times higher than staining with classical Coomassie Blue and 10 times higher than colloidal Coomassie Blue. However, the first silver staining protocols were not trouble-free. High backgrounds and silver mirrors were frequently experienced, with a subsequent decrease in sensitivity and reproducibility. This led many authors to suggest improved protocols, so that more than 100 different silver staining protocols for proteins in polyacrylamide gels can be found in the literature. However, all of them are based on the same principle (*see* [2, 3] for details) and comprise four major steps:

(a) The first step is fixation. It precipitates the proteins in the gels and removes at the same time the interfering compounds present in the 2D gels (glycine, Tris, SDS, and carrier ampholytes present in 2D gels).

(b) The second step is sensitization and aims at increasing the subsequent image formation. Numerous compounds have been proposed for this purpose. All these compounds bind

Katrin Marcus et al. (eds.), *Quantitative Methods in Proteomics*, Methods in Molecular Biology, vol. 2228,
https://doi.org/10.1007/978-1-0716-1024-4_6, © Springer Science+Business Media, LLC, part of Springer Nature 2021

to the proteins and are also able either to bind silver ion, or to reduce silver ion into metallic silver, or to produce silver sulfide [2, 3].

(c) The third step is silver impregnation. Either plain silver nitrate or ammoniacal silver can be used, but nowadays silver nitrate is more extensively used (*see* **Note 1**).

(d) The fourth last step is image development. For gels soaked with silver nitrate, the developer contains formaldehyde, carbonate, and thiosulfate. The use of the latter compound, introduced by Blum et al. [4], reduces dramatically the background and allows for thorough development of the image. When the desired image level is obtained, development is stopped by dipping the gel in a stop solution, generally containing acetic acid and an amine to reach a pH of 7. Final stabilization of the image is achieved by thorough rinsing in water to remove all the compounds present in the gel.

However, the development of downstream protein characterization methods, such as analysis by mass spectrometry starting from gel-separated proteins, has brought several constraints to the forefront. The first is the interface with mass spectrometry, and this encompasses the compatibility with enzymatic digestion and peptide extraction, as well as the absence of staining-induced peptide modifications. While the exquisite sensitivity of silver staining is unanimously recognized, its compatibility with downstream analysis appears more problematic than staining with organic dyes (e.g., Coomassie Blue). A mechanistic study [5] has shown that these problems are linked in part to the pellicle of metallic silver deposited on the proteins during staining but are mainly due to the presence of formaldehyde during silver staining. Up to now, formaldehyde is the main chemical known able to produce a silver image of good quality in protein staining, and attempts to use other chemicals have proven rather unsuccessful. However, besides a lowered peptide representation in silver-stained gels [5], formaldehyde induces a host of peptide modifications [6, 7] as well as formylation [8], the latter being most likely caused by the formic acid produced upon reaction of formaldehyde with silver ions in the image development step. In order to minimize these problems, destruction of the remaining formaldehyde by oxidation [9] is highly recommended, and this should take place as early as possible after image development and spot excision [5]. These problems are common to all silver staining protocols, although their extent is variable from one protocol to another. Some guidelines for the choice of a silver staining protocol are described in **Note 1**. To alleviate the problems linked with the use of formaldehyde, a protocol using reducing sugars as developing agents has been described [10] and is detailed herein. Typical gels stained by the

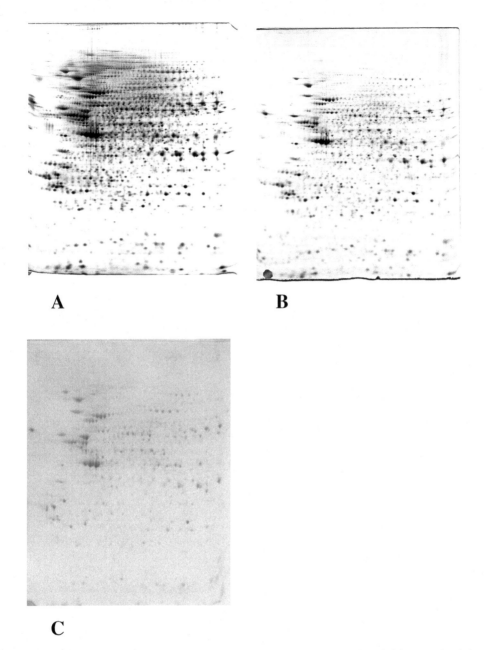

Fig. 1 200μg of proteins extracted from the murine macrophage cell line J774 were loaded on a two-dimensional gel. First dimension, 4–8 linear pH gradient; second dimension, 10% polyacrylamide gel. The protein detection method was the following: (**a**) Silver stain with formaldehyde developer, (**b**) silver stain with xylose-borate developer, (**c**) colloidal Coomassie Blue

protocols described in this chapter are shown in Fig. 1. As artifacts arising from grafting of the sugars cannot be excluded and may be confused with protein glycation occurring in vivo, pentoses such as xylose can be used for silver staining development, as they will give a peptide mass increase different than that induced by in vivo glycation.

The second constraint very prevalent in proteomics is linked to the fact that stained 2D gels are used to perform quantitative analysis. In addition to sensitivity, this puts special emphasis on stain linearity and homogeneity from run to run.

In fact, silver staining is linear over a rather limited dynamic range, as shown in Figs. 2 and 3 on 1D and 2D gels, respectively.

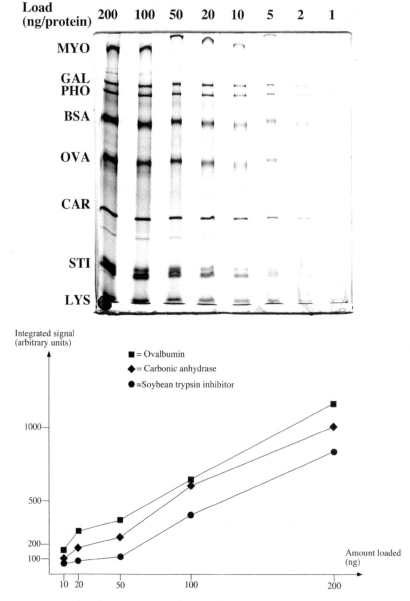

Fig. 2 Molecular mass markers (wide range from Bio-Rad) were serially diluted, loaded on a SDS-PAGE gel (10%T), and silver stained by protocol 1. The resulting image is shown on the top panel. Quantification of three of the protein bands was then achieved by the ImageJ software, and the plotted results are shown on the bottom panel. This shows the biphasic shape of the curve, with a plateau at low loads and the linear portion of the curve at higher loads

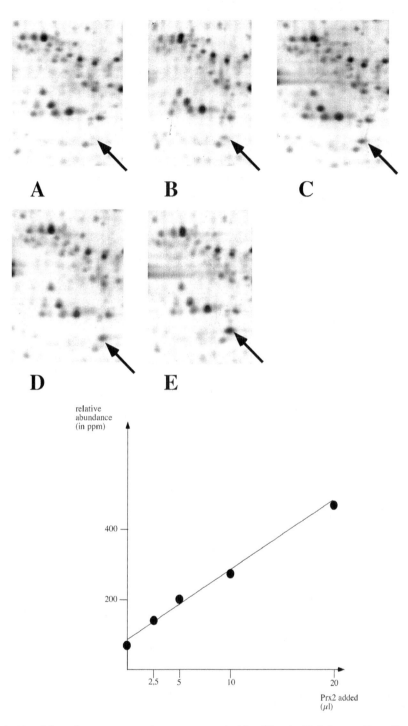

Fig. 3 A total extract from human monocytes was separated by 2D gels (4–8 linear pH gradient in the IEF dimension, 10% acrylamide in the second dimension) and silver stained (protocol 1). Prior to separation, a variable amount of a semi-purified preparation of peroxiredoxin 2 (prx2) was added to the monocyte extract. The prx2 spot is shown by an arrow in the gel excerpts shown: (**a**) no prx2 added; (**b**) 2.5μL prx2 added; (**c**) 5μL prx2 added; (**d**) 10μL prx2 added; (**e**) 20μL prx2 added. The gels were then analyzed by the Delta 2D software (Decodon) and the amount of prx2 determined (in ppm of the total spots intensities). The results are shown in the bottom part of the figure

Figure 2 shows in particular two important features of silver staining, i.e., the rather poor linearity at very low intensities and the variable slope of the dose-signal curve from one protein to the other. However, it must be emphasized that the slope of the response curve is always lesser or equal to 1. This means in turn that silver staining is a conservative technique that generally underestimates the variations in protein abundances.

Moreover, modern silver staining is no longer an erratic technique, as it was in its infancy [1]. This is shown in Fig. 4, which compares the variability of a series of 2D gels run on the same sample and stained by various methods. It can be shown that the dispersion of the signals (a measure of variability) is not greater with adequate silver staining than with Coomassie Blue. This variability can be further decreased by the use of batch apparatus for silver staining [11].

2   Materials

2.1   Equipment

1. Glass dishes or polyethylene food dishes. The latter are less expensive, have a cover, and can be easily piled up for multiple staining. They are however more difficult to clean, and it is quite important to avoid scratching of the surface, which will induce automatic silver deposition in subsequent stainings. Traces of silver are generally easily removed by wiping the plastic box with a tissue soaked with ethanol. If this treatment is not sufficient, stains are easily removed with Farmer's reducer (0.1% sodium carbonate, 0.3% potassium hexacyanoferrate (III), and 0.6% sodium thiosulfate). Thorough rinsing of the box with water and ethanol terminates the cleaning process.

2. Plastic sheets (e.g., the thin polycarbonate sheets sold by Bio-Rad for multiple gel casting) used for batch processing.

3. Reciprocal shaking platform: The use of orbital or three-dimensional movement shakers is not recommended.

2.2   Reagents and Protocols

2.2.1   Reagents

Generally speaking, chemicals are of standard pro-analysis grade:

1. Water: The quality of the water is of great importance. Water purified by ion exchange cartridges, with a resistivity greater than 15 megaohms/cm, is very adequate, while distilled water gives more erratic results.

2. Formaldehyde: Formaldehyde stands for commercial 37–40% formaldehyde. This is stable for months at room temperature. It should not be stored at 4 °C, as this increases polymerization and deposition of formaldehyde. The bottle should be discarded when a layer of polymer is visible at the bottom of the bottle.

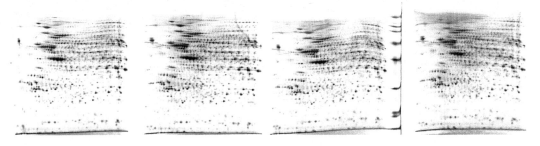

Silver stain protocol 1

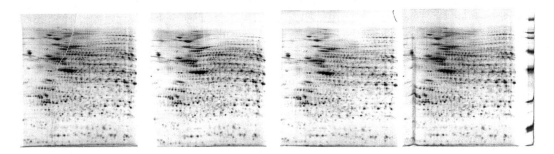

Silver stain protocol 2

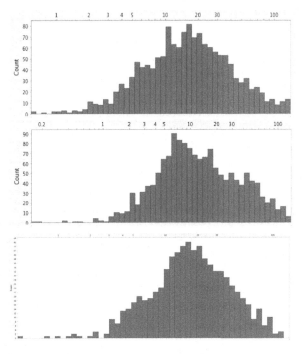

rsd silver
protocol 1
median: 20%

rsd silver
protocol 2
median: 15%

rsd colloidal
Coomassie Blue
median: 20%

Fig. 4 A total cell extract from J774 cells was separated in quadruplicate on 2D gels. One four-gel series was stained with silver and formaldehyde developer (protocol 1) and is shown on the top row. The second series was stained with silver and aldose developer (protocol 2) and is shown on the second row. The gels were then analyzed with the Delta 2D software (Decodon), and the relative standard deviation (rsd) was calculated for each spot and is then plotted as a distribution graph (bottom part of the figure). The median of the rsd was calculated and gives a measure of the variability of the process. This rsd plot is also shown for Coomassie-stained gels for comparison

3. Sodium thiosulfate solution: 10% (w/v) solution of crystalline sodium thiosulfate pentahydrate in water. Small volumes of this solution (e.g., 10 mL) are prepared fresh every week and stored at room temperature.

4. Ethanol: A technical grade of alcohol can be used, and 95% ethanol can be used instead of absolute ethanol, without any volume correction. The use of denatured alcohol is however not recommended.

5. Silver nitrate solution: 1 N silver nitrate. A 1 N silver nitrate solution (Fluka) is less expensive than solid silver nitrate and is stable for months if kept in a fridge (a black and cold place).

2.2.2 Fast Silver Staining

1. Fix solution: 5% acetic acid, 30% ethanol.

2. Sensitivity-enhancing solution: 2 mL of 10% sodium thiosulfate per liter.

3. Silver stain solution: 12.5 mL of 1 N silver nitrate per liter.

4. Development solution I: 30 g anhydrous potassium carbonate, 250 μL 37% formaldehyde, and 125 μL 10% thiosulfate per liter (*see* **Note 8**).

5. Stop solution: 40 g of Tris base and 20 mL of glacial acetic acid per liter.

2.2.3 Silver Staining with Aldose-Based Developer

1. Fix solution: 5% acetic acid/30% ethanol.

2. Sensitivity-enhancing solution: 2 mL of 10% thiosulfate per liter.

3. Silver stain solution: 12.5 mL 1 N silver nitrate per liter.

4. Development solution II: 0.1 M boric acid, 0.15 M sodium hydroxide, 2% (w/v) galactose or xylose, and 125 μL 10% thiosulfate per liter (*see* **Note 9**).

5. Stop solution: 40 g of Tris base and 20 mL of glacial acetic acid per liter.

2.2.4 Spot Destaining Prior to Mass Spectrometry

Prepare a 30 mM potassium ferricyanide solution in water. Additionally, prepare a 100 mM sodium thiosulfate solution and a 200 mM ammonium hydrogenocarbonate solution.

Caution: All solutions must be prepared the day of use.

Just before use, mix equal volumes of the potassium ferricyanide and thiosulfate solutions.

Caution: This resulting yellowish solution is stable and active for less than 30 min and must be used immediately.

3   Methods

3.1   General Practice

Batches of gels (up to four gels per box) can be stained. For a batch of three or four medium-sized gels (e.g., $160 \times 200 \times 1.5$ mm), 1 L of the required solution is used, which corresponds to a solution/gel volume ratio of at least 5. 500 mL of solution is used for one or two gels. Batch processing can be used for every step longer than 5 min, except for image development, where one gel per box is required. For steps shorter than 5 min, the gels should be dipped individually in the corresponding reagent(s).

For changing solutions, the best way is to use a plastic sheet. This is pressed on the pile of gels with the aid of a gloved hand. Inclining the entire setup allows to empty the box while keeping the gels in it. The next solution is poured with the plastic sheet in place, which prevents the flow to fold or break the gels. The plastic sheet is removed after the solution change and kept in a separate box filled with water until the next solution change. This water is changed after each complete round of silver staining.

When gels have to be handled individually, the must be manipulated with gloved hands. The use of powder-free, nitrile, gloves is strongly recommended, as standard latex gloves are often the cause of pressure marks.

Except for development or short steps, where occasional hand agitation of the staining vessel is convenient, constant agitation is required for all the steps. A reciprocal ("ping-pong") shaker is used at 30–40 strokes per minutes.

Two different silver staining protocols are detailed below. The rationale for choosing one of them according to the constraints brought by the precise 2D protocol used and the requisites of the experimentator are described in **Note 1**.

3.2   Fast Silver Staining

This protocol is based on the protocol of Blum et al. [4], with modifications [12, 13]:

1. Soak the gels in fix solution for at least 3×30 min or overnight with one solution change for 2D gels (*see* **Notes 2** and **3**).

2. Rinse in water for 3×10 min.

3. To sensitize, soak gels for 1 min (one gel at a time) in sensitivity-enhancing solution.

4. Rinse 2×1 mi in water (*see* **Note 4**).

5. Impregnate for at least 30 min in silver solution (*see* **Note 5**).

6. Rinse in water for 5–15 s (*see* **Note 6**).

7. Develop image (10–20 min) in development solution I (*see* **Note 7**).

8. Stop development (30–60 min) in stop solution.

9. Rinse with water (several changes) prior to drying or densitometry.

3.3 Silver Staining with Aldose-Based Developer

1. Soak the gels in fix solution for at least 3×30 min or overnight with one solution change for 2D gels (*see* **Notes 2** and **3**).

2. Rinse in water for 3×10 min.

3. To sensitize, soak gels for 1 min (one gel at a time) in sensitivity-enhancing solution.

4. Rinse 2×1 min in water (*see* **Note 4**).

5. Impregnate for at least 30 min in silver solution (*see* **Note 5**).

6. Rinse in water for 5–15 s (*see* **Note 6**).

7. Develop image (10–20 min) in development solution II (*see* **Notes 7** and **8**).

8. Stop development (30–60 min) in stop solution.

9. Rinse with water (several changes) prior to drying or densitometry.

3.4 Spot Destaining Prior to Mass Spectrometry

Silver staining interferes strongly with mass spectrometry analysis of spots or bands excised from stained electrophoresis gels. This interference can be reduced by destaining the spots or bands prior to the standard digestion protocols. The destaining protocols giving minimal artifacts is the ferricyanide-thiosulfate protocol of Gharahdaghi et al. [9]. This protocol can be carried out on spots or bands in microtubes (0.5 or 1.5 mL) or in 96-well plates. The use of a shaking device (plate shaker or rotating wheel for tubes) is recommended.

3.4.1 Procedure

1. Cover the spots or bands with 0.15 mL of spot destaining solution. The stain should be removed in 5–10 min.

2. Remove the solution, and rinse the spots 5×5 min with water (0.15 mL per gel piece).

3. Remove the water, and soak the gel pieces in 200 mM ammonium hydrogenocarbonate (in water) for 20 min (0.15 mL per gel piece).

4. Repeat **step 2** above.
   Process the rinsed gel pieces for mass spectrometry, or store dry at -20 °C until use.

4 Notes

1. From the rather simple theoretical bases described in the introduction, more than 100 different protocols were derived. The changes from one protocol to another are present either in the duration of the different steps or in the composition of the

solutions. The main variations concern either the concentration of the silver reagent or the nature and concentration of the sensitizers. Only a few comparisons of silver staining protocols have been published [12]. From these comparisons, selected protocols have been proposed in the former sections. The choice of a protocol will depend on the constraints of the experimental setup and of the requisites of the experimenter (speed, reproducibility, compatibility with mass spectrometry, etc.). Although they can be very sensitive for basic proteins, we have excluded from this selection protocols using silver-ammonia, as the results are fairly dependent on the ammonia/silver concentration. Furthermore, homemade gels with included thiosulfate must be used for optimal results [14], and tricine-based gels cannot be used. These restrictions have thus driven us to consider in priority protocols using plain silver nitrate or staining, as they are more robust and versatile.

2. Other fixation processes can be used. For gels running overnight, a shorter process can be used.

   For silver nitrate staining, fixation can be reduced to a single 30-min bath [15]. This will improve sequence coverage in mass spectrometry, at the expense of a strong chromatism (spots can be yellow, orange, brown, or gray), making image analysis difficult. Furthermore, ampholytes are not removed by short fixation and give a gray background at the bottom of the 2D gels. Thorough removal of ampholytes requires an overnight fixation.

3. The fixation process can be altered if needed. The figures indicated in the protocol are the minimum times. Gels can be fixed without any problem for longer periods. For example, gels can be fixed overnight, with only one solution change. For ultrarapid fixation, the following process can be used:

   1'. Fix in 10% acetic acid/40% ethanol for 10 min; then rinse for 10 min in water [15].

   2'. Rinse in 40% ethanol for 2× 10 min and then in water for 2× 10 min. Proceed to **step 3**.

4. The optimal setup for sensitization is the following: Prepare four staining boxes containing the sensitizing thiosulfate solution, water (two boxes), and the silver nitrate solution, respectively. Put the vessel containing the rinsed gels on one side of this series of boxes. Take one gel out of the vessel, and dip it in the sensitizing and rinsing solutions (1 min in each solution). Then transfer to silver nitrate. Repeat this process for all the gels of the batch. A new gel can be sensitized while the former one is in the first rinse solution, provided that the 1 min time is kept (use a bench chronometer). When several batches of gels are stained on the same day, it is necessary to prepare several batches of silver solution. However, the sensitizing and rinsing

solutions can be kept for at least three batches, and probably more.

5. Gels can be impregnated with silver for at least 30 min and at most 2 h without any change in sensitivity or background.

6. This very short step is intended to remove the liquid film of silver solution brought with the gel.

7. When the gel is dipped in the developer, a brown microprecipitate of silver carbonate should form. This precipitate must be redissolved to prevent deposition and background formation. This is simply achieved by *immediate* agitation of the box. Do not expect the appearance of the major spots before 3 min of development. The spot intensity reaches a plateau after 15–20 min of development, and then background appears. Stop development at the beginning of background development. This ensures maximal and reproducible sensitivity.

8. The developing solution II is prepared as follows: For 1 L of solution, add 0.1 mole of boric acid and 150 mL of 1 N sodium hydroxide to 500 mL water. When the boric acid is dissolved, add 20 g of galactose (or xylose) and finally the sodium thiosulfate. Complete to 1 L with water. This solution is not stable and must be prepared the day of use.

References

1. Switzer RC, Merril CR, Shifrin S (1979) A highly sensitive silver stain for detecting proteins and peptides in polyacrylamide gels. Anal Biochem 98:231–237

2. Rabilloud T (1990) Mechanisms of protein silver staining in polyacrylamide gels: a 10-year synthesis. Electrophoresis 11:785–794

3. Rabilloud T, Vuillard L, Gilly C et al (1994) Silver-staining of proteins in polyacrylamide gels: a general overview. Cell Mol Biol 40:57–75

4. Blum H, Beier H, Gross HJ (1987) The expression of the TMV-specific 30-kDa protein in tobacco protoplasts is strongly and selectively enhanced by actinomycin. Electrophoresis 8:93–99

5. Richert S, Luche S, Chevallet M et al (2004) About the mechanism of interference of silver staining with peptide mass spectrometry. Proteomics 4:909–916

6. Metz B, Kersten GF, Hoogerhout P et al (2004) Identification of formaldehyde-induced modifications in proteins: reactions with model peptides. J Biol Chem 279:6235–6243

7. Metz B, Kersten GF, Baart GJ et al (2006) Identification of formaldehyde-induced modifications in proteins: reactions with insulin. Bioconjug Chem 17:815–822

8. Oses-Prieto JA, Zhan X, Burlingame AL (2007) Formation of epsilon-formyllysine on silver-stained proteins: implications for assignment of isobaric dimethylation sites by tandem mass spectrometry. MCP 6:181–192

9. Gharahdaghi F, Weinberg CR, Meagher DA et al (1999) Mass spectrometric identification of proteins from silver-stained polyacrylamide gel: a method for the removal of silver ions to enhance sensitivity. Electrophoresis 20:601–605

10. Chevallet M, Luche S, Diemer H et al (2008) Sweet silver: a formaldehyde-free silver staining using aldoses as developing agents, with enhanced compatibility with mass spectrometry. Proteomics 8:4853–4861

11. Sinha P, Poland J, Schnolzer M et al (2001) A new silver staining apparatus and procedure for matrix-assisted laser desorption/ionization-time of flight analysis of proteins after two-dimensional electrophoresis. Proteomics 1:835–840

12. Rabilloud T (1992) A comparison between low background silver diammine and silver nitrate protein stains. Electrophoresis 13:429–439

13. Chevallet M, Luche S, Rabilloud T (2006) Silver staining of proteins in polyacrylamide gels. Nat Protoc 1:1852–1858

14. Hochstrasser DF, Merril CR (1988) 'Catalysts' for polyacrylamide gel polymerization and detection of proteins by silver staining. Appl Theor Electrophor 1:35–40

15. Shevchenko A, Wilm M, Vorm O et al (1996) Mass spectrometric sequencing of proteins silver-stained polyacrylamide gels. Anal Chem 68:850–858

Chapter 7

Differential Proteome Analysis Using 2D-DIGE

Caroline May, Frederic Brosseron, Piotr Chartowski, Kristin Fuchs, Helmut E. Meyer, Barbara Sitek, and Katrin Marcus

Abstract

Classical 2D-PAGE allows comparison and quantitation of proteomes by visualization of protein patterns using gel stains and comparative image analysis. The introduction of fluorescent reagents for protein labeling (difference in-gel electrophoresis or DIGE) has brought substantial improvement in this field. It provides multiplexing of up to three samples in one gel, higher sensitivity compared to normal protein staining methods, and a higher linear range for quantitation. This article gives detailed protocols for 2D-DIGE, including both minimal and saturation labeling.

Key words CyDye™, G-Dyes, Minimal labeling, Saturation labeling, S-Dyes, Two-dimensional difference in-gel electrophoresis (2D-DIGE), Two-dimensional polyacrylamide gel electrophoresis (2D-PAGE)

1 Introduction

The development of two-dimensional difference in-gel electrophoresis (2D-DIGE) in 1997 was a milestone substantially increasing the reproducibility of 2D-PAGE (*see* Chaps. 4 and 5) in terms of quantitation [1–3]. With this approach it was first time possible to label covalently up to three different protein samples with different spectrally resolvable fluorescent dyes and afterwards simultaneously separate in one gel. One protein, labeled with different colored fluorescent dyes, co-migrates to the same position in the gel. Different sets of fluorescent dyes for protein labeling are today commercially available. This article focuses on CyDye™s (GE Healthcare, Munich, Germany) and G-Dyes/S-Dyes (NH DyeAGNOSTICS GmbH, Halle, Germany). Both are highly similar in handling.

Technical variations based on gel-to-gel differences are minimized due to two advantages: first a more accurate quantitation by use of an internal standard as reference, which is usually a mixture of the same amounts of all samples included in the study [4]. This

Katrin Marcus et al. (eds.), *Quantitative Methods in Proteomics*, Methods in Molecular Biology, vol. 2228,
https://doi.org/10.1007/978-1-0716-1024-4_7, © Springer Science+Business Media, LLC, part of Springer Nature 2021

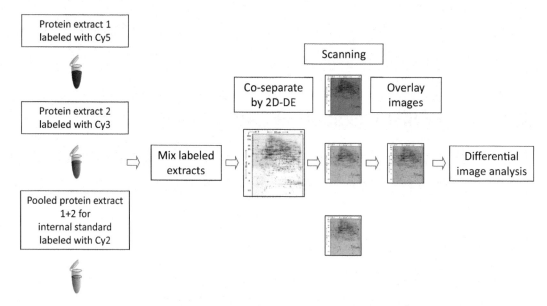

Fig. 1 *Overview of DIGE minimal labeling workflow.* Three different dyes are available for CyDye™ minimal labeling: Cy™2, Cy™3, and Cy™5. In total three different samples or two samples plus internal standard can be labeled, mixed, and simultaneously separated in one gel. Afterwards all images of a study can be compared using image analysis software, which allow an automated spot detection, background subtraction, quantitation, normalization, as well as inter-gel matching

internal standard can be used to match protein patterns of different gels, reducing quantitation errors and improving the statistical analysis. Therefore, 2D-DIGE increases not only the reproducibility of 2D-PAGE, which has to be high in order to guarantee a successful quantitation of biological differences in the proteome. Furthermore, the amount of gels is drastically reduced, because up to three samples can be run in one instead of three gels (Fig. 1) [5].

Two options exist for protein labeling on different amino acid residues: CyDye™ minimal dyes (GE Healthcare Biosciences) and Refraction-2D™ G-Dyes (NH DyeAGNOSTICS GmbH) for general differential analysis as well as CyDye™ saturation dyes (GE Healthcare Biosciences) and Saturn-2D™ S-Dyes (NH DyeAGNOSTICS GmbH) for differential analysis of scarce sample amounts.

CyDye™ minimal dyes and G-Dyes react with the ε-amino group of lysine residues [5]. Whereas GE Healthcare Bioscience provides three different dyes (Cy™2, Cy™3, and Cy™5), NH Dye AGNOSTICS offers corresponding dyes (G-Dye100, G-Dye200, G-Dye300) plus one additional dye (G-Dye400) which is measured in the near-infrared (NIR) fluorescence range. Therefore, it is possible to separate two or three samples (labeled with Cy™3 and Cy™5 or G-Dye200, G-Dye300, and G-Dye400) plus internal standard (labeled with Cy™2 or G-Dye100) in a single gel (Fig. 2). To rule out effects between treatments and dyes, a color

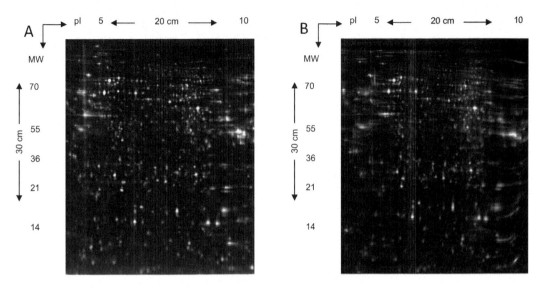

Fig. 2 *Fluorescence images of SHSY5Y cell lysates labeled with the different dyes.* Internal standard and samples from two experimental groups were co-separated, 50 µg protein of each sample and in total 150 µg on each gel. (**a**) Cy™Dye minimal labeling (internal standard with Cy™2, green; samples with Cy™3, red; and Cy™5, blue, respectively). The figure shows an overlay image of all channels. (**b**) Refraction-2D™ labeling (internal standard with G-Dye100, green; samples with G-Dye200, red; and G-Dye300, blue, respectively). The figure shows an overlay image of all channels. If the signal intensities of all channels are equal in a spot, the spot appears white. Both types of commercial dyes are comparable in spot pattern and signal intensities

switch is recommended. In contrast to the CyDye™ saturation dyes and S-Dyes, the minimal dyes only react with approximately 3% of all proteins, and only one lysine per protein is labeled in average.

CyDye™ saturation labeling allows for the analysis of scarce samples even down to 3 µg [6–8]. S-Dyes are suited for analysis of 0.5 µg, according to the manufacturers' specifications. The sensitivity is therefore about 20 times higher compared to the detection with minimal labeling [8] or G-Dyes. In contrast to minimal labeling and G-Dyes, saturation label and S-Dyes react with thiol groups of cysteine residues, and consequently cysteine residues of all proteins are labeled. Only two different dyes are available for the saturation labeling, namely, Cy™3 and Cy™5 or S-Dye200 and S-Dye300, respectively. Thus every sample has to be separated in an individual gel; nevertheless, the simultaneous separation of the internal standard results in accurate quantitation. In contrast to the minimal labeling procedure, a labeling optimization is essential for every experiment/study in order to determine the appropriate amount of dye.

Visualization and digitalization of 2D-DIGE gels are done with a fluorescence scanner or camera. In case of using the G-Dye400, you need a scanner or camera which is able to detect IR fluorophores. For subsequent image analysis, different tools are commercially available for automated spot detection, background subtraction, quantitation, normalization, as well as inter-gel matching (e.g., Delta2D, DECODON GmbH, Greifswald, Germany; Redfin, Ludesi AB, Lund, Sweden; DeCyder™ 6.5, GE Healthcare; ProgenesisSameSpots, Nonlinear Dynamics Limited, Newcastle upon Tyne, UK) [9]. Subsequently proteins of interest (e.g., differentially expressed ones) can be excised from the gels and further analyzed.

2 Materials

2.1 Technical Equipment

2.1.1 CyDye™ Minimal Labeling

1. Impervious to light reaction tubes or a dark tissue.

2.1.2 Refraction-2D™ Labeling

1. All necessary materials are included in the kits available from NH DyeAGNOSTICS GmbH.

2.1.3 CyDye™ Saturation Labeling

1. Impervious to light reaction tubes or a dark tissue.

2.1.4 Saturn-2D™ Labeling

1. All necessary materials are included in the kit available from NH DyeAGNOSTICS GmbH.

2.1.5 CA- and IPG-Based 2D-PAGE

1. Additionally to equipment for 2D-PAGE (see Chap. 5), glass plates compatible with fluorescence imaging are required and a confocal fluorescence scanner with appropriate excitation wavelength (488 nm, 532 nm, 633 nm) and emission filters.
2. Dark tissues.

2.2 Buffers and Solutions

2.2.1 CyDye™ Minimal Labeling

1. CyDye™ minimal dyes (Cy™2, Cy™3, Cy™5).
2. DIGE lysis buffer (pH 8.5): 30 mM (w/v) Tris, 7 M (w/v) urea, 2 M (w/v) thiourea, 4% (w/v) CHAPS (see Notes 1–3).
3. Dimethylformamide (DMF, anhydrous).
4. Quenching solution: 10 mM lysine.

2.2.2 Refraction-2D™ Labeling

1. All necessary solvents and solutions are included in the kits available from NH DyeAGNOSTICS GmbH.

2.2.3 CyDye™ Saturation Labeling

1. CyDye™ saturation dyes (Cy™3, Cy™5).
2. DIGE lysis buffer (pH 8.0): 30 mM (w/v) Tris, 7 M (w/v) urea, 2 M (w/v) thiourea, 4% (w/v) CHAPS (see Notes 1–3).
3. Reduction solution: 2 mM TCEP (Tris(2–carboxyethyl) phosphine hydrochloride).

4. Dimethylformamide (anhydrous; DMF).

5. 2× DIGE lysis buffer (pH 8.5): 8 M (w/v) urea, 4% (w/v) CHAPS, 2% (v/v) CA, 2% (w/v) DTT (*see* **Notes 1** and **2**).

2.2.4 Saturn-2D™
Labeling

1. All necessary solvents and solutions are included in the kit available from NH DyeAGNOSTICS GmbH.

*2.2.5 CA- and IPG-Based
2D-PAGE*

1. Solutions or buffers are used as in classical 2D-PAGE (*see* Chap. 5).

---

3 Methods

3.1 CyDye™ Minimal Labeling/ Refraction-2D™ Labeling

The labeling procedure is the same for CyDye™ minimal dyes and G-Dyes.

1. Prior to labeling, solubilize sample in DIGE buffer. The optimal pH range for minimal labeling is between pH 8.0 and 9.0 (*see* **Notes 4–7**).

2. Solubilize CyDye™ according to manufacturer's protocol in anhydrous DMF (*see* **Note 8**).

3. For labeling, add 400 pmol dyes to 50 μg protein (*see* **Notes 9–12**), mix, and incubate on ice for 30 min in the dark.

4. To stop the reaction, add 1 μL 10 mM lysine.

5. Mix the samples and incubate for another 10 min on ice in the dark.

6. Combine all three samples, mix, and place the samples on IPG strips or CA gels.

3.1.1 CyDye™ Saturation Labeling/Saturn-2D™ Labeling

The labeling procedure is the same for CyDye™ saturation dyes and S-Dyes.

1. Prior to labeling, solubilize sample in DIGE buffer. The optimal pH range for saturation labeling is between pH 7.8 and 8.2 (*see* **Notes 4–7**).

2. Before starting the differential study, optimize labeling for each type of sample. Therefore, the protein concentration must range from 0.55 to 10 mg/mL (*see* **Notes 9, 10, 13**).

3. The cysteine residues have to be reduced by using a solution with 2 mM TCEP for analytical and 20 mM TCEP for preparative gels. The optimal volume must be determined in preliminary experiments (*see* **Note 14**).

4. For labeling, add dyes (2 mM working solution for analytical and 20 mM for preparative gels in anhydrous DMF). Mix and incubate the samples at 37 °C for 30 min (*see* **Notes 2** and **12**).

5. Stop the reaction by adding DTT, e.g., using 2× DIGE lysis buffer or DTT alone (*see* **Note 15**).

6. Combine both samples and apply on the IPG strips or CA gels.

3.2 CA-
and IPG-Based
2D-PAGE

1. CA-based IEF and IPG-based 2D-PAGE should be performed as described in Chap. 5. To avoid long-time light exposure and therefore photobleaching of the dyes, a dark tissue placed above the focusing chambers is recommended (*see* **Notes 16–19**).

4 Notes

1. To ensure a high reproducibility and to minimize the technical variability of 2D gels, all solutions and buffers should be produced in large batches for one proteome study. Most of the buffers and solutions can be stored in aliquots at $-20\ ^{\circ}$C or $-80\ ^{\circ}$C.

2. Avoid heating of urea solutions above $37\ ^{\circ}$C to reduce the risk for protein carbamylation.

3. CHAPS can also be omitted.

4. Salts should be avoided during or removed after sample preparation because they may result in an inefficient labeling.

5. For all solutions and buffers, use high-purity water and p.a. reagents and never change reagents during one study.

6. Test sample pH by pipetting one droplet of the sample on a standard pH test paper. Note that salts, like additional protease or phosphatase inhibitors, might change the pH. For adjustment of pH prior to labeling, refer to the manufacturer's instructions.

7. Avoid primary amines when using minimal label and thiols when using saturation label, as these substances might interfere with the labeling.

8. To prevent degradation of NHS groups, store dyes under water-free conditions.

9. The protein concentration should be accurately determined to allow the comparison of proteomes and to ensure a defined ratio of protein to dye, leading to a maximum of sensitivity. Be aware that colorimetric protein quantitation methods are sensitive for high concentrations of urea, thiourea, and CHAPS, which are part of the DIGE lysis buffer. These ingredients pretend a higher protein concentration. Therefore it is recommended to prepare the standard dilution series in DIGE lysis buffer or to determine the protein concentration by amino acid analysis.

10. The optimal protein concentration for labeling is >5 μg/μL for minimal label/G-Dye and 0.5–10 μg/μL for saturation labels/S-Dyes.

11. The dye to protein ratio recommended by the manufacturer is 400 pmol dye for 50 μg protein. However, it is possible to use the half amount of protein (25 μg per sample) and therefore half of the amount of dye without loss of sensitivity. This can/should be tested if samples are scarce or to save costs.

12. When preparing the internal standard, calculate sufficient amount as backup to ensure that gel runs can be repeated if necessary, all with the same standard. For example, if a study includes 10 gels, prepare internal standard for 15–20 gels at least. Gels cannot be compared with each other if the standard is changed in one study.

13. If saturation label/S-Dye is used for samples such scarce that protein quantitation cannot be applied prior to labeling, try to use other ways for estimation of the sample amount. For example, prepare larger amounts of the respective sample, determine protein amount, and calculate back to the original sample or use the same number of cells for all experiments.

14. When using saturation label/S-Dye, proteins need to be first denatured and then completely reduced with TCEP to make all cysteine residues accessible for labeling. The ratio of TCEP to dye should be 1:2. Lower ratios result in insufficient labeling, while excess TCEP induces unspecific labeling on lysine residues. It is necessary to optimize the amount of dye used for saturation labeling/S-Dye, as protein in different samples might include different amounts of cysteine in their sequences.

15. Care should be taken here to ensure that the volume does not become too large. This is particularly important with the CA-based method.

16. The fluorescently labeled proteins can be detected by scanning the gels between the glass slides. Therefore, slides must be cleaned carefully. Only use high purity ethanol, not technical ethanol for cleaning.

17. After digitalization of the stained gels, crop the areas which include spacers, IPG stripes, and the running front as these show autofluorescence and therefore disturb the analysis. Consider the compatibility of the storage data format to any of the used image analysis software.

18. All CyDye™s are similar in both p*I* and MW, whereas G-Dyes/S-Dyes are only p*I* matched. However, intra-gel matching for G-Dyes/S-Dyes is not problematic using appropriate analysis software (Delta2D, Redfin, DeCyder™ 6.5, Progenesis Same Spots).

19. Protein spots can be picked "blind" from gels for mass spectrometry analysis by printing the fluorescence images and using them as templates. If proceeding this way, note that in minimal label only a portion of the proteins is labeled and therefore shows a slight mass shift to higher MW compared to the majority of unlabeled proteins.

Acknowledgments

This work was supported by the Bundesministerium für Bildung und Forschung (NGFN, FZ 01GS08143) as well as the European Regional Development Fond (ERDF) of the European Union and the Ministerium für Innovation, Wissenschaft und Forschung des Landes Nordrhein-Westfalen (ParkChip,FZ 280381102), and P.U.R.E. (Protein Research Unit Ruhr within Europe), Ministry of Innovation, Science and Research of North-Rhine Westphalia, Germany, a federal German state.
Thanks to Dr. Jan Heise (NH Dye AGNOSTICS GmbH) for providing S-Dyes and G-Dyes.

References

1. Rabilloud T, Chevallet M, Luche S et al (2010) Two-dimensional gel electrophoresis in proteomics: past, present and future. J Proteome 73 (11):2064–2077

2. Rabilloud T, Vaezzadeh AR, Potier N et al (2009) Power and limitations of electrophoretic separations in proteomics strategies. Mass Spectrom Rev 28:816–843

3. Unlu M, Morgan ME, Minden JS (1997) Difference gel electrophoresis: a single gel method for detecting changes in protein extracts. Electrophoresis 18:2071–2077

4. Alban A, David SO, Bjorkesten L et al (2003) A novel experimental design for comparative two-dimensional gel analysis: two-dimensional difference gel electrophoresis incorporating a pooled internal standard. Proteomics 3:36–44

5. Tonge R, Shaw J, Middleton B et al (2001) Validation and development of fluorescence two-dimensional differential gel electrophoresis proteomics technology. Proteomics 1:377–396

6. Helling S, Schmitt E, Joppich C et al (2006) 2-D differential membrane proteome analysis of scarce protein samples. Proteomics 6:4506–4513

7. Sitek B, Luttges J, Marcus K et al (2005) Application of fluorescence difference gel electrophoresis saturation labeling for the analysis of microdissected precursor lesions of pancreatic ductal adenocarcinoma. Proteomics 5:2665–2679

8. Shaw J, Rowlinson R, Nickson J et al (2003) Evaluation of saturation labeling two-dimensional difference gel electrophoresis fluorescent dyes. Proteomics 3:1181–1195

9. Dowsey AW, English J, Pennington K et al (2006) Examination of 2-DE in the human proteome organisation brain proteome project pilot studies with the new RAIN gel matching technique. Proteomics 6:5030–5047

Chapter 8

Quantitative Mass Spectrometry-Based Proteomics: An Overview

Svitlana Rozanova, Katalin Barkovits, Miroslav Nikolov, Carla Schmidt, Henning Urlaub, and Katrin Marcus

Abstract

In recent decades, mass spectrometry has moved more than ever before into the front line of protein-centered research. After being established at the qualitative level, the more challenging question of quantification of proteins and peptides using mass spectrometry has become a focus for further development. In this chapter, we discuss and review actual strategies and problems of the methods for the quantitative analysis of peptides, proteins, and finally proteomes by mass spectrometry. The common themes, the differences, and the potential pitfalls of the main approaches are presented in order to provide a survey of the emerging field of quantitative, mass spectrometry-based proteomics.

Key words Mass spectrometry, Proteomics, Relative quantification, Absolute quantification, Label-free, Isotope labeling, Metabolic labeling, Chemical labeling

1 Introduction

Proteomics is a powerful platform for studying both single proteins and complex protein samples. Combining gel- and chromatography-based separation techniques with subsequent mass spectrometry (MS)-based analysis and bioinformatics approaches enables processing of questions from all areas of medicine and basic science. In particular, biological processes and pathways can be examined in a more detailed manner. The key technology for the protein identification in biological samples is MS. Improvement in MS sensitivity, resolution, and mass accuracy, advances in high-performance liquid chromatography (HPLC) efficiency, progress in software development, and increasing computing power have allowed transition from "what" to "what and how much" by introducing MS-based protein quantification techniques.

Katrin Marcus et al. (eds.), *Quantitative Methods in Proteomics*, Methods in Molecular Biology, vol. 2228,
https://doi.org/10.1007/978-1-0716-1024-4_8, © The Author(s) 2021

Depending on the research question and the sample at hand, quantification of proteins on a global level as well as accurate quantification of individual proteins can be performed. MS-based proteome-wide analysis of protein changes between different conditions (i.e., healthy versus diseased) allows to uncover pathological mechanisms on the phenotypic level. In addition, quantification of a specific stimulus, ranging from changes in the amounts of a single protein or its defined post-translational modification (PTM) to the proteome-wide kinetics of the same modification between different stages of the cell cycle, can be achieved. Accurate and reliable quantification of proteins and their PTMs is playing a decisive role for new disease-related biomarkers for better diagnostics, prediction, and treatment.

The variety of questions being asked has impelled the development of a variety of quantitative MS techniques, which can be generally divided into four strategies (Fig. 1). According to the study aim and scope, two main strategies are used: (i) untargeted

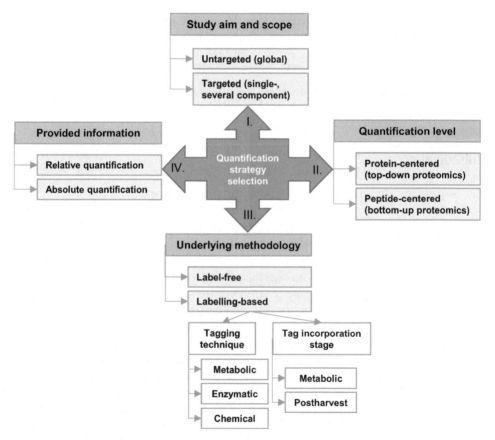

Fig. 1 Overview for the MS-based protein quantification strategies selection, based on a study scope (I); quantification level (II); methodology, i.e., label-based or label-free (III); and the data, which will be available for the further statistical analysis (IV)

(global) quantification of hundreds or thousands of proteins for protein profile comparison (Chaps. 10, 11, 14–25) and (ii) targeted (single- or several-component) quantification of only few components (Chaps. 12, 13, 26), which are selectively isolated from a sample and quantified. Depending on the quantification level, protein-centric (top-down) or peptide-centric (bottom-up) approach could be applied.

According to the underlying methodology, MS-based quantification can be further divided into two subgroups: (i) label-based quantification utilizing stable isotope labels incorporated within the peptides/proteins (Chaps. 10, 11, 14–16, 18–20, 26, 27) and (ii) label-free quantification (Chaps. 17, 21–25) in which sample retains its native isotope composition. Label-based technology comprises artificial labeling of peptides or proteins, which introduces an expectable mass difference within different experimental conditions. Depending on the way of tagging, chemical, metabolic, and enzymatic labeling are introduced. Furthermore, the methodologies for the stable isotope tagging can be generally divided into two main groups: post-harvest methods (Chaps. 10, 11, 13, 15, 16, 18, 26, 27) and metabolic labeling methods (Chaps. 14, 19, 20), which involve the incorporation of an isotopic label into the protein when the sample is still metabolically active. Depending on the information provided by these quantification methods, they are classified into (i) relative (Chaps. 10, 11, 14–22, 24, 25, 27) and (ii) absolute (Chaps. 12, 13, 26). Relative quantification yields protein quantitative ratio or relative change, by comparing the amount of single protein or whole proteomes between samples. In its turn, absolute quantification provides information regarding absolute amount or the concentration of a protein within a sample. Schematic overview of the main classification for MS-based protein quantification as well as quantification strategy selection is provided in Fig. 1. This chapter provides a general overview of MS-based protein quantification (overview is given in Table 1) and the main strategies and methodologies known at the time of writing this overview.

2   General Principles for MS-Based Quantification

MS comprises analyte ionization, their separation based on mass-to-charge ratio (m/z) with the further detection of these ions [1, 2]. Among a wide variety of ionization techniques, electrospray ionization (ESI) and matrix-assisted laser desorption/ionization (MALDI) are commonly used in proteomics research, as they enable MS analysis of highly polar and large molecules. ESI can be directly combined with separation techniques, e.g., HPLC. MALDI, in contrast, has an advantage of producing singly charged

Table 1
Methods for relative and absolute quantification

Quantification method	Description	Advantages	Disadvantages
Methods for relative quantification			
ICAT	Cysteine-specific chemical labeling on protein/peptide level (in vitro). Quantification at MS level. 2-plex	Reduced in sample complexity due to affinity enrichment of labeled peptides. Applicable to any sample (cells, animal or human tissue)	Requires cysteine-containing protein/peptides for labeling. Limited by the 2-plex multiplexing capability. Does not allow in vivo labeling
ICPL	Chemical labeling of free amino groups on the protein/peptide level (in vitro). Quantification at MS level. Up to 4-plex	Labeling of all peptides within a sample; no side reaction. Applicable to any sample (cells, animal or human tissue). Improved fragmentation efficiency during MS/MS	Peptides resulting from tryptic digestion are relatively large. Increased spectrum complexity requires complex computational analysis. Multiplexing capability is limited (up to 4-plex). Does not allow in vivo labeling
iTRAQ	Amine-specific labeling with isobaric tags on the peptide level (in vitro). Quantification at MS/MS level. Up to 8-plex	Efficient labeling enhanced signal intensity in MS and MS/MS, high multiplexing capability, simple data analysis. Applicable to any sample (cells, animal or human tissue). Commercially available	Expensive. Does not allow in vivo labeling
TMT	Amine-specific labeling with isobaric tags on the peptide level (in vitro). Quantification at MS/MS level. Up to 10-plex	Efficient labeling enhanced signal intensity in MS and MS/MS, high multiplexing capability, simple data analysis. Applicable to any sample (cells, animal or human tissue)	Expensive. Requires fragmentation with HCD or ETD. Does not allow in vivo labeling
DiLeu	Amine-specific labeling with isobaric tags on the peptide level (in vitro). Quantification at MS/MS level. Up to 12-plex	Inexpensive alternative to commercially available reagents (i.e., TMT, iTRAQ): Can be in-house synthesized. High multiplexing capability. Applicable to any sample (cells, animal or human tissue)	Additional steps for activation are required. Still in early stages of development. Does not allow in vivo labeling

(continued)

Table 1
(continued)

Quantification method	Description	Advantages	Disadvantages
IPTL	Amine-specific labeling with isobaric tags on the peptide level (in vitro), quantification at MS/MS level. Up to 3-plex	Inexpensive alternative to commercially available reagents (i.e., TMT, iTRAQ): Can be in-house synthesized. Applicable to any sample (cells, animal or human tissue). Increased robustness	Limited multiplexing capability (up to 3-plex). Does not allow in vivo labeling
^{15}N labeling	Metabolic labeling with ^{15}N-enriched media on the Protein level (in vivo and ex vivo). 2-plex	Efficient labeling. Applicable to cells and model organisms	Expensive. Complex data analysis. Limited multiplexing capability (up to 2-plex) Not suitable for clinical samples
SILAC	Metabolic labeling with amino acids containing stable heavy isotopes on the protein level (in vivo and ex vivo). Up to 5-plex	Efficient labeling, one label for (tryptic) peptide, semiautomatic data analysis. Applicable to cells but can be expanded to tissues or model organisms using internal standards (e.g., super-SILAC)	High costs, especially when applied to whole organisms. Super-SILAC experiments have reduced quantitative proteome coverage. Requires metabolically active cells to introduce labels
NeuCode	Metabolic labeling with amino acids containing stable heavy isotopes on the protein level (in vivo and ex vivo). Up to 5-plex	High multiplexing capability. When combined with PRM allows up to 30-plex. Shorter labeling time in vivo	Requires metabolically active cells to introduce labels. High-resolution mass analyzers are required
CTAP/ NANCAT	Metabolic labeling with essential amino acids on the protein level (in vivo and ex vivo), quantification at MS/MS level. 2-plex	Allows cell line-specific labeling in co-culture with near SILAC efficiency	High costs, limited multiplexing capability (up to 2-plex). Requires metabolically active cells
^{18}O labeling	Enzymatic labeling with ^{18}O-labeled water on the peptide level (in vitro). 2-plex	Low costs, simple in handling. Applicable to any sample (cells, animal or human tissue)	Incomplete labeling complicates data analysis. Limited multiplexing capability (up to 2-plex) Not suitable for in vivo labeling

(continued)

Table 1
(continued)

Quantification method	Description	Advantages	Disadvantages
Spectral counting	Relative comparison of different samples based on the number of identified peptides or acquired MS/MS spectra, respectively	No labeling required, inexpensive compared to stable isotope labeling methods. Applicable to any sample (cells, animal or human tissue), simplified sample handling, unlimited number of samples	Requires high reproducibility. Longer data acquisition time compared to isotope labeling. Lower accuracy than labeling methods Requires large sample size (spectral counts) to confidently predict small changes in expression Lower accuracy than labeling and XIC-based LFQ methods
Ion intensity (XIC)	Based on ion intensities of peptides	No labeling required, inexpensive compared to stable isotope labeling methods. Applicable to any sample (cells, animal or human tissue), simplified sample handling, unlimited number of samples	Requires high reproducibility. Longer data acquisition time compared to isotope labeling. Lower accuracy than labeling methods Requires more computationally intensive data analysis compared to spectral counting
Methods for absolute quantification			
AQUA	Based on chemically synthesized peptide standards containing stable isotopes, added in known amounts to the sample	Accurate absolute quantification	High cost; prior information about the quantified peptides is needed; only few peptides per protein are quantified; fractionation is not applicable
emPAI	Calculation of the emPAI using the number of observed and observable peptides. Spectral count based	Simple; can be applied to every sample as no labeling is required	Low accuracy
APEX	Similar to emPAI but corrected by background expectation, sampling depth, and confidence in protein identification. Spectral count based	Introduces correction factors for more accurate quantification	Correction values based on prior MS results are needed

(continued)

Table 1
(continued)

Quantification method	Description	Advantages	Disadvantages
Top3	Based on the relationship between the signal response of three most abundant tryptic peptides and the protein concentration. Intensity-based	High accuracy; no labeling is required	The sample has to be spiked with a standard protein before tryptic hydrolysis
iBAQ	Based on the relationship between the signal intensities of the precursor peptides that map to each protein and the number of theoretically observable peptides. Intensity-based	High accuracy; no labeling or spiking of standard peptides is required	Requires high-resolution instruments

ions of peptides and proteins, minimizing spectral complexity [1–3]. The sensitivity and resolution of a mass spectrometer strongly depends on the mass analyzer ability to effectively separate ions. The most commonly used mass analyzers for accurate protein quantification are quadrupole, time of flight (ToF), and Orbitrap. In its basic operation with continuous sample introduction, the mass spectrometer continuously acquires mass spectra, when the instrument operates in the full-spectrum (or full-scan MS1) mode. In this case, a three-dimensional data array, defined by time, m/z, and ion intensity (counts), is acquired [1, 2]. Coupling several mass analyzers in one mass spectrometer allows two or more sequential separations of ions with their fragmentation occurring in between and refers to as tandem mass spectrometry [1, 2]. Mostly, in tandem mass spectrometry experiments (known as MS/MS or MS2), the first mass analyzer is used to isolate a precursor ion, which then undergoes fragmentation in a collision cell to yield product ions to be sorted and weighted in a second analyzer. The number of steps can be increased to yield MSn experiments (where n refers to as the number of generations of ions being analyzed) [2]. Analysis of MS-acquired spectra allows not only molecular analytes mass/weight determination and structure/sequence elucidation but also their quantitative analysis. In the following sections, an overview of the most important analytical approaches in quantitative MS-based proteomics is discussed.

2.1　Global and Targeted Quantification

For untargeted proteomics experiments, the data-dependent acquisition (DDA) (*see* Chaps. 10, 11, 14–18, 21, 22, 27) or data-independent acquisition (DIA) (*see* Chaps. 17, 23–25) can be used. Currently, most liquid chromatography-coupled tandem mass spectrometry (LC-MS/MS) approaches rely on DDA, which comprises a selection of the most common Top N precursor ions from a full MS1 overview scan for further fragmentation and acquisition of the respective MS2 spectra [3–6]. Data derived from each MS2 scan can be analyzed with a database search algorithm [7–9]. DDA typically yields thousands of protein identifications together with the quantitative information. However, the selection of peptide precursors is stochastic; if too many peptide species co-elute and appear in a single MS1 scan, then DDA samples only the most abundant peptides, missing low-abundance ones [7]. Consequently, different subgroups of peptides could be selected for fragmentation between different samples, resulting in high variations between replicates such as lack of identification of low abundant peptides and thus a reduced number of quantifiable proteins [10–12]. Moreover, quantification based on DDA depends on the analysis of the chromatographic MS1 peak area, which is particularly susceptible to interferences, especially in complex samples [13, 14]. In spite of its drawbacks, the flexibility, scope of detection, and the relative simple setup and analysis make DDA still the preferred method within the proteomics community. Moreover, DDA allows relative quantification of peptides between samples through a variety of labeling techniques, and by label-free proteomics [8, 15, 16].

In DIA, a set of predetermined sequential mass isolation windows is used to send all precursor ions of the same mass window for simultaneous fragmentation and analysis [17]. This approach allows more reproducible and accurate protein identification and quantification, compared to DDA (*see* **Note 1**) [17–20]. As the fragmentation and subsequent analysis of all the ionized peptides occurs within a defined mass isolation window [12], theoretically the identification and quantification of all precursor peptides is possible. DIA-based quantification is performed at the MS2 level by extraction of fragment ion chromatograms, which are less susceptible to interferences than MS1-based extracted ion chromatograms [17, 21]. However, the large number of fragment ions, derived from different peptides from the same selection window, prohibits the analysis in a classical database search strategy. Moreover, as DIA allows acquisition of nearly complete MS2 data, the direct correlation between precursor and its fragment ions is lost resulting in the need for more complex data analysis algorithms (*see* Chap. 32). Most commonly spectral libraries, containing the information regarding the elution time of the peptide and its fragment ions, are used to infer the precursor peptide-fragment connection and thus allow peptide and protein identification [22–24]. At that, due to data complexity, special software tools, i.e., Spectronaut, OpenSWATH, Skyline, PeakView [25–28], or others (Chap. 32), are required. A spectral library is usually generated during preliminary profound analysis of the same samples with the same instrument by DDA (*see* Chap. 23). This requires extra time and sample expanses. Furthermore, if a peptide is not present in a spectral library, in principle, it cannot be quantified. Therefore usage of a larger spectral library does not always lead to the increase in identifications from DIA data and may also lead to a lower quantification accuracy [20, 29]. A trend toward increasing coefficients of variations (CVs) with increasing library size was reported and explained as a result of detecting more low abundant species that naturally entail a higher variation [20]. Recently, spectral library-free approaches, such as DIA-Umpire and DirectDIA, have been developed [26, 30]. The quantification accuracy for these methods is similar to the spectral library-based approaches, but with lower number of identified precursors [20, 26, 30].

While global proteomics is a strategy of choice for, e.g., the discovery of new biomarkers, their further validation requires targeted methods for sensitive, accurate, and specific protein quantification [31]. The targeted MS approaches, i.e., selected reaction monitoring (SRM), also known as multiple reaction monitoring (MRM) (*see* Chap. 26), and parallel reaction monitoring (PRM) (*see* Chaps. 12, 26), are highly reproducible and relatively fast. In contrast to the DIA approach, which is normally not performed using heavy isotope labeling, SRM and PRM offer its best performance for absolute quantification when applying isotope-labeled peptides/proteins as internal standards [32, 33].

SRM is typically performed with triple-quadrupole instruments: Q1 selects a peptide ion, Q2 fragments the peptide, and Q3 selects a specific fragment ion for detection and quantification on the MS2 level [32]. Thus, due to the two-level mass filtering, most of the co-eluting peptides are effectively excluded, making SRM a highly sensitive technique [34]. However, due to the quadrupole's low resolving power, separation of interfering near-isobaric ions that co-elute with the target peptides is limited [35]. In addition to the m/z of the peptides, SRM requires the beforehand information concerning the fragment ions to target [36].

The SRM limitations were overcome with the implementation of PRM approach, which uses high-resolution Orbitrap or time of flight as MS2 analyzers. PRM uses targeted tandem MS to simultaneously monitor product ions of a targeted peptide with high resolution and mass accuracy [35, 37]. PRM offers the same selectivity and accuracy but wider dynamic range and selectivity compared to SRM. Herewith, PRM has a longer cycle time. This approach was successfully used for the accurate quantification of specific low-abundant peptide/protein in complex biological samples [38–40].

2.2 Peptide or Protein-Centric Approach

Nowadays two distinct strategies co-exist in MS-based proteomics studies: bottom-up, peptide-centric approaches and top-down, protein-centric approaches. The bottom-up approach enables high-throughput analysis and allows identification and quantification of thousands of proteins in complex samples, due to their proteolytic digestion into shorter peptides before analysis [41, 42]. This simplifies MS/MS sequencing, since peptides are easier fragmented in the mass spectrometer than intact proteins. However, the protein digestion at the very first stage of the experiment discards the connectivity between peptides and proteins, complicating computational analysis and biological interpretation of the data [42, 43]. Despite the fact that MS/MS spectra analysis comprises peptide spectrum matches (PSMs) determination, most of the search engines and methods return protein lists. These lists will vary according to the underlying model and the implied independence assumptions [43, 44]. Moreover, the inconsistences are connected with so-called protein inference, when two or more proteins share the same peptide [42, 44, 45]. In this case, accurate identification is only possible, if discriminating (unique) peptides are identified and the applied database was adequate for the sample. Otherwise, all of the possible proteins with shared peptides will be assigned as protein group members (according to the inference algorithm) [42, 43]. Protein inference is automatically performed only by few peptide search engines, i.e., X!Tandem and Mascot, and these can only employ peptide identifications found by the respective algorithms. Most of the widely used programs, for example, MS-GF+ and Comet, return only the spectra identifications without any inference [43]. Thus, identification and quantification of a large part of proteins and proteoforms may be restricted by the presence of non-unique low-quality or incorrectly identified peptides [42, 44]. In attempt to solve this issue, programs for protein inference from PSMs (e.g., PIA, ProteinProphet, Scaffold, and IDPicker) were developed [43, 46, 47]. PIA, for example, is able to combine PSMs from different search engine runs and turn these into consistent results [43]. Additionally, protein quantification could be complicated due to the necessary proliferation of peptide-level information up to the protein-level. Thus, the peptides are usually quantified first, and then these data are transferred to the protein level [45]. Hence, quantification and analysis on peptide level is preferable because of better differential abundance detection and higher accuracy [43, 44].

Distinguishing and quantification of different proteoforms, protein products of a single gene derived from genetic variation, alternative splicing of RNA transcripts, and PTMs (e.g., phosphorylation, glycosylation, acetylation), is essential for biological processes and pathology understanding [48]. In this context, the top-down MS approach is especially attractive for relative quantification of protein modifications because MS of a digest mixture may not detect the peptide carrying a given particular isoform modification. Moreover, addition of smaller PTMs (i.e., phosphorylation) has much less effect on ionization/detection of intact proteins compared to peptides. Thus, top-down approach may be employed for the quantification of modified and unmodified protein species in biological samples [49–52]. The top-down MS comprises introduction of intact proteins into the mass spectrometer and analysis of both intact and fragment ions masses. This approach allows up to 100% sequence coverage and full characterization of proteoforms [53]. Despite its attractivity, due to technical difficulty of proteome-wide analysis at the intact protein level, reproducible and accurate protein quantification by top-down MS remains challenging. Mainly, quantification is restricted due to the much lower efficiency of ionization and detection of large proteins [54]. For complex samples such as human plasma, target protein is usually enriched prior to top-down MS analysis, e.g., by immunoprecipitation [54, 55]. Successful protein quantification with top-down approach may be also achieved when using targeted MS [56] or when analyzing relatively small molecules (<30 kDa) [55, 57].

2.3 Label-Free and Label-Based Quantification by Mass Spectrometry

MS is a technique for the measurement of the mass-to-charge ratios of charged particles and does not allow in itself their quantification. Different peptides and proteins exhibit a wide range of physicochemical properties, i.e., size, charge, and hydrophobicity; therefore, their mass-spectrometric responses cannot be used for quantitative comparison between different molecular species within one sample. For accurate quantification, it is therefore generally required to compare each individual molecule between experiments (includes methods for label-free quantification), or within a single experiment, when molecules differ only in their isotopic composition and have identical physical and chemical properties (label-based quantification) [58].

Label-based methodology comprises in vivo (metabolic) (*see* Chaps. 14, 19, 20) or in vitro (post-harvested) peptide or protein labeling with heavy, non-radioactive, isotopes (*see* Chaps. 10, 11, 13, 15, 16, 18, 26, 27). Owing to the natural occurrence of certain stable heavy isotopes, each peptide/protein contains a certain proportion of these; the isotope pattern seen in the mass spectrometer thus reflects the natural abundances of these heavy isotopes within the peptide. Artificial incorporation of heavy isotopes (e.g., ^{13}C, ^{15}N, ^{18}O, ^{2}H) produces a mass-to-charge shift of the peptide's peaks in the mass spectrum. Therefore, heavy isotope-labeled peptides, with the exception of deuterium incorporation, are chemically identical to its native counterpart, and therefore the two peptides behave identically during LC-MS analysis (*see* **Note 2**). Given that a mass spectrometer can recognize the mass difference between the labeled and unlabeled forms of a peptide, quantification is achieved by comparing their respective signal intensities

[58]. Isotope labels can be introduced as an internal standard into amino acids metabolically, chemically, or enzymatically or, alternatively, as an external standard using spiked synthetic peptides [59]. Thus labeling techniques enable comparing two or more samples within the same mass spectrum with a high accuracy and reproducibility of quantitative measurements. It allows avoiding missing value problem of label-free approaches: the signal absence rather testifies to peptide signal being below the detection limit, than not being picked up by chance [60]. The major limitations of the approach are additional steps in sample preparation, high costs, and the limited number of samples, analyzed within one experiment [58]. Using the MS1 spectrum for quantification could be another limitation of the methodology [61]. The higher the number of samples, the more complex is the MS1 spectrum for analysis, due to overlapping peaks and identical properties of precursor ions [61]. In practice, this limits the number of samples that can be compared in a single experiment [60]. Another limitation of MS1-based quantification is that the number of ions that can be accumulated in the most commonly used high-resolution analyzer, the Orbitrap, is limited [62, 63]. The number of ions for low-abundance peptides can therefore be very small if some very high-abundant peptides co-elute at the same time in the MS1 spectrum, resulting in less precise quantification due to poor ion statistics [60]. This limitation has been somewhat alleviated by ion-mobility separation or BoxCar [60]. Moreover, quantification information can be also obtained from the MS/MS fragment ion spectrum. In this context, more than one independent spectrum will be available for analysis. However, quantification based on isobaric labels can be complicated due to co-isolated peptides, creating reporter tags, which, if superimposed on the reporter tags from the selected precursor ion, gives an inaccurate link between peptide quantity and identity [64]. Double isolation method (MS3) was suggested to address this problem. A disadvantage with the MS3 isolation is that it results in lower number of proteins being quantified [65].

Label-free quantification relies on comparison of peptides and proteins in their natural state in consecutive experiments (*see* Chaps. 17, 21–25). It requires highly reproducible sample-handling and analysis protocols [58, 66]. Since all samples in a label-free study are separately analyzed by LC-MS/MS, even running all samples in a single sequence in the same instrument can lead to variation even for technical replicate of the same sample. In order to consider the data bias and to make data more comparable, normalization is required [58, 67]. Label-free quantification enables analysis of virtually unlimited number of samples without introduction of any labels, thus keeping costs low and minimizing the sample preparation steps. This approach is highly preferable for the biomarker research [5, 6, 18]. On the other hand, poor reproducibility may require analysis of many technical replicates and may lead to the low accuracy of the quantitative measurements

[58]. Thus, label-free approach has been shown to give the largest dynamic range and the highest proteome coverage for identification but the lower quantification accuracy and reproducibility compared to labeling-based strategies [68].

2.4 Quantitative Cross-Linking/Mass Spectrometry

Cross-linking MS (XL-MS) is an advanced technique to study single proteins, protein complexes, and protein–protein interaction networks [69, 70] (*see* Chap. 27). The method is based on the ability of a cross-linker to convert 3D proximity of amino acid residues into covalent bonds [71, 72]. The bridgeable distance between residues depends on the cross-linker used. Thus the commonly used bis[sulfosuccinimidyl] suberate (BS3) links residues up to 25–30 Å apart [71]. Following proteolytic digestion of the proteins, cross-linked peptides are identified using LC-MS and database search [71].

Quantitative XL-MS allows comparison of cross-linked peptides across experimental conditions and varying biological states and may apply both label-free and label-based approaches [73]. Successful label-based XL-MS was reported when using isotope-labeled cross-linker combined to the software tool, e.g., XiQ, xTract MaxQuant, and Skyline [73–76]. The majority of the methods are designed for quantification based on MS1 signal [73]. Combination with isobaric tandem mass tags (TMT) allowed quantification from the reporter ion signal in MS3 spectra [77]. The main limitations of the isotope labeling-based XL-MS are costs, complex sample preparation, and reduced data coverage [72]. Advantages of label-free quantitation, allowing non-limiting numbers of samples, were presented recently with an MS1- and MS2-based XL-MS workflow using Skyline [72, 73]. Since cross-linking sample preparation procedure is more sophisticated than in normal proteomics, one might expect a larger variance. However, it has been shown that the reproducibility of label-free XL-MS is in line with the reproducibility of general quantitative proteomics [72]. Recently, DIA-based XL-MS have been demonstrated to be capable of detecting changing abundances of cross-linked peptides in complex mixtures when combined to the Spectronaut software [78].

3 Methods for Protein/Peptide Quantification

3.1 Relative Quantification

Relative quantification provides calculation of abundance ratios between peptides and proteins by comparing their signals originating from different samples (*see* Chaps. 10, 11, 14–22, 24, 25, 27). Usually performed in "discovery" (non-targeted) mode, it allows quantitative profiling of tens of thousands of peptides from thousands of proteins within a single experiment without a prior information. It can be based upon heavy isotope labeling or label-free.

3.1.1 Stable Isotope Labeling-Based Methods

Chemical Labeling

The methods for relative quantification by chemical labeling rely on the chemical reaction (without enzymatic catalysis) between a reagent and the peptides (or proteins) in the sample of interest in vitro (i.e., after isolation of the protein/peptide from the biological sample). The reagent used bears different numbers of stable heavy isotopes and thus produces a mass shift in the MS spectrum (e.g., dimethyl labeling) or MS/MS spectrum (in case of isobaric reagents, e.g., iTRAQ, TMT).

One of the first commercially available reagents for the chemical labeling was ICAT (isotope-coded affinity tags) [79]. ICAT is a protein-specific non-isobaric chemical label, which consists of three moieties: a sulfhydryl-reactive group for coupling to the analyte cysteines, an affinity group for isolation of the tagged species (peptides), and a linker in light (with natural isotope distribution) and heavy (containing eight deuterium (^{2}D) atoms instead of ^{1}H) form. Two samples to be compared are labeled with light or heavy ICAT reagent and subsequently mixed. The ratio of the peak intensities of light−/heavy-labeled peptide pairs obtained by MS correlates with their abundance. In the original version, deuterium labeling of the linker was used, but due to differential elution of light- and heavy-labeled peptides, the method was improved by using ^{13}C labeling [80]. Significant disadvantages of the approach are the side reactivity of the biotin tag and its inability to label peptides lacking cysteine.

Another non-isobaric labeling method based on the same principle is ICPL (isotope-coded protein labeling). The advantage of ICPL is their reactivity toward free amines (lysine side chains and N-terminus), allowing labeling of virtually all peptides present in the samples [81]. This approach can be also performed on the protein level, reducing influence of sample processing variabilities. However in this case, trypsin is unable to cleave at ICPL-modified lysine residues, resulting in longer peptides that are difficult to fragment. Moreover, a part of proteins is identified with no lysine-containing peptides, restricting their usage for quantification [82].

An important group of reagents used for relative quantification comprises the isobaric chemical labels [59, 83]. These rely on isobaric labeling of peptides from different samples, which upon fragmentation give rise to different reporter ions in the MS/MS spectrum. The iTRAQ (isobaric tags for relative and absolute quantification) labels each contain an amine-reactive group, a balance group, and a reporter group. Overall, different reagents have the same molecular weight and upon labeling produce identical mass shifts. Different samples are labeled with reagents containing different distributions of heavy isotopes between the balance and reporter groups and are subsequently mixed. Identical peptides from the samples to be compared co-elute and are detected as a single precursor ion. The iTRAQ labels are designed in such a way that, upon fragmentation, different reagents give rise to reporter ions with identical chemical composition but different molecular weights, owing to their different isotope compositions. Their intensities are proportional to the relative abundances of the labeled peptide originating from the different samples. A major advantage of this method is that it is capable of "multiplexing"; it enables analysis of up to eight samples within a single experiment. The disadvantage of the method is that, similar to other isobaric techniques, ratio compression due to background interference occurs. This problem can be solved through high-resolution sample fractionation and additional isolation and fragmentation (MS3) [83, 84].

A very similar approach is the labeling with tandem mass tags (TMTs), which consist of an amine-reactive, a balance, and a reporter group, which are released upon fragmentation during MS/MS, and the intensity of which is used to calculate relative peptide amounts between the samples [85] (*see* Chaps. 10, 15, 16, 18). The multiplexing capability of TMT reagents was recently improved using differential stable isotope incorporation (^{15}N, ^{13}C) across the reporter and mass balancing regions of the tag, giving rise to 6-plex and 10-plex sets for relative quantitative profiling of multiple conditions [86]. Importantly, quantitation of TMT-labeled peptides requires fragmentation using higher-energy collision dissociation (HCD) or electron transfer dissociation (ETD), because TMT reporter ions are not visible in ion traps following collision-induced dissociation (CID). Accurate quantification for all isobaric labeling strategies is only possible when a single precursor ion is selected for MS2 analysis. Thus, co-eluting peptides will lead to the underestimation of actual protein abundance differences. This challenge can be overcome by additional fragmentation [84]. Even though multiple software packages support relative quantification using TMT reagents, for data analysis, followed by multiple fragmentation, Proteome Discoverer or custom scripts are required.

A cost-effective alternative to commercially available TMT and iTRAQ is the N,N-dimethyl leucine (DiLeu) reagent. DiLeu has the same labeling efficacy, protein coverage, and quantitation accuracy, as well as higher fragmentation efficacy compared to iTRAQ [87]. Due to the inclusion of eight new reporter isotopologues that differ in mass from the existing four reporters by intervals of 6 mDa, DiLeu yields a 12-plex isobaric set that preserves the synthetic simplicity and quantitative performance of the original implementation [88].

Another approach for quantification at the MS/MS level is IPTL (isobaric peptide termini labeling) [89] (*see* Chap. 11). This uses isobaric sequential labeling of the C- and N-termini of the analyzed peptides with deuterated and non-deuterated succinic anhydride. Upon fragmentation, either the N-terminal or the C-terminal label is lost, which results in differentially labeled C- and N-terminal fragment ion series, respectively. These appear as fragment ion pairs in MS/MS, and their relative intensities can be used for quantification. An advantage of this strategy is that the quantification is based on several data points per MS/MS spectrum, although this complicates data analysis enormously. An improved duplex IPTL tags N- and C-terminal with succinic anhydride and dimethyl, respectively, and does not require peptide purification between steps [90]. Triplex-IPTL comprises dimethylation of both peptide termini using different stable isotopes of formaldehyde and cyanoborohydride [91].

A significant advantage of all chemical labeling methods is that they can be applied to practically any type of sample (cell culture, tissues, body fluids, etc.), in contrast to metabolic labeling as discussed below. However, it is crucial to optimize labeling conditions (*see* **Note 3**).

Metabolic Labeling

Metabolic labeling with stable heavy isotope labels (*see* Chaps. 14, 19, 20) introduces the label at the earliest time point in an experiment, i.e., during cell growth and duplication. This is achieved by feeding organisms with special media containing a subset of the metabolites in heavy-labeled form. Metabolic labeling ensures lower deviations in quantification, as the samples to be compared can be mixed at a very early stage during the experiment. Metabolic labeling can easily be achieved in cell culture, but scaling up to whole organisms such as *Drosophila, Caenorhabditis elegans,* and even mice is also possible.

Labeling with ^{15}N-containing media (*see* Chap. 20) has been used successfully for protein quantification in the yeast [92], mammalian cells [93], *C. elegans, Drosophila melanogaster* [94], *Arabidopsis thaliana* [59], and rat [95]. Very high levels of isotope incorporation can be achieved by this method, but the mass difference between labeled and unlabeled samples depends on the number of ^{15}N atoms present in different peptides and determines a significant challenge for data analysis and quantification. Moreover, highly enriched ^{15}N sources are required in order to avoid complex isotope distributions of partially labeled peptides [59]. A computationally simpler method AACT (amino acid-coded mass tagging) also known as SILAC (stable isotope labeling with amino acids in cell culture) was developed to address these issues [95, 96]. SILAC (*see* Chaps. 14, 19) takes advantage of the fact that organisms are naturally (or genetically manipulated to be) auxotrophic for certain amino acids. These amino acids can therefore be provided in labeled and unlabeled form to growth media and would be used by the organism for building proteins in vivo. SILAC experiments usually employ lysine and arginine containing different numbers of the heavy isotopes ^{13}C, ^{15}N, and ^{2}H. Using trypsin for protein digestion ensures that each resulting peptide will contain at least one labeled amino acid (except for the C-terminal peptide of the protein). By comparison of the intensities of the precursor isotope envelopes of non-labeled and labeled peptides, quantitative information can be easily obtained. It can be applied not only in cell culture but also to whole organisms such as *Drosophila* [97] or mice (SILAM) [98]. However, application in animals comprises high costs and sophisticated procedures of animal feeding, limiting the method's common usage. As with most other label-based approaches, when metabolic labeling is applied, nearly 100% incorporation of the label should be aimed at. Incomplete labeling results in inaccurate quantification. Additionally, any changes or stress in the experimental organism due to the artificial growth medium should be taken into account (e.g., when using dialyzed fetal bovine serum for mammalian cells). Another important consideration when using SILAC is the metabolic conversion of the isotopically labeled amino acids within the cell. This can lead to incorrect quantification if, for example, the pathway leading from arginine to proline is stimulated when the concentration of the added arginine is not carefully adjusted, or if the conversion is not corrected for [99] (*see* **Note 4**). In the case of affinity interaction pull-downs using SILAC in vitro, careful adherence to identical conditions for preparation of heavy and light cell extracts is important for obtaining reliable results [100]. A significant disadvantage of metabolic labeling methods is their inability to be used for tissues and body fluids from organisms that cannot be labeled easily (e.g., human patients). To overcome this issue, super-SILAC, isotopically labeled internal SILAC standards were introduced; this allowed successful quantification in tumor tissue samples [101].The SILAC approach limitation is that it is impractical for quantification of up to three samples. 5-plex SILAC was suggested to compare five different cellular conditions within a single experiment. In addition to Lys and Arg, stable isotopically labeled Tyr ($^{13}C_6$ and $[^{13}C_9,^{15}N_1]$) was combined to introduce the necessary peptide mass shifts [102].

Another approach for SILAC multiplicity improvement up to 6-plex is neutron encoding (NeuCode). This method is based on labeling with six lysine isotopologues with varying combinations of ^{13}C, ^{15}N, and ^{2}H substitution that differ in mass by 6 mDa [103]. While using for the targeted proteome analysis through the combination of PRM and NeuCode labeling, the multiplicity rises up to 30 channels of quantitative information in a single MS experiment [104]. NeuCode can be also used both in cell culture and in mammals with shorter labeling times, compared to standard approach (~2 weeks for cultured cells and 3–4 weeks for mammals) [105].

Cell line-specific labeling using amino acid precursors (CTAP) was developed to differentiate the proteome of individual cell populations in co-culture [106]. This method utilizes the inability of vertebrate cells to synthesize certain amino acid required for growth and homeostasis. Transgenic expression of enzymes that synthesize essential amino acids would allow vertebrate cells to overcome auxotrophy by producing their own amino acids from supplemented precursors. These precursors can be isotopically labeled, allowing cell of origin of proteins to be determined by label status identified with MS/MS [61]. The similar method is nitrilase-activatable noncanonical amino acid tagging (NANCAT) that exploits an exogenous nitrilase to enzymatically convert the nitrile-substituted precursors to their corresponding noncanonical amino acids (ncAAs), L-azidohomoalanine (AHA) or homopropargylglycine (HPG), in living cells. Only cells expressing the nitrilase can generate AHA or HPG in cellulo and metabolically incorporate them into the nascent proteins. Subsequent click-labeling of the azide- or alkyne-incorporated proteins with fluorescent probes or with affinity tags enables visualization and proteomic profiling of nascent proteomes, respectively [107].

Enzymatic Labeling

Heavy stable isotopes can be incorporated during enzymatic proteolysis of proteins. Performing proteolysis in heavy ($H_2^{18}O$) or light ($H_2^{16}O$) water incorporates, respectively, two ^{18}O or ^{16}O atoms at the C-terminus of the generated peptides, resulting in a mass shift of 4 Da between heavy- and light-labeled peptides [108, 109]. This label can also be incorporated after digestion in a second incubation step with a protease. This method ensures near-complete labeling, in contrast to ICAT it does not favor cysteine-containing proteins; it does not require enrichment of labeled peptides and unlike metabolic labeling it can be used for human specimens. It is also less expensive compared to other stable isotope labeling techniques. Acid-catalyzed back-exchange at extreme pH conditions can occur [110] (*see* **Note 5**); however, the mild conditions used during ESI or MALDI analyses do not influence the stability of the introduced label. Incomplete labeling by incorporation of only one ^{18}O atom can complicate data analysis and needs to be taken into account [111].

3.1.2   *Label-Free Quantification (LFQ)*

Label-based approaches for proteomic quantification usually come at higher cost, require additional steps of sample preparation, and are characterized by limited multiplicity. Therefore, it is not surprising that the use of label-free methods increased during the last few decades. As mentioned above, label-free quantitative approaches (*see* Chaps. 17, 21–25) rely on the comparison of different features between independent LC-MS or LC-MS/MS measurements. They fall into two general categories: (i) spectral counting (SC), methods that involve counting the number of identified peptides or acquired fragment spectra, and (ii) methods that involve comparing precursors' ion intensities, determined by the extracted ion chromatogram (XIC).

Spectral Counting-Based LFQ

Spectral counting-based LFQ relies on the practical observation that more abundant peptides are more likely to be observed and detected in an MS experiment. These approaches use the number of peptides or the number of fragment spectra observed for a particular protein in the analysis. Protein abundance index (PAI), calculated as a ratio of the number of peptides identified for a protein to a total number of peptides, the protein could theoretically produce, was suggested for LFQ [112]. However, Liu et al. found a linear correlation over two orders of magnitude between the number of spectra and the relative protein abundance, whereas no correlation between the relative protein amounts and the number of peptides and the sequence coverage was observed [11]. Though spectral counting is a relatively simple and reliable technique and is easily implemented, normalization and careful statistical evaluation are still needed for accurate quantification. This accuracy can decrease significantly for proteins with only a few observable peptides, as well as when the quantitative changes between experiments are small [113]. Furthermore, since larger proteins give rise to more peptides than do smaller ones, additional normalization factors can be applied to improve the results of quantification [114]. Thus, several approaches were implemented. The first is the normalized spectral abundance factor (NSAF), calculated as a ratio of the spectral counts for a given protein to its length. This value is then normalized by dividing it by the sum of all the ratios obtained for each protein identified in the experiment. Additionally this method improves the minimal fold change detectible by SC [114]. The protein abundance factor normalizes the total number of non-redundant spectra to the molecular mass of the intact protein [115]. The last one is the absolute protein expression algorithm, a machine learning classification system, that corrects spectral counts for the likelihood that a spectrum might be detected [116]. The SC disability to quantify smaller abundance changes can be improved by increasing the scoring requirements for spectrum identification; however, the accurate quantification of low-abundance proteins will be restricted [117]. Another disadvantage of the SC is connected with the peptides that can be assigned to more than one protein. The normalized spectral abundance factor bears with the problem by dividing shared spectra proportionally between the possible contributing proteins based on the distribution of the other identified unique peptides [118].

Intensity-Based LFQ

Signal intensities of ions after ESI correlate with ion concentrations [119, 120]. The extracted peak areas from chromatograms in LC-MS measurements specific for certain ions (extracted ion chromatograms, XIC) can therefore be used for relative quantification of specific peptides and proteins between different samples. Quantitation is an area under the curve or peak height calculation for each peptide that elutes from the LC column at an expected retention time. The method allows measurements with high precision and wide dynamic range, especially when high-resolution mass spectrometers are used. It can also be applied to MALDI measurements combined with offline LC separation. However, the following important considerations should be taken into account: First, the variation between measurements of the peak intensities of peptides from the same sample (technical replicates) should be recorded, and appropriate normalization should be applied. Secondly, and more critically, variation of the LC retention time and/or m/z values of identical peptides between measurement runs should be considered. Any variability in this respect requires alignment of individual ion chromatograms for correct quantification and elimination of any global drift in retention time. Practical normalization strategies may include the addition of identical amounts of standard protein in different sample or normalization, based on a priori information about a protein that does not change quantitatively between the samples compared [121]. Reproducibility of LC separation, stability of the electrospray ion source, and the use of computational algorithms for comparison, retention time alignment, and statistical evaluation of several LC-MS datasets in a single procedure are therefore crucial (*see* **Note 6**).

When carried out using high-resolution mass spectrometers, intensity-based LFQ is more sensitive and accurate than SC [122]. Moreover it allows lower accurately distinguishing fold changes, compared to SC [123].

3.2 Absolute Quantification

Absolute quantification is used to determine the absolute amount (mass, mole number, or copy number) of proteins in a mixture or complex (*see* Chaps. 12, 13, 26). This is very informative, but label-based methods are usually relatively laborious, and label-free ones are less accurate. Absolute quantification is generally performed at the peptide level, although top-down absolute quantification has been introduced [124].

3.2.1 Label-Based Absolute Quantification

The arguably most widely used method for absolute quantification (AQUA) employs peptides labeled with heavy stable isotopes (AQUA peptides) as added, internal standards [125] (*see* Chaps. 12, 26). This method can be used for accurate profiling and absolute quantification of proteins within a complex sample, for monitoring changes in post-translational modification [125, 126], and for determining the stoichiometry of subunits within a protein complex [127]. Being a targeted approach, the method requires a priori information about the peptides and proteins that are subject to analysis. The specific characteristics of the targeted precursor ion (elution time, m/z value, charge state), optimum fragmentation conditions (collision energy), and resulting fragmentation pattern are determined in prior measurements. Peptides labeled with heavy stable isotopes (^{13}C- and ^{15}N-labeled amino acids), identical in sequence to the peptides of interest naturally present in the sample, are synthesized chemically. These two peptides have identical physicochemical properties but present a specific mass shift in the mass spectrum. The AQUA peptides are added to the protein digest or peptide sample at known concentrations and analyzed in SRM mode. The co-eluting analytes—i.e., the endogenous and the mass-shifted labeled peptides—are selected for fragmentation on the basis of their (already determined) elution time and m/z value. The intensities of the fragment ions of the peptide of interest are compared with those of the AQUA peptide, and this reflects directly their quantitative relationship. As the amount of the added peptide is known, the amount of the sample peptide can be deduced. The AQUA approach allows very specific, targeted detection of the peptides of interest, thereby minimizing the variability and the influence of background noise. Even in complex samples, several hundred peptides can be targeted within a single LC–MS/MS experiment [128]. As the method is strictly hypothesis-driven, it allows the selection of peptides with optimal chromatographic performance and ionization efficiency (i.e., good "detectability"), which do not undergo uncontrolled modification in vitro (e.g., oxidation of methionine) and which are unique to the protein of interest. Such peptides are called prototypic peptides and can be identified or predicted for particular proteomic platform using peptide libraries and public databases [113, 129, 130]. AQUA strategy suffers from quantification uncertainties: peptide standard spiking in occurs after sample preparation and enzymatic proteolysis. Moreover, any losses of the peptides—for example, during storage—would directly influence quantification results. There are several critical aspects that should be considered when an AQUA experiment is being planned such as incomplete proteolytic digestion, exact amount of AQUA peptide, application of AQUA peptides, and number of applied AQUA peptides for each protein to be quantified (*see* **Note 7**).

In order to simplify the quantification of several peptides per protein, heavy-labeled standard proteins can be used instead of individual peptides. Several approaches have been developed in that direction, including PSAQ (protein standard absolute quantification) [131], absolute SILAC [132], absolute NeuCode [104], and FLEXIQuant (full-length expressed stable isotope-labeled proteins for absolute quantification) [133]. Protein Epitope Signature

Tags (PrESTs), developed in the course of the Human Protein Atlas project [134], can be combined with SILAC allowing accurate and streamlined quantification of the absolute or relative amount of a large amount of proteins of interest at a time, in a wide variety of applications [135].

QconCAT (concatenated signature peptides encoded by Qcon-CAT genes) uses artificial, labeled standard proteins assembled from diverse peptides belonging to different proteins [136]. It utilizes synthetic DNA that encodes a concatenated series of peptides of interest, which are expressed in *Escherichia coli* grown in stable isotope-labeled media. After purification and quantification, peptides are introduced to cell lysates at digestion. Similarly to AQUA, this approach does not allow sample fractionation before analysis.

3.2.2 Label-Free Absolute Quantification

Development of the label-free absolute quantification has several advantages: (i) omitting the time-consuming and often costly step of introducing standard peptides and (ii) the opportunity to compare virtually unlimited numbers of samples. On the other hand, they entail the disadvantages of lower accuracy and the requirement for high reproducibility. One of the first label-free approaches used for absolute quantification was the emPAI (exponentially modified PAI) [137], calculated as $emPAI = 10^{PAI} - 1$. It is proportional to the protein content in a protein mixture and, therefore, can be used for the estimation of absolute amounts of proteins. This approach still has a lower accuracy compared to other methods [123].

An approach termed APEX (absolute protein expression) based on spectral counting can also be used for profiling absolute protein quantities per cell [116]. Important features of APEX are the correction factors that it introduces, providing a relationship of direct proportionality between the numbers of observed and expected peptides.

In addition to the above discussed spectral counting-based absolute quantification, several intensity-based methods have also been reported. As incomplete digestion is a critical issue when one is performing absolute quantification of peptides or proteins [138], an alternative approach (generally known as "Top3") has been developed that deals with this problem. In this approach the quantities of the three most abundant tryptic peptides are averaged [139]. It is generally assumed that some parts of the protein are completely digested and, therefore, the three most abundant peptides reflect the protein concentration. The protein sample is therefore spiked with a known amount of standard protein, and, after digestion, the average MS signal response of the standard protein is used to calculate a universal signal response factor (ion counts per mole of protein). This factor is then applied to calculate the concentration of the proteins in the sample to be analyzed [138].

In the intensity-based absolute quantitation (iBAQ) algorithm, the summed intensities of the precursor peptides that map to each protein are divided by the number of theoretically observable peptides, which is considered to be all tryptic peptides between 6 and 30 amino acids in length [140]. This operation converts a measure that is expected to be proportional to mass (intensity) into one that is proportional to molar amount (iBAQ). Interestingly, iBAQ and dividing Top3 by the number of identified peptides gave the most accurate quantitation. Here iBAQ shows less bias when calculating the abundance of smaller proteins [141]. Relative iBAQ (riBAQ), which is the iBAQ (calculated by MaxQuant) for a protein or protein group divided by all non-contaminant, non-reversed iBAQ values for a replicate, is an equivalent to normalized molar intensity [142, 143].

Total protein approach (TPA) is an approach for determination of protein copy numbers per cell without protein standards [144]. The method is based on the observation that the 3000 most abundant proteins of the cell already constitute >99% of the proteome mass. Thus, using intensity values for each protein, a fractional value of the MS signal (LFQ intensity) of a protein compared with the total MS signal is a good proxy of the percentage of its protein mass to total protein mass. This can then be converted into numbers of molecules per cell by measuring or estimating the volume and protein content of the analyzed cells [144]. Comparison analysis showed similarity of the data obtained by the TPA and SILAC-PrEST approaches [144]. Recently, DIA-based label-free absolute quantification method was reported. It uses the TPA algorithm (DIA-TPA) [145].

Nowadays MALDI-MS is still considered by many researchers as a non-quantitative technique (*see* Chap. 13). However MALDI-mass spectrometry imaging (MSI) permits label-free in situ analysis of chemical compounds directly from the surface of two-dimensional biological tissue slices [146]. MALDI-MSI is a key label-free technology for quantitative analysis of drugs, metabolites, and formulations as well as biomarkers in tissues [147]. It allows both protein relative and absolute quantification. Absolute quantitative MSI experiments deliver concentration levels of analytes, which are expressed in units of moles or mass quantity of compound per mass/volume or area of tissue. Relative quantification is achieved by visualizing and estimating relative concentration values of the analyte across the tissue or whole body section in comparison with other compounds [146]. Achieving reliable relative quantification by MALDI-MSI requires the application of a signal intensity normalization method, baseline correction (subtraction), and mass spectral realignment (recalibration) [148].

4 Notes

1. Quantification performance of DIA is superior to DDA, especially in terms of reproducibility and accuracy. Quantification accuracy is decreased when considering low protein/peptide amounts in DDA, but not in DIA. However, for DIA-based quantification several issues should be taken into account. The larger the spectral library, the higher the coefficient of variations on peptide and protein level. In DIA small fold changes might entail the risk of more false positive discoveries. For DIA quantification on peptide level is preferable because of better differential abundance detection and higher accuracy [20].

2. For relative quantification using stable isotopes, the quantitative correspondence does not always apply exactly when deuterium is used as a label, as labeling with deuterium can affect retention time in LC [80].

3. Relative quantification using stable isotope chemical or enzymatic labeling: The labeling procedure has to be optimized ensuring ideal labeling; 100% label incorporation should be aimed at, which might not be achievable for all approaches. Additionally, side reactions should be avoided to prevent erroneous quantification results.

4. Relative quantification using metabolic labeling: In general, large-scale SILAC experiments use both isotope-coded arginine and lysine to obtain labeling of all possible tryptic peptides, thereby maximizing quantitative coverage of all potential peptides in a given experiment. Quantification using SILAC may be disturbed by the fact that the isotopically labeled amino acid arginine is a metabolic precursor of proline and as such might be converted to labeled proline. As with other labeling approaches, complete incorporation of the heavy label should be aimed at (which should be limited only by the isotopic enrichment of the commercially available labeling sources).

5. Relative quantification using enzymatic labeling: Under extreme pH conditions in $H_2 16O$ buffers, acid-catalyzed back-exchange could result in partial loss of the ^{18}O label. Therefore, it is recommended that the enzymatic reactions are stopped by addition of protease inhibitors or freezing of the reaction mixture, rather than by acidifying with 10% TFA.

6. Label-free quantification: The most crucial parameter in label-free quantification is the consistent reproducibility of the LC separation, ionization, and mass measurements of the peptides. All variations of peptide intensities, as well as LC retention times, should be recorded between technical replicates and used for normalization and alignment between runs.

7. Absolute quantification: First, when peptides from protease digests are to be quantified, complete digestion of the protein sample must be guaranteed. Missed protease cleavages affecting the targeted peptide will result in an artificial decrease in the amounts observed in quantification. Additionally, AQUA peptides are usually obtained in known absolute amounts in lyophilized form and therefore have to be dissolved quantitatively. As it is advisable to add standard peptides after rather than before digestion, any variability and losses during the prior sample preparation should be minimized. Finally, for reliable quantification results, several peptides per targeted protein should be monitored, in order to provide more than one reference value per protein.

Acknowledgments

The authors were supported by P.U.R.E. (Protein Research Unit Ruhr within Europe) and ProDi (Center for protein diagnostics), Ministry of Innovation, Science and Research of North-Rhine Westphalia, Germany, and by the H2020 project NISCI, (GA no. 681094) and by the ISF program of the European Union.

References

1. Niessen WMA, Falck D (2015) Introduction to mass spectrometry, a tutorial. In: Kool J, Niessen WMA (eds) Analyzing biomolecular interactions by mass spectrometry. Wiley-VCH, Weinheim

2. Hoffmann ED, Stroobant V (2001) Mass spectrometry: principles and applications. Wiley, New York

3. El-Aneed A, Cohen A, Banoub J (2009) Mass spectrometry, review of the basics: electrospray, MALDI, and commonly used mass analyzers. Appl Spectrosc Rev 44(3):210–230. https://doi.org/10.1080/05704920902717872

4. Domon B, Aebersold R (2010) Options and considerations when selecting a quantitative proteomics strategy. Nat Biotechnol 28 (7):710–721. https://doi.org/10.1038/nbt.1661

5. McDonald WH, Yates JR (2002) Shotgun proteomics and biomarker discovery. Dis Markers 18:99–105. https://doi.org/10.1155/2002/505397

6. Sajic T, Liu Y, Aebersold R (2015) Using data-independent, high-resolution mass spectrometry in protein biomarker research: perspectives and clinical applications. Proteomics Clin Appl 9(3–4):307–321. https://doi.org/10.1002/prca.201400117

7. Meyer JG (2019) Fast proteome identification and quantification from data-dependent acquisition-tandem mass spectrometry (DDA MS/MS) using free software tools. Methods Protoc 2(1):8. https://doi.org/10.3390/mps2010008

8. Cox J, Hein MY, Luber CA et al (2014) Accurate proteome-wide label-free quantification by delayed normalization and maximal peptide ratio extraction, termed MaxLFQ. Mol Cell Proteomics 13(9):2513–2526. https://doi.org/10.1074/mcp.M113.031591

9. Bateman NW, Goulding SP, Shulman NJ et al (2014) Maximizing peptide identification events in proteomic workflows using data-dependent acquisition (DDA). Mol Cell Proteomics 13(1):329–338. https://doi.org/10.1074/mcp.M112.026500

10. Michalski A, Cox J, Mann M (2011) More than 100,000 detectable peptide species elute in single shotgun proteomics runs but the majority is inaccessible to data-dependent LC-MS/MS. J Proteome Res 10 (4):1785–1793. https://doi.org/10.1021/pr101060v

11. Liu H, Sadygov RG, Yates JR 3rd (2004) A model for random sampling and estimation of relative protein abundance in shotgun proteomics. Anal Chem 76(14):4193–4201. https://doi.org/10.1021/ac0498563

12. Bruderer R, Bernhardt OM, Gandhi T et al (2015) Extending the limits of quantitative proteome profiling with data-independent acquisition and application to acetaminophen-treated three-dimensional liver microtissues. Mol Cell Proteomics 14 (5):1400–1410. https://doi.org/10.1074/mcp.M114.044305

13. Bondarenko PV, Chelius D, Shaler TA (2002) Identification and relative quantitation of protein mixtures by enzymatic digestion followed by capillary reversed-phase liquid chromatography-tandem mass spectrometry. Anal Chem 74(18):4741–4749. https://doi.org/10.1021/ac0256991

14. Zhu W, Smith JW, Huang CM (2010) Mass spectrometry-based label-free quantitative proteomics. J Biomed Biotechnol 2010:1. https://doi.org/10.1155/2010/840518

15. Ong SE, Mann M (2006) A practical recipe for stable isotope labeling by amino acids in cell culture (SILAC). Nat Protoc 1 (6):2650–2660. https://doi.org/10.1038/nprot.2006.427

16. Wiese S, Reidegeld KA, Meyer HE et al (2007) Protein labeling by iTRAQ: a new tool for quantitative mass spectrometry in proteome research. Proteomics 7 (3):340–350. https://doi.org/10.1002/pmic.200600422

17. Gillet LC, Navarro P, Tate S et al (2012) Targeted data extraction of the MS/MS spectra generated by data-independent acquisition: a new concept for consistent and accurate proteome analysis. Mol Cell Proteomics 11(6):O111.016717. https://doi.org/10.1074/mcp.O111.016717

18. Muntel J, Xuan Y, Berger ST et al (2015) Advancing urinary protein biomarker discovery by data-independent acquisition on a quadrupole-orbitrap mass spectrometer. J Proteome Res 14(11):4752–4762. https://doi.org/10.1021/acs.jproteome.5b00826

19. Bruderer R, Bernhardt OM, Gandhi T et al (2017) Optimization of experimental parameters in data-independent mass spectrometry significantly increases depth and reproducibility of results. Mol Cell Proteomics 16(12):2296–2309. https://doi.org/10.1074/mcp.RA117.000314

20. Barkovits K, Pacharra S, Pfeiffer K et al (2020) Reproducibility, specificity and accuracy of relative quantification using spectral library-based data-independent acquisition. Mol Cell Proteomics 19(1):181–197. https://doi.org/10.1074/mcp.RA119.001714

21. Weisbrod CR, Eng JK, Hoopmann MR et al (2012) Accurate peptide fragment mass analysis: multiplexed peptide identification and quantification. J Proteome Res 11 (3):1621–1632. https://doi.org/10.1021/pr2008175

22. Bilbao A, Varesio E, Luban J et al (2015) Processing strategies and software solutions for data-independent acquisition in mass spectrometry. Proteomics 15(5–6):964–980. https://doi.org/10.1002/pmic.201400323

23. Shao W, Lam H (2017) Tandem mass spectral libraries of peptides and their roles in proteomics research. Mass Spectrom Rev 36 (5):634–648. https://doi.org/10.1002/mas.21512

24. Bruderer R, Bernhardt OM, Gandhi T et al (2016) High-precision iRT prediction in the targeted analysis of data-independent acquisition and its impact on identification and quantitation. Proteomics 16(15–16):2246–2256. https://doi.org/10.1002/pmic.201500488

25. Li S, Cao Q, Xiao W et al (2017) Optimization of acquisition and data-processing parameters for improved proteomic quantification by sequential window acquisition of all theoretical fragment ion mass spectrometry. J Proteome Res 16(2):738–747. https://doi.org/10.1021/acs.jproteome.6b00767

26. Röst HL, Rosenberger G, Navarro P et al (2014) OpenSWATH enables automated, targeted analysis of data-independent acquisition MS data. Nat Biotechnol 32(3):219–223. https://doi.org/10.1038/nbt.2841

27. Egertson JD, MacLean B, Johnson R et al (2015) Multiplexed peptide analysis using data-independent acquisition and skyline. Nat Protoc 10(6):887–903. https://doi.org/10.1038/nprot.2015.055

28. Bruderer R, Sondermann J, Tsou CC et al (2017) New targeted approaches for the quantification of data-independent acquisition mass spectrometry. Proteomics 17(9). https://doi.org/10.1002/pmic.201700021

29. Govaert E, Van Steendam K, Willems S et al (2017) Comparison of fractionation proteomics for local SWATH library building. Proteomics 17(15–16). https://doi.org/10.1002/pmic.201700052

30. Tsou CC, Avtonomov D, Larsen B et al (2015) DIA-umpire: comprehensive computational framework for data-independent acquisition proteomics. Nat Methods

12(3):258–264, 257 p following 264. https://doi.org/10.1038/nmeth.3255

31. Smith RD (2012) Mass spectrometry in biomarker applications: from untargeted discovery to targeted verification, and implications for platform convergence and clinical application. Clin Chem 58(3):528–530. https://doi.org/10.1373/clinchem.2011.180596

32. Lange V, Picotti P, Domon B et al (2008) Selected reaction monitoring for quantitative proteomics: a tutorial. Mol Syst Biol 4:222. https://doi.org/10.1038/msb.2008.61

33. Rauniyar N (2015) Parallel reaction monitoring: a targeted experiment performed using high resolution and high mass accuracy mass spectrometry. Int J Mol Sci 16 (12):28,566–28,581. https://doi.org/10.3390/ijms161226120

34. Gallien S, Duriez E, Demeure K et al (2013) Selectivity of LC-MS/MS analysis: implication for proteomics experiments. J Proteome 81:148–158. https://doi.org/10.1016/j.jprot.2012.11.005

35. Peterson AC, Russell JD, Bailey DJ et al (2012) Parallel reaction monitoring for high resolution and high mass accuracy quantitative, targeted proteomics. Mol Cell Proteomics 11(11):1475–1488. https://doi.org/10.1074/mcp.O112.020131

36. Prakash A, Tomazela DM, Frewen B et al (2009) Expediting the development of targeted SRM assays: using data from shotgun proteomics to automate method development. J Proteome Res 8(6):2733–2739. https://doi.org/10.1021/pr801028b

37. Gallien S, Duriez E, Crone C et al (2012) Targeted proteomic quantification on quadrupole-orbitrap mass spectrometer. Mol Cell Proteomics 11(12):1709–1723. https://doi.org/10.1074/mcp.O112.019802

38. Keshishian H, Addona T, Burgess M et al (2007) Quantitative, multiplexed assays for low abundance proteins in plasma by targeted mass spectrometry and stable isotope dilution. Mol Cell Proteomics 6 (12):2212–2229. https://doi.org/10.1074/mcp.M700354-MCP200

39. Zhao C, Trudeau B, Xie H et al (2014) Epitope mapping and targeted quantitation of the cardiac biomarker troponin by SID-MRM mass spectrometry. Proteomics 14(11):1311–1321. https://doi.org/10.1002/pmic.201300150

40. Gauthier MS, Perusse JR, Awan Z et al (2015) A semi-automated mass spectrometric immunoassay coupled to selected reaction monitoring (MSIA-SRM) reveals novel relationships between circulating PCSK9 and metabolic phenotypes in patient cohorts. Methods 81:66–73. https://doi.org/10.1016/j.ymeth.2015.03.003

41. McDonald WH, Yates JR 3rd (2003) Shotgun proteomics: integrating technologies to answer biological questions. Curr Opin Mol Ther 5(3):302–309

42. Nesvizhskii AI, Aebersold R (2005) Interpretation of shotgun proteomic data: the protein inference problem. Mol Cell Proteomics 4 (10):1419–1440. https://doi.org/10.1074/mcp.R500012-MCP200

43. Uszkoreit J, Perez-Riverol J, Eggers B et al (2019) Protein inference using PIA workflows and PSI standard file formats. J Proteome Res 18(2):741–747. https://doi.org/10.1021/acs.jproteome.8b00723

44. Gerster S, Kwon T, Ludwig C et al (2014) Statistical approach to protein quantification. Mol Cell Proteomics 13(2):666–677. https://doi.org/10.1074/mcp.M112.025445

45. Perez-Riverol Y, Sanchez A, Ramos Y et al (2011) In silico analysis of accurate proteomics, complemented by selective isolation of peptides. J Proteome 74(10):2071–2082. https://doi.org/10.1016/j.jprot.2011.05.034

46. Serang O, Noble W (2012) A review of statistical methods for protein identification using tandem mass spectrometry. Stat Interface 5 (1):3–20. https://doi.org/10.4310/sii.2012.v5.n1.a2

47. Ma ZQ, Dasari S, Chambers MC et al (2009) IDPicker 2.0: improved protein assembly with high discrimination peptide identification filtering. J Proteome Res 8(8):3872–3881. https://doi.org/10.1021/pr900360j

48. Smith LM, Kelleher NL (2018) Proteoforms as the next proteomics currency. Science 359 (6380):1106–1107. https://doi.org/10.1126/science.aat1884

49. Kelleher NL, Lin HY, Valaskovic GA et al (1999) Top down versus bottom up protein characterization by tandem high-resolution mass spectrometry. J Am Chem Soc 121 (4):806–812. https://doi.org/10.1021/ja973655h

50. Auclair JR, Salisbury JP, Johnson JL et al (2014) Artifacts to avoid while taking advantage of top-down mass spectrometry based detection of protein S-thiolation. Proteomics 14(10):1152–1157. https://doi.org/10.1002/pmic.201300450

51. Doll S, Burlingame AL (2015) Mass spectrometry-based detection and assignment of protein posttranslational modifications.

ACS Chem Biol 10(1):63–71. https://doi.org/10.1021/cb500904b

52. Donnelly DP, Rawlins CM, DeHart CJ et al (2019) Best practices and benchmarks for intact protein analysis for top-down mass spectrometry. Nat Methods 16(7):587–594. https://doi.org/10.1038/s41592-019-0457-0

53. Smith LM, Kelleher NL, Consortium for Top Down Proteomics (2013) Proteoform: a single term describing protein complexity. Nat Methods 10:186–187, United States. https://doi.org/10.1038/nmeth.2369

54. Chen B, Brown KA, Lin Z et al (2018) Top-down proteomics: ready for prime time? Anal Chem 90(1):110–127. https://doi.org/10.1021/acs.analchem.7b04747

55. Cheon DH, Yang EG, Lee C et al (2017) Low-molecular-weight plasma proteome analysis using top-down mass spectrometry. Methods Mol Biol 1619:103–117. https://doi.org/10.1007/978-1-4939-7057-5_8

56. Chen Y, Mao P, Wang D (2018) Quantitation of intact proteins in human plasma using top-down parallel reaction monitoring-MS. Anal Chem 90(18):10,650–10,653. https://doi.org/10.1021/acs.analchem.8b02699

57. Ntai I, Toby TK, LeDuc RD et al (2016) A method for label-free, differential top-down proteomics. Methods Mol Biol 1410:121–133. https://doi.org/10.1007/978-1-4939-3524-6_8

58. Bantscheff M, Lemeer S, Savitski MM et al (2012) Quantitative mass spectrometry in proteomics: critical review update from 2007 to the present. Anal Bioanal Chem 404(4):939–965. https://doi.org/10.1007/s00216-012-6203-4

59. Ong SE, Mann M (2005) Mass spectrometry-based proteomics turns quantitative. Nat Chem Biol 1(5):252–262. https://doi.org/10.1038/nchembio736

60. Meier F, Geyer PE, Virreira Winter S et al (2018) BoxCar acquisition method enables single-shot proteomics at a depth of 10,000 proteins in 100 minutes. Nat Methods 15(6):440–448. https://doi.org/10.1038/s41592-018-0003-5

61. Pappireddi N, Martin L, Wuhr M (2019) A review on quantitative multiplexed proteomics. Chembiochem 20(10):1210–1224. https://doi.org/10.1002/cbic.201800650

62. Eliuk S, Makarov A (2015) Evolution of orbitrap mass spectrometry instrumentation. Annu Rev Anal Chem (Palo Alto, Calif) 8:61–80. https://doi.org/10.1146/annurev-anchem-071114-040325

63. Zubarev RA, Makarov A (2013) Orbitrap mass spectrometry. Anal Chem 85(11):5288–5296. https://doi.org/10.1021/ac4001223

64. Sandberg A, Branca RM, Lehtio J et al (2014) Quantitative accuracy in mass spectrometry based proteomics of complex samples: the impact of labeling and precursor interference. J Proteome 96:133–144. https://doi.org/10.1016/j.jprot.2013.10.035

65. Altelaar AF, Frese CK, Preisinger C et al (2013) Benchmarking stable isotope labeling based quantitative proteomics. J Proteome 88:14–26. https://doi.org/10.1016/j.jprot.2012.10.009

66. Piehowski PD, Petyuk VA, Orton DJ et al (2013) Sources of technical variability in quantitative LC-MS proteomics: human brain tissue sample analysis. J Proteome Res 12(5):2128–2137. https://doi.org/10.1021/pr301146m

67. Valikangas T, Suomi T, Elo LL (2018) A systematic evaluation of normalization methods in quantitative label-free proteomics. Brief Bioinform 19(1):1–11. https://doi.org/10.1093/bib/bbw095

68. Li Z, Adams RM, Chourey K et al (2012) Systematic comparison of label-free, metabolic labeling, and isobaric chemical labeling for quantitative proteomics on LTQ Orbitrap Velos. J Proteome Res 11(3):1582–1590. https://doi.org/10.1021/pr200748h

69. Sinz A (2018) Cross-linking/mass spectrometry for studying protein structures and protein-protein interactions: where are we now and where should we go from here? Angew Chem Int Ed Engl 57(22):6390–6396. https://doi.org/10.1002/anie.201709559

70. Yugandhar K, Gupta S, Yu H (2019) Inferring protein-protein interaction networks from mass spectrometry-based proteomic approaches: a mini-review. Comput Struct Biotechnol J 17:805–811. https://doi.org/10.1016/j.csbj.2019.05.007

71. Rappsilber J (2011) The beginning of a beautiful friendship: cross-linking/mass spectrometry and modelling of proteins and multiprotein complexes. J Struct Biol 173(3):530–540. https://doi.org/10.1016/j.jsb.2010.10.014

72. Muller F, Fischer L, Chen ZA et al (2018) On the reproducibility of label-free quantitative cross-linking/mass spectrometry. J Am Soc Mass Spectrom 29(2):405–412. https://doi.org/10.1007/s13361-017-1837-2

73. Chavez JD, Eng JK, Schweppe DK et al (2016) A general method for targeted quantitative cross-linking mass spectrometry. PLoS One 11(12):e0167547. https://doi.org/10.1371/journal.pone.0167547

74. Fischer L, Chen ZA, Rappsilber J (2013) Quantitative cross-linking/mass spectrometry using isotope-labelled cross-linkers. J Proteome 88:120–128. https://doi.org/10.1016/j.jprot.2013.03.005

75. Walzthoeni T, Joachimiak LA, Rosenberger G et al (2015) xTract: software for characterizing conformational changes of protein complexes by quantitative cross-linking mass spectrometry. Nat Methods 12(12):1185–1190. https://doi.org/10.1038/nmeth.3631

76. Chen ZA, Fischer L, Cox J et al (2016) Quantitative cross-linking/mass spectrometry using isotope-labeled cross-linkers and MaxQuant. Mol Cell Proteomics 15(8):2769–2778. https://doi.org/10.1074/mcp.M115.056481

77. Yu C, Huszagh A, Viner R et al (2016) Developing a multiplexed quantitative cross-linking mass spectrometry platform for comparative structural analysis of protein complexes. Anal Chem 88(20):10301–10308. https://doi.org/10.1021/acs.analchem.6b03148

78. Muller F, Kolbowski L, Bernhardt OM et al (2019) Data-independent acquisition improves quantitative cross-linking mass spectrometry. Mol Cell Proteomics 18(4):786–795. https://doi.org/10.1074/mcp.TIR118.001276

79. Gygi SP, Rist B, Gerber SA et al (1999) Quantitative analysis of complex protein mixtures using isotope-coded affinity tags. Nat Biotechnol 17(10):994–999. https://doi.org/10.1038/13690

80. Tao WA, Aebersold R (2003) Advances in quantitative proteomics via stable isotope tagging and mass spectrometry. Curr Opin Biotechnol 14(1):110–118. https://doi.org/10.1016/s0958-1669(02)00018-6

81. Schmidt A, Kellermann J, Lottspeich F (2005) A novel strategy for quantitative proteomics using isotope-coded protein labels. Proteomics 5:4–15

82. Paradela A, Marcilla M, Navajas R et al (2010) Evaluation of isotope-coded protein labeling (ICPL) in the quantitative analysis of complex proteomes. Talanta 80(4):1496–1502. https://doi.org/10.1016/j.talanta.2009.06.083

83. Ow SY, Salim M, Noirel J et al (2011) Minimising iTRAQ ratio compression through understanding LC-MS elution dependence and high-resolution HILIC fractionation. Proteomics 11(11):2341–2346. https://doi.org/10.1002/pmic.201000752

84. Ting L, Rad R, Gygi SP et al (2011) MS3 eliminates ratio distortion in isobaric multiplexed quantitative proteomics. Nat Methods 8(11):937–940. https://doi.org/10.1038/nmeth.1714

85. Thompson A, Schafer J, Kuhn K et al (2003) Tandem mass tags: a novel quantification strategy for comparative analysis of complex protein mixtures by MS/MS. Anal Chem 75(8):1895–1904. https://doi.org/10.1021/ac0262560

86. Werner T, Becher I, Sweetman G et al (2012) High-resolution enabled TMT 8-plexing. Anal Chem 84(16):7188–7194. https://doi.org/10.1021/ac301553x

87. Xiang F, Ye H, Chen R et al (2010) N, N-dimethyl leucines as novel isobaric tandem mass tags for quantitative proteomics and peptidomics. Anal Chem 82(7):2817–2825. https://doi.org/10.1021/ac902778d

88. Frost DC, Greer T, Li L (2015) High-resolution enabled 12-plex DiLeu isobaric tags for quantitative proteomics. Anal Chem 87(3):1646–1654. https://doi.org/10.1021/ac503276z

89. Koehler CJ, Strozynski M, Kozielski F et al (2009) Isobaric peptide termini labeling for MS/MS-based quantitative proteomics. J Proteome Res 8(9):4333–4341

90. Koehler CJ, Arntzen MO, Strozynski M et al (2011) Isobaric peptide termini labeling utilizing site-specific N-terminal succinylation. Anal Chem 83:4775–4781

91. Koehler CJ, Arntzen MO, de Souza GA et al (2013) An approach for triplex-isobaric peptide termini labeling (triplex-IPTL). Anal Chem 85:2478–2485

92. Krijgsveld J, Ketting RF, Mahmoudi T et al (2003) Metabolic labeling of C. elegans and D. melanogaster for quantitative proteomics. Nat Biotechnol 21(8):927–931. https://doi.org/10.1038/nbt848

93. Nelson CJ, Huttlin EL, Hegeman AD et al (2007) Implications of 15N-metabolic labeling for automated peptide identification in Arabidopsis thaliana. Proteomics 7(8):1279–1292. https://doi.org/10.1002/pmic.200600832

94. Wu CC, MacCoss MJ, Howell KE et al (2004) Metabolic labeling of mammalian organisms with stable isotopes for quantitative proteomic analysis. Anal Chem 76(17):4951–4959. https://doi.org/10.1021/ac049208j

95. Ong SE, Blagoev B, Kratchmarova I et al (2002) Stable isotope labeling by amino acids in cell culture, SILAC, as a simple and accurate approach to expression proteomics. Mol Cell Proteomics 1(5):376–386. https://doi.org/10.1074/mcp.m200025-mcp200

96. Chen X, Smith LM, Bradbury EM (2000) Site-specific mass tagging with stable isotopes in proteins for accurate and efficient protein identification. Anal Chem 72(6):1134–1143. https://doi.org/10.1021/ac9911600

97. Sury MD, Chen JX, Selbach M (2010) The SILAC fly allows for accurate protein quantification in vivo. Mol Cell Proteomics 9 (10):2173–2183. https://doi.org/10.1074/mcp.M110.000323

98. Kruger M, Moser M, Ussar S et al (2008) SILAC mouse for quantitative proteomics uncovers kindlin-3 as an essential factor for red blood cell function. Cell 134 (2):353–364. https://doi.org/10.1016/j.cell.2008.05.033

99. Van Hoof D, Pinkse MW, Oostwaard DW et al (2007) An experimental correction for arginine-to-proline conversion artifacts in SILAC-based quantitative proteomics. Nat Methods 4(9):677–678. https://doi.org/10.1038/nmeth0907-677

100. Nikolov M, Stutzer A, Mosch K et al (2011) Chromatin affinity purification and quantitative mass spectrometry defining the interactome of histone modification patterns. Mol Cell Proteomics 10(11):M110.005371. https://doi.org/10.1074/mcp.M110.005371

101. Geiger T, Cox J, Ostasiewicz P et al (2010) Super-SILAC mix for quantitative proteomics of human tumor tissue. Nat Methods 7 (5):383–385. https://doi.org/10.1038/nmeth.1446

102. Tzouros M, Golling S, Avila D et al (2013) Development of a 5-plex SILAC method tuned for the quantitation of tyrosine phosphorylation dynamics. Mol Cell Proteomics 12(11):3339–3349. https://doi.org/10.1074/mcp.O113.027342

103. Merrill AE, Hebert AS, MacGilvray ME et al (2014) NeuCode labels for relative protein quantification. Mol Cell Proteomics 13 (9):2503–2512. https://doi.org/10.1074/mcp.M114.040287

104. Potts GK, Voigt EA, Bailey DJ et al (2016) Neucode labels for multiplexed, absolute protein quantification. Anal Chem 88 (6):3295–3303. https://doi.org/10.1021/acs.analchem.5b04773

105. Overmyer KA, Tyanova S, Hebert AS et al (2018) Multiplexed proteome analysis with neutron-encoded stable isotope labeling in cells and mice. Nat Protoc 13(1):293–306. https://doi.org/10.1038/nprot.2017.121

106. Gauthier NP, Soufi B, Walkowicz WE et al (2013) Cell-selective labeling using amino acid precursors for proteomic studies of multicellular environments. Nat Methods 10 (8):768–773. https://doi.org/10.1038/nmeth.2529

107. Li Z, Zhu Y, Sun Y et al (2016) Nitrilase-Activatable noncanonical amino acid precursors for cell-selective metabolic labeling of proteomes. ACS Chem Biol 11 (12):3273–3277. https://doi.org/10.1021/acschembio.6b00765

108. Desiderio DM, Kai M (1983) Preparation of stable isotope-incorporated peptide internal standards for field desorption mass spectrometry quantification of peptides in biologic tissue. Biomed Mass Spectrom 10(8):471–479. https://doi.org/10.1002/bms.1200100806

109. Mirgorodskaya OA, Kozmin YP, Titov MI et al (2000) Quantitation of peptides and proteins by matrix-assisted laser desorption/ionization mass spectrometry using (18)O-labeled internal standards. Rapid Commun Mass Spectrom 14(14):1226–1232. https://doi.org/10.1002/1097-0231(20000730)

110. Schnolzer M, Jedrzejewski P, Lehmann WD (1996) Protease-catalyzed incorporation of 18O into peptide fragments and its application for protein sequencing by electrospray and matrix-assisted laser desorption/ionization mass spectrometry. Electrophoresis 17 (5):945–953. https://doi.org/10.1002/elps.1150170517

111. Johnson KL, Muddiman DC (2004) A method for calculating 16O/18O peptide ion ratios for the relative quantification of proteomes. J Am Soc Mass Spectrom 15 (4):437–445. https://doi.org/10.1016/j.jasms.2003.11.016

112. Rappsilber J, Ryder U, Lamond AI et al (2002) Large-scale proteomic analysis of the human spliceosome. Genome Res 12 (8):1231–1245. https://doi.org/10.1101/gr.473902

113. Mallick P, Schirle M, Chen SS et al (2007) Computational prediction of proteotypic peptides for quantitative proteomics. Nat Biotechnol 25(1):125–131. https://doi.org/10.1038/nbt1275

114. Zybailov B, Mosley AL, Sardiu ME et al (2006) Statistical analysis of membrane proteome expression changes in Saccharomyces cerevisiae. J Proteome Res 5(9):2339–2347. https://doi.org/10.1021/pr060161n

115. Powell DW, Weaver CM, Jennings JL et al (2004) Cluster analysis of mass spectrometry data reveals a novel component of SAGA. Mol

Cell Biol 24(16):7249–7259. https://doi.org/10.1128/mcb.24.16.7249-7259.2004

116. Lu P, Vogel C, Wang R et al (2007) Absolute protein expression profiling estimates the relative contributions of transcriptional and translational regulation. Nat Biotechnol 25 (1):117–124. https://doi.org/10.1038/nbt1270

117. Zhou JY, Schepmoes AA, Zhang X et al (2010) Improved LC-MS/MS spectral counting statistics by recovering low-scoring spectra matched to confidently identified peptide sequences. J Proteome Res 9 (11):5698–5704. https://doi.org/10.1021/pr100508p

118. Zhang Y, Wen Z, Washburn MP et al (2010) Refinements to label free proteome quantitation: how to deal with peptides shared by multiple proteins. Anal Chem 82 (6):2272–2281. https://doi.org/10.1021/ac9023999

119. Voyksner RD, Lee H (1999) Investigating the use of an octupole ion guide for ion storage and high-pass mass filtering to improve the quantitative performance of electrospray ion trap mass spectrometry. Rapid Commun Mass Spectrom 13(14):1427–1437

120. Wiener MC, Sachs JR, Deyanova EG et al (2004) Differential mass spectrometry: a label-free LC-MS method for finding significant differences in complex peptide and protein mixtures. Anal Chem 76 (20):6085–6096. https://doi.org/10.1021/ac0493875

121. Wilm M (2009) Quantitative proteomics in biological research. Proteomics 9 (20):4590–4605. https://doi.org/10.1002/pmic.200900299

122. Ahrne E, Molzahn L, Glatter T et al (2013) Critical assessment of proteome-wide label-free absolute abundance estimation strategies. Proteomics 13(17):2567–2578. https://doi.org/10.1002/pmic.201300135

123. Dowle AA, Wilson J, Thomas JR (2016) Comparing the diagnostic classification accuracy of iTRAQ, peak-area, spectral-counting, and emPAI methods for relative quantification in expression proteomics. J Proteome Res 15(10):3550–3562. https://doi.org/10.1021/acs.jproteome.6b00308

124. Waanders LF, Hanke S, Mann M (2007) Top-down quantitation and characterization of SILAC-labeled proteins. J Am Soc Mass Spectrom 18(11):2058–2064. https://doi.org/10.1016/j.jasms.2007.09.001

125. Gerber SA, Rush J, Stemman O et al (2003) Absolute quantification of proteins and phosphoproteins from cell lysates by tandem MS. Proc Natl Acad Sci U S A 100 (12):6940–6945. https://doi.org/10.1073/pnas.0832254100

126. Kirkpatrick DS, Gerber SA, Gygi SP (2005) The absolute quantification strategy: a general procedure for the quantification of proteins and post-translational modifications. Methods 35(3):265–273. https://doi.org/10.1016/j.ymeth.2004.08.018

127. Schmidt C, Lenz C, Grote M et al (2010) Determination of protein stoichiometry within protein complexes using absolute quantification and multiple reaction monitoring. Anal Chem 82(7):2784–2796. https://doi.org/10.1021/ac902710k

128. Stahl-Zeng J, Lange V, Ossola R et al (2007) High sensitivity detection of plasma proteins by multiple reaction monitoring of N-glycosites. Mol Cell Proteomics 6 (10):1809–1817. https://doi.org/10.1074/mcp.M700132-MCP200

129. Picotti P, Lam H, Campbell D et al (2008) A database of mass spectrometric assays for the yeast proteome. Nat Methods 5 (11):913–914. https://doi.org/10.1038/nmeth1108-913

130. Picotti P, Rinner O, Stallmach R et al (2010) High-throughput generation of selected reaction-monitoring assays for proteins and proteomes. Nat Methods 7(1):43–46

131. Brun V, Dupuis A, Adrait A et al (2007) Isotope-labeled protein standards: toward absolute quantitative proteomics. Mol Cell Proteomics 6(12):2139–2149. https://doi.org/10.1074/mcp.M700163-MCP200

132. Hanke S, Besir H, Oesterhelt D et al (2008) Absolute SILAC for accurate quantitation of proteins in complex mixtures down to the attomole level. J Proteome Res 7 (3):1118–1130. https://doi.org/10.1021/pr7007175

133. Singh S, Springer M, Steen J et al (2009) FLEXIQuant: a novel tool for the absolute quantification of proteins, and the simultaneous identification and quantification of potentially modified peptides. J Proteome Res 8(5):2201–2210. https://doi.org/10.1021/pr800654s

134. Ponten F, Schwenk JM, Asplund A et al (2011) The human protein atlas as a proteomic resource for biomarker discovery. J Intern Med 270(5):428–446. https://doi.org/10.1111/j.1365-2796.2011.02427.x

135. Zeiler M, Straube WL, Lundberg E et al (2012) A protein epitope signature tag (PrEST) library allows SILAC-based absolute

quantification and multiplexed determination of protein copy numbers in cell lines. Mol Cell Proteomics 11(3):O111.009613. https://doi.org/10.1074/mcp.O111.009613

136. Pratt JM, Simpson DM, Doherty MK, Rivers J, Gaskell SJ, Beynon RJ (2006) Multiplexed absolute quantification for proteomics using concatenated signature peptides encoded by QconCAT genes. Nat Protoc 1 (2):1029–1043. https://doi.org/10.1038/nprot.2006.129

137. Ishihama Y, Oda Y, Tabata T et al (2005) Exponentially modified protein abundance index (emPAI) for estimation of absolute protein amount in proteomics by the number of sequenced peptides per protein. Mol Cell Proteomics 4(9):1265–1272. https://doi.org/10.1074/mcp.M500061-MCP200

138. Brownridge P, Beynon RJ (2011) The importance of the digest: proteolysis and absolute quantification in proteomics. Methods 54 (4):351–360. https://doi.org/10.1016/j.ymeth.2011.05.005

139. Silva JC, Gorenstein MV, Li GZ, Vissers JP et al (2006) Absolute quantification of proteins by LCMSE: a virtue of parallel MS acquisition. Mol Cell Proteomics 5 (1):144–156. https://doi.org/10.1074/mcp.M500230-MCP200

140. Schwanhausser B, Busse D, Li N et al (2011) Global quantification of mammalian gene expression control. Nature 473 (7347):337–342. https://doi.org/10.1038/nature10098

141. Wilhelm M, Schlegl J, Hahne H et al (2014) Mass-spectrometry-based draft of the human proteome. Nature 509(7502):582–587. https://doi.org/10.1038/nature13319

142. Krey JF, Wilmarth PA, Shin JB et al (2014) Accurate label-free protein quantitation with high- and low-resolution mass spectrometers. J Proteome Res 13(2):1034–1044. https://doi.org/10.1021/pr401017h

143. Shin JB, Krey JF, Hassan A et al (2013) Molecular architecture of the chick vestibular hair bundle. Nat Neurosci 16(3):365–374. https://doi.org/10.1038/nn.3312

144. Wisniewski JR, Ostasiewicz P, Dus K et al (2012) Extensive quantitative remodeling of the proteome between normal colon tissue and adenocarcinoma. Mol Syst Biol 8:611. https://doi.org/10.1038/msb.2012.44

145. He B, Shi J, Wang X, Jiang H et al (2019) Label-free absolute protein quantification with data-independent acquisition. J Proteome 200:51–59. https://doi.org/10.1016/j.jprot.2019.03.005

146. Rzagalinski I, Volmer DA (2017) Quantification of low molecular weight compounds by MALDI imaging mass spectrometry—a tutorial review. Biochim Biophys Acta Proteins Proteom 1865(7):726–739. https://doi.org/10.1016/j.bbapap.2016.12.011

147. Schulz S, Becker M, Groseclose MR et al (2019) Advanced MALDI mass spectrometry imaging in pharmaceutical research and drug development. Curr Opin Biotechnol 55:51–59. https://doi.org/10.1016/j.copbio.2018.08.003

148. Norris JL, Cornett DS, Mobley JA et al (2007) Processing MALDI mass spectra to improve mass spectral direct tissue analysis. Int J Mass Spectrom 260(2–3):212–221. https://doi.org/10.1016/j.ijms.2006.10.005

Chapter 9

Tandem Mass Tags for Comparative and Discovery Proteomics

Oliver Pagel, Laxmikanth Kollipara, and Albert Sickmann

Abstract

Relative or comparative proteomics provides valuable insights about the altered protein abundances across different biological samples in a single (labeled) or series (label-free) of LC–MS measurement(s). Chemical labeling of peptides using isobaric mass tags for identification and quantification of different proteomes simultaneously has become a routine in the so-called discovery proteomics in the past decade. One of the earliest isobaric tags-based technologies is TMT (tandem mass tags), which relies on the comparison of the unique "reporter ions" intensities for relative peptide/protein quantification. This differential labeling approach has evolved over time with respect to its multiplexing capability, i.e., from just 2 samples (TMTduplex) to 10 samples (TMT10plex) and a nowadays of up to 16 samples (TMTpro 16plex). Here, we describe a straightforward protocol to perform relatively deep proteome quantitative analyses using TMT10plex.

Key words TMT, Multiplexing, LC–MS/MS, Relative quantitative proteomics

Abbreviations

ACN	Acetonitrile
DTT	1,4-Dithiothreitol
FA	Formic acid
HPLC	High-performance liquid chromatography
IAA	2-Lodoacetamide
MS	Mass spectrometer/mass spectrometry
SILAC	Stable isotope labeling of amino acids in cell cultures
SPE	Solid-phase extraction
TEAB	Triethylammonium bicarbonate
TFA	Trifluoroacetic acid
TMT	Tandem mass tags

Katrin Marcus et al. (eds.), *Quantitative Methods in Proteomics*, Methods in Molecular Biology, vol. 2228, https://doi.org/10.1007/978-1-0716-1024-4_9, © Springer Science+Business Media, LLC, part of Springer Nature 2021

1   Introduction

Besides mere identification, relative quantification of proteins that are derived from different backgrounds (e.g., wild type vs. mutant/ treated) is of great interest to biologists and allied researchers worldwide. Liquid chromatography–mass spectrometry (LC–MS)-based proteomics has been benefited immensely by the usage and application of stable isotopes (^{2}H, ^{13}C, ^{15}N, ^{18}O) in the labeling technologies, which allow both absolute and relative quantification of proteins [1, 2]. Among such approaches, chemical labeling with isobaric mass tags has been widely used recently in comparative proteomics studies that provide a comprehensive profiling of different proteomes simultaneously when combined with appropriate fractionation methods [3]. This knowledge (discovery-based) can be utilized and transferred to more specific (targeted-based) MS-based techniques (*see* Chaps. 8, 11, 12, 25) that focus on (i) investigation, (ii) identification, and (iii) validation of biomarkers in, e.g., clinical cancer samples [4]. Furthermore, isobaric tagging methods are applicable to both cell culture and non-cell culture-derived biological samples, such as mammalian tissues [5] and body fluids [6] when compared to metabolic labeling techniques, e.g., SILAC [7] (*see* Chaps. 8, 13, 18).

Tandem mass tags (TMT) [8, 9] (*see* also Chaps. 14, 15, 17) are a group of isobaric chemical tags that were first introduced in the year 2003 as TMTduplex for quantifying relative abundances of proteins in two different samples simultaneously based on the unique "reporter ion" signal intensities in the low m/z region. Since then a relatively high degree of multiplexing has been demonstrated for this differential labeling approach, i.e., from TMTduplex to TMT10plex [10] (*see* Chaps. 8, 14, 17) and the very recent TMTpro 16plex (*see* Chap. 15) [11]. The higher multiplexing possibility is mainly attributed to different combinations of the light and heavy stable isotope ($^{12/13}$C, $^{14/15}$N) variants that are incorporated in the reporter and balancer groups of each TMT reagent. When combined with off-line fractionation strategies (e.g., C18 high-pH reversed phase) and fast scanning high-resolution MS settings, this approach provides relatively deep proteome quantitative analysis of ten (TMT10plex) different samples in a single experiment.

2   Materials

All chemical solutions and buffers should be prepared using ultrapure deionized water or LC–MS grade water. All HPLC solvents, such as acetonitrile (ACN), formic acid (FA), and trifluoroacetic acid (TFA) must be of LC–MS grade (e.g., Biosolve). To minimize sample losses, always use protein LoBind (Eppendorf) tubes for all protein/peptide and enzyme preparations.

2.1 Cell Lysis of Biological Samples and Protein Concentration Estimation

1. Biological sample: Cell culture, e.g., HeLa.

2. Cell lysis buffer: 1% (w/v) sodium dodecyl sulfate (SDS), 50 mM Tris–HCl and 150 mM sodium chloride (NaCl), pH 7.8 (adjust with HCl). Add one tablet of protease inhibitors cocktail complete Mini, EDTA-free (Sigma Aldrich) to 10 mL of buffer (*see* **Note 1**).

3. Endonuclease and cofactor: Benzonase (>99% purity, activity 25.0 U/μL; Novagen) stored at −20 °C and always keep it on ice during usage. 1 M magnesium chloride ($MgCl_2$) solution. Stock can be stored at −40 °C.

4. Lab equipment: Benchtop centrifuge (Eppendorf), ultrasonic water bath, ultrasonic processor (Vibra-Cell), and a vortex mixer.

5. Determination of protein concentration: Pierce Bicinchoninic acid (BCA) protein assay kit (Thermo).

2.2 Reduction and Alkylation (Carbamido-methylation)

1. Reducing agent: 2 M dithiothreitol (DTT) solution. Stock can be stored at −40 °C.

2. Alkylating agent: 0.5 M iodoacetamide (IAA) solution. Freshly prepared (*see* **Note 2**).

2.3 Sample Cleanup and on-Filter Enzymatic Digestion

1. Buffers: 8.0 M urea in 100 mM Tris–Cl, pH 8.5 and 50 mM NH_4HCO_3, pH 7.8. Prepare both buffers freshly (*see* **Note 3A, B**). Store at 4 °C prior usage.

2. Centrifugal devices: 30 kDa molecular weight cutoff (MWCO) spin filters (Nanosep with Omega Membrane; PALL).

3. Centrifuge: Benchtop centrifuge (Eppendorf).

4. Enzyme: Sequencing grade modified trypsin (Promega). Reconstitute lyophilized trypsin in 50 mM acetic acid (provided in the kit) to get a final concentration of 1 mg/mL (*see* **Note 4**).

5. Digestion buffer: 0.2 M guanidine hydrochloride (GuHCl), 2 mM calcium chloride ($CaCl_2$), trypsin solution [1 mg/mL] and 50 mM NH_4HCO_3, pH 7.8. Freshly prepared.

6. Stop digestion: 10% TFA.

7. pH indicator: pH test strip/paper.

2.4 Digestion Control by Monolithic Reversed-Phase Chromatography (RPC)

1. HPLC: UltiMate 3000 rapid separation liquid chromatography (RSLC) HPLC equipped with a dedicated ultraviolet (UV) detector using a 3 nL flow cell (Thermo Fisher Scientific) at 214 nm (peptide bond) for quality control or similar HPLC system.

2. HPLC columns: PepSwift reversed-phase (RP) monolithic columns (both Thermo Fisher Scientific), trap column 200 μm inner diameter (ID), 5 mm length, and capillary column 200 μm ID, 5 cm length.

3. HPLC buffer A: 0.1%TFA.

4. HPLC buffer B: 0.08% TFA, 84% ACN.

2.5 Solid-Phase Extraction Cartridges (SPEC) for Sample Cleanup

1. SPEC sorbent material: *C18 AR 4 mg* columns (*Agilent*).

2. Lab equipment: Vacuum manifold system and vacuum centrifuge (SpeedVac).

3. Wetting/activating buffer: 100% ACN.

4. Equilibration and wash buffer: 0.1% TFA.

5. Elution buffer: 60% ACN in 0.1% TFA.

2.6 TMT Labeling

1. TMT reagents: 10plex Isobaric Label Reagent Set, 1×0.8 mg (126, 127N, 127C, 128N, 128C, 129N, 129C, 130N, 130C, 131) from Thermo Scientific, Rockford, IL, USA.

2. Dissolution buffer: 100 mM triethylammonium bicarbonate (TEAB), pH 8.5.

3. pH indicator: pH test strip/paper,

4. Reagent dilution solvent: 100% anhydrous ACN.

5. Reaction quenching solvent: 5% (v/v) hydroxylamine solution.

6. Thermomixer.

2.7 Sample Fractionation Using High-pH C18 RP Chromatography

1. HPLC: UltiMate 3000 RSLC HPLC equipped with a dedicated UV detector using a 3 nL flow cell (Thermo Fisher Scientific) at 214 nm or similar HPLC system.

2. HPLC column: BioBasic C18 0.5 mm ID, 15 cm length, 5 μm particle size, and 300 Å pore size (Thermo Scientific).

3. HPLC buffer A: 10 mM ammonium formate (NH_4HCO_2), pH 8.0.

4. HPLC buffer B: 84% ACN in 10 mM NH_4HCO_2, pH 8.0.

2.8 NanoLC-NanoESI MS and MS/MS

1. HPLC: UltiMate 3000 nanoRSLC HPLC (Thermo Fisher Scientific) equipped with a dedicated UV detector using a 3 nL flow cell at 214 nm or a similar HPLC system.

2. HPLC columns: Acclaim PepMap C18, nanoViper columns (both Thermo Fisher Scientific), i.e., trap column 100 μm ID $\times$ 2 cm length, 5 μm particle size, 100 Å pore size, and analytical column 75 μm ID, 50 cm length, 3 μm particle size, 100 Å pore size.

3. HPLC loading buffer: 0.1% TFA.

4. HPLC solvent A: 0.1% FA.

5. HPLC solvent B: 0.1% FA, 84% ACN.

6. LC-MSinterface: Nano-electrospray ionization (nanoESI).

7. Emitter: PicoTip with distal coating 20 μm ID, 10 μm tip ID, 5 cm length.

8. Mass spectrometer (MS): Q Exactive High Field (HF) (Thermo Fisher Scientific) or other MS that can acquire both survey MS and tandem MS/MS scans with high mass accuracy ($\leq$1 ppm) and at high resolution ($\geq$60,000 at m/z 200) (*see* **Note 5**).

2.9 Data Analysis

1. Data analysis software: Proteome Discoverer (version 1.4, Thermo Scientific) with the following nodes.

2. Search algorithm: Mascot [12] (version 2.6, Matrix Science).

3. Quantification: Reporter ions quantifier.

4. False discovery rate (FDR) estimation: Percolator [13].

3 Methods

To minimize technical variation and to ensure reliable quantitative results, all samples should be processed simultaneously and in an identical manner. Additionally, quality control (QC) measures must be taken at critical steps of the TMT sample preparation workflow (*see* Fig. 1).

3.1 Sample Lysis and Determination of Protein Concentration

1. Pre-warm (37 °C) 1% SDS buffer and use 100 μL to lyse, e.g., 10^6 HeLa cells. Promote homogenization by mechanical/shearing forces induced by slow pipetting (*see* **Note 6**).

2. To degrade all forms of DNA and RNA (genetic material), add 3 μL of benzonase per 200 μL of cell lysate plus 2 mM $MgCl_2$. Do not vortex. Mix by slowly pipetting up/down. Incubate samples at 37 °C for 30 min (*see* **Note 7**). Alternatively, use an ultrasonic probe for 30 s (amplitude, 30%; pulse, 1 s/1 s). Place the sample containing LoBind Eppendorf tube on ice during the whole process to avoid protein degradation caused due to the heat generated by ultrasonication.

3. Clarify the lysates by centrifugation. Use a benchtop centrifuge and spin down the sample tubes at 18,000 $\times$ g at room temperature (RT) (*see* **Note 8**) for 30 min.

4. Collect the supernatant in a new LoBind Eppendorf tube and determine the protein concentration using BCA assay according to the manufacturer's instructions (*see* **Note 9**).

3.2 Carbamido-methylation, On-Filter Sample Cleanup, and Proteolytic Digestion

1. Reduce disulfide bonds by incubating the cell lysates with 10 mM of DTT at 56 °C for 30 min. After incubation, cool down the tubes to RT.

2. Promptly add the alkylating agent (IAA) to a final concentration of 30 mM and incubate at RT for 30 min in the dark.

3. Quench excess of IAA by adding same volume (used for reduction) of DTT and incubate samples at RT for 15 min.

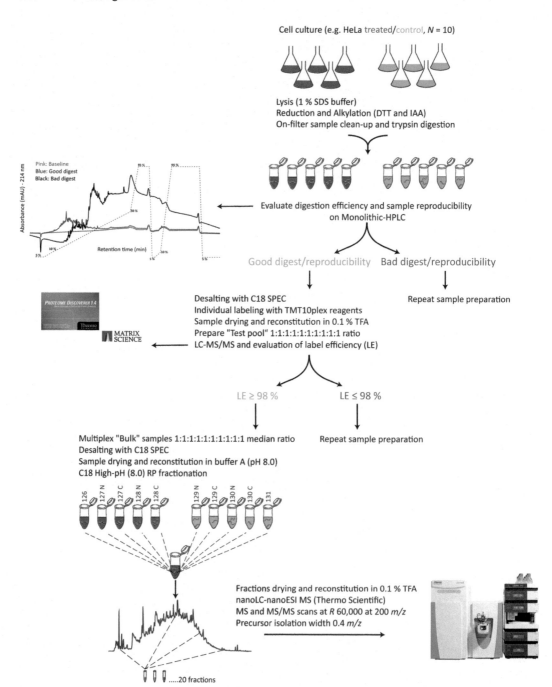

Fig. 1 Schematic of TMT workflow. Up to ten different biological samples can be differentially labeled using 10-plex isobaric reagents that allow simultaneous comparison of relative protein abundances between samples. Quality control checks at critical time points of the sample preparation, e.g., digestion efficiency and labeling efficiency, ensure reliable and accurate quantification results. Furthermore, pre-fractionation using high-pH RP HPLC prior to LC–MS/MS analysis minimizes sample complexity and enables deep proteome coverage

4. Spin-filter based sample cleanup and digestion [14, 15] (*see* **Note 10**). Precondition the 30 kDa MWCO spin-filters (*see* **Note 11**) with 100 μL of freshly prepared [16] 8.0 M urea in 100 mM Tris–Cl, pH 8.5 buffer (UB). Centrifuge at 13,800 × *g* at RT for 5 min. Discard the flow through from the collection tube.

5. Transfer an aliquot of cell lysate (carbamidomethylated) corresponding to a defined amount of protein, e.g., 150 μg (*see* **Note 12**), in a LoBind Eppendorf tube and dilute with UB such that the final concentration of SDS is ≤10 mM (*see* **Note 13**).

6. After dilution, place the mixture carefully onto the spin-filters without touching the membrane surface and centrifuge the devices at 13,800 × *g* at RT for 20 min (*see* **Note 14**). Discard the flow through from the collection tube.

7. Add 100 μL of UB onto the spin-filters and centrifuge under same conditions as mentioned above for 15 min. Repeat this step twice (in total three times). Discard the flow through from the collection tube.

8. Add 100 μL of 50 mM NH_4HCO_3, pH 7.8 onto the spin-filters and centrifuge under same conditions as mentioned above for 15 min. Repeat this step twice (in total three times). Discard the flow through from the collection tube.

9. To the concentrated proteins, add 100 μL of trypsin digestion buffer comprising 0.2 M GuHCl, 2 mM $CaCl_2$, trypsin [1 mg/mL] in 1:20 (w/w) ratio of enzyme to substrate and 50 mM NH_4HCO_3, pH 7.8 (*see* **Note 15**).

10. Incubate the spin-filters at 37 °C for 14 h.

11. For recovering peptides, transfer the spin-filters onto a new 2 mL LoBind Eppendorf tube and centrifuge the devices at 13,800 × *g* at RT for 15 min. Perform successive washing steps with 50 μL each of 50 mM NH_4HCO_3, pH 7.8, and ultrapure water, respectively, using similar centrifugation conditions as above.

12. Stop the enzyme activity by adding 10% TFA to a final concentration of 1%. Check the pH using pH indicator strip/paper. pH ≤3.0 is optimal.

3.3 Digestion Control with RP Monolithic Chromatography

1. Take an aliquot of 500 ng of acidified tryptic digest of each sample.

2. Measure all samples on a RP monolithic HPLC system to evaluate the digestion efficiency and reproducibility (*see* **Note 16**) [17].

3. If the samples look properly digested, clean, and reproducible, then proceed with Subheading 3.5; otherwise, consider

repeating sample preparation process from the beginning (*see* Fig. 1) or try a cleanup by solid-phase extraction as in Subheading 3.4.

3.4 SPEC Sample Cleanup (Desalting)

1. To obtain reproducible results, use a vacuum manifold system for peptides desalting. Use C18 AR 4 mg material, 100 μL (Agilent) (*see* **Note 17**).

2. Activation: three times with 100 μL of 100% ACN.

3. Equilibration: three times with 100 μL of 0.1% TFA.

4. Sample loading: Place the sample on the material and collect flow through (FT) in a new tube. Reload FT on the SPEC once.

5. Washing: three times with 100 μL of 0.1% TFA.

6. Peptides elution: three times with 100 μL of 60% ACN in 0.1% TFA.

7. Dry the eluates in the SpeedVac.

8. Store the dried peptides at −40 °C until further use.

3.5 TMT Labeling

1. Reconstitute each peptides pellet (~100 μg) (*see* **Note 18**) in 100 μL of 100 mM TEAB, pH 8.5. Control the pH using a pH strip/paper.

2. Proceed with labeling according to the manufacturer's instructions (TMT10plex, Thermo Scientific).

3. After incubation, quench the reaction with 8 μL of 5% (v/v) hydroxylamine solution and place the samples on a thermo-mixer with agitation (550 rpm) at 25 °C for 15 min.

4. Snap freeze all samples and completely dry them in the SpeedVac.

5. Reconstitute each sample in 300 μL of 0.1% TFA. Vortex for complete solubilization.

6. Take equal volume (e.g., 2 μL) from each TMT channel and pool in a LoBind Eppendorf tube. Label it as "Test pool" and use this for checking the label efficiency and to adjust (or normalize) sample amounts prior to bulk multiplexing (*see* **Note 19**).

7. Store the remainder "Bulk" solutions promptly at −80 °C until further use.

8. Measure 1 μL of the "Test pool" sample on a nano-HPLC coupled to Q Exactive HF using the LC–MS method as described in the Subheading 3.8.

3.6 Data Analysis, Evaluation of Labeling Efficiency, and Multiplexing

1. Perform database search of the MS raw data of the "Test pool" sample using Proteome Discoverer (PD) software version 1.4.

2. Use a human UniProt "target" database (*see* **Note 20**) as PD which generates random decoy hits *on the fly*.

3. Use the spectrum selector node for precursor ion selection to process MS raw data with default settings.

4. Use the "Reporter ion quantifier" node and select TMT as the quantification method. Open the "Quantification method editor" window and check "Show the raw quan values" under "Ratio calculation" tab.

5. Use Mascot as search algorithm and set the following parameters: precursor and fragment ion tolerances of 10 ppm and 0.02 Da for MS and MS/MS, respectively; trypsin as enzyme with a maximum of two missed cleavages; carbamidomethylation of Cys as fixed modification; and TMT on Lys and oxidation of Met as variable modifications.

6. Use the Percolator node to filter the data at FDR of 1%.

7. From PD software, export the peptide-spectrum matches (PSMs) list to Microsoft Excel (*see* **Note 21**).

8. For checking labeling efficiency, calculate the percentage by taking into account the number of modified (represented as "k") and unmodified (represented as "K") Lys in all identified PSMs to the total number of Lys residues. Typically, a labeling efficiency of $\geq$98% is acceptable.

9. To normalize sample amounts before multiplexing differentially labeled TMT samples, individually sum up the raw reporter ion intensity values of each TMT channel. Calculate the median across all channels summed intensity values.

10. Then, divide each summed up value of each TMT channel by the previously calculated median to obtain normalization factor (NF) per individual channel.

11. Use the NFs from above to get the correction factors and pool the differently TMT labeled "Bulk" samples accordingly ensuring a 1:1:1:1:1:1:1:1:1:1 median ratio of multiplexed sample.

12. Desalt the multiplexed sample using SPEC as described in Subheading 3.4.

3.7 Off-Line High-pH C18 RP Fractionation

To minimize sample complexity and to obtain deep proteome coverage, we recommend pre-fractionation by high-pH C18-based RP (high-pH RP) chromatography of the TMT multiplexed sample. For better orthogonality, high-pH RP fractionation is typically carried out at pH 10 [18, 19] followed by low-pH RP LC–MS/MS analysis. However, pH 10 conditions could lead to the hydrolysis of siloxane groups in the silica-based RP column and thereby hamper the column stability [20]. Thus, we suggest high-pH RP fractionation at pH 8.0.

1. Reconstitute the dried multiplexed sample in buffer A. Take an aliquot corresponding to 50 μg of peptides and proceed with high-pH RP fractionation.

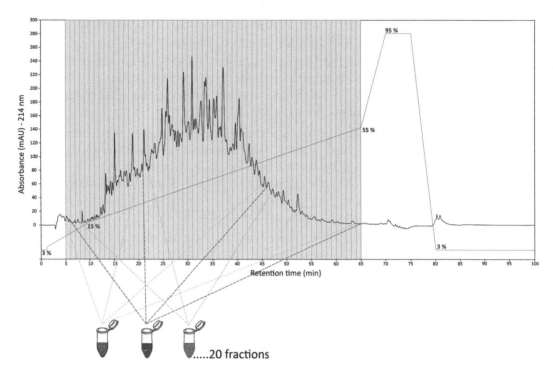

Fig. 2 Pre-fractionation of a TMT multiplexed sample on a high-pH C18-based RP HPLC system. The peptides are separated using a binary gradient (buffer A, 10 mM NH_4HCO_2; buffer B, 10 mM; NH_4HCO_2, 84% ACN; both pH, 8.0) ranging from 3 to 55% buffer B in 50 min. In total, 20 fractions are collected at 1 min intervals in a concatenation mode (represented in the gray area of the UV chromatogram). Separation of the peptides on high-pH RP columns in combination with concatenation mode of collecting the eluted fractions has been demonstrated to provide better orthogonality and increases proteome coverage [31]

2. Perform peptide fractionation on a BioBasic C18, 0.5 mm × 15 cm column using an UltiMate 3000 nanoRSLC HPLC (Thermo Fisher Scientific) system with the following gradients, 3–15% B in 10 min, 15–55% B in 55 min, 55–95% B in 5 min, 95% B hold for 5 min, and 95–3% B in 5 min, and finally re-equilibrate the column with 3% B for 20 min.

3. Collect 20 fractions at 1 min intervals from HPLC retention time 5–65 min in a concatenation mode (*see* Fig. 2).

4. Finally, dry the collected fractions in the SpeedVac.

3.8 LC–MS/MS Analysis

3.8.1 LC Conditions

1. Reconstitute each high-pH RP fraction ($N = 20$; from Subheading 3.7) in an appropriate volume of HPLC loading buffer (0.1% TFA) (*see* **Note 22**).

2. Using a UltiMate 3000 nanoRSLC HPLC (Thermo Fisher Scientific) system, load each fraction with 0.1% TFA and a flow rate of 20 μL/min for 5 min onto the trap column followed by separation of peptides on the analytical column using a binary gradient ranging from 3 to 38% B in 120 min at a flow rate of 250 nL/min at 60 °C.

1. Operate the Q Exactive HF MS in data-dependent acquisition mode.

2. Acquire MS survey scans at a resolution of 60,000 with a target value of 1×10^6 ions and maximum injection time of 120 ms. Acquire MS/MS scans of 15 most abundant ions (Top 15) at a resolution of 60,000 using 2×10^5 ions as target value and a maximum fill time of 200 ms. Use a normalized collision energy of 33% or a stepped collision energy of 30+/−3% and a dynamic exclusion of 30 s. Set the first fixed mass 100 m/z to allow a good signal intensity of the first reporter ion (i.e., 126) and select only precursor ions with charge state between +2 and +5 for MS/MS fragmentation. To minimize the potential precursor co-isolation, use a quadrupole isolation width of 0.4 m/z (*see* **Note 23**).

4 Notes

1. Prepare 1 mL aliquots of the 1% SDS lysis buffer containing complete mini EDTA-free protease inhibitors cocktail tablet. The aliquots can be stored up to 12 weeks at −40 °C from the date of preparation.

2. IAA solution is unstable and photosensitive. Therefore, prepare the stock solution freshly, protect from exposure to light (e.g., wrap the tube with aluminum foil), and use it ≤1 h after preparation. Furthermore, sulfhydryl group containing and not slightly alkaline (pH 7.5–8.0) buffers should be avoided. Notably, excess of IAA or non-buffered IAA reagent can promote unwanted side reactions including alkylation of primary amines (Lys, N-termini), thioethers (Met), imidazoles (His), and carboxylates (Asp, Glu) [21].

3. (A) In aqueous buffers, urea dissociates into ammonium and isocyanate upon heating and over time. Under basic pH conditions, isocyanate covalently reacts with primary amines leading to artificial carbamylation of proteins [16]. This unwanted reaction drastically hampers the labeling efficiency of isobaric tags that primarily target Lys and N-termini. (B) NH_4HCO_3 buffer is volatile, and it is prone to pH instability over time.

4. Reconstitute lyophilized trypsin (Promega) in 50 mM acetic acid solution to a concentration of 1 mg/mL and prepare aliquots. These aliquots can be stored up to 1 year from the date of preparation at −80 °C. To prevent denaturation, avoid multiple freeze-thaw cycles of enzyme solutions.

5. In TMT10plex, the ^{13}C and ^{15}N isotopologues and their isobaric variants differ by a mere 6.32 mDa mass. Therefore, high-resolution MS/MS analysis is recommended to get precise

quantitative measurements [10]. With 60,000 resolving power at 200 m/z setting, both ^{13}C and ^{15}N isotopologues are fully resolved using a Q Exactive HF MS.

6. Upon adding lysis buffer to cells, the lysate turns viscous due to the release of genetic material into the surrounding environment. Proceed directly with Benzonase treatment or ultrasonication procedure.

7. After incubation with Benzonase at 37 °C for 30 min, check the viscosity of the lysates by pipetting up/down. If the lysates are still viscous, then add same volumes of Benzonase and 1 M $MgCl_2$ as before and incubate the samples under same conditions.

8. SDS precipitates at temperatures ≤4 °C. Perform centrifugation at RT.

9. Besides BCA, other colorimetric assays, e.g., Bradford or modified Lowry can be used for protein concentration determination. Regardless of the method used, the assay should provide comparably accurate concentration values since this is critical for calculating the amount of protein for labeling and amount of trypsin for digestion.

10. Besides on-filter-based sample preparation [14, 15], cold organic solvent (e.g., acetone [22], ethanol [23], trichloroacetic acid [24])-based protein precipitation followed by in solution digestion (trypsin) protocol can be employed for generating peptides [25].

11. Depending on the purpose, use either 10, 30, or 50 kDa MWCO spin-filters. Typically, 30 kDa MWCO membrane spin-filters are most widely used in the proteomics field [26].

12. In our experience, processing and digesting >150 μg of protein sample with 30 kDa MWCO spin-filters (P/N OD030C35; PALL) yielded irreproducible results. This might be due to the clogging of the membrane of the spin-filter with higher protein amounts. Therefore, we recommend using ≤150 μg protein per sample when using this type of spin-filter. For processing samples containing >150 μg protein, use organic solvent precipitation and subsequent in-solution digestion as mentioned in **Note 10**.

13. For low-concentrated protein samples, dilute them with UB to get ≤10 mM SDS and place the mixture on to the spin-filter in a batch mode, e.g., if the final volume of lysate and UB combination is 700 μL, centrifuge twice with 350 μL sample. In the first place though, plan your experiment such that the lysate protein concentration is at least 1.5 mg/mL to avoid multiple loadings.

14. Typically, centrifugation at $13,800 \times g$ for 15–20 min at RT should be sufficient to pass the liquid through the spin-filter. Centrifuge for additional 10 min in case there is still liquid present on the spin-filter. Turn the position/angle of the tubes by $180°$ if necessary. Place the next buffer solution only after all the liquid from the previous centrifugation step has passed through.

15. For a TMT10plex experiment ($N = 10$), prepare a stock trypsin digestion buffer for 11 samples (i.e., 1100 µL @ 100 µL per sample) in order to compensate for the errors due to the usage of different range of pipettes.

16. It is most important to evaluate the enzymatic digestion efficiency as this step is critical for the subsequent quantitative analysis [17]. The monolithic HPLC system provides a rapid and direct comparison of the samples as it can be used to measure both proteins and peptides. Monolithic columns are more robust, sensitive, and fast compared to other methods that are used for digestion control, e.g., SDS-PAGE followed by *Coomassie* staining.

17. Desalt the acidified peptides with appropriate C18 SPE cartridges (e.g., 1–5% of peptides weight: C18 bed weight) using a vacuum manifold system. Adjust solvents volumes used for desalting according to the column specifications, i.e., size/capacity of the used C18 SPEC.

18. Typically, each TMT reagent (0.8 mg) is suitable to label 100–200 µg of peptides and to achieve high labeling efficiencies ($\geq98\%$). However, the TMT label-to-peptides ratio (w/w) can also be adjusted (or downscaled) especially, for samples with lower starting amounts of peptides [27, 28].

19. To obtain accurate quantification results, near-complete labeling of all samples with TMT reagents is a prerequisite as well as diligent sample amount normalization. The amount of TMT label used, changes in the labeling buffer pH, or unequal starting amounts of peptides in different samples can lead to inefficient labeling or systematic bias and thereby effect the overall quantitative analysis.

20. Note down the date and the source of the download, as well as the number of target (forward) sequences (source: https://www.uniprot.org/).

21. Before exporting the PSMs list from the PD software, apply the data reduction filters such as high confidence corresponding to an FDR $\leq1\%$ on the PSM level and peptide search engine rank of 1.

22. For the low pH-RP LC–MS/MS analysis, use either complete/one-half/one-third of each high pH-RP fraction based on the peak intensities observed in the UV chromatogram (214 nm) of pH 8.0 HPLC system.

23. Pre-fractionation (high-pH RP chromatography) of a TMT10plex multiplexed sample and using narrower precursor isolation windows (e.g., 0.4 m/z) for MS/MS analysis should minimize the so-called co-isolation interference problem that causes "ratio compression or distortion" leading to inaccurate protein quantification [29, 30].

Acknowledgments

OP, LK, and AS acknowledge the support by the Ministerium für Kultur und Wissenschaft des Landes Nordrhein-Westfalen, the Regierende Bürgermeister von Berlin—inkl. Wissenschaft und Forschung, and the Bundesministerium für Bildung und Forschung.

References

1. Bantscheff M, Lemeer S, Savitski MM et al (2012) Quantitative mass spectrometry in proteomics: critical review update from 2007 to the present. Anal Bioanal Chem 404 (4):939–965

2. Bantscheff M, Schirle M, Sweetman G et al (2007) Quantitative mass spectrometry in proteomics: a critical review. Anal Bioanal Chem 389(4):1017–1031

3. Rauniyar N, Yates JR (2014) Isobaric labeling-based relative quantification in shotgun proteomics. J Proteome Res 13:5293

4. Faria SS, Morris CFM, Silva AR et al (2017) A timely shift from shotgun to targeted proteomics and how it can be groundbreaking for cancer research. Front Oncol 7:13

5. Plubell DL, Wilmarth PA, Zhao Y et al (2017) Extended multiplexing of tandem mass tags (TMT) labeling reveals age and high fat diet specific proteome changes in mouse epididymal adipose tissue. Mol Cell Proteomics 16 (5):873–890

6. Moulder R, Bhosale SD, Goodlett DR et al (2018) Analysis of the plasma proteome using iTRAQ and TMT-based isobaric labeling. Mass Spectrom Rev 37(5):583–606

7. Ong S-E, Blagoev B, Kratchmarova I et al (2002) Stable isotope labeling by amino acids in cell culture, SILAC, as a simple and accurate approach to expression proteomics. Mol Cell Proteomics 1(5):376–386

8. Thompson A, Schäfer J, Kuhn K et al (2003) Tandem mass tags: a novel quantification strategy for comparative analysis of complex protein mixtures by MS/MS. Anal Chem 75 (8):1895–1904

9. Thompson A, Schäfer J, Kuhn K et al (2003) Tandem mass tags: a novel quantification strategy for comparative analysis of complex protein mixtures by MS/MS. Anal Chem 75 (18):4942–4942

10. McAlister GC, Huttlin EL, Haas W et al (2012) Increasing the multiplexing capacity of TMTs using reporter ion isotopologues with isobaric masses. Anal Chem 84(17):7469–7478

11. Thompson A, Wölmer N, Koncarevic S et al (2019) TMTpro: design, synthesis, and initial evaluation of a proline-based isobaric 16-Plex tandem mass tag reagent set. Anal Chem 91:15,941

12. Perkins DN, Pappin DJ, Creasy DM et al (1999) Probability-based protein identification by searching sequence databases using mass spectrometry data. Electrophoresis 20 (18):3551–3567

13. Kall L, Canterbury JD, Weston J et al (2007) Semi-supervised learning for peptide identification from shotgun proteomics datasets. Nat Methods 4(11):923–925

14. Manza LL, Stamer SL, Ham A-JL et al (2005) Sample preparation and digestion for proteomic analyses using spin filters. Proteomics 5 (7):1742–1745

15. Wisniewski JR, Zougman A, Nagaraj N et al (2009) Universal sample preparation method for proteome analysis. Nat Meth 6 (5):359–362. http://www.nature.com/ nmeth/journal/v6/n5/suppinfo/nmeth. 1322_S1.html

16. Kollipara L, Zahedi RP (2013) Protein carbamylation: in vivo modification or in vitro artefact? Proteomics 13(6):941–944

17. Burkhart JM, Schumbrutzki C, Wortelkamp S et al (2012) Systematic and quantitative comparison of digest efficiency and specificity reveals the impact of trypsin quality on MS-based proteomics. J Proteome 75 (4):1454–1462

18. Toll H, Oberacher H, Swart R et al (2005) Separation, detection, and identification of peptides by ion-pair reversed-phase high-performance liquid chromatography-electrospray ionization mass spectrometry at high and low pH. J Chromatogr A 1079 (1–2):274–286

19. Gilar M, Olivova P, Daly AE et al (2005) Orthogonality of separation in two-dimensional liquid chromatography. Anal Chem 77(19):6426–6434

20. Xie F, Smith RD, Shen Y (2012) Advanced proteomic liquid chromatography. J Chromatogr A 1261:78–90

21. Suttapitugsakul S, Xiao H, Smeekens J et al (2017) Evaluation and optimization of reduction and alkylation methods to maximize peptide identification with MS-based proteomics. Mol BioSyst 13(12):2574–2582

22. Buxton TB, Crockett JK, Moore WL 3rd et al (1979) Protein precipitation by acetone for the analysis of polyethylene glycol in intestinal perfusion fluid. Gastroenterology 76(4):820–824

23. Mellanby J (1907) The precipitation of the proteins of horse serum. J Physiol 36 (4–5):288–333

24. Arnold U, Ulbrich-Hofmann R (1999) Quantitative protein precipitation from guanidine hydrochloride-containing solutions by sodium deoxycholate/trichloroacetic acid. Anal Biochem 271(2):197–199

25. Solari FA, Kollipara L, Sickmann A et al (2016) Two birds with one stone: parallel quantification of proteome and phosphoproteome using iTRAQ. In: Reinders J (ed) Proteomics in systems biology: methods and protocols. Springer New York, New York, NY, pp 25–41

26. Wiśniewski JR, Zielinska DF, Mann M (2011) Comparison of ultrafiltration units for proteomic and N-glycoproteomic analysis by the filter-aided sample preparation method. Anal Biochem 410(2):307–309

27. Zecha J, Satpathy S, Kanashova T et al (2019) TMT labeling for the masses: a robust and cost-efficient, in-solution labeling approach. Mol Cell Proteomics 18:1468–1478

28. Erdjument-Bromage H, Huang F-K, Neubert TA (2018) Sample preparation for relative quantitation of proteins using tandem mass tags (TMT) and mass spectrometry (MS). Methods Mol Biol 1741:135–149

29. Savitski MM, Mathieson T, Zinn N et al (2013) Measuring and managing ratio compression for accurate iTRAQ/TMT quantification. J Proteome Res 12(8):3586–3598

30. Bai B, Tan H, Pagala VR et al (2017) Deep profiling of proteome and phosphoproteome by isobaric labeling, extensive liquid chromatography, and mass spectrometry. Methods Enzymol 585:377–395

31. Yang F, Shen Y, Camp DG et al (2012) High pH reversed-phase chromatography with fraction concatenation as an alternative to strong-cation exchange chromatography for two-dimensional proteomic analysis. Expert Rev Proteomics 9(2):129–134

Chapter 10

An Approach for Triplex-IPTL

Christian J. Koehler and Bernd Thiede

Abstract

Isobaric peptide termini labeling (IPTL) is an approach for quantitative proteomics based on crosswise isotopic labeling of peptides at the N- and C-terminus. The labeling reagents are chosen in isotopic variations that the resulting mass of all labels per peptide is isobaric, but the individual label on each peptide terminus is different. Therefore, the quantitative difference of the peptide signal can be determined by the fragment ions of the corresponding MS2 spectra. Here, we describe an approach for triplex-IPTL to allow the comparison of three proteomes. This approach is based on digestion of the proteins by endoproteinase Lys-C, followed by three combinations of selective dimethylation of the peptide N-termini and subsequent dimethylation of the lysine residues at the C-termini. Data analysis is performed using Mascot for database searches and the freely available software package IsobariQ for quantification.

Key words Chemical labeling, Dimethylation, Isobaric labeling, IsobariQ, IPTL, Mass spectrometry, Quantitative proteomics

1 Introduction

Quantitative proteome profiling is most commonly performed by label-free quantification (LFQ) (*see* Chaps. 8, 16, 20–24), stable isotopic labeling with amino acids in cell culture (SILAC) (*see* Chaps. 8, 13, 18), and reporter ion-based isobaric labeling methods (TMT and iTRAQ) [1] (*see* Chaps. 8, 9, 14, 15, 17, 28). IPTL was introduced as an alternative approach which is based on relative protein quantification by isobaric derivatization of both peptide termini with complementary isotopically labeled reagents [2]. The mixed isotopic labeling results in isobaric precursor masses and provides several quantification data points per peptide in MS2 spectra. These data enable the statistical treatment of the quantitative data of each peptide spectrum match (PSM). In addition, the presence of pairs of fragment ions in MS2 spectra with reverse quantification ratios increases the confidence of database hits and/or aids in the assignment of ions when de novo sequencing needs to be performed. Since its introduction, several different labeling approaches for IPTL were established; most of them are

Katrin Marcus et al. (eds.), *Quantitative Methods in Proteomics*, Methods in Molecular Biology, vol. 2228,
https://doi.org/10.1007/978-1-0716-1024-4_10, © Springer Science+Business Media, LLC, part of Springer Nature 2021

based on differential labeling of the amine groups at the N- and C-terminus of the peptides after Lys-C digestion. Recently, an IPTL approach for tryptic peptides was reported as well [3]. Moreover, IPTL seems to be particularly useful for the analysis of ubiquitinylation sites [4]. Here, we present an approach for the comparison of three proteomes using triplex-IPTL [5]. After Lys-C digestion, different combinations of isotopically labeled formaldehyde and sodium cyanoborohydride are used for site-specific labeling of the amine group at the N-terminus followed by the dimethylation of C-terminal lysines. Subsequently, the procedure for database searches of IPTL data using Mascot and quantitative data analysis using IsobariQ [6] is described.

2   Materials

2.1 Protein Digestion with Endoproteinase Lys-C

1. Endoproteinase Lys-C, sequencing grade (e.g., Roche Applied Science, Sigma-Aldrich).
2. 1 pmol/μL solution of a standard protein (e.g., bovine serum albumin (BSA)).
3. 25 mM Tris–HCl, pH 8.5, 1 mM EDTA.
4. SPE C18 cartridges or pipette tips.
5. HPLC grade (or better) water.
6. HPLC grade (or better) acetonitrile.

2.2 Dimethylation of Alpha-N-Termini

1. 4% (v/v) formaldehyde (CH_2O, C^2H_2O, $^{13}CH_2O$, $^{13}C^2H_2O$) (e.g., Cambridge Isotopes, C/D/N Isotopes, Sigma-Aldrich) (in water).
2. 1% (v/v) acetic acid, pH 2.8 (in water).
3. 600 mM sodium cyanoborohydride ($NaBH_3CN$) and 600 mM sodium cyanoborodeuteride (NaB^2H_3CN) (37 mg/mL).
4. 1% (v/v) ammonium hydroxide (NH_4OH) (in water).
5. 5% (v/v) formic acid (in water).
6. SPE C18 cartridges or pipette tips.

2.3 Dimethylation of Lysine Residues

1. 4% (v/v) formaldehyde (CH_2O, C^2H_2O, $^{13}CH_2O$, $^{13}C^2H_2O$) (e.g., Cambridge Isotopes, C/D/N Isotopes, Sigma-Aldrich), all in water.
2. 200 mM triethylammonium bicarbonate (TEAB).
3. 600 mM sodium cyanoborohydride ($NaBH_3CN$) and 600 mM sodium cyanoborodeuteride (NaB^2H_3CN) (37 mg/mL).
4. 1% (v/v) ammonium hydroxide (NH_4OH) (in water).
5. 5% (v/v) formic acid (in water).
6. SPE C18 cartridges or pipette tips.

2.4 MALDI-MS Analysis	1. 20 mg/mL α-cyano-4-hydroxycinnamic acid in 0.3% aqueous trifluoroacetic acid/acetonitrile (1/1, v/v).
2.5 LC-ESI-MS Analysis	1. Solvent A: 0.1% (v/v) formic acid in water. 2. Solvent B: 0.1% (v/v) formic acid, 90% (v/v) acetonitrile in water. 3. Reversed phase C18 precolumn and analytical column.
2.6 Data Analysis	1. PC. 2. Protein database search engine Mascot from Matrix Sciences (London, UK). 3. Quantification software IsobariQ (https://www.mn.uio.no/ibv/english/research/sections/bmb/ research-groups/enzymology-and-protein-structure-and-function/proteomics-thiede/software/) [6].

3   Methods

3.1 General Practice

IPTL is performed by crosswise peptide termini labeling to produce isobaric peptides. Three different states of a protein sample can be compared and distinguished after mass spectrometry data acquisition. An outline of the triplex-IPTL approach is presented in Fig. 1. First, the proteins are digested with endoproteinase Lys-C to generate peptides with lysines at the C-terminal end. The three different samples are subsequently dimethylated by N-terminal specific reductive amination resulting in a light L (C_2H_4), a medium M1 ($C_2{}^2H_4$), and a heavy H ($^{13}C_2{}^2H_6H_{-2}$) labeled sample, respectively (Table 1). The second chemical modification of each of the three samples is performed to dimethylate the free amine group of lysines by reductive amination resulting in a heavy label H ($^{13}C_2{}^2H_6H_{-2}$), a medium label M2 ($^{13}C_2{}^2H_2H_2$), and a light L (C_2H_4) labeled sample, respectively (Table 1). The labeled samples are combined that the N-terminal light label is modified with a heavy lysine label (L-H), the N-terminal medium label is modified with a second variant of the medium label at the lysine (M1-M2), and the N-terminal heavy label is modified with a light lysine label (H-L). The total mass of the isobaric label $^{13}C_2C_2{}^2H_6H_2$ of the doubly labeled peptides with single lysines is 64.10697 Da in all three cases. The isobaric peptides co-elute during reversed-phase LC-separation, and corresponding peptides of the three states result in single peaks in MS1 mode. The relative quantitative abundance of the peptides derived from the three different states can be detected by the ion intensities of peptide fragment ions in the MS2 spectrum, which occur in triplets with 4 Da mass shifts. The light b-ion series and the heavy y-ion series (L-H), the heavy b-ion series and

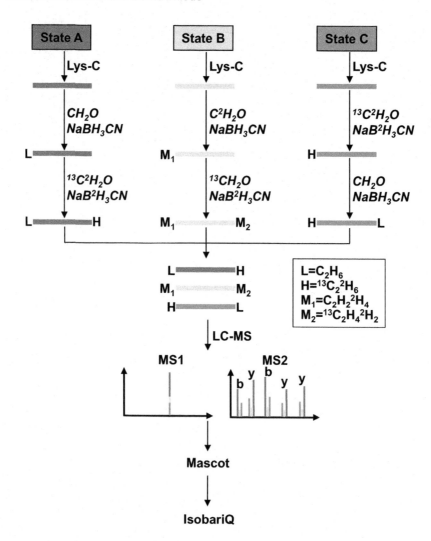

Fig. 1 Flowchart of the triplex-IPTL approach

the light y-ion series (H-L), and the two medium b- and y-ion series (MS1-MS2), respectively, derive each from one of the three samples. Data analysis can be performed by a database search using Mascot and quantitative analysis using IsobariQ. The scoring algorithm in Mascot considers matching peaks in a MS2 spectrum to determine the score, while non-matching peaks cause a penalty and thus a lower score. In triplex-IPTL-labeled peptides, three peptide sequences coexist of which the most intense will be identified by Mascot. To overcome the penalty of the other two ion series, satellite neutral losses can be defined in Mascot which are not considered for scoring and thereby improving the score for IPTL data. In triplex-IPTL, three satellite neutral losses need to be specified for the light (L) and the heavy (H) modification and two for the two medium (M1 and M2) modifications (Table 2).

Table 1
Dimethylation labeling scheme for triplex-IPTL

Label name	Short name	Formaldehyde variant	Sodium cyanoborohydride variant	Resulting label	Label mass [Da]
L-H					
1. Light	L	CH_2O	$NaBH_3CN$	C_2H_4	28.03130
2. Heavy	H	$^{13}C^2H_2O$	NaB^2H_3CN	$^{13}C_2{}^2H_6H_{-2}$	36.07567
M1-M2					
1. Medium 1	M1	C^2H_2O	$NaBH_3CN$	$C_2{}^2H_4$	32.05641
2. Medium 2	M2	$^{13}CH_2O$	NaB^2H_3CN	$^{13}C_2{}^2H_2H_2$	32.05056
H-L					
1. Heavy	H	$^{13}C^2H_2O$	NaB^2H_3CN	$^{13}C_2{}^2H_6H_{-2}$	36.07567
2. Light	L	CH_2O	$NaBH_3CN$	C_2H_4	28.03130

Table 2
Modifications and corresponding neutral loss specifications in Mascot for triplex-IPTL

Modification	Light (L)	Medium (M1)	Medium (M2)	Heavy (H)
Delta				
Monoisotopic mass (Da)	28.031300	32.056407	32.050563	36.075670
Composition	H(4) C(2)	2H(4) C(2)	2H(2) H(2) 13C(2)	2H(6) H(−2) 13C(2)
Specificity—neutral loss				
Satellite to match L	–	2H (4) H(−4)	2H(2) H(−2)	2H(6) H(−6)
	–	–	13C(2) C(−2)	13C(2) C(−2)
Satellite to match M1	2H(−4) H(4)	–	–	2H(2) H(−2)
	–	–	–	13C(2) C(−2)
Satellite to match M2	2H(−2) H(2)	–	–	2H(4) H(−4)
	13C(−2) C(2)	–	–	–
Satellite to match H	2H(−6) H(6)	2H(−2) H(2)	2H(−4) H(4)	–
	13C(−2) C(2)	13C(−2) C(2)	–	–

The Mascot configuration for the light modification is shown in Fig. 2. Finally, the Mascot dat-file can be used for IPTL quantification using IsobariQ.

3.2 Protein Digestion with Endoproteinase Lys-C

1. Reconstitute 5 µg Lys-C in 50 µL water.

2. Add 950 µL of 25 mM Tris-HCl, pH 8.5, 1 mM EDTA to generate a stock Lys-C solution.

Name

Title	IPTL_3P_L
Fullname	Triplex_di-Methylation light variant

Delta | Specificity | Ignore Masses | Misc | References

Specificity

Specificity Site [N-term ⌄]   Position [Any N-term ⌄]   [Copy] [Delete] [Hide Details]

Classification [Post-translational ⌄] Hidden ☐   Group [3]

Notes [   ]

Neutral loss ○ Scoring ◉ Satellite ○ Peptide ○ Required Peptide   [Delete]
Monoisotopic: **-4.025107** Average: **-4.0246**

Composition [2H(-4) H(4)   ] Symbols [13C ⌄] [1 ⌄] [Add]

Neutral loss ○ Scoring ◉ Satellite ○ Peptide ○ Required Peptide   [Delete]
Monoisotopic: **-4.019263** Average: **-3.9976**

Composition [13C(-2) 2H(-2) C(2) H(2)   ] Symbols [13C ⌄] [1 ⌄] [Add]

Neutral loss ○ Scoring ◉ Satellite ○ Peptide ○ Required Peptide   [Delete]
Monoisotopic: **-8.044370** Average: **-8.0223**

Composition [13C(-2) 2H(-6) C(2) H(6)   ] Symbols [13C ⌄] [1 ⌄] [Add]

Neutral loss ◉ Scoring ○ Satellite ○ Peptide ○ Required Peptide   [Delete]

Composition [   ] Symbols [13C ⌄] [1 ⌄] [Add]

[New Neutral Loss]

Specificity Site [K ⌄]   Position [Anywhere ⌄]   [Copy] [Delete] [Hide Details]

Classification [Post-translational ⌄] Hidden ☐   Group [1]

Notes [   ]

Neutral loss ○ Scoring ◉ Satellite ○ Peptide ○ Required Peptide   [Delete]
Monoisotopic: **-4.025107** Average: **-4.0246**

Composition [2H(-4) H(4)   ] Symbols [13C ⌄] [1 ⌄] [Add]

Neutral loss ○ Scoring ◉ Satellite ○ Peptide ○ Required Peptide   [Delete]
Monoisotopic: **-4.019263** Average: **-3.9976**

Composition [13C(-2) 2H(-2) C(2) H(2)   ] Symbols [13C ⌄] [1 ⌄] [Add]

Neutral loss ○ Scoring ◉ Satellite ○ Peptide ○ Required Peptide   [Delete]
Monoisotopic: **-8.044370** Average: **-8.0223**

Composition [13C(-2) 2H(-6) C(2) H(6)   ] Symbols [13C ⌄] [1 ⌄] [Add]

Neutral loss ◉ Scoring ○ Satellite ○ Peptide ○ Required Peptide   [Delete]

Composition [   ] Symbols [13C ⌄] [1 ⌄] [Add]

[New Neutral Loss]

Fig. 2 Mascot modification specificity for the light modification C_2H_4 (L). Apart from the two specificities (N-term and K), satellite neutral loss definitions for the corresponding labels M1, M2, and H are shown. Similar modification specificities must be defined for the medium and heavy labels (M1, M2, and H) according to Table 1

3. Add an appropriate amount of the Lys-C stock solution to all samples. Include three tubes containing each 10 μL (10 pmol) of the standard protein solution.

4. Mix by vortexing.

5. Incubate for 16 h at 37 °C under continuous shaking (e.g., at 1000 rpm in a thermomixer).

6. Purify the digestion products using SPE C18 using the instructions provided by the manufacturer (*see* **Note 1**).

7. Evaporate the solvent from all sample tubes by vacuum drying.

8. Check the efficiency of the Lys-C digest by analyzing one of the three aliquots of the digest of the control protein using either MALDI-MS or ESI-MS analysis.

3.3 Dimethylation of Peptides

To achieve a triplex labeling of IPTL peptides, correct combinations of isotopically labeled formaldehyde and sodium cyanoborohydride must be used. The protocol for the modification of the N-termini (*see* Subheading 3.3.1) and the lysines (*see* Subheading 3.3.2) is used with its isotopic variants of the reagents according to the labeling scheme in Table 1.

3.3.1 Dimethylation of Peptide N-Termini

1. Add 50 μL of 1% acetic acid, pH 2.8 to the dried samples.

2. Add 2 μL of 4% formaldehyde (CH_2O (L), C^2H_2O (M1), $^{13}C^2H_2O$ (H)), respectively, to each of the three samples and mix thoroughly.

3. Add 2 μL of 600 mM sodium cyanoborohydride (L, M1) or cyanoborodeuteride (H) and mix thoroughly (*see* **Notes 2 and 3**).

4. Incubate for 1 h at 37°C under continuous shaking (e.g., at 1000 rpm in a thermomixer).

5. Add 8 μL of 1% ammonium hydroxide (in water), mix thoroughly, and wait for 1 min (*see* **Note 4**).

6. Add 4 μL of 5% formic acid (in water) to the samples and mix thoroughly and wait for 1 min.

7. Purify the derivatized peptides using SPE C18 cartridges or pipette tips (*see* Subheading 3.2, **step 6** and **Note 1**).

8. Evaporate samples to dryness.

3.3.2 Dimethylation of Lysine Residues

1. Add 30 μL of 200 mM TEAB solution to the dried samples and mix thoroughly.

2. Add 2 μL of 4% formaldehyde ($^{13}C^2H_2O$ (H), $^{13}CH_2O$ (M2), CH_2O (L)) to the corresponding sample (*see* **Note 2**).

3. Add 2 μL of 600 mM sodium cyanoborohydride (H, M2) or cyanoborodeuteride (L) and mix thoroughly (*see* **Notes 2 and 3**).

4. Incubate for 1 h at 37 °C under continuous shaking (e.g., at 1000 rpm in a thermomixer).

5. Add 8 μL of 1% ammonium hydroxide solution (in water), mix thoroughly, and wait for 1 min (*see* **Note 4**).

6. Add 4 μL of 5% formic acid (in water) to the samples and mix thoroughly and wait for 1 min.

7. Purify the derivatized peptides using SPE C18 cartridges or pipette tips (*see* Subheading 3.2, **step 6** and **Note 1**).

8. Evaporate solvent from all samples by vacuum drying.

9. Check complete derivatization of the control protein digest by MALDI-MS (*see* Subheading 3.4) or ESI-MS (*see* Subheading 3.5) analysis using the aliquots taken in **step 8** from Subheading 3.2 (unmodified), **step 8** from Subheading 3.3.1 (N-terminally dimethylated), and **step 8** from Subheading 3.3.2 (fully dimethylated) after purification using SPE C18 pipette tips (*see* **Note 5**).

10. Combine samples (L-H, M1-M2, and H-L).

11. Analyze by LC-ESI-MS.

3.4 MALDI-MS

1. Mix 0.5 μL of the α-cyano-4-hydroxycinnamic acid stock solution with an equal volume of sample.

2. Record peptide mass fingerprints.

3. Check for the completeness of the derivatization reactions by comparing the peptide mass fingerprints of the Lys-C digest with those of the N-terminal dimethylated and fully dimethylated Lys-C digest (*see* **Note 5**).

3.5 LC-ESI Mass Spectrometry

1. Dissolve the dried peptides in 10 μL of 1% formic acid and 2% acetonitrile.

2. Inject 5 μL into the LC system.

3. Separate peptides by RP-C18 HPLC by applying a linear gradient from 3% of solvent B to 40% of solvent B (*see* **Note 6**).

4. Record MS1 and MS2 spectra (*see* **Notes 7** and **8**).

3.6 Protein Identification Using Mascot and Quantification Using IsobariQ

1. Process raw data to mgf files (*see* **Note 9**).

2. Perform a database search with Mascot using the following search parameters: Lys-C as enzyme with no missed cleavage sites. Apply all modifications for the triplex-IPTL labeling to variable (six in total: L, H for the N-terminus and lysines, M1 for the N-terminus, and M2 for lysines). Follow Table 2 for the correct Mascot modification configuration to yield the best results. Apply the automatic decoy database to determine the false discovery rate.

3. Copy the Mascot result file (dat-file) to a local folder.

4. Launch the IsobariQ application (*see* **Note 10**). Select "Import Data…" from the file menu; in the Import Data Wizard, select the IPTL technique and give the path to the correct Mascot

dat-file. If the data was acquired on a mass spectrometer utilizing electron transfer dissociation as fragmentation technique, select the check box for ETD. See the IsobariQ manual for detailed description about other settings. Click "Next," choose "Standard IPTL," and navigate in the drop-down boxes so that the different IPTL labels match the local Mascot names. For example, "N-terminal light label" should be set to "IPTL_3P_L (N-term)," while "N-terminal heavy label" should be set to "IPTL_3P_H (N-term)." The naming may be slightly different depending on the naming of the modifications on the Mascot servers. Click "Next" and "Finish." IsobariQ processes the Mascot dat-file and displays the Mascot results in a table (Fig. 3) where every protein can be clicked to display its assigned peptides and their individual scores.

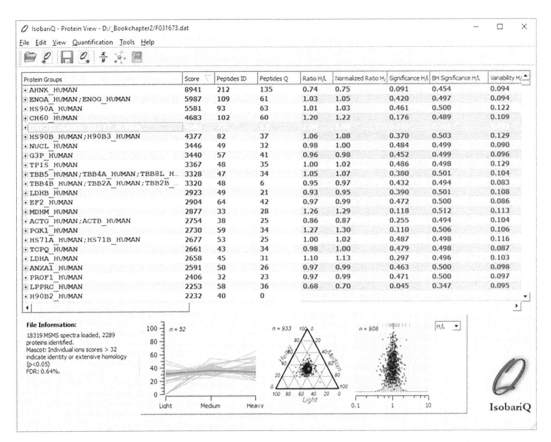

Fig. 3 Main window of IsobariQ. All identified proteins and their respective peptides, scores, quantitative information, ratio, normalized ratio, significance, Benjamini-Hochberg corrected significance, variability, etc., are shown. Every peptide may have several quantification points which enable a robust and accurate protein ratio to be inferred. The protein ratio is calculated as the median of the individual peptide ratios to minimize the effect of outliers. A ternary graph displays the distribution of all quantified proteins according to their H/M/L abundances, and histogram plots the ratios. Additionally, a profile plot displays ratio trends of the selected protein over the three states. The results can be saved as XML or exported to a spreadsheet application for further analysis

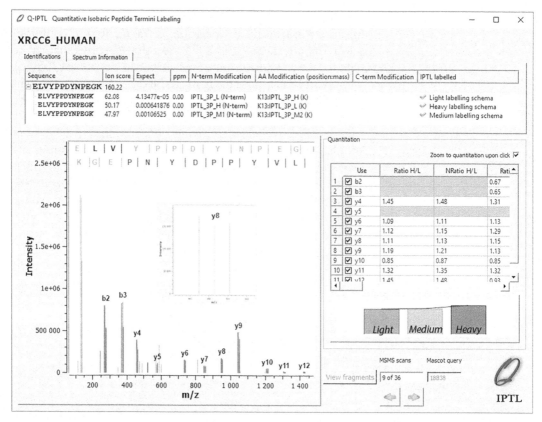

Fig. 4 IsobariQ, Q-IPTL module. When a protein has been double-clicked in the protein view, all MS2 spectra assigned to this protein are displayed here. All Mascot hits assigned to a given MS2 spectrum are displayed in the top panel, and when a triplex-IPTL label of the same sequence occurs, IsobariQ creates an IPTL triplex and recalculates the ion scores to take all three sequences of the spectrum into account. The annotated MS2 spectrum is shown in the bottom panel. When quantified, all the quantification events are listed in the quantification table to the right where the user can select which ratios to include or exclude for this particular MS2 spectrum. Therein, the extreme ratios are highlighted for easy assessment of possible outliers

Inspection of the data can be performed by double-clicking on a protein. This will load this protein with all its assigned peptides and MS2 spectra into the quantification module Q-IPTL (Fig. 4). The first MS2 spectrum identifying this protein is shown in the bottom panel, and all Mascot hits to this MS2 spectrum are shown in the top panel with detailed information about sequence, ion score, modifications, and ppm error. This is the same information as given by Mascot when hovering over a query number with the mouse in the web view. By clicking on the sequences, the spectrum updates its annotations accordingly, and the sequences can thus be manually validated. For triplex-IPTL-labeled peptides, three identifications for the same peptide should be found, L-H, M1-M2, and H-L, respectively. As a control, clicking on these three individually should change the b- and y-ion series annotation. Clicking on an IPTL-triplex will show annotation with all three sequences.

5. In the menu of the main window, click on "Quantify all proteins" and select normalization strategy (Fig. 3). We recommend using variance stabilizing normalization (VSN) as this minimizes the variance heterogeneity seen in proteomic data sets. This is the phenomenon that the statistical spread of ratios is larger for low intense proteins than for high intense (see the IsobariQ manual for more details on normalization). IsobariQ will now quantify every MS2 spectrum individually, calculate peptide and protein ratios, and perform statistical tests. In the Q-IPTL module, detailed information about the quantification events per peptide can be viewed and validated. Manual alterations, for example, removal of a possible outlier, will cause IsobariQ to recalculate the protein values on the fly.

6. For further analysis, the protein list and all of its quantification information can be exported to a spreadsheet application via a tab-separated values (tsv) file. From the menu choose "Export results...," and give the file a unique name ending with .tsv. In the spreadsheet application, the protein list can be opened and processed for post-quantification analysis. The results and all manual alteration can also be stored in an XML format: In the menu click "Save As...," and give the file a unique name. This saving enables the user to go back to the data at a later time point and do further alterations or reanalysis.

4 Notes

1. Different companies offer cartridges or pipette tips for SPE C18 with different loading capacities and slightly different protocols.

2. Use correct isotopic version for your labeling Scheme (L-H, M1-M2, H-L). Work under a fume hood, and comply with the health and safety regulations that apply in your country for handling this hazardous compound.

3. The solutions of $NaBH_3CN$ must be prepared fresh shortly before use.

4. Ammonium hydroxide can be purchased as 28%–30% solution. Ammonium hydroxide must be rapidly pipetted to avoid evaporation.

5. Peptide mass fingerprinting of a control protein (e.g., BSA) should be performed to check if the Lys-C digest and the chemical reactions have been complete. Start with enough control protein (e.g., 10 pmol) to ensure that the reactions have been complete even with a high amount of sample. Theoretical peptide masses of protein digests can be calculated, e.g., at http://web.expasy.org/peptide-mass/. After derivatization of the Lys-C digest, mass increases of 28.03 Da (N-terminal

dimethylation, L), 32.05 Da (N-terminal dimethylation, M1 and M2), 36.08 Da (N-terminal dimethylation, H), and 64.11 Da (N-terminal and lysine dimethylation, L-H, M1-M2, H-L) must be detected in comparison to the Lys-C digest of the unmodified control protein. Furthermore, the detected masses of the precedent spectra must disappear. Peptide mass fingerprinting can easily be performed using MALDI-MS. If a MALDI mass spectrometer is not available, the control sample can also be analyzed using ESI-MS.

6. In average, longer peptides are produced using Lys-C in comparison to digestion of proteins with trypsin. Therefore, a steeper gradient should be used (e.g., 40% of solvent B at the end of the separating gradient) than for tryptic peptides.

7. Peptide fragmentation by collision-induced dissociation (CID), higher energy collisionally activated dissociation (HCD), and electron transfer dissociation (ETD) can be used [7].

8. The mass differences within the MS2 spectra according to IPTL labeling are dependent on the charge of the fragments: $m/z = 4$ for $z = 1$, $m/z = 2$ for $z = 2$, $m/z = 1.33$ for $z = 3$.

9. Mgf-files can be generated using different software routines depending on the instrument used and the type of raw data. For Orbitrap instruments we recommend using msconvert from ProteoWizard available free of charge from http://proteowizard.sourceforge.net/download.html/.

10. IsobariQ can be downloaded from https://www.mn.uio.no/ibv/english/research/sections/ bmb/research-groups/enzymology-and-protein-structure-and-function/proteomics-thiede/software/.

References

1. Ankney JA, Muneer A, Chen X (2018) Relative and absolute quantitation in mass spectrometry-based proteomics. Annu Rev Anal Chem 11:40–77

2. Koehler CJ, Strozynski M, Kozielski F et al (2009) Isobaric peptide termini labeling for MS/MS-based quantitative proteomics. J Proteome Res 8:4333–4341

3. Waldbauer J, Zhang L, Rizzo A, Muratore D (2017) diDO-IPTL: a peptide labeling strategy for precision quantitative proteomics. Anal Chem 89:11498–11504

4. Cao T, Zhang L, Zhang Y et al (2017) Site-specific quantification of protein ubiquitination on MS2 fragment ion level via isobaric peptide labeling. Anal Chem 89:11468–11475

5. Koehler CJ, Arnzten MØ, de Souza GA, Thiede B (2013) An approach for triplex-isobaric peptide termini labeling (Triplex-IPTL). Anal Chem 85:2478–2485

6. Arntzen MØ, Koehler CJ, Barsnes H, Berven FS et al (2011) IsobariQ: software for isobaric quantitative proteomics using IPTL, iTRAQ, and TMT. J Proteome Res 10:913–920

7. Koehler CJ, Arnzten MØ, Treumann A, Thiede B (2012) Comparison of data analysis parameters and MS/MS fragmentation techniques for quantitative proteome analysis using isobaric peptide termini labeling (IPTL). Bioanal Anal Chem 404:1103–1114

Chapter 11

Targeted Protein Quantification Using Parallel Reaction Monitoring (PRM)

Katalin Barkovits, Weiqiang Chen, Michael Kohl, and Thilo Bracht

Abstract

Targeted proteomics represents an efficient method to quantify proteins of interest with high sensitivity and accuracy. Targeted approaches were first established for triple quadrupole instruments, but the emergence of hybrid instruments allowing for high-resolution and accurate-mass measurements of MS/MS fragment ions enabled the development of parallel reaction monitoring (PRM). In PRM analysis, specific peptides are measured as representatives of proteins in complex samples, with the full product ion spectra being acquired, allowing for identification and quantification of the peptides. Ideally, corresponding stable isotope-labeled peptides are spiked into the analyzed samples to account for technical variation and enhance the precision. Here, we describe the development of a PRM assay including the selection of appropriate peptides that fulfill the criteria to serve as unique surrogates of the targeted proteins. We depict the sequential steps of method development and the generation of calibration curves. Furthermore, we present the open-access tool CalibraCurve for the determination of the linear concentration ranges and limits of quantification (LOQ).

Key words Targeted proteomics, Parallel reaction monitoring, Stable isotope-labeled synthetic peptides, Calibration curve, Limit of quantification

---

1 Introduction

Shotgun proteomics is ideal for the identification of hundreds to thousands of proteins from a complex sample and allows discovery-based experiments aiming to identify proteins specific to treatment conditions, genotypes, or diseases in an unbiased way. The method most commonly used for data generation in mass spectrometry (MS)-based proteomics is data-dependent acquisition (DDA) (*see* Chaps. 8–10, 13–17, 20, 21, 26). In DDA mode, an overview scan is initially performed from which the Top N most intense precursor ions are selected for fragmentation. Although DDA is a powerful technique for identifying proteins in a sample, it, unfortunately, lacks sensitivity and reproducibility and the stochastic selection of

Katrin Marcus et al. (eds.), *Quantitative Methods in Proteomics*, Methods in Molecular Biology, vol. 2228,
https://doi.org/10.1007/978-1-0716-1024-4_11, © Springer Science+Business Media, LLC, part of Springer Nature 2021

the precursor ions leads to complementary identifications in repetitive experiments [1]. For quantification, however, it is essential to consistently measure all proteins of interest in all samples to obtain sufficient information for statistical analysis. Targeted proteomics offers the possibility to quantify proteins with increased sensitivity, accuracy, and reproducibility. Multiple reaction monitoring (MRM), also referred to as selected reaction monitoring (SRM), and parallel reaction monitoring (PRM) (*see* Chaps. 8, 25) are targeted proteomics approaches aiming to quantify up to hundreds of proteins in one experiment [2, 3]. These approaches have in common that selected peptides, which are exclusive representatives of proteins of interest, are measured in a predefined m/z range and optionally also in a predefined retention time window. MRM experiments are performed with triple quadrupole mass spectrometers where the scan mode comprises three main steps: (1) specific precursor ion selection in the first quadrupole, (2) fragment ion generation by a specific collision energy-induced fragmentation in the second quadrupole, and (3) specific product ion selection in the third quadrupole (Fig. 1a). The major limitation of MRM-based quantification is that the monitored product ions need to be defined and evaluated in advance. In contrast to MRM, no preselection of fragment ions is required for PRM experiments, which are performed with high-resolution and accurate-mass (HR/AM) instruments such as quadrupole orbitrap (Q-Orbitrap) and quadrupole time-of-flight (Q-TOF) mass spectrometers. Similar to MRM, the PRM scan mode also comprises three main steps. The first two steps are in principle identical to MRM; however, in the third step, MS/MS data acquisition is performed using HR/AM scanning, which allows the separation of co-isolated background ions from the target MS/MS product ions. This allows a complete MS/MS spectrum to be acquired including all possible product ions, which confirm the correct identity of the target peptide. Contrarily, in MRM only three to five product ions are monitored, and the correct identification of target peptides needs to be verified during method development [4, 5] (Fig. 1b). Various studies have shown that MRM and PRM have comparable linearity, dynamic range, and precision of quantification of target peptides, with PRM having the following advantages: (1) minimal pre-development time, (2) simple data analysis, and (3) similar performance features [3, 6, 7].

Targeted proteomics methods have been established to determine the amount of selected endogenous proteins measuring peptides that are unique surrogates of the target protein within a complex sample. For both PRM and MRM experiments, key information about the target peptides is required, which can be obtained from DDA analyses performed in advance or by searching data repositories. The selection of target peptides is crucial for a successful quantitative experiment because only an optimally selected set of

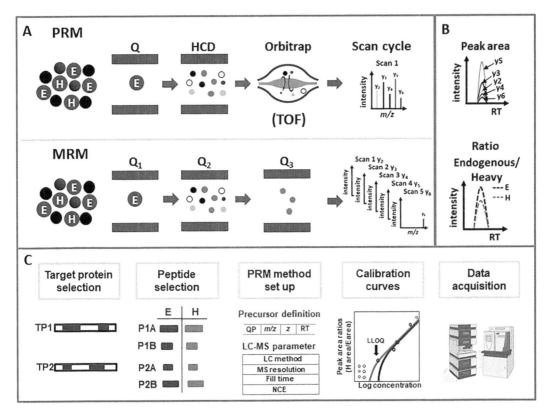

Fig. 1 *Principle of data collection, evaluation, and workflow development of targeted proteomics quantification strategies.* (**a**) Data collection in PRM and MRM approaches requires three main steps: (1) selection of the precursor ion in the quadrupole; (2) fragmentation of the precursor ion (in the HCD cell for PRM, in the second quadrupole (Q_2) for MRM); and (3) detection of the MS/MS product ions (in the Orbitrap for PRM and through selection by the third quadrupole (Q_3) for MRM). Due to the high-resolution and accurate-mass scan in PRM, all product ions are acquired within one scan cycle. MRM, in contrast, requires multiple scan cycles for multiple product ions, as only one product ion is acquired within one scan. (**b**) For both approaches, data analysis is based on the peak areas of the acquired MS/MS product ions. The presence of fragment ions is used to confirm the identity of the target peptide, and the peptide-specific distribution of fragment ions is used for quality control. Ratios of endogenous peptides (E) and stable isotope-labeled standard peptides (heavy, H) can be used for quantification. (**c**) For the development of PRM methods, specific proteins are targeted and peptides that serve as unique surrogates for the targeted proteins need to be selected. MS parameters can be optimized depending on the characteristics of the respective peptides and samples. Calibration curves need to be generated to determine the linear range and the lower and upper limits of quantification (LLOQ, ULOQ) for each peptide before data acquisition is conducted

peptides allows reliable quantification of the proteins of interest. For precise quantification of the endogenous peptides, reference peptides can be spiked into the samples and the ratios of endogenous and standard peptide intensities can be used for quantification. Here, synthetic peptides, which have the identical sequence as the target peptides but are labeled using stable isotopes, have been established as the method of choice (stable isotope-labeled synthetic (SIS) peptides). While in PRM fragment ion spectra ensure the correct identity of the target peptides, SIS peptides also warrant the correct identity of the analyzed peptides in MRM. Another important step in targeted proteomics method development is the generation of calibration curves to determine the concentration ranges in which endogenous peptides can be reliably quantified. Calibration curves can be generated using SIS peptides, and in addition to the linear range of the assay, the corresponding lower and upper limits of quantification [LLOQ/ULOQ] can be calculated (Fig. 1c).

Here, we present the experimental setup of a PRM experiment, including a tutorial for the selection of the targeted peptides. We describe the generation of calibration curves and present an open-source software named CalibraCurve [8], which uses several accuracy measures (i.e., precision and trueness) for determination of the linear concentration range and LOQs.

2 Materials

Prepare all reagents using ultrapure water (resistivity at 25 °C > 18.18 MΩ cm). Purchase analytical grade reagents unless the use of LC–MS grade reagents is indicated.

2.1 Stable Isotope-Labeled Synthetic (SIS) Peptides

Stable isotope-labeled synthetic (SIS) peptides can be purchased at various manufacturers or synthesized in-house if the required equipment is available. SIS peptides are usually labeled C-terminally with stable isotope-labeled amino acids (either $[^{13}C_6, {}^{15}N_2]$-lysine or $[^{13}C_6, {}^{15}N_4]$-arginine), ensuring that all tryptic peptides carry exactly one heavy-labeled amino acid. Non-tryptic peptides or peptides representing the protein C-terminus might also be labeled using alternative amino acids [9] (*see* **Note 1**).

2.2 LC–MS/MS Analysis

1. An Ultimate 3000 RSLC nano UHPLC System coupled online to a Q Exactive HF™ mass spectrometer. Both LC and MS instruments are from Thermo Fisher Scientific (USA).

2. Trap column: Acclaim PepMap100 C18 Nano-Trap Column (particle size 5 μm, pore size 100 Å, I.D. 100 μm, length 2 cm; Thermo Fisher Scientific, USA).

3. Analytical column: Acclaim PepMap RSLC, C18, 2 μm, 100 Å, 75 μm × 50 cm, nanoViper (Thermo Fisher Scientific, USA).

4. Loading solvent: 0.1% (v/v) TFA in water.

5. Gradient solvent A: 0.1% (v/v) formic acid (FA, LC–MS grade) in water.

6. Gradient solvent B: 0.1% (v/v) FA (LC–MS grade), 84% (v/v) acetonitrile (ACN, LC–MS gradient grade) in water.

2.3   Data Analysis

1. The Skyline software (www.skyline.ms) is a freely available, open-source application for the performance of targeted proteomics assay development and analysis.

2. CalibraCurve [8] is available from the software section of our website (www.ruhr-uni-bochum.de/mpc/). CalibraCurve is implemented both as an R script and as a KNIME workflow. Both versions are included in the software download. Comprehensive instructions for software preparation and customization via advanced settings are given in the manual that is included in the download.

3   Methods

3.1   Selection of Stable Isotope-Labeled Synthetic (SIS) Peptides

1. If available, consult in-house generated DDA MS data, ideally measured on the same instrument, that will be used for PRM assays. If peptides representing proteins of interest were detected before, include these ones for further selection procedure. If quantitative or semi-quantitative data is available (relative abundances, peptide spectrum matches), also consider these for the selection process and give preference to high abundant peptides.

2. Carefully obtain information on the protein candidates of interest from a public database, e.g., UniProt (www.uniprot.org). Relevant information might be the occurrence of isoforms that share sequences with the target isoform, natural amino acid variants, and known PTMs. Some criteria like carrying glycosylation or being part of a truncated sequence might qualify a peptide for direct exclusion from the assay. Other features like phosphorylation at tyrosine, serine, and threonine residues or the occurrence of several isoforms might justify the exclusion of peptides and should be considered for data interpretation.

3. Perform in silico digestion of the protein candidates either by editing the corresponding FASTA files manually or using software like Skyline or online tools like the ExPASy peptide cutter (www.expasy.org). Predictions of trypsin digestion efficiency might be considered but should be compared to experimental observations.

4. Peptides used for parallel reaction monitoring should have a minimum of six amino acids. For shorter peptides, the likelihood to be a unique representative of a respective protein is extremely low. Peptides longer than 20 amino acids might be difficult to synthesize.

5. Peptides likely to produce missed cleavages, ending with consecutive lysine and arginine residues (-KR, -RR, -KK, -RK,

etc.), should be excluded. Peptides carrying internal K or R residues must be excluded, as they represent peptides with missed cleavage sites.

6. Although the cleavage properties of trypsin are to some extent controversially discussed [10], peptides with K and R followed by proline should be excluded. Even if KP/RP cleavages occur, the digestion efficiencies might be precarious and not suitable for robust PRM assays.

7. Peptides containing methionine residues, which are prone to oxidation, should not be considered. Cysteine residues are likely to form disulfide bonds or be modified, and chemical alkylation of cysteine residues during sample preparation might be incomplete and bias quantification; thus, peptides containing cysteine should be excluded as well. Preference should be given to those peptides not containing tryptophan, which is prone to oxidation [11], as well as glutamine and asparagine, which both can undergo deamidation [12]. In addition, N-terminal glutamates and glutamines, which can be cyclized to pyroglutamate, might also be avoided [13].

8. Experimental depositories might be consulted to review which peptides have been experimentally observed by other researchers. PeptideAtlas (www.peptideatlas.org) represents a convenient database to assess such data and additionally receive information about observation frequencies. In addition, specific datasets might be downloaded via ProteomeXchange (www.proteomexchange.org), as not all deposited data is automatically available from PeptideAtlas.

9. The uniqueness of selected peptide in front of a respective background proteome needs to be verified. An appropriate feature is implemented in the Skyline software but can also be performed using BLAST search (www.ncbi.nlm.nih.gov).

3.2 SIS Peptide Concentration Correction

If the respective instrumentation is available, the precise concentration and purity of SIS peptides should be determined carefully. For in-house synthesized SIS peptides, the determination of these characteristics is mandatory. For commercially purchased SIS peptides, the manufacturers' specifications can be considered but might also be additionally verified.

1. Dissolve the SIS peptides in 0.1% TFA, 5% ACN in water to approximately 100 pmol/μL (replace TFA by FA, if no trap column is used).

2. Measure the accurate concentration of SIS peptides using quantitative amino acid analysis as described before [14].

3. Measure the purity of SIS peptide by the capillary zone electrophoresis as described elsewhere [15].

4. Calculate the corrected SIS peptide concentration by multiplying concentration and purity.

3.3 PRM Analysis of SIS Peptides

1. Prepare a master mix stock solution of SIS peptides in 0.1% TFA by mixing the peptides in equal molarities of 10 pmol/µL for each SIS peptide. Aliquot the solution, and store aliquots at −80 °C.

2. Dilute the master mix in 0.1% TFA to a working solution containing 50 fmol/µL of each SIS peptide.

3. Inject the master mix into the LC–MS system, with the mass spectrometer operated in PRM mode (*see* **Note 2**).

4. Peptide separation is achieved using an appropriate LC–MS method. Dependent on the number of analyzed peptides, a gradient between 30 min and 2 h might be chosen (e.g., a 98 min linear gradient from 5 to 40% solvent B at a flow rate of 400 nL/min).

5. In the MS method editor, enter an inclusion list including mono-isotopic precursor m/z, charge states (z), polarity, start and end time for scheduling, and peptide sequence information in the comment field (*see* **Note 3**). Other instrument parameters for PRM analysis are listed in Table 1 (*see* **Note 4**).

6. Evaluate measured data in the Skyline software. Stable isotope labels can be selected accordingly using the peptide settings option of the Skyline software. At the used molarity of 50 fmol,

Table 1
Exemplary LC–MS instrument parameters for PRM analysis using a Q Exactive HF instrument

Parameters for PRM	Settings on a Q Exactive HF
Run time	Same as LC run time
Polarity	Positive
Default charge state	2
Resolution	17,500
AGC target	1e6
Maximum inject time	50 ms (*see* **Note 4**)
Loop count	1
Isolation window	0.5–2 m/z
Isolation offset	0
Fixed first mass	100.0 m/z
Spectrum data type	Profile

all SIS peptides should show very good peaks of precursor ions and product ions in the extracted ion chromatogram (sufficient intensity, symmetric peak shape, reproducible product ion pattern). SIS peptides with poor peaks in the extracted ion chromatogram should be excluded from further assay development.

3.4 Calibration Curve of SIS Peptides

If the availability of an appropriate complex matrix is limited (*see* **Note 5**), a dilution series of the SIS peptides might first be measured in the absence of a complex background. Dependent on their performance, SIS peptides might be excluded from further assay development.

1. Prepare a dilution series using SIS peptide molarities from 50 to 0.001 fmol using the dilution pattern: 1:5:2:5:2:5:2:5:2:5.

2. If in addition to SIS peptides the corresponding light synthetic peptides are available, spike in constant molarities of 50 fmol per peptide into all samples to serve as an internal standard (*see* **Note 6**).

3. Measure the dilution series in PRM mode using five replicates per dilution. Start with the lowest dilution to account for possible carryover of high abundant peptides.

4. Analyze the result using Skyline, and export either the peak areas or the heavy to light ratios, the latter if light peptides were used in the analysis.

5. Adapt and/or check the settings that are required for the basic analysis of the Skyline export (*see* **Note 7**) with the Calibra-Curve software. CalibraCurve is used for the determination of linear concentration ranges/LOQs.

6. Adjust advanced settings, if necessary (e.g., for customization of the CalibraCurve results).

7. Run CalibraCurve: The R script can be executed from the command line using the following command:

   Rscript --vanilla path/to/directory/CalibraCurve_v2.0.R.
   The KNIME implementation of CalibraCurve provides the same functionality for users not familiar with command line software and can be used alternatively.

8. Evaluate the results. CalibraCurve computes four result files for each input file:

   (a) A calibration curve

   (b) A so-called Response factor plot

   (c) A plain text file containing the calculated accuracy measures, the linear ranges, and LOQ values.

   (d) A plain text file with measures calculated for the linear fit

9. Identify inconsistencies (*see* **Note 8**).

3.5 Measurement of SIS Peptides in a Complex Matrix

1. Generate a complex peptide matrix by tryptic digestion of an appropriate amount of proteins (*see* **Note 5**), e.g., using the SP3 protocol [15].

2. Spike the SIS peptide mix with 50 fmol each into 200 ng complex matrix.

3. Perform triplicate LC–MS analysis in PRM mode using the optimized method. The inclusion list contains both heavy SIS and light endogenous peptides.

4. Evaluate the results in Skyline, and review if the corresponding endogenous light peptides are detectable.

5. Inspect the product ion peak profiles for interferences. Peptides which are present in the complex matrix might co-elute with the target peptides and produce product ions of similar m/z that interfere with the target product ions. Product ions that are affected by interferences need to be excluded from quantification.

3.6 Peptide-Specific Optimization of Normalized Collision Energies (NCEs)

If DDA analysis of the respective sample was performed beforehand and the target peptides were already identified, the corresponding NCEs might also be used for PRM analysis. Optional, the NCEs can be optimized specifically for each peptide.

1. Set up multiple MS methods with different NCEs such as 20, 22.5, 25, 27.5, 30, 32.5, and 35.

2. Measure the SIS peptide master mix in triplicates using MS methods with the different NCEs.

3. Use Skyline to process the MS *.raw data files. The combined peak areas of the top five product ions can be used to determine the best NCE for each peptide.

3.7 Calibration Curves of SIS Peptides in a Complex Matrix

1. Prepare a dilution series of SIS peptides analogous to Subheading 3.4 but using the complex peptide matrix as a background.

2. Measure the dilution series in PRM mode with five replicates.

3. Analyze the result with Skyline, and export the heavy/light ratios (*see* **Note 6**).

4. Import the heavy/light ratios into CalibraCurve, and process data as described in Subheading 3.4.

3.8 Use Optimized PRM Assay to Measure Real Samples

1. Digest the real samples using your in-house protocol.

2. Spike the fixed amount of SIS peptides into 100 ng protein digest. SIS peptides need to be spiked into the samples in concentrations/molarities, yielding intensities that are close to the observed levels of the corresponding endogenous peptides. If the endogenous peptides' intensities do not robustly

reach the LLOQ, it is still important to spike in the SIS peptides within at least the LLOQ concentration.

3. For certain kinds of samples, it may also be relevant to consider the ULOQ. If the intensities of endogenous peptides are above the ULOQ, the samples have to be diluted accordingly to reach the linear range of the respective peptide (for some instruments, ion accumulation times might also be adapted to the requirements). The SIS peptides should be spiked into the samples within the linear range.

4. Analyze samples in PRM mode with the optimized method, optimally in technical triplicates. The corresponding inclusion list contains both heavy and light peptides.

5. Evaluate the LC–MS results using the Skyline software.

6. Calculate the amount of endogenous light peptide, its concentration, and coefficient of variation.

7. Explore the biological meaning of endogenous peptide/protein between different conditions.

4   Notes

1. Many providers offer SIS peptides in different purification states. Charges for peptides of >95% purity are naturally higher than for non-purified crude peptides. If reference peptides were selected without being experimentally observed *in-house* before, solely based on the general peptide selection criteria, crude peptides might be ordered first to assess their characteristics in LC–MS analysis. Peptides proving to be well suited for PRM might subsequently be ordered in purified states.

2. A full MS measurement should be performed as well to enable complete monitoring of the total ion intensity on MS level. If some targeted peptides are not detected, the MS spectrum might be inspected manually to review the existence of the respective SIS peptides.

3. Duplicate test runs with an inclusion list without start and end retention time should be analyzed every time the LC methods or conditions such as solvents or separation columns are changed. From the results, the start and end retention time of each peptide can be retrieved. Depending on the peak width, the start time could be retention time −2 min and the end time could be retention time +2 min. If no specific NCE is used for individual peptides, the respective parameter in the inclusion list can be omitted. Consequently, all peptides will be analyzed using the overall NCE as defined in the MS method. After the optimization of NCEs, the appropriate NCE for each peptide can be entered in the inclusion list.

4. The maximum injection time might be increased to improve the sensitivity of peptide detection and eventually lower the LLOQ. Accordingly, the resolution should be increased as well, since the time to scan is longer. The number of data points per peak needs to be monitored so that at least eight data points per peak are realized to warrant accurate quantification.

5. The choice of an appropriate complex matrix for PRM method development is critical. Ideally, a complex matrix consists of a mixture of the samples which are supposed to be measured or a representative set of analogous samples. However, especially in clinical applications when scarce samples like biopsies or micro-dissected tissue are analyzed, the sample material is limited. In such cases, the experimentalists might make use of a mixture of representative cell lines, which warrant the unlimited availability of a complex matrix of constant quality. However, proteins representing the extracellular matrix or originating from miscellaneous cell types might be underrepresented. In such cases, tissue from animal models might serve as an alternative. Here, species-associated differences in amino acid sequences might or might not cause interferences, which might be present or absent in the samples eventually analyzed.

6. The analysis of peptide abundances based on heavy/light ratios offers higher precision compared to the analysis of peak areas. Variations in total ion intensities that might be caused by pipetting errors or variations in the electrospray performance are already considered when the analyses are based on heavy/light ratios. Also, the calculation of calibration curves and the determination of the LOQ are more precise when heavy/light ratios are used. However, in some cases, endogenous peptides might not be detected reliably in the complex matrices used for measurements of the calibration curves. In such cases light peptides might also be spiked into the samples representing the SIS peptide dilution series, to serve as an internal standard accounting for technical variations. Light peptides are generally cheaper than the corresponding stable isotope-labeled ones, and no high purity is required for the respective application.

7. The CalibraCurve data input format requires two columns: the first column contains the expected concentrations and the second column either the measurement values or heavy/light ratios. Further columns can be included in the input in order to provide additional information.

8. CalibraCurve result files and the log file contain several opportunities to identify inconsistencies that should be considered. For example, warnings are given that indicate concentration levels within the calculated linear range that show accuracy values that are outside the specified threshold values. In

particular, users of the software are encouraged to evaluate the calibration and response factor plots thoroughly in order to decide whether deviations of specific concentration levels are negligible or re-measurements are required. A specific issue may occur if in one sample several concentration ranges exist that pass the threshold that is required for the coefficient of variation of the measurements. For example, imagine a sample with ten concentration levels. Levels 3–5 and Levels 7–9 pass the threshold criteria, but Level 6 does not. CalibraCurve provides several opportunities to deal with this situation, which are explained in detail in the software manual.

Acknowledgments

A part of this study was funded by P.U.R.E. (Protein Research Unit Ruhr within Europe) and ProDi (Center for protein diagnostics), Ministry of Innovation, Science and Research of North-Rhine Westphalia, Germany; and i:DSem-Verbundprojekt: Electronic Patient Path (EPP)' [FKZ 031 L 0025], a project of the German Federal Ministry of Education and Research (BMBF). In addition, the work was supported by the H2020 project NISCI (GA no. 681094) and by the ISF program of the European Union.

References

1. Barkovits K, Linden A, Galozzi S et al (2018) Characterization of cerebrospinal fluid via data-independent acquisition mass spectrometry. J Proteome Res 17(10):3418–3430. https://doi.org/10.1021/acs.jproteome.8b00308

2. Hoofnagle AN, Becker JO, Oda MN et al (2012) Multiple-reaction monitoring-mass spectrometric assays can accurately measure the relative protein abundance in complex mixtures. Clin Chem 58(4):777–781. https://doi.org/10.1373/clinchem.2011.173856

3. Peterson AC, Russell JD, Bailey DJ et al (2012) Parallel reaction monitoring for high resolution and high mass accuracy quantitative, targeted proteomics. Mol Cell Proteomics 11(11):1475–1488. https://doi.org/10.1074/mcp.O112.020131

4. Sherman J, McKay MJ, Ashman K et al (2009) How specific is my SRM?: the issue of precursor and product ion redundancy. Proteomics 9(5):1120–1123. https://doi.org/10.1002/pmic.200800577

5. Duncan MW, Yergey AL, Patterson SD (2009) Quantifying proteins by mass spectrometry: the selectivity of SRM is only part of the problem. Proteomics 9(5):1124–1127. https://doi.org/10.1002/pmic.200800739

6. Ronsein GE, Pamir N, von Haller PD et al (2015) Parallel reaction monitoring (PRM) and selected reaction monitoring (SRM) exhibit comparable linearity, dynamic range and precision for targeted quantitative HDL proteomics. J Proteome 113:388–399. https://doi.org/10.1016/j.jprot.2014.10.017

7. Hoffman MA, Fang B, Haura EB et al (2018) Comparison of quantitative mass spectrometry platforms for monitoring kinase ATP probe uptake in lung cancer. J Proteome Res 17(1):63–75. https://doi.org/10.1021/acs.jproteome.7b00329

8. Kohl M, Stepath M, Bracht T et al (2020) CalibraCurve: a tool for calibration of targeted MS-based measurements. Proteomics 22: e1900143. https://doi.org/10.1002/pmic.201900143

9. Bracht T, Schweinsberg V, Trippler M et al (2015) Analysis of disease-associated protein expression using quantitative proteomics-fibulin-5 is expressed in association with hepatic fibrosis. J Proteome Res 14(5):2278–2286. https://doi.org/10.1021/acs.jproteome.5b00053

10. Rodriguez J, Gupta N, Smith RD et al (2008) Does trypsin cut before proline? J Proteome

Res 7(1):300–305. https://doi.org/10.1021/pr0705035

11. Perdivara I, Deterding LJ, Przybylski M et al (2010) Mass spectrometric identification of oxidative modifications of tryptophan residues in proteins: chemical artifact or post-translational modification? J Am Soc Mass Spectrom 21(7):1114–1117. https://doi.org/10.1016/j.jasms.2010.02.016

12. Geiger T, Clarke S (1987) Deamidation, isomerization, and racemization at asparaginyl and aspartyl residues in peptides. Succinimide-linked reactions that contribute to protein degradation. J Biol Chem 262(2):785–794

13. Hoofnagle AN, Whiteaker JR, Carr SA et al (2016) Recommendations for the generation, quantification, storage, and handling of peptides used for mass spectrometry-based assays. Clin Chem 62(1):48–69. https://doi.org/10.1373/clinchem.2015.250563

14. Taleb RSZ, Moez P, Younan D et al (2019) Protein biomarker discovery using human blood plasma microparticles. Methods Mol Biol 1959:51–64. https://doi.org/10.1007/978-1-4939-9164-8_4

15. Percy AJ, Chambers AG, Yang J et al (2014) Advances in multiplexed MRM-based protein biomarker quantitation toward clinical utility. Biochim Biophys Acta 1844(5):917–926. https://doi.org/10.1016/j.bbapap.2013.06.008

Chapter 12

Quantitative Approach Using Matrix-Assisted Laser Desorption/Ionization Time-of-Flight (MALDI-ToF) Mass Spectrometry

Brooke A. Dilmetz, Peter Hoffmann, and Mark R. Condina

Abstract

Quantitation using mass spectrometry (MS) is a routine approach for multiple analytes, including small molecules and peptides. Electrospray-based MS platforms are typically employed, as they provide highly reproducible outputs for batch processing of multiple samples. Quantitation using matrix-assisted laser desorption/ionization (MALDI) time-of-flight (ToF) mass spectrometry, while less commonly adopted, offers the ability to monitor analytes at significantly higher throughput and lower cost compared with ESI MS. Achieving accurate quantitation using this approach requires the development of appropriate sample preparation, spiking of appropriate internal standards, and acquisition to minimize spot-to-spot variability. Here we describe the preparation of samples for accurate quantitation using MALDI-ToF MS. The methodology presented shows the ability to quantitate perfluorooctanesulfonic acid (PFOS) from contaminated water.

Key words MALDI-ToF, PFOS, Quantitation, Monoclonal antibodies

1 Introduction

Matrix-assisted laser desorption/ionization (MALDI) time-of-flight (ToF) mass spectrometry (MS) can be used for quantitation of multiple analytes, including small molecules [1–9], peptides and proteins [10–19]. This can be done as purified or enriched analytes from complex samples (including blood plasma or tissue lysates) or, in the case of small molecule therapeutics, on tissue samples prepared for MALDI-MS imaging (MALDI-MSI) acquisition. Liquid chromatography (LC) coupled with electrospray (ESI) MS (LCMS) is the routine approach for quantitative workflows as the approach is highly reproducible, highly sensitive, and selective across a larger dynamic range compared with MALDI-MS. The

Peter Hoffmann and Mark R. Condina are equally contributed to last author.

Katrin Marcus et al. (eds.), *Quantitative Methods in Proteomics*, Methods in Molecular Biology, vol. 2228,
https://doi.org/10.1007/978-1-0716-1024-4_12, © Springer Science+Business Media, LLC, part of Springer Nature 2021

combination with LC minimizes complexity of the sample prior to introduction into the MS and maintains the reproducible signal response for multiple samples. Quantitation using MALDI-MS is a desirable complementary approach as it allows significantly higher throughput at lower cost per sample compared with LCMS. Achieving repeatable and reproducible results for an analyte requires the selection of the correct matrix, matrix concentration, optimized sample preparation, target plate, and concentration range of isotopic standards.

Per- and polyfluoroalkyl substances (PFAS) such as PFOS are a diverse group of compounds that are used in a wide range of consumer products and industrial processes [1, 20]. These compounds are highly resistant to degradation and bio-accumulative, leading to their persistence in the environment [1, 8, 21–23]. PFAS compounds have been regarded as a new class of environmental contaminants and banned from use in some countries. Additionally, exposure of rodents to PFAS compounds has been shown to have adverse effects (e.g., increased incidence of tumors and changes in lipid metabolism) on multiple organs and cause developmental changes [24, 25]. The widespread use of PFAS compounds in the manufacturing industry and emerging concerns of their impact to the environment and human health have resulted in the need to develop rapid and cost-effective techniques for their analysis.

We here describe a method for quantitation using MALDI-ToF MS. The method outlines the sample and calibration preparation to quantify PFOS, using α-cyano-4-hydroxycinnamic acid (CHCA) matrix. This quantitative method can be applied to other PFAS compounds such as perfluorooctanoic acid (PFOA) and perfluorohexane sulfonate (PFHxS); however, matrix composition and concentration would need to be evaluated to ensure no interfering matrix peaks are present at these masses. A MALDI-ToF MS approach for PFOS monitoring allows high sampling of water and soil from contaminated areas, which can be used as a screening method to identify key areas/samples that are of high importance for further analysis using LCMS. This will reduce costs to industries that require high-throughput analysis and high sampling of potentially contaminated areas.

2   Materials

Prepare all solutions using ultrapure water (sensitivity of 18 MΩ-cm at 25 °C) and LCMS grade reagents. Prepare and store all reagents at room temperature (unless indicated otherwise). Diligently follow all waste disposal regulations when disposing waste materials.

2.1 MALDI-ToF MS Calibrants and Matrix	1. Red phosphorus ((w/v), 3 mg/0.3 mL acetonitrile) (*see* **Note 1**). 2. Matrix: 10 mg/mL CHCA matrix in 70% acetonitrile/0.1% trifluoroacetic acid (TFA) [26]. A 0.5 mg/mL dilution (in 90% acetonitrile/0.1% TFA) is used for measurements.
2.2 PFOS Standards	1. PFOS (L-PFOS, sodium perfluoro-1-octanesulfonate) and labelled PFOS (M-PFOS, sodium perfluoro-1-[1,2,3,4-$^{13}C_4$]-octane sulfonate) (Wellington Laboratories, Guelph, Canada) in methanol (1.2 mL) (*see* **Note 2**).
2.3 MALDI-ToF MS	1. MTP AnchorChip 800 μm 384 raster MALDI-ToF MS target (Bruker Daltonics, Bremen, Germany). 2. UltrafleXtreme MALDI-ToF/ToF MS (P.N. 276601.00538, Bruker Daltonics). 3. flexControl (version 3.4, build 76) and flexAnalysis software (version 3.4, build 76) (Bruker Daltonics).
2.4 Solid-Phase Extraction (SPE)	1. C18 SPE columns (PN 60108–305, Thermo Fisher, Waltham, USA).

3 Methods

Carry out all procedures at room temperature unless otherwise specified.

3.1 Minimization of Contamination

1. To minimize and prevent contamination and carryover of PFAS, many precautions should be undertaken. Plastic and glassware used for all analyses were pre-washed three times with methanol, followed by three times with ultrapure water, and dried.

2. Prior to use, the MALDI target must be cleaned using a modified procedure to minimize contamination of PFAS. The MALDI target requires rinsing with methanol, followed by sonication in methanol for 10 min in a glass or ceramic dish. The target is then rinsed with acetone and subsequently sonicated for 10 mins with isopropanol. Finally, the target is then rinsed with ultrapure water and dried using high-purity nitrogen.

3. All MALDI-ToF MS analyses included matrix-only controls, as well as MPFAS only controls, with experiments discarded if PFAS mass peaks were detected in these controls.

4. All SPE processed samples included the analysis of a blank column prepared with the same solutions (excluding PFAS) to ensure no contamination.

Table 1
Dilution steps of L-PFOS and M-PFOS

Final concentration (ng/μL)	Initial PFOS concentration (ng/μL)	Volume of PFOS solution (μL)	Volume of M-PFOS 50 ng/ μL solution in μL	Volume of methanol (μL)
0	0	0	0	100
0.1	1	10	10	90
1	50	10	10	500
5	50	10	10	100
10	50	10	10	50

3.2 Calibration Preparation

1. Prepare calibration standard solutions of L-PFOS and M-PFOS using methanol. The calibration concentration range should cover the anticipated concentration range expected from the sample. The respective mass-labelled M-PFOS is added prior to spotting on the MALDI target as an internal control. For L-PFOS, prepare a calibration standard curve from 0.1 to 10 ng/μL, as outlined in Table 1. M-PFOS is spiked in at 1 ng/μL to normalize for spot-to-spot variations in intensity (*see* **Note 3**).

2. For L-PFOS MALDI analysis, all samples were spotted onto the target in replicates of 10 in HPLC grade methanol. 1 μL of sample is spotted onto the target and allowed to dry at room temperature before being overlaid with 1 μL of CHCA matrix. Include matrix-only controls to ensure the absence of overlapping peaks for the analyte of interest (in this case, L-PFOS and M-PFOS) (*see* **Note 4**).

3. Between replicate spots, deposit 1 μL red phosphorus onto the dedicated calibration spot position of the MTP AnchorChip 800 μm 384 MALDI-ToF target.

3.3 MALDI-ToF MS Acquisition and Analysis

1. For a Bruker ultrafleXtreme MALDI-ToF MS system, running flexControl 3.4, load a reflectron negative ion method, with the detection range of m/z 120–3000 Da. The laser repetition rate is 2000 Hz, acquiring 10,000 shots in 100 shot increments using a small spiral movement within the 800 μm raster (*see* **Note 4**).

2. Upon acquisition, the obtained spectra were re-calibrated using the red phosphorus signals obtained from the respective calibrant position using a multipoint quadratic near neighbor calibration routine and peaks annotated using centroid peak picking.

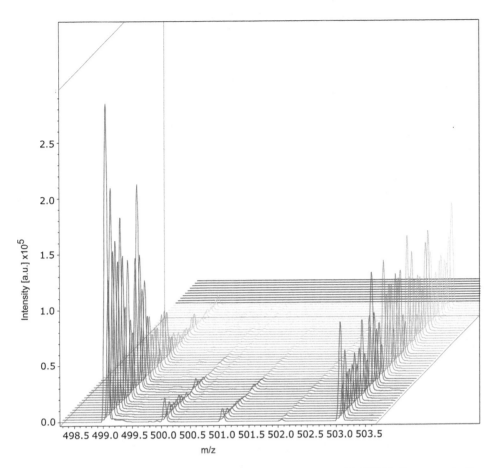

Fig. 1 *Representative MALDI-ToF MS spectra of PFOS standards spotted with CHCA in the concentration range of 0.1–10 ng/µL.* Different concentrations of L-PFOS (498.9 *m/z*) at 0.1–10 ng/µL were spiked with M-PFOS (502.9 *m/z*) at 1 ng/µL in replicates of 10. Samples were overlaid with CHCA matrix. Samples were acquired using an ultrafleXtreme MALDI-ToF MS in reflectron negative mode from 120 to 3000 *m/z*. Spectra were analyzed in flexAnalysis 4.0 (Bruker Daltonics). The heavy labelled standard does not always show consistent intensity when acquired with different samples due to the inherent limited dynamic range associated with MALDI-ToF MS [11, 29]

3. In flexAnalysis 3.4 (Bruker Daltonics), load and display spectra acquired for the calibration curve. Ensure the intensity values for L-PFOS (K-PFOS 498.9 m/z) and M-PFOS ($502.9\ m/z$) $[M + H]^-$ are present and no contaminating peaks are observed in the matrix-only sample spots, as shown in Fig. 1:

4. Export obtained peaklists from flexAnalysis 3.4 (Bruker Daltonics) to Excel, and calculate the ratio of L-PFOS/M-PFOS. The intensity of the monoisotopic peak of L-PFOS and M-PFOS is used for quantitation of all replicates. Plot this ratio to generate a calibration curve, such as outlined in Fig. 2:

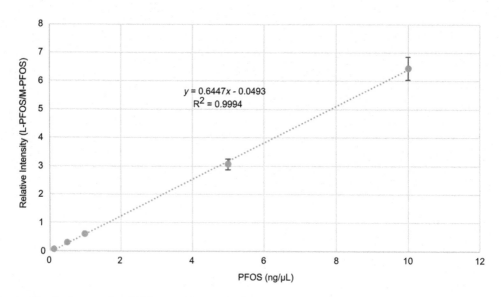

Fig. 2 *Standard curve for PFOS spotted with CHCA in the concentration range of 0.1–10 ng/μL.* Curve generated using ratio of unlabelled (L-PFOS) to labelled (M-PFOS) intensity ($r^2 = 0.9994$). The M-PFOS was spiked in at 1 ng/μL

Table 2
Quantification of PFOS spiked into ultrapure water at drinking water guideline concentrations, analyzed by MALDI-ToF MS

Prepared concentration (μg/L)	Calculated concentration (μg/L)	Error (%)
0.07	$0.091 \pm 1.06 \times 10^{-4}$	30.0

3.4 Extraction and Analysis of Contaminated Samples

1. Regulation guidelines state that PFOS concentrations should be below 0.07 μg/L [27]. To test the feasibility of this workflow for the screening of water samples, ultrapure water (500 mL) was spiked with 0.07 μg/L of PFOS.

2. The SPE C18 material was pre-conditioned by running 2 × 5 mL of methanol, followed by 2 × 5 mL ultrapure water through the cartridge under vacuum.

3. The sample was then passed through the cartridge, eluted with 2 × 2 mL methanol, and dried in a vacuum concentrator (*see* **Note 5**). Samples were re-dissolved in 50 μL of methanol containing 5 μL of 50 ng/μL of M-PFOS.

4. Samples were spotted in 10× replicates and overlaid with CHCA matrix.

5. L-PFOS concentrated from ultrapure water was quantified using MALD-ToF MS as shown in Table 2. This procedure can be applied to environmental samples (*see* **Note 6**).

3.5 Quantification of PFAS

1. The intensity of the monoisotopic peak from L-PFOS and M-PFOS was exported into Excel.

2. The ratio of L-PFOS/M-PFOS was determined for each replicate and averaged. To compute the concentration (x) of the sample, re-arrange the equation of the line (*see* Fig. 2 for an example) in terms of concentration, and substitute the relative intensity value (y). After determining the concentration, consider any dilution factors to give the final concentration.

4 Notes

1. The red phosphorus does not dissolve. It requires being crushed in the acetonitrile using an Eppendorf tip, followed by vortexing for 30 s and sonication for 5 min. The calibrant was centrifuged at $14,000 \times g$ for 2 min and the supernatant spotted onto the target.

2. The standards are supplied in 100% methanol. To prevent evaporation, we recommend aliquoting samples (e.g., 50 µL) into an amber glass vial that contains a glass insert. The samples should then be vacuum concentrated to dryness. The aliquot should be resuspended in the same volume of methanol aliquoted (e.g., 50 µL) on the day of use.

3. Repeatability and reproducibility for the method were assessed based on Food and Drug Administration Guidelines [28].

4. The method employs a higher number of shots to ensure high sampling of the entire spot and to minimize spot-spot variation.

5. All SPE extractions included analysis of a blank column, with the same solutions run through (excluding added PFAS) to ensure no contamination.

6. Environmental water samples need to be filtered through a 0.2 µm filter prior to clean-up using a C18 SPE column. After elution with methanol, the sample is dried using vacuum concentration, re-dissolved in methanol containing 1 ng/µL of M-PFOS, and spotted in 10× replicates.

References

1. Cao D, Wang Z, Han C et al (2011) Quantitative detection of trace perfluorinated compounds in environmental water samples by matrix-assisted laser desorption/ionization-time of flight mass spectrometry with 1,8-bis (tetramethylguanidino)-naphthalene as matrix. Talanta 85(1):345–352

2. Chumbley CW, Reyzer ML, Allen JL et al (2016) Absolute quantitative MALDI imaging mass spectrometry: a case of rifampicin in liver tissues. Anal Chem 88(4):2392–2398

3. Cohen LH, Gusev AI (2002) Small molecule analysis by MALDI mass spectrometry. Anal Bioanal Chem 373(7):571–586

4. Ling L, Li Y, Wang S et al (2018) DBDA as a novel matrix for the analyses of small molecules and quantification of fatty acids by negative ion MALDI-TOF MS. J Am Soc Mass Spectrom 29(4):704–710

5. Shariatgorji M, Nilsson A, Goodwin RJ et al (2014) Direct targeted quantitative molecular imaging of neurotransmitters in brain tissue sections. Neuron 84(4):697–707

6. Takai N, Tanaka Y, Saji H (2014) Quantification of small molecule drugs in biological tissue sections by imaging mass spectrometry using surrogate tissue-based calibration standards. Mass Spectrom (Tokyo) 3(1):A0025

7. Wang P, Giese RW (2017) Recommendations for quantitative analysis of small molecules by matrix-assisted laser desorption ionization mass spectrometry. J Chromatogr A 1486:35–41

8. Yang C, Lee HK, Zhang Y et al (2019) In situ detection and imaging of PFOS in mouse kidney by matrix-assisted laser desorption/ionization imaging mass spectrometry. Anal Chem 91(14):8783–8788

9. Schulz S, Becker M, Groseclose MR et al (2019) Advanced MALDI mass spectrometry imaging in pharmaceutical research and drug development. Curr Opin Biotechnol 55:51–59

10. Ahn SH, Kang JW, Moon JH et al (2015) Quick quantification of proteins by MALDI. J Mass Spectrom 50(3):596–602

11. Anderson NL, Razavi M, Pearson TW et al (2012) Precision of heavy-light peptide ratios measured by maldi-tof mass spectrometry. J Proteome Res 11(3):1868–1878

12. Gutierrez JA, Dorocke JA, Knierman MD et al (2005) Quantitative determination of peptides using matrix-assisted laser desorption/ionization time-of-flight mass spectrometry. Biotechniques (Suppl):13–17. https://doi.org/10.2144/05386su02

13. Jiang J, Parker CE, Hoadley KA et al (2007) Development of an immuno tandem mass spectrometry (iMALDI) assay for EGFR diagnosis. Proteomics Clin Appl 1(12):1651–1659

14. Li H, Popp R, Frohlich B et al (2017) Peptide and protein quantification using automated immuno-MALDI (iMALDI). J Vis Exp (126):55,933. https://doi.org/10.3791/55933

15. Mason DR, Reid JD, Camenzind AG et al (2012) Duplexed iMALDI for the detection of angiotensin I and angiotensin II. Methods 56(2):213–222

16. Moon JH, Park KM, Ahn SH et al (2015) Investigations of some liquid matrixes for analyte quantification by MALDI. J Am Soc Mass Spectrom 26(10):1657–1664

17. Popp R, Li H, Borchers CH (2018) Immuno-MALDI (iMALDI) mass spectrometry for the analysis of proteins in signaling pathways. Expert Rev Proteomics 15(9):701–708

18. Spainhour JC, Janech MG, Schwacke JH et al (2014) The application of Gaussian mixture models for signal quantification in MALDI-TOF mass spectrometry of peptides. PLoS One 9(11):e111016

19. Szajli E, Feher T, Medzihradszky KF (2008) Investigating the quantitative nature of MALDI-TOF MS. Mol Cell Proteomics 7(12):2410–2418

20. Schultz MM, Barofsky DF, Field JA (2004) Quantitative determination of fluorotelomer sulfonates in groundwater by LC MS/MS. Environ Sci Technol 38(6):1828–1835

21. Giesy JP, Kannan K, Jones PD (2001) Global biomonitoring of perfluorinated organics. ScientificWorldJournal 1:627–629

22. Kannan K, Tao L, Sinclair E et al (2005) Perfluorinated compounds in aquatic organisms at various trophic levels in a Great Lakes food chain. Arch Environ Contam Toxicol 48(4):559–566

23. Fromme H, Tittlemier SA, Volkel W et al (2009) Perfluorinated compounds—exposure assessment for the general population in Western countries. Int J Hyg Environ Health 212(3):239–270

24. Biegel LB, Hurtt ME, Frame SR et al (2001) Mechanisms of extrahepatic tumor induction by peroxisome proliferators in male CD rats. Toxicol Sci 60(1):44–55

25. Onishchenko N, Fischer C, Wan Ibrahim WN et al (2011) Prenatal exposure to PFOS or PFOA alters motor function in mice in a sex-related manner. Neurotox Res 19(3):452–461

26. Zhang X, Shi L, Shu S et al (2007) An improved method of sample preparation on AnchorChip targets for MALDI-MS and MS/MS and its application in the liver proteome project. Proteomics 7(14):2340–2349

27. PFOS and PFOA—fact sheets; 2017. https://www.health.nsw.gov.au/environment/factsheets/Pages/pfos.aspx

28. Hubbard WK (1997) International conference on harmonisation; guideline on the validation of analytical procedures: methodology; availability. Daily J U S Government 62:27,464–27,467

29. Bucknall M, Fung KYC, Duncan MW (2002) Practical quantitative biomedical applications of MALDI-TOF mass spectrometry. J Am Soc Mass Spectrom 13(9):1015–1027

Chapter 13

Application of SILAC Labeling in Phosphoproteomics Analysis

Markus Stepath and Thilo Bracht

Abstract

The analysis of disease-related changes in the phosphorylation status of cellular signal transduction networks is of major interest to biomedical researchers. Mass spectrometry-based proteomics allows the analysis of phosphorylation in a global manner. However, several technical challenges need to be addressed when the phosphorylation of proteins is analyzed. Low-abundant phosphopeptides need to be enriched before analysis, thereby introducing additional steps in sample preparation. Consequently, the applied quantification strategies should be robust towards elaborate sampling handling, rendering label-based quantification strategies the methods of choice in many experiments. Here, we present a protocol for SILAC labeling and the subsequent isolation of phosphopeptides using TiO_2 affinity chromatography. We outline the corresponding LC–MS/MS analysis and the essential steps of data processing.

Key words Phosphoproteomics, SILAC, Phosphorylation, Quantification, Signaling network dynamics

1 Introduction

Cellular signal transduction processes and their disease-associated modulations are of major interest to biomedical research. The activity of cellular signal transduction networks is mainly mediated by the phosphorylation status of proteins, which is tightly controlled by the activity of kinases and phosphatases. Mass spectrometry-based (phospho-)proteomics (*see* also Chaps. 8, 14–16) has been demonstrated to be a powerful tool for analyzing protein phosphorylation in a global manner and assess functional changes in signal transduction networks. However, several technical challenges have to be addressed when protein phosphorylation is investigated. Phosphorylated peptides are usually low abundant in complex samples and need to be enriched prior to analysis. Here, various approaches were established and evaluated in the past, including also exceptional ones like the isolation using molecularly imprinted polymers [1]. The most frequently used enrichment

Katrin Marcus et al. (eds.), *Quantitative Methods in Proteomics*, Methods in Molecular Biology, vol. 2228,
https://doi.org/10.1007/978-1-0716-1024-4_13, © Springer Science+Business Media, LLC, part of Springer Nature 2021

strategies are based on ion metal affinity chromatography (IMAC [2]) or metal oxide affinity chromatography (MAOC), the latter mostly utilizing TiO_2-functionalized matrices [3]. These chromatography techniques can be applied in sophisticated 2D-LC approaches but also in simple batch enrichment protocols, as presented here. On the one hand, enrichment of phosphopeptides is urgently needed for a comprehensive phosphoproteome analysis, but additional steps in sample handling increase the technical variation between individual samples, on the other hand. For this reason, label-based quantification approaches allowing for multiplex analysis have emerged as methods of choice for phosphoproteomics applications. Especially, if additional pre-fractionation is performed to increase the analysis depth, label-based methods are advisable. Here, metabolic stable isotope labeling of amino acids in cell culture (SILAC) (*see* Chaps. 8, 13, 18) or chemical labeling using tandem mass tags (TMT) (*see* Chaps. 8, 9, 14, 15, 17) or isobaric tags for relative and absolute quantification (iTRAQ) are widely used techniques. Chemical labeling strategies require thoughtful experimental designs and strategies for data analysis [4]. In addition, labeling reagents can be costly, especially for higher plex capacities. SILAC is typically limited to cell culture experiments, but a detailed documentation is available in the literature and online, making it applicable in a straightforward manner. Additional costs for stable isotope-labeled amino acids are moderate. SILAC has been demonstrated to outperform label-free quantification and TMT for the analysis of phosphorylation dynamics in a cell culture experiment [5]. In the following, we present a protocol based on SILAC labeling and quantification and the enrichment of phosphopeptides using TiO_2 affinity chromatography. We describe the mass spectrometric measurement and present a basic outline of the corresponding data analysis. The described workflow refers to an experiment that was recently published by our group [5]. We inhibited EGFR signaling in the colorectal cancer cell line DiFi using the therapeutic antibody cetuximab and performed sampling of cells at three different time points. In this protocol, the experimental set up is meant to illustrate general considerations and supposed to be adjusted according to the experimentalists needs.

2   Materials

2.1   SILAC Medium and Cell Culture Experiment

1. Mammalian cell line used here: Colorectal cancer cell line DiFi [6].

2. DMEM, high glucose, no glutamine, no lysine, no arginine.

3. DPBS (Dulbecco's phosphate-buffered saline), w/o calcium and magnesium.

4. Dialyzed fetal calve serum (FCS) (*see* **Note 1**).

5. Labeled amino acids.

   (a) "Light": L-Arginine HCl (Arg0), L-Lysine HCl (Lys0).

   (b) "Medium Heavy": L-Arginine-13C6 HCl (Arg6), L-Lysine-4,4,5,5-d4 HCl (Lys4).

   (c) "Heavy": L-Arginine-13C6,15N4 HCl (Arg10), L-Lysine-13C6,15N2 HCl (Lys8).

6. 200 mM L-Glutamine.

7. L-Proline.

8. Penicillin-Streptomycin.

9. EGFR inhibitor antibody: cetuximab (5μg/mL, Erbitux, Merck).

2.2 Cell Lysis and Protein Extraction

1. DPBS w/o calcium and magnesium.

2. RIPA buffer.

3. Protease inhibitor: Complete Protease Inhibitor Cocktail (Roche).

4. Phosphatase inhibitors: PhosSTOP™ (Roche).

5. Acetone (ice-cold).

6. Lysis buffer: Urea Buffer (pH = 8.5 with HCl), 7 M Urea, 2 M Thiourea, 30 mM Tris Base, 0.1% NaDoc (sodium deoxycholate) (see **Note 2**).

7. Sonication probe.

8. Protein assay.

2.3 Protein Digestion and Peptide Purification

1. Reducing agent: 250 mM dithiothreitol (DTT, stock solution in H_2O).

2. Alkylating agent: 550 mM iodoacetamide (IAA, stock solution in H_2O).

3. Trypsin NB Sequencing Grade, modified from porcine pancreas (SERVA Electrophoresis, Heidelberg, Germany).

4. 50 mM Ammonium bicarbonate (NH_4HCO_3)

5. Oasis HLB PLUS Extraction Cartridge (Waters).

6. H_2O (from Millipore desalting system).

7. Trifluoroacetic acid (TFA).

8. Formic acid (FA).

9. Acetonitrile (ACN).

10. Washing buffer: 0.1% TFA in H_2O.

11. Elution buffer: 5% FA, 90% ACN in H_2O.

12. Disposable syringes: 20, 5 and 1 mL (see **Note 3**).

2.4 Phosphopeptide Enrichment

1. TiO$_2$ beads (5μm, GL Science Inc., Tokyo, Japan).
2. LoBind tubes.
3. Glycolic acid.
4. 24.5% Ammonium hydroxide.
5. Formic acid (FA).
6. Filter for TiO$_2$ Beads: C8 membrane for SPE (3 M Empore).
7. Loading Buffer: 80% ACN (v/v), 5% TFA (v/v), 1 M Glycolic acid.
8. Washing Buffer A: 80% ACN, 1% TFA (v/v).
9. Washing Buffer B: 20% ACN, 0.5% TFA (v/v).
10. Elution Buffer: ~1% Ammonium hydroxide (pH ~11.3).
11. H$_2$O (from Millipore desalting system).
12. 0.1% TFA.

2.5 Stage Tip Purification

1. Octadecyl C18 extraction disks (2215, Empore™).
2. Preparation solution: Methanol.
3. Cleaning buffer: 75% ACN/25% H$_2$O/0.05% TFA (v/v).
4. Washing buffer: 0.05% TFA in H$_2$O (pH $\geq$2.0 & $\leq$4).
5. Elution buffer: 50% ACN/0.5% AcOH in H$_2$O.
6. 200μL tip adapter.

2.6 LC–MS/MS Analysis

1. ESI-MS instrument, e.g., Q Exactive Orbitrap mass spectrometer (Thermo Fisher Scientific).
2. Nano-flow capillary HPLC system, e.g., Ultimate 3000 RSLCnano (Dionex, Thermo Fisher Scientific).
3. Trap column: Acclaim PepMap, 100, 100μm × 2 cm, nanoViper, C18, 5μm, 100 Å (Thermo Fisher Scientific).
4. Analytical column: Acclaim PepMap RSLC, 75μm × 50 cm, nanoViper, C18, 2μm, 100 Å (Thermo Fisher Scientific).
5. Solvent A: 0.1% FA in H$_2$O.
6. Solvent B: 0.1% FA, 84% ACN in H$_2$O.

2.7 Raw Data Processing (Peptide Identification and Quantification) and Data Analysis

1. Protein sequence data in FASTA format for your analyzed species, e.g., from the public comprehensive protein sequence databases UniProt (https://www.uniprot.org/).
2. Protein identification and quantification software such as MaxQuant (freely available and documented here: https://maxquant.net/maxquant/).

Data analysis software such as Perseus (freely available and documented here: https://maxquan.net/perseus/ or R (freely available and documented here: https://www.r-project.org/).

3   Methods

3.1   General Considerations and Remarks

During the planning phase of the (phospho-)SILAC experiment (*see* Fig. 1a), the following aspects should be considered: (1) the number of replicates (biological or technical), (2) the labeling design, and (3) the cell culture model. We suggest at least four biological replicates for each experimental condition. For phosphoproteomics analysis, which typically includes additional sample preparation steps (e.g., phosphopeptide enrichment via IMAC or

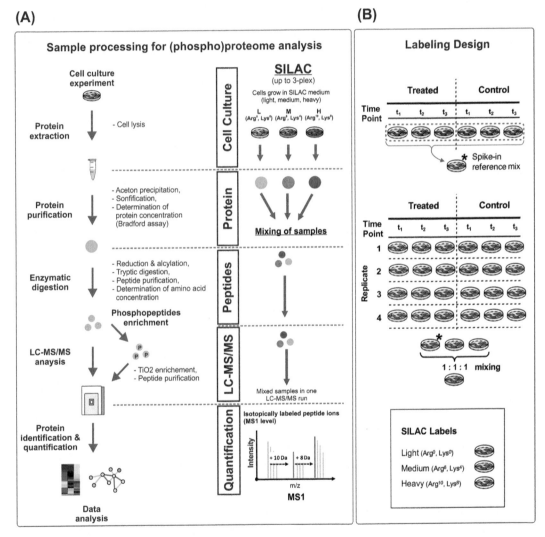

Fig. 1 Illustration of the SILAC quantification strategy in mass spectrometry-based proteomics. (**a**) Schematic representation of a workflow for mass spectrometry-based analysis of (phospho)proteome samples. (**b**) Schematic representation of the applied labeling strategy. SILAC samples were metabolically labeled, and the experimental replicates were mixed directly after sample generation and subsequently processed in combination

TiO$_2$) and relies on single peptide quantifications, a higher technical variability can be expected. Here, it might be beneficial for subsequent statistical data analysis to increase number of replicates. Cells can be metabolically labeled with either light (L), medium (M), or heavy (H) arginine and lysine isotopes in a SILAC triplex experiment. Therefore, the compared experimental conditions require differential labeling. In addition, when only two conditions are analyzed (e.g., treated and untreated), a reference mix of samples using the remaining label can be spiked into each SILAC triplex. We recommend using a light-labeled reference mix containing all experimental conditions in equal amounts to cover a broad portion of the analyzed (phospho)proteome. This reference mix can be used for normalization and direct comparison of SILAC triplex samples. To monitor and compensate for possible effects of the labeling on biology or accuracy [7], we recommend a labeling swap for half of the samples (excluding the reference mix). The requirements for the cellular model are certainly to be auxotroph for the amino acids used for metabolic labeling and for practical reasons the doubling time during proliferation. Under optimal conditions five to six doublings (with each doubling the existing fraction of non-labeled peptides is reduced by a factor of two) are required to reach the recommended labeling degree of >95%. However, stable isotope labeling can also affect the cellular metabolism, eventually leading to experimental bias between differentially labeled samples. Such effects need to be controlled carefully, especially in experiments requiring long culture times [8]. In this context, arginine-to-proline conversion is described for various cell lines and can lead to incomplete SILAC labeling. L-proline might be added to reduce arginine-to-proline conversion and incomplete SILAC labeling [7, 9]. Completeness of SILAC labeling should be tested by LC–MS/MS before continuing the experiment.

In general, it is highly advised to use LC–MS grade reagents and solutions for the sample preparation and LC–MS/MS analysis. Furthermore, in phosphoproteomics fast working under cooled conditions is recommended whenever possible. While in standard proteomics analysis workflows protein amounts of ng to µg per sample are sufficient, in phosphoproteomics considerable larger protein amounts of mg to g are necessary to obtain a sufficient number of phosphorylated peptides after enrichment and clean-up steps for LC–MS/MS analysis.

3.2 SILAC Labeling and Cell Culture Experiment

1. Prepare sterile cell culture media: DMEM high glucose medium supplemented with 10% dialyzed FCS, 4 mM L-glutamine, 1.7 mM L-proline (optional), and 100 U/mL penicillin/streptomycin.

2. Supplement the prepared cell culture media with light, medium, or heavy arginine and lysine isotopes. For the light (L), medium (M), and heavy (H)-labeled versions, first

reconstitute the respective amino acids (Lys0 and Arg0 for L; Lys4 and Arg6 for M; Lys8 and Arg10 for H) in, e.g., PBS and sterilities these stock solutions by filtration. Subsequently add the required amount of amino acid stock solutions. The final amount of Arg and Lys must be equal for the three different SILAC media. Finally add PBS or the respective solvent used to reconstitute the L-Proline, L-glutamine, and amino acids to adjust concentrations. Here, final concentrations are based on the original composition of the light DMEM medium with ~0.398 mM for Arg and ~ 0.798 mM for Lys. Consider the different molar masses of the isotopes for each amino acid.

3. Culture your mammalian cell line (here DiFi) in the prepared SILAC media for 5–6 cell doublings to achieve a labeling degree of >95% (*see* **Note 4**).

4. Here, we describe a time course experiment with four experimental replicates for each time point (0 h, 3 h, 24 h) and treatment condition (cetuximab treated or untreated control). The experiment is performed with each SILAC medium (L, M, H) in parallel, and the label swap has to be considered (*see* Fig. 1b). In theory, duplicates would be sufficient in this experimental labeling design; however we suggest performing the experiment in at least triplicates with each SILAC medium to compensate for potential sample losses or handling errors.

5. Seed cells in 75 cm² flasks (4.8 × 10E6 cells per flask) with 14 mL working volume.

6. After a cultivation of 72 h, treat cells with cetuximab (5μg/mL) at a final concentration of 5μg/mL and leave control groups untreated.

7. Harvest the cells after 0, 3, and 24 h at 80–90% confluency.

3.3 Cell Lysis and Protein Extraction

1. Supplement ice-cold RIPA buffer with protease and phosphatase inhibitors according to manufacturer's instructions.

2. Perform the following steps with pre-cooled solutions, and work under cooled conditions (on ice) and fast.

3. Discard cell culture medium and wash cells twice with ice-cold DPBS. Remove remaining DPBS completely!

4. Next, add 800μL (sufficient to cover the surface of the cell culture flasks used) of the ice-cold supplemented RIPA buffer and let cells incubate for 1 min.

5. Scrape cells from the flasks and transfer the lysates into 2 mL protein LoBind tubes.

6. Let lysates agitate gently at 4 °C for 30 min in an overhead rotator.

7. Add −20 °C acetone to a final 50% (v/v) to precipitate proteins over night at −20 °C (*see* **Note 5**).

8. Centrifuge precipitated proteins (10 min, 14,000 × *g*, 4 °C) and remove supernatant.

9. Let remaining acetone evaporate for a few minutes.

10. Supplement ice-cold urea lysis buffer with phosphatase inhibitors. Dissolve the pellet in 200μL ice-cold lysis urea buffer (*see* **Note 6**).

11. Vortex pellet for 30 s and mix gently at 4 °C until the pellets have dissolved.

12. Shortly spin down samples in the LoBind tubes (5 s).

13. Homogenize lysates with a sonication probe (10 s, 3 cycles, 60% power) to destroy slime and poorly soluble substances. Repeat sonication one to three times and place samples on ice between each cycle.

14. Next, centrifuge homogenized lysates (10 min, 14,000 × *g*, 4 °C) and transfer supernatants into new 1.5 mL LoBind tubes. Avoid to aspirate the insoluble pellets.

15. Determine protein concentration via protein assay, e.g., Bradford assay or amino acid analysis (*see* also Chap. 2).

16. Add urea lysis buffer to the individual samples to adjust protein concentrations (e.g., 6μg/μL) if possible, since this is easy for batch processing of samples in the following steps. Scale according to the lowest concentrated sample.

3.4 Mixing of SILAC Samples

1. Prepare reference mix by combining equal amounts of each of the six experimental condition (i.e., time point and treatment) using the light-labeled samples (*see* **Note 7**).

2. Mix SILAC samples at a final protein ratio of 1:1:1 with medium and heavy labels being swapped for half of the samples (here: use 55μL (6μg/μL) per label equivalent to ~1 mg protein per SILAC triplex sample).

3.5 Protein in-Solution Digestion

1. Use 1 mg proteins of each SILAC triplex sample and reduce with 5 mM (final concentration) DTT for 30 min at 37 °C in a thermomixer (600 rpm) (*see* **Note 8**).

2. Next, alkylate with 15 mM (final concentration) IAA for 30 min at room temperature in the dark.

3. Lower the total urea concentration to 1.5 M by adding 50 mM ammonium bicarbonate (*see* **Note 9**).

4. Prepare trypsin freshly in 50 mM ammonium bicarbonate (0.5μg/μL).

5. Add trypsin solution to your samples at a trypsin: protein ratio of 1:50 (w/w) and mix gently.

6. Spin down samples shortly.

7. Let samples digest for 14 h at 37 °C (overnight).

8. Quench trypsin by acidification with TFA to a final concentration of 1% (v/v).

9. Centrifuge at 15.7 × g for 5 min at room temperature.

10. Transfer the purified supernatants carefully into new 1.5 mL LoBind tube. Avoid aspirating the sedimented sodium deoxycholate pellet.

3.6 Peptide Purification on Oasis HLB Cartridges

1. Prepare purification buffers and store at 4 °C. 30 mL per buffer are sufficient for six samples.

2. Keep cartridges, samples, and buffers on ice during the purification.

3. Equilibrate cartridge material with 5 mL elution buffer (*see* **Note 10**).

4. Next, wash the cartridges with 5 mL water. Conditioned cartridges can be stored in cooled water.

5. Before applying the peptide samples, flush cartridges with air to remove remaining solvents.

6. Load samples slowly onto the cartridges and collect the eluting flow through in the original tube.

7. Repeat the procedure twice.

8. Wash unbound material and contaminants using 5 mL Washing Buffer with fast flow.

9. Next flush with 5 mL water with fast flow.

10. Flush cartridges with air to remove remaining solvents.

11. Finally, elute the peptides in 1.5 mL elution buffer into a new 2 mL LoBind tube. Elute complete peptide amount by flushing the cartridges with air.

12. Determine peptide concentration by quantitative amino acid analysis [10] or estimate concentrations based on the amount of protein input and yield of purification.

13. Reduce sample volumes to dryness by vacuum centrifugation.

14. Samples can be stored at −80 °C until further processing.

3.7 Phosphopeptide Enrichment

Here phosphopeptides are enriched using 100μg starting material and TiO$_2$ beads according to modified protocols [3, 11] (*see* **Notes 11** and **12**). The protocol is subdivided into the preparation of TiO$_2$ removal tips (Subheading 3.7.1.), the TiO$_2$ beads (Subheading 3.7.2.), and the phosphopeptide enrichment steps (Subheading 3.7.3.):

3.7.1 Preparation of TiO₂ Removal Tips

1. To prepare the TiO_2 removal tips, use 200µL pipetting tips and transfer the desired amount of stamped out C8 membrane material on the bottom of each tips. Use only little mechanical pressure to compress the disks. Usually two disks with ~0.7 mm diameter are sufficient per tip. Prepare a few tips in excess.

2. Next, place the C8 packed TiO_2 removal tips with appropriate adapter into a 1.5 or 2 mL collection vial (*see* **Note 13**).

3. Place collections vials together with adapter and tips in a centrifuge.

4. Add 100µL 50% ACN/50% H_2O (v/v) into each tip and let centrifuge for 10–30 s at $2.3 \times g$.

5. Quality control (QC) step: Check liquid levels in each tip and sort out empty tips or tips with considerable higher liquid levels than the average level of our tips. Proceed with the tips that passed this quality check.

6. Let remaining packed tips centrifuge at $2.3 \times g$ until liquid has run through completely and store tips at RT.

7. Prepare TiO_2 beads in desired amounts (*see* **Note 14**).

3.7.2 Preparation of TiO₂ Beads

1. To prepare the TiO_2 beads, weigh a desired amount (e.g., 60 mg) and place in a 1.5 mL tube.

2. Add 200µL ACN and mix thoroughly.

3. Centrifuge shortly to sediment the TiO_2 beads.

4. Aspirate and discard the ACN supernatant.

5. Repeat washing procedure with ACN twice.

6. Next, add 1 mL ACN to the washed beads and prepare aliquots of the desired amount. Alternatively, the beads must be pipetted later directly from the slurry (*see* **Note 15**).

3.7.3 Phosphopeptide Enrichment

1. Reconstitute 100µg of the dried and purified peptides in 100µL 0.1% TFA. Mix the peptide solution thoroughly.

2. Take a few µL (2µL) from the dissolved peptides and store them for the proteome analysis by LC–MS/MS. Here, 500 ng per LC–MS/MS analysis are sufficient.

3. Add 1 mL cold loading buffer to the remaining peptide solution and mix.

4. Combine the peptides dissolved in loading buffer with the pre-aliquoted 0.6 mg TiO_2 beads per sample.

5. Incubate reconstituted peptides and beads for 10 min on a thermo shaker at 1000 rpm and 4 °C.

6. Centrifuge samples for 1 min at $15.7 \times g$ and 4 °C.

7. Discard supernatant without beads.

8. Wash beads with 200µL loading buffer thoroughly, followed by centrifugation (1 min, $15.7 \times g$, 4 °C).

9. Aspirate and discard supernatant without beads.

10. Wash beads with 200µL Washing Buffer A thoroughly, followed by centrifugation (1 min, $15.7 \times g$, 4 °C).

11. Aspirate and discard supernatant without beads.

12. Wash beads with 200µL Washing Buffer B thoroughly, followed by centrifugation (1 min, $15.7 \times g$, 4 °C).

13. Aspirate and discard supernatant without beads.

14. Dry beads to remove remaining solvents in a vacuum centrifuge.

15. Finally add with 50µL elution buffer to the beads and mix (*see* **Note 16**).

16. Incubate for 10 min on a thermo shaker at 1000 rpm and 4 °C.

17. In the meantime, prepare elution vials (2 mL LoBind tubes) containing 10µL FA and place them into a cooled (4 °C) centrifuge together with the previously prepared C8 packed TiO_2 removal tips and the adapters (*see* **Note 17**).

18. Shortly spin down the incubated TiO_2 beads

19. Add and mix gently with 50µL ACN.

20. Transfer complete volume of the samples into the TiO_2 removal tips placed in the centrifuge.

21. Subsequently centrifuge samples for 5 min at $2.3 \times g$ and 4 °C. Check if samples eluted completely and none of the C8 restraints is broken (i.e., TiO_2 beads were collected in the elution vial).

22. Evaporate enriched samples with a vacuum centrifuge and store at −80 °C.

3.8 Stage Tip Purification

1. For the final purification of the TiO_2-enriched phosphopeptides, we used self-packed stage tips according to a modified protocol [12] (*see* **Note 18**).

2. Prepare buffers and work under cooled conditions whenever possible (*see* **Note 19**).

3. Prepare self-house packed C18 stage tips (*see* **Note 20**).

4. Reconstitute phosphopeptides with 50µL ice-cold 0.05% TFA (pH 2–4).

5. Place the self-packed C18 stage tips, adapters, and 2 mL waste vials into a pre-cooled centrifuge (4 °C).

6. Activate the tips by adding 40µL methanol into each tip and centrifuge 5 min at $0.3 \times g$.

7. Next, wash the tips by adding 40μL 75% ACN into each tip and centrifuge 5 min at 0.3 × *g*.

8. Assure that solutions passed the packed filters completely.

9. Add 40μL 0.05% TFA into the tips and perform a quality control step during this equilibration step.

10. QC step (1/5): Sort out the tips that do not perform well in order to reduce technical variation and optimize peptide recovery and handling time.

11. QC (2/5) step: Therefore, centrifuge the tips at 0.3 × *g* for 30–60 s. Check the liquid levels of 0.05% TFA solution in the tips. Mark the tips that contain considerable less or more volume than the average as the outliers.

12. QC step (3/5): Continue centrifugation for the 5–6 min and check the volumes again.

13. QC step (4/5): Mark also the tips that still have surpassing volume of solution in it.

14. QC step (5/5): Sort out the marked tips (*see* **Note 21**).

15. Exchange the 2 mL waste vial with a 2 mL LoBind tube to collect the flow-through during sample loading and repeat the following sample loading step two to three times for increased sensitivity and reproducibly.

16. Otherwise, keep the waste vial and load samples only once.

17. Add 40μL 0.05% TFA into the tip and centrifuge for a few minutes (5 min) at 0.3 × *g* to wash out unbound material.

18. After this step, exchange the 2 mL waste vial or collection vial with the final 2 mL elution vial.

19. Pipet 60μL elution buffer (50% ACN, 0.5% AcOH) into each tip and centrifuge for 5 min at 0.3 × *g*.

20. Reduce the volume in a vacuum centrifuge.

21. Samples can be stored at −80 °C until further processing (e.g., LC–MS/MS analysis).

3.9 LC–MS/MS Analysis

1. In the described workflow, LC–MS/MS analysis is performed on a Q Exactive Orbitrap mass coupled to an Ultimate 3000 RSLCnano HPLC system (*see* **Notes 22** and **23**).

2. Dilute the peptide samples in 17μL 0.1% TFA. Use 500 ng per proteome sample and the total yield of each TiO_2 phosphopeptide enrichment for loading and pre-concentration on the trap column.

3. Perform the peptide separation on an analytical column at a flow rate of 400 nL/min with a 180 min linear gradient from 5 to 40% solvent B (solvent A: 0.1% FA, solvent B: 0.1% FA, 84% ACN).

3.10   Raw Data Processing (Peptide Identification and Quantification) and Data Analysis

1. We will only cover the essential first steps of data analysis. The described generic data analytical workflow is adapted to the materials used and the respective bioanalytical question, i.e., investigation of changes on proteome and phosphoproteome level upon EGFR inhibition. Spectra of the generated *.raw files are searched with the MaxQuant software and Andromeda search algorithm against the human UniProtKB/Swiss-Prot database to identify and quantify peptides and protein abundancies. Downstream data analysis (e.g., filtering, normalization, integration, statistical testing, network analysis, visualization, etc.) is conducted with the Perseus software [13, 14].

2. Select parameters as follows: Trypsin/P as enzyme in specific digestion mode with a maximum of two missed cleavage sides allowed. Set fixed modification to carbamidomethyl at cysteine (C) resides. Include methionine (M) oxidation and phosphorylation at serine (S), threonine (T), and tyrosine (Y) as variable modifications with a maximum number of five modifications per peptide. Set false discovery rate (FDR) thresholds to 0.01 on peptide spectrum matching and on protein level. Enable identification of co-fragmented peptides (second peptides) and match between run (MBR) features. Set the minimum score for modified peptides to 40. Set Arg6/Lys4 as medium and Arg10/Lys8 as heavy labels. For improved quantification apply the re-quantify option (mode: "Match from to"). Set the minimum ratio count for protein quantification to two, considering unmodified oxidized (M) and phosphorylated (STY) peptides.

3. After the peptide identification and quantification step using MaxQuant, import the output file proteinGroups.txt into Perseus for proteome analysis. Here, we mainly work with protein ratio intensities (ratios M/L, H/L).

4. Pre-process the data by filtering out contaminants, observations only identified by site, and reversed hits for protein data.

5. Consider further filtering of protein groups with ≥ 1 unique peptide and ≥ 2 peptides in total.

6. $\log_2$-transform intensity ratios (M/L, H/L) and inspect distribution and centering of your data for each sample via histograms.

7. Next, inspect missing values and consider applying a threshold for proteins you want to consider as quantified. Here, only proteins were used for further analysis if detected in at least 50% of replicates in each of the six experimental groups. Alternatively, you could use missing value imputation approaches.

8. Normalize intensity ratios by replicate-wise median subtraction and inspect distribution and centering of your data for each sample via histograms.

9. Perform further data analysis: statistical analysis, functional annotation, enrichment, and protein interaction network analysis.

10. Next, perform phosphosite (p-site) analysis. Therefore, import the "Phospho (STY)Sites.txt" file into Perseus software.

11. Filter p-sites for contaminants, reversed hits and a localization probability ≥75% (i.e., class I p-sites).

12. Data analysis regarding p-sites can be conducted analogously to the analysis of proteome data.

13. In addition p-sites can be normalized to protein abundancies.

14. To determine TiO$_2$ phosphopeptide enrichment efficiency, import the "modificationSpecificPeptides.txt" file into Perseus and compare amount/intensity of phosphorylated to non-phosphorylated peptides before and after enrichment.

4   Notes

1. Dialysis is an important step to remove small unlabeled (native, light) amino acids from the serum.

2. NaDoc is a cost-effective RapiGest alternative to increase solubility of proteins.

3. Check if syringe material loses polymers on contact with the reagents of the buffers, especially acetonitrile. Syringes without rubber coating are preferred.

4. The incorporation of medium and heavy labeled Arg and Lys cells can be assessed using LC–MS/MS analysis. Samples from labeled cells need to be collected, digested, and measured separately. For LC–MS/MS and data analysis, use settings analogue as described in Subheading 3.10. Additionally, select variable proline modifications (medium-labeled samples, Pro5; heavy-labeled samples, Pro6). Assess intensities of labeled and unlabeled peptides for different peptide species (content of Arg, Lys) separately and in combination.

5. This step is necessary to remove detergents (e.g., SDS) of the RIPA buffer. In addition, the precipitated proteins can be dissolved in a defined volume to increase concentration, which has practical advantages during sample processing.

6. Phosphatase inhibitors should be added freshly.

7. Each SILAC triplex contains an untreated control and its corresponding cetuximab-treated sample, both labeled with

medium and heavy isotopes, respectively (*see* Fig. 1b). In addition, a reference mix of all light-labeled samples is spiked into each SILAC triplex. To perform the subsequent phosphoproteomics workflow in addition to the proteomics workflow, an absolute protein amount of ~1 mg is intended for each of the resulting 12 SILAC triplex samples.

8. Here we use lower temperatures than other protocols during the reduction step since the lysis buffer contains high amounts of urea. 37 °C are sufficient for reduction and prevent unwanted side modifications. In general, reduction and alkylation steps might be omitted at all to reduce sample processing times and technical variation.

9. For high trypsin activity and prevention of trypsin denaturation, the urea concentrations should be <2 M. Some proteases like LysC can also operate at higher concentrations.

10. It is acceptable for the cartridges to run dry, but in any case, they should only be flushed in one flow direction as recommended by the manufacturer.

11. In order to verify the digestion efficiency and determined concentrations on the proteome level, a master mix of the samples might be analyzed before proceeding with the phosphopeptide enrichment.

12. Work under cooled conditions and prepare buffers fresh if possible. Either prepare TiO_2 removal tips or carefully aspirate enriched peptides without beads in the later steps. We suggest using the tip-based technique.

13. Emptied Mini-Prep columns with drilled and emptied lid (adapter to fix the tips above the waste vials) have proven very useful for this purpose. Be careful to always keep liquid levels of collection or waste vials below the bottom of the tips.

14. A widely recommended ratio of peptides to TiO_2 beads (m/m) is 1–6. However, the optimal ratio also depends on the specific characteristics of your samples and the applied TiO_2 beads.

15. We recommend to aliquot the beads beforehand. Here, we aim for 0.6 mg TiO_2 quantities to enrich 100µg peptides each. Ensure a good mixing during aliquot preparation and pipette volumes of sufficient size. If necessary, perform aliquotation as dilution series. For example, first prepare 200µL aliquots corresponding to 12 mg TiO_2 beads. Then produce 50µL aliquots, followed by 10µL aliquots with 0.6 mg TiO_2 beads each. Reduce ACN volumes to dryness by vacuum centrifugation and store TiO_2 aliquots at RT.

16. Check the pH and adjust where appropriate if the elution buffer (~1% ammonia (pH 11.3)) is not prepared freshly.

17. The FA is used to rapidly neutralize the high pH of the Elution buffer, which is important for the stability of the phosphopeptides.

18. Alternatively, commercially available purification tools, such as ZipTips, can be applied. However, when larger sample amounts have to be processed (>10), we found the stage tip protocol to be more effective, as it represents a fast and cost-effective technique with high and reproducible peptide yield.

19. Check pH of the Washing Buffer (0.05% TFA in H_2O) to be within the range of 2.0–4.0.

20. Prepare self-house packed C18 stage tips as described previously for the C8 TiO_2 removal tips (see Subheading 3.7.). But here use the C18 material. Consider the peptide capacity of the C18 membrane. From our experience, three disks are sufficient for the purification of TiO_2-enriched phosphopeptides on the basis of 100µg starting material.

21. If the solutions pass very slowly through the packed tips, this is typically caused by tight packing. In this case, you can increase either the centrifugation time or the rotation speed (e.g., $0.3 \times g$ - $0.4 \times g$) or increase both. If this does not work, you should consider packing new columns.

22. Avoid letting dry out the disks between equilibration and sample addition. Check if the solution passed through the tip before going further in protocol in all following steps.

23. If possible, avoid the usage of metal needles whenever phosphopeptides are analyzed. Use glass needles instead. We tested different settings for the normalized collision energy (NCE) of the HCD fragmentation in the range from 27 to 35. We found lower NCEs (22, 27) with superior performance for phosphopeptide identification and phosphosite localization.

Acknowledgments

The work was supported by a grant from the German Cancer Aid (70111971). A part of this study was funded by P.U.R.E. (Protein Research Unit Ruhr within Europe); by Ministry of Innovation, Science and Research of North Rhine-Westphalia, Germany; and by de.NBI, a project of the German Federal Ministry of Education and Research (BMBF) [FKZ 031 A 534 A].

References

1. Chen J, Shinde S, Koch M-H et al (2015) Low-bias phosphopeptide enrichment from scarce samples using plastic antibodies. Sci Rep 5:11,438

2. Ruprecht B, Koch H, Domasinska P et al (2017) Optimized enrichment of phosphoproteomes by Fe-IMAC column chromatography. Methods Mol Biol 1550:47–60

3. Thingholm TE, Jørgensen TJD, Jensen ON et al (2006) Highly selective enrichment of phosphorylated peptides using titanium dioxide. Nat Protoc 1:1929–1935

4. Brenes A, Hukelmann J, Bensaddek D et al (2019) Multibatch TMT reveals false positives, batch effects and missing values. Mol Cell Proteomics 18:1967–1980

5. Stepath M, Zu B, Maghnouj A et al (2019) Systematic comparison of label-free, SILAC, and TMT techniques to study early adaption toward inhibition of EGFR signaling in the colorectal cancer cell line DiFi. J Proteome Res 19:926–937

6. Olive M, Untawale S, Coffey RJ et al (1993) Characterization of the DiFi carcinoma cell line dereives from a familial adenomatous polyposis patient. Vitr Cell Dev Biol 29A:239–248

7. Park S-S, Wu WW, Zhou Y et al (2012) Effective correction of experimental errors in quantitative proteomics using stable isotope labeling by amino acids in cell culture (SILAC). J Proteome 75:3720–3732

8. Lehmann W (2016) A timeline of stable of isotopes and mass spectrometry. Mass Spectrom Rev 36:58–85

9. Bendall SC, Hughes C, Stewart MH et al (2008) Prevention of amino acid conversion in SILAC experiments with embryonic stem cells. Mol Cell Proteomics 7:1587–1597

10. Taleb RSZ, Moez P, Younan D, Eisenacher M et al (2019) Protein biomarker discovery using human blood plasma microparticles. Methods Mol Biol 1959:51–64

11. Dehghani A, Gödderz M, Winter D (2018) Tip-based fractionation of batch-enriched phosphopeptides facilitates easy and robust phosphoproteome analysis. J Proteome Res 17:46–54

12. Rappsilber J, Mann M, Ishihama Y (2007) Protocol for micro-purification, enrichment, pre-fractionation and storage of peptides for proteomics using stagetips. Nat Protoc 2:1896–1906

13. Tyanova S, Cox J (2018) Perseus: a bioinformatics platform for integrative analysis of proteomics data in cancer research. Methods Mol Biol 1711:133–148

14. Tyanova S, Temu T, Sinitcyn P et al (2016) The Perseus computational platform for comprehensive analysis of (prote)omics data. Nat Methods 13:731–740

Chapter 14

Relative Quantification of Phosphorylated and Glycosylated Peptides from the Same Sample Using Isobaric Chemical Labelling with a Two-Step Enrichment Strategy

Ivan Silbern, Pan Fang, Yanlong Ji, Lenz Christof, Henning Urlaub, and Kuan-Ting Pan

Abstract

Post-translational modifications (PTMs) are essential for the regulation of all cellular processes. The interplay of various PTMs on a single protein or different proteins comprises a complexity that we are far from understanding in its entirety. Reliable strategies for the enrichment and accurate quantification of PTMs are needed to study as many PTMs on proteins as possible. In this protocol we present a liquid chromatography-tandem mass spectrometry (LC/MS/MS)-based workflow that enables the enrichment and quantification of phosphorylated and N-glycosylated peptides from the same sample. After extraction and digestion of proteins, we label the peptides with stable isotope-coded tandem mass tags (TMTs) and enrich N-glycopeptides and phosphopeptides by using zwitterionic hydrophilic interaction chromatography (ZIC-HILIC) and titanium dioxide (TiO_2) beads, respectively. Labelled and enriched N-glycopeptides and phosphopeptides are further separated by high pH (basic) reversed-phase chromatography and analyzed by LC/MS/MS. The enrichment strategies, together with quantification of two different PTM types from the same sample, allow investigation of the interplay of those two PTMs, which are important for signal transduction inside the cell (phosphorylation), as well as for messaging between cells through decoration of the cellular surface (glycosylation).

Key words Post-translational modifications, Phosphorylation, N-Glycosylation, Isobaric labelling, TMT, TiO_2, ZIC-HILIC

1 Introduction

Phosphorylation and N-glycosylation are among the most abundant post-translational protein modifications (PTMs). It is estimated that up to 25% and 20% of proteins curated in the Swiss-Prot database [1] can be phosphorylated or glycosylated, respectively [2]. Phosphorylation, which in mammals occurs mainly on serine, threonine, or tyrosine residues, can induce protein conformational changes that modulate protein–protein interactions and enzyme activity. Phosphorylation signalling thereby regulates key cellular

Katrin Marcus et al. (eds.), *Quantitative Methods in Proteomics*, Methods in Molecular Biology, vol. 2228,
https://doi.org/10.1007/978-1-0716-1024-4_14, © Springer Science+Business Media, LLC, part of Springer Nature 2021

processes, including the survival, migration, and division of cells [3]. Aberrant kinase or phosphatase activity that leads to sustained proliferative activity is one of the hallmarks of cancer [4]. In comparison to phosphorylation, protein N-linked glycosylation (N-glycosylation), an attachment of a glycan to the side chain amide nitrogen of an asparagine residue within the Asn-X-Ser/Thr sequence motif, is chemically more complex and less well understood. N-glycosylation assists protein folding and trafficking and governs cell–cell interactions [5]. Altered glycosylation is associated with a variety of diseases, including cancer [6], diabetes [7], inflammatory diseases [8–10], and neurodegenerative disorders [11, 12]. Despite the known biological importance of both modifications, much less is known about the interactions between them. A few reports have shown that changes in N-glycosylation in cells affect cellular phosphorylation signalling [13–15]. For example, glycosylation influences protein phosphorylation events by modulating the stability of receptor tyrosine kinases [13].

Phosphopeptides and N-glycopeptides are low abundant in complex samples. Despite recent advances in mass spectrometry-based PTM analysis, efficient enrichment of phosphopeptides and N-glycopeptides remains necessary in a complex proteome. For phosphopeptide enrichment, approaches such as metal oxide affinity chromatography (MOAC [16–19]) and immobilized metal ion affinity chromatography (IMAC [20–23]) are widely used. The enrichment is performed at low pH and with a high concentration of organic solvent, to prevent deprotonation of the carboxyl groups of aspartic and glutamic acids and of the peptide C terminus. Under these conditions, phosphate groups can still bind to the metal oxide or metal ion on the solid phase because of their different pK_a values. In contrast, glycopeptide enrichment utilizes either the affinity of glycan-binding proteins (or lectins) to specific glycans or chromatographic materials that can retain hydrophilic glycopeptides. Chromatographic techniques commonly used for this include hydrophilic interaction chromatography (HILIC [24, 25]), electrostatic repulsion/hydrophilic interaction chromatography (ERLIC [26, 27]), boronic acid chromatography [28], and combinations of hydrophilic, reversed-phase, and anion exchange chromatography (e.g., MAX [29]).

Although methods have been developed for simultaneous or sequential enrichment of phosphopeptides and glycopeptides [30–34], none of them can completely separate glyco- and phosphopeptides owing to their similar chemical properties. Recently, Cho et al. [35] optimized a two-step workflow that utilizes IMAC and MAX to enrich a substantial number of phospho- and glycopeptides from a complex proteomic sample. Here we present an alternative strategy that uses ZIC-HILIC [25, 36] and TiO_2 beads [37, 38] to capture N-glycopeptides and phosphopeptides, respectively. This workflow incorporates tandem mass tag (TMT) labelling and allows

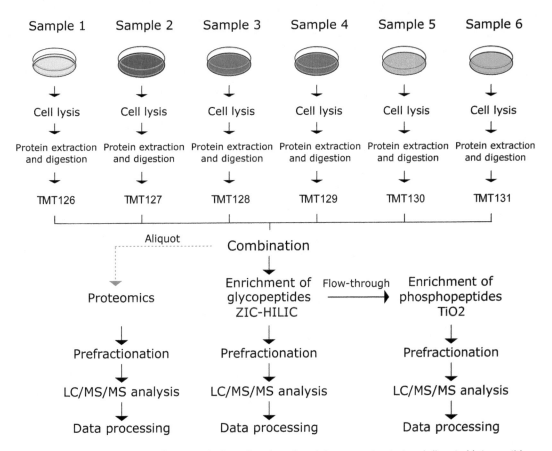

Fig. 1 Experimental workflow. Samples 1–6 are lysed, and proteins are extracted and digested into peptides. Each peptide sample is labelled with an individual isobaric tag. Equal peptide amounts from each sample are combined in a "pooled" sample. If required, an aliquot of the pooled sample can be used for a general proteomics workflow. The sample is subjected to glycopeptide enrichment using ZIC-HILIC beads. The flow-through (unbound fraction) is subjected further to phosphopeptide enrichment using TiO_2 beads. Glycopeptide, phosphopeptide, and general proteomics sample are individually prefractionated using basic reverse-phase chromatography. The obtained fractions are measured in a mass spectrometer, and acquired data are subjected to data processing tools

robust and time-efficient preparation of TMT-labelled phosphorylated and N-glycosylated peptides in a cell-wide manner. Specifically, this workflow, which is suitable for any cultured cells, comprises (1) cell lysis; (2) protein extraction and cleanup; (3) proteolysis; (4) TMT labelling; (5) glycopeptide enrichment; (6) phosphopeptide enrichment; (7) basic reversed-phase prefractionation; and (8) LC/MS/MS analysis (Fig. 1). We also note critical details of each step to assist reproduction of the workflow in all laboratories.

2   Materials

Unless otherwise noted, highly pure LC–MS grade water and solvents are used.

2.1   Laboratory Equipment

1. Ultrasonication device, e.g., Bioruptor (Diagenode).

2. Table-top centrifuge with cooling option, e.g., Heraeus™ Pico with 24 × 1.5/2.0 mL rotor (Thermo Fisher Scientific).

3. Magnetic rack with neodymium magnets suitable for 1.5/2 mL Eppendorf tubes, e.g., DynaMag-2 Magnet (Thermo Fisher Scientific).

4. Thermomixer (e.g., Eppendorf) for 1.5 or 2 mL tubes with shaking function.

5. Rotation wheel suitable for 2 mL Eppendorf tubes.

6. Vortex shaker.

7. Centrifugal vacuum concentrator, e.g., Savant SPD121P SpeedVac Concentrator (Thermo Fisher Scientific).

8. Universal pH indicator paper.

9. Ultrasonication bath, e.g., Sonorex (Bandelin).

10. HPLC system for off-line fractionation, e.g., Agilent 1100 series HPLC system with UV detector.

11. Reversed-phase chromatography HPLC column suitable for peptide fractionation under conditions of basic pH, e.g., Waters XBridge C18 column (3.5 μm particles, 1.0 mm inner diameter, 150 mm length).

12. Online UHPLC system coupled to a high-resolution mass spectrometer, e.g., Dionex UltiMate 3000 UHPLC system (Thermo Scientific) equipped with an in-house-packed C18 column (ReproSil-Pur 120 C18-AQ, 1.9 μm pore size, 75 μm inner diameter, 30 cm length, Dr. Maisch GmbH).

13. Quadrupole-linear ion trap-Orbitrap mass spectrometer, e.g., Orbitrap Fusion Tribrid mass spectrometer (Thermo Fisher Scientific), or alternatively high-resolution quadrupole-Orbitrap or quadrupole-time-of-flight mass spectrometers.

2.2   Cell Lysis and Protein Extraction and Digestion

1. Ethylenediaminetetraacetic acid (EDTA, Sigma).

2. 4-(2-Hydroxyethyl)-1-piperazineethanesulfonic acid (HEPES, Sigma).

3. Sodium dodecyl sulfate (SDS, Sigma).

4. 1× phosphate buffer (PBS) for cell cultures.

5. Lysis buffer, 4% (w/w) SDS, 100 mM HEPES, 1 mM EDTA pH 8.5.

6. Protease and phosphatase inhibitor cocktail, e.g., Halt™ Protease and Phosphatase Inhibitor Cocktail (Thermo Fisher Scientific).

7. Colorimetric assay for protein concentration determination with SDS tolerance up to 4%, e.g., Pierce™ BCA Protein Assay Kit (Thermo Fisher Scientific).

8. 0.5 M Tris(2-carboxethyl)phosphine (TCEP, Thermo Fisher Scientific).

9. 800 mM 2-chloroacetamide (CAA, Merck) in water.

10. Magnetic beads for protein aggregation capture and cleanup: e.g., Sera-Mag SpeedBeads with hydrophilic surface (GE Healthcare, cat. no 45152101010250, Magnetic Carboxylate Modified) and Sera-Mag SpeedBeads with hydrophobic surface (GE Healthcare, cat. no. 65152105050250, Magnetic Carboxylate Modified). Mix the beads at 1:1 ratio, wash three times with 2 volumes of water, and resuspend in water.

11. 80% (v/v) ethanol, freshly prepared. Use absolute ethanol (absolute ethanol for analysis, Merck) for preparation.

12. Trypsin (lyophilized, mass spectrometry grade), e.g., sequencing grade modified trypsin (Promega). Dissolve in a suitable amount of trypsin buffer according to the manufacturer's instructions. Prepare freshly.

13. 50 mM triethylammonium bicarbonate buffer (TEAB) in water. Prepare freshly from 1 M stock (Merck).

2.3 TMT Labelling Reaction

1. Isobaric reagents for peptide labelling, 0.8 mg per tag, e.g., TMT6plex or TMT10plex (Thermo Fisher Scientific).

2. 50 mM TEAB, prepared freshly from 1 M stock (Merck).

3. 5% (v/v) hydroxylamine solution, prepared freshly from 50% (v/v) hydroxylamine solution in water.

4. Trifluoroacetic acid, LC–MS grade (TFA, Thermo Fisher Scientific).

2.4 Glycopeptide Enrichment

1. Loading and washing buffer (75% (v/v) acetonitrile, 1% (v/v) TFA).

2. Zwitterionic hydrophilic interaction liquid chromatography 5 μm beads (ZIC-HILIC beads, Ultimate™ HILIC Amphion II, Welch).

3. Filter paper (a coffee filter can be used).

4. Disposable insulin syringe (see Note 1) with plastic plunger and a pipette tip adapter.

5. 200 μL plastic pipette tips.

2.5 Phosphopeptide Enrichment

1. Acetonitrile.

2. Triethylammonium bicarbonate buffer, 1 M (TEAB, Sigma).

3. Trifluoroacetic acid, LC–MS grade (TFA, Thermo Fisher Scientific).

4. Trifluoroethanol (TFE, Sigma).

5. Resuspension buffer, 100 mM TEAB containing 10% (v/v) TFE.

6. KCl, 3.2 M in water.

7. KH_2PO_4, 150 mM in water.

8. Loading buffer, 80% (v/v) acetonitrile, 6% TFA (v/v) in water.

9. Washing buffer, 60% (v/v) acetonitrile, 1% TFA (v/v) in water.

10. Transfer buffer, 80% (v/v) acetonitrile, 0.5% acetic acid.

11. Elution buffer, 40% (v/v) acetonitrile, 3.75% NH_4OH in water.

12. TiO_2 beads, 10 μm (GL Sciences).

13. Empty spin columns with 5–10 μm frit, e.g., empty Micro Spin Columns (Harvard Apparatus).

14. Disposable insulin syringe (*see* **Note 1**) with plastic plunger and a pipette tip adapter.

2.6 Peptide Desalting Using C18 Spin Columns

1. Spin column filled with C18 material and binding capacity up to 100 μg of peptides, e.g., Micro Spin Columns (Harvard Apparatus).

2. Conditioning buffer, 80% (v/v) acetonitrile, 0.1% (v/v) TFA in water.

3. Washing buffer, 2% (v/v) acetonitrile, 0.1% (v/v) TFA in water.

4. Elution buffer, 60% (v/v) acetonitrile in water.

2.7 High-pH Reversed-Phase Fractionation

1. Buffer A: 10 mM ammonium hydroxide in water, pH ~10.

2. Buffer B: 10 mM ammonium hydroxide and 80% (v/v) acetonitrile in water, pH ~10.

3. Sample loading buffer: 2% Buffer B plus 98% Buffer A (i.e., 1.6% (v/v) acetonitrile plus 10 mM NH_4OH in water, pH ~10).

2.8 LC/MS/MS Analysis

1. Sample loading buffer: 2% (v/v) acetonitrile and 0.1% (v/v) formic acid in water.

2. Mobile phase A: 0.1% (v/v) formic acid in water.

3. Mobile phase B: 80% (v/v) acetonitrile plus 0.08% (v/v) formic acid in water.

3 Methods

3.1 Cell Lysis and Protein Extraction and Digestion

This protocol is based on publications by Hughes et al. [39, 40] and Batth et al. [41].

1. Wash the cells of interest from the cell culture medium with PBS buffer, and remove the remaining buffer by aspiration. For our studies, we use 10 million cells per condition which corresponds to ~1 mg of extracted protein amount. Lyse the cells by adding lysis buffer. Adjust the volume of the lysis buffer based on the amount of material to lyse. As a guiding value, 200 μL can be used to lyse cell pellet corresponding to $5–10 \times 10^6$ cells or 20 mg of wet tissue sample (*see* **Notes 2** and **3**).

2. Sonicate the cell lysate for 10 min using 15 s on/15 s off cycle at the maximum output level.

3. Clear the lysate by centrifugation in a table-top centrifuge at $14,000 \times g$ for 15 min at RT.

4. Determine the protein concentration of the lysate using the BCA Protein Assay Kit (*see* **Note 4**).

5. Add TCEP and CAA simultaneously to reach final concentrations of 10 mM TCEP and 40 mM CAA. Incubate at 55 °C for 30 min to reduce protein disulfide bridges and to alkylate the reduced cysteine side chains.

6. Add the mixed magnetic beads at a beads-to-protein ratio of 10:1 (w/w) and the minimum required final bead concentration of 0.5 μg/μL (*see* **Note 5**).

7. Add acetonitrile to the protein lysate to reach a final concentration of at least 50% (v/v) (*see* **Note 6**).

8. Incubate samples for 10 min at room temperature (RT) in the thermomixer at 1000 rpm, and then place the sample tubes in the magnetic rack for 2 min at room temperature (RT).

9. Discard the liquid phase, and rinse the beads three times with 80% ethanol. For each wash, take the samples off the rack, and allow them to remain for 2 min at RT on the bench before placing them back in the magnetic rack.

10. After the last rinsing step, allow the beads to air-dry (~1 min), and resuspend them in a suitable amount of 50 mM triethylammonium bicarbonate buffer. 50–100 μL is a suitable volume for peptide recovery, as it allows beads' resuspension without excessive sample dilution.

11. Add sequencing grade trypsin at an enzyme-to-protein ratio of 1:20 to 1:50 (w/w) (*see* **Note 7**).

12. Incubate at 37 °C for 4 h in a thermomixer. When using 1.5 or 2 mL Eppendorf tubes, mix at 800 rpm. Adjust the rotation

speed to allow gentle mixing of the bead suspension. It is not recommended to use speeds above 1000 rpm, as liquid splashes might cause sample loss at the tube walls.

13. Spin down the beads by short centrifugation (~5 s). Place the samples on a magnetic rack, let the beads aggregate on the tube wall under the magnetic force for 2 min, and then transfer the supernatant into a fresh vessel.

3.2 TMT Labelling Reaction

Consult Zecha et al. [42] for an optimized use of TMT reagents.

1. Warm up the TMT reagents to RT.

2. Use acetonitrile to dissolve TMT reagents. Add 41 μL of acetonitrile per 0.8 mg of the labelling reagent (*see* **Note 8**).

3. Add labelling reagent to the sample. The smallest recommended ratio of labelling reagent to initial protein amount is 1:1 (w/w) [42]. We recommend increasing this ratio to 2:1 (using a single TMT vial (0.8 mg), one can achieve sufficient labelling of up to 400–800 μg of peptides). Adjust the estimated concentrations of peptide and labelling reagent to approx. 2 mg/mL and 10 mM, respectively. Dilute the sample using 50 mM TEAB, or concentrate the sample in a vacuum concentrator if required. Check the pH of the reaction mixture by spotting a small droplet onto a piece of pH indicator paper. The pH should remain around 8 (*see* **Note 9**).

4. Incubate the sample at RT for 1 h.

5. Quench the labelling reaction by adding 5% hydroxylamine to give a final concentration of 1%. Incubate for further 15 min (*see* **Note 10**).

6. Combine the samples labelled with different TMT reagents (*aka* channels). Take the same peptide amount (refer to the starting protein amount) from each sample for pooling.

7. Add 10% trifluoroacetic acid to give a final TFA concentration of 1%.

8. Dry the samples in a vacuum concentrator (*see* **Note 11**).

3.3 Glycopeptide Enrichment

1. Redissolve the dried samples in the loading buffer (75% acetonitrile (v/v) and 1% TFA (v/v)) to maintain the estimated peptide concentration at 2–5 mg/mL (*see* **Note 12**).

2. Weigh out ZIC-HILIC beads at a peptide-to-beads ratio of 1:50 (w/w).

3. Wash the beads three times with the loading buffer (*see* **Note 13**).

4. Insert a piece of filter paper (e.g., coffee filter) into a 200 μL plastic pipette tip.

5. Load the ZIC-HILIC beads suspension onto the pipette tip with the filter paper. The ZIC-HILIC beads should be retained by the filter paper in the tip and form a small chromatographic column.

6. Load the sample and press the liquid through the column using a syringe (*see* **Note 1**) or by centrifugation. When using a centrifuge, adjust the speed so that the liquid passes through the column within 2–3 min and a thin liquid layer is left above the column material at the end of the centrifugation.

7. Reload the flow-through, and press it through the column as before. Repeat this step four times in all.

8. Collect the flow-through, and evaporate it to dryness in a vacuum concentrator (*see* **Note 11**). Store at −20 °C until it is used for phosphopeptide enrichment.

9. After the last loading step, wash the column three times using loading buffer. In each washing step, fill the pipette tip almost completely (~200 μL), and slowly press the liquid through the column using a syringe (*see* **Note 1**) or a table-top centrifuge at 4500 rpm. Make sure that the liquid passes through the beads within 2–3 min.

10. Elute the glycopeptides retained on the column by applying 100 μL of 0.1% (v/v) TFA in water. Collect the eluate in a fresh tube. Repeat the elution step and combine the eluates (*see* **Note 14**).

11. Evaporate the eluates to dryness in a centrifugal vacuum concentrator device (*see* **Note 11**). Store the dried glycopeptides at −20 °C before use.

3.4 Phosphopeptide Enrichment

This protocol is based on publications by Humphrey et al. [37, 38].

1. Redissolve the dried unbound fraction (flow-through) after glycopeptide enrichment in 500 μL of 100 mM TEAB 10% (v/v) TFE buffer. Sonicate briefly in a sonication bath (1–3 min). If the estimated peptide amount is lesser than 1 mg, use 200 μL of the buffer, and reduce other volumes accordingly.

2. Add 75 μL of 3.2 M KCl, 27.5 μL of 150 mM KH_2PO_4, 400 μL of acetonitrile, and 47.5 μL of TFA to reach final concentrations of 228 mM KCl, 3.9 mM KH_2PO_4, 38% (v/v) acetonitrile, and 4.5% (v/v) TFA. The final volume should add up to ~1050 μL.

3. Mix using a vortex shaker, and clear the mixture by centrifugation at maximum speed (14,000 × g) in a table-top centrifuge for 15 min. Transfer the supernatant to a fresh 2 mL Eppendorf tube.

4. Prepare TiO$_2$ beads:

   (a) Weigh out the beads at a beads-to-protein ratio of 10:1 (w/w).

   (b) Wash three times with 400 μL of loading buffer.

   (c) Discard the liquid phase, and resuspend the beads in 100 μL of loading buffer per sample.

   (d) Sonicate in a sonication bath (3 min).

5. Add TiO$_2$ bead suspension to the cleared sample (*see* **Note 15**).

6. Incubate at 40 °C for 20 min in a thermomixer with shaking at 1000 rpm. Adjust the shaking speed to keep the beads suspended and evenly distributed in the solution.

7. Centrifuge the sample in a table-top centrifuge for 2 min at 2000 rpm at RT, and then transfer the liquid phase (flow-through) into a fresh tube. Use an aliquot of the flow-through for a general proteome analysis (dry in a vacuum concentrator to remove acetonitrile and TFA, dissolve in water, and remove salts using spin columns with C18 material).

8. Resuspend TiO$_2$ beads in 800–1000 μL of washing buffer.

9. Incubate for 2 min at RT on a rotation wheel.

10. Centrifuge for 2 min at 2000 rpm at RT; discard the liquid phase.

11. Repeat **steps 8, 9**, and **10** another three times, respectively.

12. After the final wash, add 100 μL of transfer buffer.

13. Gently resuspend the TiO$_2$ beads by pipetting up and down. Use a pipette tip with a wider opening (e.g., 1000 μL tips), or pre-cut a smaller pipette tip accordingly.

14. Load the TiO$_2$ beads onto an empty spin column with a 5 μm frit. Alternatively, a 200 μL pipette tip with filter paper as described above (Subheading 3.3, **step 5**) can be used.

15. Make sure that the TiO$_2$ beads are completely transferred to the spin column by rinsing the tube with another 100 μL of transfer buffer.

16. Centrifuge at 500 rpm for 1 min. Reload the beads that have passed through the filter, and repeat the centrifugation step. Make sure that all the beads are retained by the filter in the spin column.

17. Repeat the centrifugation step, and discard the flow-through (transfer buffer).

18. Elute using 40 μL of elution buffer.

19. Centrifuge at 500 rpm for 1 min at 4 °C. Collect the flow-through.

20. Repeat the elution step once more.

21. Use a syringe (*see* **Note 1**) to press out the remaining liquid carefully.

22. Immediately acidify the collected flow-through using 10% (v/v) TFA in water (*see* **Note 16**). Check the pH of the solution by spotting a small amount onto pH indicator paper. Adjust the pH to 4–6.

23. Concentrate the sample in a centrifugal vacuum concentrator device to remove acetonitrile and trifluoroacetic acid (*see* **Notes 11** and **17**).

24. Proceed to the desalting step (Subheading 3.5) (*see* **Note 18**).

3.5 Peptide Desalting Using C18 Spin Columns (See Notes 19 and 20)

1. Condition the C18 material by applying 400 μL of acetonitrile (this fills the tip almost completely); let it run through under gravity.

2. Refill the column with another portion of acetonitrile, and use a centrifuge to facilitate the column flow: adjust the centrifugation speed to allow at least 2 min for interaction between the liquid and solid phases. Importantly, never let the column material dry, and always leave a liquid layer above the column material (*see* **Notes 21** and **22**).

3. Repeat the conditioning using 80% acetonitrile (v/v), 0.1% TFA (v/v) in water.

4. Equilibrate the column three times using 400 μL of 2% acetonitrile (v/v), 0.1% TFA (v/v) in water. Increase the centrifugation speed to compensate for the water's viscosity (~800–1000 rpm).

5. Dissolve or dilute the sample in 0.1% TFA in water.

6. Load the sample onto the column, and adjust the centrifugation speed to allow the sample to interact with the C18 material for at least 2 min.

7. Repeat the previous step by reloading the flow-through.

8. Wash the column three times using 2% acetonitrile (v/v), 0.1% TFA (v/v) in water.

9. Elute the peptides retained on the C18 material by applying 150 μL of 60% (v/v) acetonitrile in water. Collect the eluate into a fresh tube.

10. Repeat the elution step. Expel the remaining liquid with a syringe (*see* **Note 1**), and combine the collected eluates.

11. Dry the eluate in a centrifugal vacuum concentrator (*see* **Note 11**).

3.6 Basic Reversed-Phase Fractionation

Off-line fractionation using reversed-phase chromatography under basic conditions is a non-mandatory step. However, given its high resolution and orthogonality to reversed-phase chromatography under acidic conditions, it can significantly improve the depth of the analysis. For a more detailed description and further reading, please see earlier publications [43–45].

1. Prepare and degas mobile phases for off-line high-pH reversed-phase chromatography. Buffer A: 10 mM ammonium hydroxide in water, pH 10. Buffer B: 10 mM ammonium hydroxide 80% (v/v) acetonitrile in water, pH 10 (see **Note 23**). Sonicate buffers in a sonication bath to remove excess gas content.

2. Fill and purge the HPLC system with freshly prepared buffers. If the system shows significant pressure fluctuations while purging, repeat purging until the pressure has stabilized.

3. Equilibrate a C18 column with a mobile-phase mixture of 2% buffer B and 98% buffer A. Follow the manufacturer's specifications for optimum column flow and pressure. If a new column is being used, additional steps may be needed to prepare the column for first-time use.

4. Dissolve the enriched and dried phosphopeptides or glycopeptides in a suitable volume of 2% buffer B/98% buffer A mixture. Sonicate the sample briefly in a sonication bath (1–3 min) to facilitate peptide solubilization. When planning the volume in which to resuspend the sample, consider the volume of the sample loop of the column. For example, if the installed sample loop volume is 50 μL, one should not inject more than 45 μL; otherwise one will risk losing a part of the sample (see **Note 24**).

5. Inject the sample. For elution, apply the following gradient at a flow rate of 60 μL/min (adjust the flow rate and the gradient accordingly, depending on the sample and the chromatography system used in order to achieve better chromatographic separation): loading at 2% buffer B for 5 min and linear increase to 34% B over 37 min, to 60% B over 8 min, and to 90% B over 1 min. Then wash at 90% B for 5.5 min, return to 2% B for 0.5 min, and allow equilibration at 2% B for 7 min (64 min total run time) (see **Note 25**).

6. Collect fractions every minute starting from minute 7 till minute 57 (adjust the time for collecting fractions according to the gradient used and the chromatogram; see **Note 26**). Pool the fractions collected into 12 major fractions using the following scheme: major fraction 1 = 1st + 13th + 25th + 37th + ... primary fractions; major fraction 2 = 2nd + 14th + 26th + ..., 3 = 3rd + 15th + 27th + ...; etc. [43] (see **Note 27**).

7. Evaporate the major fractions to dryness in a vacuum concentrator (see **Note 11**). Store at −20 °C before use.

3.7 LC/MS/MS Analysis

The dried samples are dissolved in sample loading buffer and loaded directly onto the LC–MS system (*see* **Note 28**). The LC settings should be optimized in advance for TMT-labelled phosphopeptides and N-glycopeptides (*see* **Note 29**). Peptides and phosphopeptides are first concentrated in a trap column and then separated by capillary column chromatography using a 60–90 min gradient. For glycopeptide analysis, we omit the trap column and use a prolonged 3 h gradient. As for regular LC/MS/MS analysis, a common top-N data-dependent acquisition method can be used [46, 47] (*see* **Notes 30** and **31**). Alternatively, an MS3 method with synchronous precursor selection (SPS-MS3) developed by McAlister et al. [48] has shown improved quantitative accuracy in multiplexed quantification using isobaric tags (*see* **Note 32**) [49]. SPS-MS3 has also been optimized for quantitative phosphoproteomics [47, 50].

3.8 Data Processing

Conventional protein database search engines, including *Andromeda* [51], *Mascot* [52], *Sequest* [53], and *Comet* [54], are capable of phosphopeptide identification. In either case, one has to specify the TMT labelling as a fixed modification on peptide N-termini and lysine residues and place the phosphorylation on serine, threonine, and tyrosine (or other amino acid residues in special applications [55]) as variable modifications (*see* **Notes 33** and **34**). For the analysis of N-glycoproteomics data, there are a number of freely available and commercial software packages; for a review, please consult [56]. For example, *pGlyco 2.0* [57] and *GPQuest* [58] are freely available software solutions, while *Byonic* (Protein Metrics) [59] is a wildly used commercial tool and can be integrated into the Proteome Discoverer platform (Thermo Fisher Scientific) (*see* **Note 35**).

4 Notes

1. When using a syringe to facilitate the flow-through of a spin column (glycopeptide enrichment, C18 desalting protocols), it is advisable to use a syringe with a plastic plunger without a rubber tip, since the rubber part might give rise to contamination, such as nylon peaks, that produces contaminant signals (e.g., m/z 453) in the mass spectrometer. Small syringes for the insulin injections are normally sufficient.

2. Test the lysis procedure on the cells/tissue of interest. Depending on the material, it might require the use of homogenizers or mechanical shredding for tissues. Adjust lysis buffer volume according to the sample size.

3. Because of the downstream TMT labelling, avoid buffers containing primary amines (urea, Tris, etc.). Despite the protein

cleanup, remaining substances can interfere with the labelling procedure.

4. When estimating protein concentration, make sure that the assay is compatible with the reagents used in the lysis buffer. According to the manufacturer's instructions, Pierce BCA Protein Assay Kit can tolerate SDS concentrations up to 5%. The assay is incompatible with reducing reagents, unless a reducing agent-compatible modification of the BCA Assay is used. In this case, one can add TCEP and CAA directly to the lysis buffer.

5. According to publications by Hughes et al. [39, 40], other types of beads can be used for the protein cleanup as well. According to Batth et al. [41], proteins can be captured on microparticles independent of the surface chemistry of the microparticles.

6. After adding the acetonitrile to the beads, it is advisable to not resuspend the beads by using a pipette tip as the beads can stick to the wall of the tip.

7. If tryptic digestion for 4 h is not sufficient, increase the sample incubation time or use a LysC/Trypsin mixture.

8. TMT is highly reactive and is quickly hydrolyzed and deactivated in aqueous solutions. Dissolve TMT using water-free acetonitrile, and avoid contamination with primary amines. Although this is not advised by the manufacturer, dissolved TMT reagent can be aliquoted and stored at $-80\ ^\circ$C for several months.

9. A basic pH facilitates the labelling reaction.

10. It is advisable to check labelling efficiency by injecting a small part of the desalted sample into the mass spectrometer and allowing TMT as a variable modification for the peptide search. The percentage of the peptides carrying a TMT modification among all peptides identified by the search engine should be above 95%.

11. In our experience snap-freezing of the sample in the liquid nitrogen often facilitates the process of vacuum drying.

12. Clear the peptide sample from any undissolved particles by centrifugation before loading it on the beads.

13. Prevent complete drying of the beads during each loading or washing step.

14. At the elution step, try to press out and collect all remaining buffer in the column by increasing centrifuge force or duration.

15. TiO_2 beads precipitate quickly. Resuspend the beads thoroughly before any step in which beads are transferred. Ensure that the beads do not precipitate during the incubation step.

16. Minimize peptide-handling time at basic pH, since failure to do this can lead to hydrolysis of the phosphorylated moiety. Acidify the eluate with 10% TFA, work at 4 °C, or even snap-freeze the eluate in liquid nitrogen.

17. The desalting step is needed to remove excess ammonium salts. Concentration in a vacuum concentrator (*see* **Note 11**) before the desalting step is required to decrease the concentrations of acetonitrile and TFA, as these can otherwise interfere with peptide retention by the C18 material.

18. It is advisable to proceed with the phosphopeptide analysis as soon as possible, as phosphopeptides might degrade faster than non-modified peptides.

19. The protocol describes desalting procedure using Harvard Apparatus Micro Spin Columns. However, other solutions can equally well be employed. In this case, consider the column's binding capacity and follow the manufacturer's instructions for use of the column.

20. In our experience, the Harvard Apparatus Micro Spin Columns can be used to desalt peptides in the range of 10–100 μg.

21. Adjust the centrifugation speed to allow at least 2 min for interaction between the liquid and solid phases in each step. Commonly, we use 400–500 rpm for acetonitrile-based solutions (activation and elution steps) and 800–1000 rpm for water-based solutions (equilibration and column washing steps).

22. Importantly, never let the column material dry, and always leave a liquid layer above the column material, except for the last elution step.

23. Use pH indicator paper strips to check the pH of the mobile phase. Normally no additional pH adjustment is required.

24. If possible, use a small aliquot of the same sample or an aliquot of the general proteome to test the performance of the chromatography system, and adjust the gradient to achieve the best separation. Be aware that glycopeptides and phosphopeptides might appear more hydrophilic than their non-modified counterparts. TMT labelling often results in increased hydrophobicity of peptides.

25. Monitor elution of peptides using light absorption at 214 nm. Adjust the time frame for the sample collection accordingly.

26. Depending on the chromatographic separation and available sampling device, other fractionation schemes might be employed. Consider that the peptide amount collected per concatenated fraction should be suitable for injection into the mass spectrometer.

27. The idea of concatenated fractions is to combine peptides with different hydrophobicity within one fraction while keeping peptides with similar hydrophobicity in separate fractions. It can be achieved if primary fractions are combined with other primary fractions separated by a certain time window. In this case, neighboring primary fractions contribute to different final concatenated fractions [45].

28. The SPS-MS3 method requires the availability of a hybrid quadrupole-linear ion trap-Orbitrap mass spectrometer capable of acquiring MS3 scans, such as Orbitrap Fusion or Orbitrap Lumos (Thermo Fisher Scientific). However, the analysis can also be carried out on the quadrupole-Orbitrap or quadrupole-time-of-flight mass spectrometers without MS3 mode by using MS2 product ion scans only. The latter might result in greater sensitivity because of the shorter duty cycle. However, SPS-MS3 provides more accurate quantification results because it minimizes precursor co-fragmentation [49].

29. It is advisable to make a test injection with a small amount of sample because one tends to underestimate the intensity of the fractionated sample. Dilute the sample if needed, and adjust the LC gradient to achieve the best separation.

30. Use of a trapping column might be justified when one is analyzing phosphopeptide samples, as it can significantly decrease the loading time and prolong the lifetime of the analytical column. However, especially for glycopeptides, it might result in the loss of hydrophilic peptide species.

31. Consider that reporter ions from the TMT10plex kit require higher resolution—up to 60,000 at 200 m/z in MS2 or MS3 scan (depending on which scan is used for quantification). Reporter ions from TMT6plex can be normally resolved when one is using 15,000 resolution at 200 m/z.

32. For MS2-based quantification, elevated collision energy is recommended as compared with conventional methods to improve the signal intensity of reporter ions. Also, a narrower precursor isolation window helps to reduce precursor co-fragmentation.

33. It is advisable to use reporter ion intensities from MS2 scans whose precursors contributed to at least 75% of the total intensities in the isolation window.

34. Check the specificity of the enrichment: compare the number of identified modified peptides with the total number of identified peptides by the search engine. For phosphopeptides, a good specificity value would lie above 80%. Check the washing steps and the beads-to-peptide ratio if the specificity is significantly lower.

35. Depending on the experimental setup, it may be advisable, or even necessary, to correct for possible changes in protein expression under the various conditions tested. Use either an aliquot of the initial sample after TMT labelling before PTM enrichment or an aliquot of the flow-through after the enrichment step. Use a single-shot injection or prefractionated sample, depending on its complexity. Correct intensities of modified peptides for observed changes in protein abundances.

References

1. Bairoch A, Apweiler R (2000) The SWISS-PROT protein sequence database and its supplement TrEMBL in 2000. Nucleic Acids Res 28(1):45–48

2. Khoury GA, Baliban RC, Floudas CA (2011) Proteome-wide post-translational modification statistics: frequency analysis and curation of the swiss-prot database. Sci Rep 1:90

3. Ardito F, Giuliani M, Perrone D, Troiano G et al (2017) The crucial role of protein phosphorylation in cell signaling and its use as targeted therapy. Int J Mol Med 40(2):271–280

4. Hanahan D, Weinberg RA (2011) Hallmarks of cancer: the next generation. Cell 144 (5):646,674

5. Varki A, Lowe JB (2009) Biological roles of glycans. In: Varki A (ed) Essentials of glycobiology, 2nd edn. Cold Spring Harbor Laboratory Press, Cold Spring Harbor (NY)

6. Stowell SR, Ju T, Cummings RD (2015) Protein glycosylation in cancer. Annu Rev Pathol 10:473–510

7. Thanabalasingham G, Huffman JE, Kattla JJ et al (2013) Mutations in HNF1A result in marked alterations of plasma glycan profile. Diabetes 62(4):1329–1337

8. Vučković F, Krištić J, Gudelj I et al (2015) Association of systemic lupus erythematosus with decreased immunosuppressive potential of the IgG glycome. Arthritis Rheumatol 67:2978–2989

9. Trbojević-Akmačić I, Ventham NT, Theodoratou E et al (2015) Inflammatory bowel disease associates with proinflammatory potential of the immunoglobulin G glycome. Inflamm Bowel Dis 21:1237–1247

10. Parekh RB, Dwek RA, Sutton BJ et al (1985) Association of rheumatoid arthritis and primary osteoarthritis with changes in the glycosylation pattern of total serum IgG. Nature 316:452–457

11. Hwang H, Zhang J, Chung KA et al (2010) Glycoproteomics in neurodegenerative diseases. Mass Spectrom Rev 29:79–125

12. Lauc G, Pezer M, Rudan I et al (2016) Mechanisms of disease: the human N-glycome. Biochim Biophys Acta 1860 (8):1574–1582

13. Itkonen HM, Mills IG (2013) N-linked glycosylation supports cross-talk between receptor tyrosine kinases and androgen receptor. PLoS One 8(5):e65016

14. Carrascal MA, Silva M, Ramalho JS et al (2018) Inhibition of fucosylation in human invasive ductal carcinoma reduces E-selectin ligand expression, cell proliferation, and ERK 1/2 and p38 MAPK activation. Mol Oncol 12 (5):579–593

15. Li CW, Lim SO, Xia W et al (2016) Glycosylation and stabilization of programmed death ligand-1 suppresses T-cell activity. Nat Commun 7:12,632

16. Larsen MR, Thingholm TE, Jensen ON et al (2005) Highly selective enrichment of phosphorylated peptides from peptide mixtures using titanium dioxide microcolumns. Mol Cell Proteomics 4(7):873–886

17. Liang SS, Makamba H, Huang SY et al (2006) Nano-titanium dioxide composites for the enrichment of phosphopeptides. J Chromatogr A 1116(1–2):38–45

18. Sugiyama N, Masuda T, Shinoda K et al (2007) Phosphopeptide enrichment by aliphatic hydroxy acid-modified metal oxide chromatography for nano-LC-MS/MS in proteomics applications. Mol Cell Proteomics 6 (6):1103–1109

19. Thingholm TE, Jørgensen TJ, Jensen ON et al (2006) Highly selective enrichment of phosphorylated peptides using titanium dioxide. Nat Protoc 1(4):1929

20. Posewitz MC, Tempst P (1999) Immobilized gallium (III) affinity chromatography of phosphopeptides. Anal Chem 71(14):2883–2892

21. Andersson L, Porath J (1986) Isolation of phosphoproteins by immobilized metal (Fe3 +) affinity chromatography. Anal Biochem 154(1):250–254

22. Villén J, Gygi SP (2008) The SCX/IMAC enrichment approach for global phosphorylation analysis by mass spectrometry. Nat Protoc 3(10):1630

23. Zhou H, Ye M, Dong J et al (2013) Robust phosphoproteome enrichment using monodisperse microsphere-based immobilized titanium (IV) ion affinity chromatography. Nat Protoc 8(3):461

24. Ruhaak LR, Huhn C, Waterreus WJ et al (2008) Hydrophilic interaction chromatography-based high-throughput sample preparation method for N-glycan analysis from total human plasma glycoproteins. Anal Chem 80(15):6119–6126

25. Mysling S, Palmisano G, Højrup P et al (2010) Utilizing ion-pairing hydrophilic interaction chromatography solid phase extraction for efficient glycopeptide enrichment in glycoproteomics. Anal Chem 82(13):5598–5609

26. Zhang H, Guo T, Li X et al (2010) Simultaneous characterization of glyco-and phosphoproteomes of mouse brain membrane proteome with electrostatic repulsion hydrophilic interaction chromatography. Mol Cell Proteomics 9(4):635–647

27. Alpert AJ (2008) Electrostatic repulsion hydrophilic interaction chromatography for isocratic separation of charged solutes and selective isolation of phosphopeptides. Anal Chem 80(1):62–76

28. Xu Y, Wu Z, Zhang L et al (2008) Highly specific enrichment of glycopeptides using boronic acid-functionalized mesoporous silica. Anal Chem 81(1):503–508

29. Yang W, Shah P, Hu Y et al (2017) Comparison of enrichment methods for intact N-and O-linked glycopeptides using strong anion exchange and hydrophilic interaction liquid chromatography. Anal Chem 89(21):11,193–11,197

30. Yan J, Li X, Yu L et al (2010) Selective enrichment of glycopeptides/phosphopeptides using porous titania microspheres. Chem Commun 46(30):5488–5490

31. Palmisano G, Parker BL, Engholm-Keller K et al (2012) A novel method for the simultaneous enrichment, identification, and quantification of phosphopeptides and sialylated glycopeptides applied to a temporal profile of mouse brain development. Mol Cell Proteomics 11(11):1191–1202

32. Zhang Y, Wang H, Lu H (2013) Sequential selective enrichment of phosphopeptides and glycopeptides using amine-functionalized magnetic nanoparticles. Mol BioSyst 9(3):492–500

33. Xie Y, Deng C (2017) Designed synthesis of a "One for Two" hydrophilic magnetic amino-functionalized metal-organic framework for highly efficient enrichment of glycopeptides and phosphopeptides. Sci Rep 7(1):1162

34. Zou X, Jie J, Yang B (2017) Single-step enrichment of N-Glycopeptides and Phosphopeptides with novel multifunctional Ti4+−immobilized dendritic polyglycerol coated chitosan nanomaterials. Anal Chem 89(14):7520–7526

35. Cho KC, Chen L, Hu Y et al (2018) Developing workflow for simultaneous analyses of phosphopeptides and glycopeptides. ACS Chem Biol 14(1):58–66

36. Boersema PJ, Divecha N, Heck AJ et al (2007) Evaluation and optimization of ZIC-HILIC-RP as an alternative MudPIT strategy. J Proteome Res 6(3):937–946

37. Humphrey SJ, Azimifar SB, Mann M (2015) High-throughput phosphoproteomics reveals in vivo insulin signaling dynamics. Nat Biotechnol 33(9):990

38. Humphrey SJ, Karayel O, James DE et al (2018) High-throughput and high-sensitivity phosphoproteomics with the EasyPhos platform. Nat Protoc 13(9):1897

39. Hughes CS, Foehr S, Garfield DA et al (2014) Ultrasensitive proteome analysis using paramagnetic bead technology. Mol Syst Biol 10(10):757

40. Hughes CS, Moggridge S, Müller T et al (2019) Single-pot, solid-phase-enhanced sample preparation for proteomics experiments. Nat Protoc 14(1):68

41. Batth TS, Tollenaere MA, Rüther P et al (2019) Protein aggregation capture on microparticles enables multipurpose proteomics sample preparation. Mol Cell Proteomics 18(5):1027–1035

42. Zecha J, Satpathy S, Kanashova T et al (2019) TMT labeling for the masses: a robust and cost-efficient, in-solution labeling approach. Mol Cell Proteomics 18(7):1468–1478

43. Wang Y, Yang F, Gritsenko MA et al (2011) Reversed-phase chromatography with multiple fraction concatenation strategy for proteome profiling of human MCF10A cells. Proteomics 11(10):2019–2026

44. Yang F, Shen Y, Camp DG et al (2012) High-pH reversed-phase chromatography with fraction concatenation for 2D proteomic analysis. Expert Rev Proteomics 9(2):129–134

45. Song C, Ye M, Han G et al (2010) Reversed-phase-reversed-phase liquid chromatography approach with high orthogonality for

multidimensional separation of phosphopeptides. Anal Chem 82(1):53–56

46. Stadlmann J, Taubenschmid J, Wenzel D et al (2017) Comparative glycoproteomics of stem cells identifies new players in ricin toxicity. Nature 549(7673):538–542

47. Hogrebe A, von Stechow L, Bekker-Jensen DB et al (2018) Benchmarking common quantification strategies for large-scale phosphoproteomics. Nat Commun 9(1):1045

48. McAlister GC, Nusinow DP, Jedrychowski MP et al (2014) MultiNotch MS3 enables accurate, sensitive, and multiplexed detection of differential expression across cancer cell line proteomes. Anal Chem 86(14):7150–7158

49. Fang P, Ji Y, Silbern I et al (2020) A streamlined pipeline for multiplexed quantitative site-specific N-glycoproteomics. Nat Commun 11 (1):5268

50. Erickson BK, Jedrychowski MP, McAlister GC et al (2015) Evaluating multiplexed quantitative phosphopeptide analysis on a hybrid quadrupole mass filter/linear ion trap/orbitrap mass spectrometer. Anal Chem 87 (2):1241–1249

51. Cox J, Neuhauser N, Michalski A et al (2011) Andromeda: a peptide search engine integrated into the MaxQuant environment. J Proteome Res 10(4):1794–1805

52. Perkins DN, Pappin DJ, Creasy DM et al (1999) Probability-based protein identification by searching sequence databases using mass spectrometry data. Electrophoresis 20 (18):3551–3567

53. Eng JK, McCormack AL, Yates JR (1994) An approach to correlate tandem mass spectral data of peptides with amino acid sequences in a protein database. J Am Soc Mass Spectrom 5 (11):976–989

54. Eng JK, Jahan TA, Hoopmann MR (2013) Comet: an open-source MS/MS sequence database search tool. Proteomics 13(1):22–24

55. Potel CM, Lin MH, Heck AJ et al (2018) Widespread bacterial protein histidine phosphorylation revealed by mass spectrometry-based proteomics. Nat Methods 15(3):187

56. Hu H, Khatri K, Zaia J (2017) Algorithms and design strategies towards automated glycoproteomics analysis. Mass Spectrom Rev 36 (4):475–498

57. Liu MQ, Zeng WF, Fang P et al (2017) pGlyco 2.0 enables precision N-glycoproteomics with comprehensive quality control and one-step mass spectrometry for intact glycopeptide identification. Nat Commun 8(1):438

58. Toghi Eshghi S, Shah P et al (2015) GPQuest: a spectral library matching algorithm for site-specific assignment of tandem mass spectra to intact N-glycopeptides. Anal Chem 87 (10):5181–5188

59. Bern M, Kil YJ, Becker C (2012) Byonic: advanced peptide and protein identification software. Curr Protoc Bioinformatics 40 (1):13–20

Chapter 15

High-Throughput Profiling of Proteome and Posttranslational Modifications by 16-Plex TMT Labeling and Mass Spectrometry

Kaiwen Yu, Zhen Wang, Zhiping Wu, Haiyan Tan, Ashutosh Mishra, and Junmin Peng

Abstract

Mass spectrometry (MS)-based proteomic profiling of whole proteome and protein posttranslational modifications (PTMs) is a powerful technology to measure the dynamics of proteome with high throughput and deep coverage. The reproducibility of quantification benefits not only from the fascinating developments in high-performance liquid chromatography (LC) and high-resolution MS with enhanced scan rates but also from the invention of multiplexed isotopic labeling strategies, such as the tandem mass tags (TMT). In this chapter, we introduce a 16-plex TMT-LC/LC-MS/MS protocol for proteomic profiling of biological and clinical samples. The protocol includes protein extraction, enzymatic digestion, PTM peptide enrichment, TMT labeling, and two-dimensional reverse-phase liquid chromatography fractionation coupled with tandem mass spectrometry (MS/MS) analysis, followed by computational data processing. In general, more than 10,000 proteins and tens of thousands of PTM sites (e.g., phosphorylation and ubiquitination) can be confidently quantified. This protocol provides a general protein measurement tool, enabling the dissection of protein dysregulation in any biological samples and human diseases.

Key words Mass spectrometry, Proteomics, Proteome, Posttranslational modifications, Phosphorylation, Ubiquitination, Ubiquitin, Tandem mass tag, Liquid chromatography, Database

1 Introduction

Mass spectrometry has become a versatile tool for analyzing proteins from small-scale assays to genome-wide studies, including protein identification, quantification, and posttranslational modifications, protein interaction networks, and systems biology of cells and tissues [1–4]. Although the top-down approach of directly analyzing intact proteins is dramatically improved [5], the

Kaiwen Yu and Zhen Wang are the co-first authors.

Katrin Marcus et al. (eds.), *Quantitative Methods in Proteomics*, Methods in Molecular Biology, vol. 2228,
https://doi.org/10.1007/978-1-0716-1024-4_15, © Springer Science+Business Media, LLC, part of Springer Nature 2021

bottom-up approach of analyzing proteolytic peptides is a sensitive and robust method with high proteome coverage [1, 6]. During the last few years, implementation of enhanced two-dimensional liquid chromatography and high-resolution mass spectrometry has dramatically improved the bottom-up proteomics platform, allowing proteome analysis at a scale comparable to transcriptomics analysis [7, 8]. Deep posttranslational modification (PTM) analysis (e.g., phosphoproteome and ubiquitinome) of mammalian tissues has also been accomplished [9, 10] (*see* also Chaps. 8, 13, 14, 16). Maturation of bottom-up proteomics enables a wide range of applications to profile disease tissues, such as cancer [11–14] and Alzheimer's disease (AD) [15, 16], to gain fundamental insights into molecular pathogenesis for novel therapeutic strategies.

In addition to protein identification, a number of MS-based techniques have been developed for quantitative untargeted proteomics. Label-free quantification can be achieved by data-dependent acquisition (DDA) methods (*see* Chaps. 8–10, 13–17, 20, 21, 26), including the comparison of spectral counts [17] or signal intensities of precursor ions [18], as well as the data-independent acquisition (DIA) method [19] (*see* Chaps. 8, 16, 22–24), such as SWATH-MS-based comparison of signal intensities of product ions [20]. The label-free method is straightforward, affordable, and capable of generating reproducible data with one-dimensional LC-MS/MS, especially when sample processing is highly automated [21]; however, the method variation increases with two-dimensional LC fractionation of unlabeled peptides. To overcome the limitation, stable isotope labeling methods have been introduced to differentially label proteins or digested peptides [2], followed by pooling prior to LC fractionation to reduce experimental variation. The chemical labeling of proteins/peptides can occur in vivo by Stable Isotope Labeling with Amino acids in Cell culture (SILAC) [2] (*see* Chaps. 8, 13, 18), or in vitro by tagging proteolytic peptides (*see* Chaps. 8–10, 12, 14, 15, 17, 25, 26, 28), using TMT [22], Isobaric Tags for Relative and Absolute Quantitation (iTRAQ) [23], and DiLeu labeling methods [24]. SILAC metabolic labeling process, however, can also introduce experimental variation and cannot be readily applied to clinical specimens. Instead, the TMT isobaric labeling method has emerged as a popular quantitative proteomics strategy [25] and is under continuous active development.

Like the existing TMT10/11-plex reagents, the newly released 16-plex TMT (TMTpro) reagents consist of a mass reporter, a mass balance group, and a reactive group [26] (Fig. 1). The digested peptides from different samples are differentially labeled with the TMT reagents, mixed, and then processed as a single sample for LC/LC-MS/MS analysis. The same peptides in different samples that carry different isobaric tags are chemically identical and non-distinguishable during LC separation and display the same

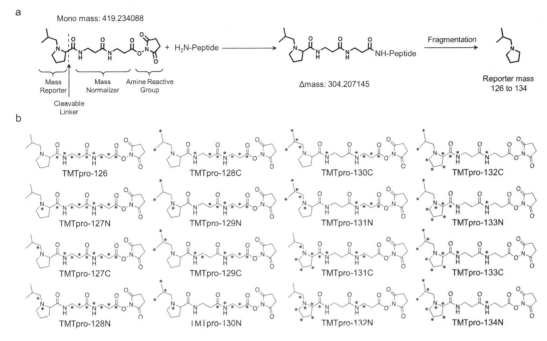

Fig. 1 Structures of the 16-plex TMT reagents. (**a**) Unlabeled structure of the 16-plex TMT reagent, the mass of the reagent, mass shift after labeling, and the mass of the reporter ion are shown. (**b**) Heavy isotope-labeled structures of the TMT16 reagents

peaks in the precursor ion scans by mass spectrometry. Importantly, during the fragmentation of precursor ions (e.g., MS2 or MS3), the reporter ions of these isobaric tags are generated to show different mass, and their intensities are used for quantification. This minimizes missing values in the same batch of up to 16 samples and allows precise quantification due to reduced experimental variation [27]. A well-known caveat of the TMT method is that the noise levels due to co-eluted interfering ions often lead to ratio compression, underestimating the protein difference, particularly in complex protein samples [28]. The distorted ratios can be rescued by multiple strategies including pre-MS extensive fractionation [29], MS settings with small MS2 isolation window [29] or gas-phase isolation [30], or post-MS corrections by subtracting interference [29, 31]. In addition, complement reporter ion clusters can be accurately quantified during MS analysis [32], while multistage MS3-based technique can almost eliminate the effect of ratio compression [28].

In this chapter, we introduce a detailed 16-plex TMT (TMT16 hereafter) method for profiling the whole proteome and the ubiquitinome, as an example of PTM analysis. Similar to other PTM profiling, ubiquitinome profiling often involves peptide-level enrichment by using an antibody recognizing the ubiquitin remnant motif di-GG tag on lysine residues of substrates after tryptic

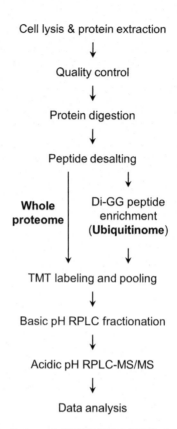

Fig. 2 Experimental design of TMT-LC/LC-MS/MS for profiling the whole proteome and ubiquitinome

digestion (K-ε-GG motif) [10, 33, 34], enabling measurements of over 10,000 ubiquitination sites in mammalian cells and tissues [35, 36]. The TMT-LC/LC-MS/MS method includes (in order) protein extraction from cells and tissues, protein digestion, peptide desalting, PTM enrichment, 16-plex TMT labeling, basic pH reverse-phase (RP) LC fractionation, and acidic pH RPLC-MS/MS (Fig. 2). Additionally, the computational data processing is discussed in detail for large-scale analysis (Fig. 3).

2   Materials

2.1   Protein Extraction from Cells and Tissues

1. Human postmortem brain tissues of normal control or Alzheimer's disease: frontal cortex with a well-characterized pathology record, stored at −80 °C. The samples are used as an example experiment in this method. Other tissues or cells may be processed similarly.

2. Lysis buffer: 8 M urea (*see* **Note 1**), 50 mM [4-(2-hydroxyethyl)-1-piperazineethanesulfonic acid] (HEPES), pH 8.5, 0.5% sodium deoxycholate, 1× PhosSTOP phosphatase

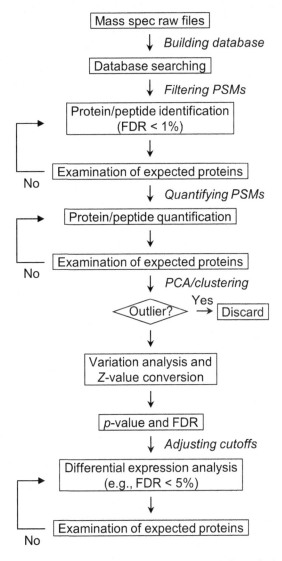

Fig. 3 Data processing strategies for large-scale proteomic analysis

inhibitor cocktail (Roche), and 1 mM dithiothreitol (DTT). For ubiquitinome analysis, DTT is replaced by 50 μM (final) PR-619 ((2,6-diamino-5-thiocyanatopyridin-3-yl) thiocyanate) and 10 mM (final) iodoacetamide (IAA) (*see* **Note 2**).

3. 0.5 mm diameter glass beads.

4. Bullet Blender (Next Advance).

5. Protein quantification standard: 20 mg/mL bovine serum albumin (BSA).

6. BCA Protein Assay Kit (Thermo Fisher Scientific).

7. NuPAGE Novex 4–12% Bis-Tris Gel (Life Technologies).

8. GelCode Blue Stain (Thermo Fisher Scientific).

2.2 Protein In-Solution Digestion

1. Digestion enzymes: Lys-C (FUJIFILM Wako) and trypsin (Promega).

2. Dilution buffer: 50 mM HEPES, pH 8.5.

3. Peptide disulfide bond reduction: 1 M DTT in 50 mM HEPES.

4. Peptide cysteine alkylation: 1 M IAA in 50 mM HEPES (*see* **Note 3**).

5. Acetonitrile (ACN, HPLC grade).

2.3 Peptide Desalting

1. Desalting columns: 500 mg Sep-Pak C18 desalting cartridge (Waters, ~1 mL of bed volume), 200 mg Sep-Pak C18 desalting cartridge (Waters, ~400 μL of bed volume), Ultra Micro-Spin C18 columns (The Nest Group, 25 μL of bed volume), C18 StageTips (Thermo Fisher Scientific) (*see* **Note 4**).

2. Vacuum manifold.

3. Trifluoroacetic acid (TFA).

4. Methanol (HPLC grade).

5. Equilibration and wash buffer: 0.1% TFA.

6. Elution buffer: 60% ACN, 0.1% TFA.

7. SpeedVac concentrator.

2.4 Enrichment of Di-GG Peptides and Di-GG Peptides Depletion Test

1. PTMScan® ubiquitin remnant motif (K-ε-GG) antibody (Cell Signaling Technology).

2. Immuno-affinity purification (IAP) buffer: 50 mM [3-(*N*-morpholino)propanesulfonic acid] (MOPS), pH 7.2, 10 mM sodium phosphate, and 50 mM NaCl.

3. Wash buffer: phosphate-buffered saline (PBS).

4. Elution buffer: 0.15% TFA.

5. C18 StageTips (Thermo Fisher Scientific).

6. 1% formic acid.

7. 40% ACN.

8. 5% formic acid.

2.5 TMT16 Labeling of Peptides and Desalting

1. Peptide labeling buffer: 50 mM HEPES, pH 8.5.

2. 16-plex TMT Isobaric Mass Tagging Kit (Thermo Fisher Scientific).

3. Anhydrous ACN.

4. Quenching solution: 5% hydroxylamine.

5. Desalting columns: 500 mg Sep-Pak C18 desalting cartridge (Waters, ~1 mL of bed volume), 200 mg Sep-Pak C18 desalting cartridge (Waters, ~400 μL of bed volume), C18 StageTips (Thermo Fisher Scientific).

6. Methanol (HPLC grade).

7. Equilibration and wash buffer: 5% ACN, 0.1% TFA (*see* **Note 5**).

8. Elution buffer: 60% ACN, 0.1% TFA.

2.6 Offline Basic pH Reverse-Phase Liquid Chromatography Fractionation

1. Waters XBridge C18 column, 3.5 μm particle size, 4.6 mm × 25 cm or 2.1 mm × 15 cm (*see* **Note 6**).

2. Buffer A: 10 mM ammonium formate; adjust pH to 8.0 by 28% ammonium hydroxide.

3. Buffer B: 10 mM ammonium formate, 90% ACN; adjust pH to 8.0 by 28% ammonium hydroxide.

4. Wash solution: 25% isopropanol, 25% methanol, 25% ACN, and 25% water.

5. High-performance liquid chromatography (HPLC) system (e.g., Agilent 1220 Infinity LC System).

6. Fraction collector (e.g., Gilson FC 203B).

2.7 Acidic pH RPLC-MS/MS Analysis

1. New Objective empty column: 75 μm I.D. × 50 cm, 15 μm tip orifice.

2. 1.9 μm particle size C18 resin (Dr. Maisch GmbH, Germany).

3. Butterfly portfolio column heater.

4. HPLC system (e.g., Waters ACQUITY UPLC or Dionex Ultimate 3000 RSLCnano System).

5. Tandem MS instrument (e.g., Thermo Q Exactive HF or Fusion).

6. Sample loading solution: 5% formic acid.

7. Sample vials with 0.2 mL bottom-spring polypropylene inserts.

8. Buffer A: 3% dimethyl sulfoxide and 0.2% formic acid.

9. Buffer B: 65% ACN, 3% dimethyl sulfoxide, and 0.2% formic acid.

2.8 MS Data Analysis

1. Database downloaded from the UniProt website (https://www.uniprot.org/downloads).

2. Proteomics software suite for data processing: JUMP software suite [29, 37, 38] or Proteome Discoverer (Thermo Scientific) (*see* also Chap. 28).

3. A computer cluster for data processing.

3 Methods

3.1 Protein Extraction from Cells and Tissues

3.1.1 Whole Proteome

1. Cut the frozen brain tissue (~10 mg) from a defined brain region (e.g., prefrontal cortex) on a metal plate on dry ice (*see* **Note 7**). Weigh the tissue, and transfer immediately to a pre-cooled Eppendorf tube on dry ice followed by the addition of glass beads (~20% of the volume of lysate).

2. Add freshly prepared lysis buffer (10 μL per mg tissue) into the tube. Generally, 1 mg tissue yields 50–100 μg total proteins. The final protein concentration is 5–10 μg/μL. For whole proteome analysis, usually 100 μg protein per sample is sufficient.

3. Seal the tube caps with parafilm, and put the tubes in a Bullet Blender in 4 °C cold room. Lyse the tissues under speed 8 with the cycle time of 30 s on and 5 s off until the sample is homogenized (~5 cycles).

4. Make aliquots for the lysates (*see* **Note 8**). Small aliquots (e.g., 10 μL) can be used for the analysis of protein concentration and the evaluation of protein quality (i.e., Western blot validation of positive control proteins) (*see* **Note 9**). Large aliquots (e.g., 50 μL) are for whole proteome analysis. Freeze the aliquots immediately on dry ice, and store at −80 °C until further use.

5. Measure the protein concentrations by the BCA assay (protocol provided by the manufacturer). Considering that the BCA assay may be affected by non-protein reducing components in the tissue lysate (e.g., DTT in lysis buffer), we often confirm the protein concentration by a short SDS gel staining method [39], in which a serial dilution of standard protein is used for the establishment of a standard curve (e.g., BSA titrations of 0.1, 0.3, 1, 3, and 10 μg).

3.1.2 Ubiquitinome

Due to the nature of low abundance/stoichiometry of ubiquitination on proteins and high activity of deubiquitinating enzymes (DUBs) in cells, lysis protocol is modified to limit DUB activities and maximize recovery of ubiquitinated proteins/peptides [40, 41].

1. Lyse ~40 mg brain tissue per sample with DUB inhibitors (e.g., PR-619 and IAA) as described in Subheading 3.1.1. Dissolve lyophilized Lys-C in lysis buffer (1 μg/μL), and keep at 4 °C for use in next step.

2. After making small aliquots (e.g., 10 μL) for protein quality control analysis, add 10 μg Lys-C to the remaining tissue lysates (at least 2 mg protein per sample) immediately, and incubate at room temperature. The rapid digestion eliminates the impact of the DUB activities.

3. Determine the protein amount in the cell lysates as described in Subheading 3.1.1 using the small aliquots. Then add additional Lys-C as well as 10% ACN to the lysates at a final protein/Lys-C ratio of 100:1 (w/w), and incubate for 3 h at room temperature. Store the digested lysates at −80 °C.

3.2 Protein In-Solution Digestion

1. For whole proteome samples, add Lys-C into each lysate at a protein/Lys-C ratio of 100:1 (w/w). Then, add ACN to a final concentration of 10%. Incubate at room temperature for 3 h [42]. For ubiquitinome samples, directly go to the next step.

2. Dilute the lysates with 50 mM HEPES (pH 8.5) to lower the urea concentration to 2 M. Add trypsin into the diluted lysates at a protein/trypsin ratio of 50:1 (w/w), and incubate at room temperature overnight.

3. For whole proteome samples, reduce the digested peptides with 1 mM DTT for 2 h and alkylate cysteine residues with 10 mM IAA for 30 min in the dark. Quench the alkylation by 30 mM DTT and incubating at room temperature for another 30 min. For ubiquitinome samples, directly quench the alkylation by 30 mM DTT.

4. Terminate the digestion by adding TFA to 0.5% (v/v). Check the pH of the mixture to ensure the pH is lower than 3.

3.3 Peptide Desalting

To limit the loss of peptides during the desalting process, we choose C18 desalting cartridges or spin columns with binding capacity matched with the amount of peptides to be desalted (*see* **Note 4**).

3.3.1 Whole Proteome

1. Centrifuge the tryptic peptides at $21,000 \times g$ for 10 min to remove any insoluble materials.

2. Wash Ultra MicroSpin C18 desalting columns with 5× bed volumes of elution buffer (60% ACN, 0.1% TFA) three times by spinning at $500 \times g$ for 30 s.

3. Equilibrate the columns with 5× bed volumes of equilibration and wash buffer (0.1% TFA) three times by spinning at $500 \times g$ for 30 s.

4. Transfer the supernatant of samples to the C18 desalting columns, and spin at $100 \times g$ for 3 min.

5. Wash the columns three times with 5× bed volumes of equilibration and wash buffer (0.1% TFA) by spinning at $500 \times g$ for 30 s.

6. Elute peptides with 3× bed volumes of elution buffer (60% ACN, 0.1% TFA) by spinning at $100 \times g$ for 3 min.

7. Dry the eluted peptides in SpeedVac, and store the peptides at −80 °C for further TMT labeling (*see* **Note 11**).

3.3.2 Ubiquitinome

1. Centrifuge the tryptic peptides at $21,000 \times g$ for 10 min to remove insoluble materials.

2. Condition 500 mg Sep-Pak C18 desalting cartridges on vacuum manifold by washing the cartridges with $5\times$ bed volumes of methanol twice, $5\times$ bed volumes of elution buffer (60% ACN, 0.1% TFA) twice, and $5\times$ bed volumes of equilibration and wash buffer (0.1% TFA) twice.

3. Transfer the supernatant of samples to the C18 cartridges. Adjust the vacuum pressure of manifold to let samples flow through the cartridges at the slowest speed (dropwise). Keep the flow-through until the desalting is done successfully. Wash the cartridge at least three times with $5\times$ bed volumes of equilibration and wash buffer (0.1% TFA).

4. Elute peptides with $3\times$ bed volumes of elution buffer (60% ACN, 0.1% TFA).

5. Dry the eluted peptides in SpeedVac, and store the peptides at $-80\ °C$ for di-GG peptides enrichment (*see* **Note 11**).

3.4 Enrichment of Di-GG Peptides

This section is only for ubiquitinome analysis. Skip this section, and move to Subheading 3.5 for whole proteome analysis (*see* **Note 10**).

1. Dissolve each of the 16 desalted peptide samples in 1 mL of ice-cold IAP buffer by vortexing, and adjust pH to 7.2 if necessary (*see* **Note 11**).

2. Centrifuge at $21,000 \times g$ for 10 min to remove insoluble materials. Aliquot 1 μL as input for testing the efficiency of di-GG peptide enrichment (*see* **Note 12**).

3. Tap the tube of ubiquitin remnant motif (K-ε-GG) antibody to mix the beads well, and transfer 40 μL of slurry into an Eppendorf tube (*see* **Note 13**).

4. Wash the slurry three times with 1 mL cold IAP buffer, and settle the beads down by spinning at $2,000 \times g$ for 1 min and remove the supernatant.

5. Transfer peptide solution into the first tube that contains antibody slurry, and put the mixture on a 360-degree rotator and rotate for 2 h at $4\ °C$.

6. Centrifuge at $2,000 \times g$ for 2 min to remove supernatant, and keep the supernatant at $-80\ °C$ as flow-through for the depletion test (*see* **Note 12**).

7. Spin the remaining beads again at $2,000 \times g$ for 2 min to remove residual solution.

8. Wash the beads twice with 1 mL cold IAP buffer and thrice with 1 mL cold PBS by inverting the tubes six times, and then remove supernatant after centrifuging at $2,000 \times g$ for 2 min.

9. Add 50 μL of elution buffer into beads, and gently tap the bottom of the tube to mix the solution with beads. Remain at room temperature for 5 min (*see* **Note 14**).

10. Centrifuge at 2,000 × *g* to settle the beads, and transfer the eluent to another tube.

11. Repeat **steps 9** and **10** and combine both eluents.

12. Desalt input and flow-through by C18 StageTips (*see* Subheading 3.3.1), and analyze the eluents by LC-MS/MS for di-GG peptide depletion test (*see* **Note 12**).

13. Based on the depletion test result, if there is still K48 chain and K63 chain in the flow-through, a second round of enrichment may be necessary. In general, we find that ~240 μg of antibody is enough to deplete K48 chain peptide in 4 mg of total peptides from cell lysate. However, it is necessary to perform an individual pilot depletion test for each sample.

3.5 TMT16 Labeling of Peptides and Desalting

3.5.1 Whole Proteome

1. Resuspend the desalted peptides with 50 mM HEPES, pH 8.5, examine the pH by pH paper, and take ~1 μg unlabeled peptides for TMT labeling efficiency test.

2. Dissolve the TMT16 reagents in anhydrous acetonitrile (10 μg/μL) immediately before use, and make aliquots if necessary.

3. For the labeling of the whole proteome sample, add TMT16 reagents to each sample at a reagent/substrate ratio of 1.5:1 (w/w), and incubate at room temperature for 30 min. Take ~1 μg labeled peptides for TMT labeling efficiency test, and keep the remaining peptides at −80 °C without quenching the reaction.

4. Perform the labeling efficiency test. Desalt all the aliquots of labeled and unlabeled samples with C18 StageTips (*see* Subheading 3.3.1) followed by analyzing the labeled and unlabeled samples using LC-MS/MS (*see* **Note 15**). If the TMT labeling is complete, the unlabeled peptide peaks should not be seen in the labeled samples.

5. If the labeling is completed, quench the reaction by adding the quenching solution (5% hydroxylamine), and incubate for 15 min at room temperature. If not, additional TMT reagents can be used to further label the peptide samples.

6. Take half of each TMT-labeled sample, and pool all the 16 samples to make a mixture "mix1." For whole proteome samples, take ~1 μg pooled peptides for LC-MS/MS analysis after desalting. Calculate and compare the average intensity of each TMT16 reporter ion. If the discrepancies of intensity among samples are larger than 5%, adjust the intensity by adding individual remaining TMT-labeled samples into "mix1"

according to the calculated average intensity to make a "mix2." Repeat the adjustment until all samples are mixed equally.

7. Desalt the pooled TMT16-labeled peptides (*see* Subheading 3.3.2). Note that the wash buffer (5% ACN, 0.1% TFA) for TMT16-labeled peptides contains a higher level of ACN than the regular wash buffer (0.1% TFA) for unlabeled peptides, because the 5% ACN is required to remove the hydrophobic derivatives of quenched TMT16 reagents.

3.5.2 Ubiquitinome

As global ubiquitinome varies dramatically under different stress conditions such as heat shock, proteasome inhibition, etc., the normalization of sample loading bias cannot be performed the same as for the whole proteome analysis. Here we only adjust the mixing amount for the replicates in the 16-plex samples to reduce experimental variations that occur in sample preparation steps for the ubiquitinome samples.

1. Add 800 µg TMT16 reagent to the di-GG peptides that are enriched from 4 mg peptides, and incubate at room temperature for 30 min. Take 1% unlabeled and labeled di-GG peptides for TMT labeling efficiency test. In the meantime, keep the remaining peptides at −80 °C.

2. For TMT16 mixing ratio adjustment, check the discrepancies of intensity only for the replicate samples, and adjust mixing volume as described in Subheading 3.5.1.

3.6 Offline Basic pH RPLC Fractionation

1. Prepare a Waters XBridge C18 column (3.5 µm particle size, 4.6 mm × 25 cm).

2. Dissolve the mixture of TMT16-labeled peptides with basic pH RPLC buffer A (10 mM ammonium formate, pH 8.0).

3. Centrifuge the peptides at $21,000 \times g$ for 10 min to remove precipitates.

4. Wash the LC sample loop with methanol, water, and buffer A sequentially using 3× loop volume.

5. Inject 100 µL of wash solution (25% isopropanol, 25% methanol, 25% ACN, and 25% water) to clean the column, and then equilibrate the column with 95% buffer A for 1 h with a flow rate of 0.4 mL/min.

6. The whole proteome samples are fractionated using the following gradient: 5% buffer B for 10 min, 5–15% buffer B for 2 min, 15–45% buffer B for 148 min, and 45–95% buffer B for 5 min, at the flow rate of 0.4 mL/min. Collect the fractions every 1 min by a fraction collector, and concatenate to 40 fractions in 4 cycles.

7. All the fractions are dried in SpeedVac and stored at −80 °C for further LC-MS/MS analysis.

8. For ubiquitinome analysis, switch to a 2.1 mm × 15 cm C18 column and flow rate of 0.2 mL/min. The di-GG peptides are fractionated in a 40 min gradient with 15–45% buffer B. Collect the fractions every 1 min by a fraction collector, and concatenate to 10 fractions in 4 cycles.

3.7 Acidic pH RPLC-MS/MS Analysis

1. Pack the LC-MS/MS columns with 1.9 μm C18 resin. The fractions of whole proteome are analyzed on 75 μm × 15 cm C18 columns, and the fractions of ubiquitinome samples are analyzed on 75 μm × 50 cm C18 columns. Heat the column at 65 °C with a butterfly portfolio heater to reduce the backpressure.

2. Inject 100 ng of rat brain or BSA tryptic peptides to evaluate the performance of LC/MS system.

3. Dissolve the basic pH RPLC fractions in the sample loading solution (5% formic acid). Centrifuge the samples at 21,000 × g for 5 min, and transfer 5 μL supernatant to inserts.

4. For whole proteome analysis, load ~1 μg for each fraction. Peptides are eluted in a 60 min gradient of 18–45% buffer B at 0.25 μL/min flow rate. The gradient range may be slightly adjusted according to specific samples and different LC-MS/MS settings. The mass spectrometer settings include MS1 scans (60,000 resolution, scan range 450–1,600 m/z, 1×10^6 AGC, and 50 ms maximal ion time) and 20 data-dependent MS2 scans (60,000 resolution, scan range starting from 120 m/z, 1×10^5 AGC, ~110 ms maximal ion time, 32% normalized collision energy (NCE), 1.0 m/z isolation window, 0.2 m/z isolation offset, exclude isotopes on, and 10 s dynamic exclusion) (*see* **Note 16**).

5. For ubiquitinome analysis, reconstitute the dried di-GG peptides in 2 μL of 5% formic acid, and load all 2 μL onto the column. Elute peptides by 18–45% buffer B in 150 min at ~0.125 μL/min flow rate. Operate mass spectrometer in data-dependent mode with a survey scan in Orbitrap (60,000 resolution, scan range 450–1,600 m/z, 1×10^6 AGC target, ~50 ms maximal ion time) and 20 data-dependent MS2 scans (60,000 resolution, scan range starting from 120 m/z, 1×10^5 AGC target, ~110 ms maximal ion time, 32% NCE, 1.0 m/z isolation window, 0.2 m/z isolation offset, exclude isotopes on, and 10 s dynamic exclusion).

3.8 MS Data Analysis

3.8.1 Database Search

The MS raw files are converted to mzXML files and searched against the human protein database using our in-house developed JUMP pipeline [29, 37, 38]. The database is generated by combining Swiss-Prot, TrEMBL, and UCSC databases, removing redundancy (83,955 entries for human proteins) and adding customized protein sequences which are not contained in the reference databases, including but not limited to protease-cleaved proteins, proteins with single nucleotide polymorphisms, and all possible contaminants. Search parameters include 10 ppm mass tolerance for precursor ions and 15 ppm for fragment ions, full trypticity with two maximal missed cleavages, and three maximal modification sites. TMT16 modification on Lys residues and N termini (+304.20715 Da) and carbamidomethylation of Cys residues (+57.02146 Da) are used as static modifications, while Met oxidation (+15.99492 Da) is set as dynamic modification. For ubiquitinome analysis, the maximal missed cleavage sites are set to three, and di-GG modification on Lys residues (+114.04293 Da) is set as an additional dynamic modification.

3.8.2 Peptide-Spectrum Match (PSM) Filtering

The resulting PSMs are first filtered by precursor ion mass accuracy and then grouped by peptide length, tryptic ends, modifications, miscleavage sites, and precursor ion charge state. The data are further filtered with the JUMP-based matching scores (Jscore and ΔJn) to reduce false discovery rate (FDR) below 1% for either proteins (whole proteome analysis) or peptides (ubiquitinome analysis), based on the target-decoy strategy [43, 44]. The shared peptides from numerous homologous proteins are matched to the protein with the top PSM number according to the rule of parsimony. If some positive control proteins are missing at the filtering steps, the FDR may be reduced to 5% with manual examination to see if these proteins/peptides can be salvaged.

3.8.3 TMT-Based Quantification

Proteins or peptides quantification is performed in the following steps based on our previous method [29]: (i) Extract TMT reporter ion intensities of each PSM. (ii) Correct the raw intensities based on isotopic distribution of each labeling reagent (e.g., TMT16-126 generates 92.6%, 7.2%, and 0.2% of 126, 127C, and 128C m/z ions, respectively). (iii) Discard PSMs of very low intensities (e.g., minimum intensity of 1000 and median intensity of 5000). (iv) Minimize sample loading bias by normalization with the trimmed median intensity of all PSMs. (v) Calculate the mean-centered intensities across samples (e.g., relative intensities between each sample and the mean). (vi) Summarize protein relative intensities by averaging related PSMs. (vii) Finally derive protein absolute intensities by multiplying the relative intensities by the grand mean of three most highly abundant PSMs. In addition, y1 ion-based correction is also used for TMT quantification analysis [29]. The intensities of positive control proteins should be checked and matched to Western blot (*see* also Chap. 3) results to see if there is any error of sample mislabeling.

In general, the analysis of quantification is carried out with data without missing values. However, PSMs with missing values in specific labeling channels may be examined under certain conditions (e.g., knockout of positive control proteins/genes in the samples). Our JUMP software generates a separate table containing all PSMs with the missing values for manual evaluation. We then evaluate the quality of samples by statistical methods such as PCA or clustering analysis to determine if there are any outliers in the samples or mislabeling of the samples. After removing all the outliers, protein fold change and p values of different comparisons are calculated based on the protein signal intensities. The $\log_2$(fold change) values are further transformed to z scores to reduce the impact of ratio compression [29], which is based on the calculated experimental standard deviation between replicate samples. Finally, the differential expression analysis is performed by applying different combinations of cutoff values (e.g., FDR (Benjamini-Hochberg corrected p) <0.05, and $z > 2$). The cutoffs are often adjusted to include known positive proteins that are dysregulated in the samples.

4   Notes

1. The solution form of urea is not stable during longtime storage and can decompose into ammonium cyanate. Ammonium cyanate reacts with Lys residues to induce protein carbamylation. Heating accelerates this process. Thus, lysis buffer should be prepared in fresh form or stored at −80 °C if necessary. All digestion steps are carried out at room temperature instead of 37 °C.

2. Some of the DUBs are still active in 8 M urea and can remove ubiquitin modifications within short period of time. Add DUB inhibitors (PR-619 and IAA) in the lysis buffer just before tissue lysis if ubiquitinome analysis is performed. Also, adding Lys-C to cell lysate immediately after protein extraction can minimize the activities of DUBs. Skip these steps if only whole proteome analysis is performed.

3. IAA modification on lysine residues may introduce ubiquitination artifact mimics [45]. This reaction does not occur at room temperature [46], and the pseudo GG peptides are not recognized by the K-ε-GG motif antibody [47]. Thus, IAA is used in the protocol. Or use other IAA alternatives such as iodoacetic acid or 2-chloroacetamide instead in ubiquitinome analysis [48].

4. Selecting appropriate cartridges or columns can improve the recovery of peptides during desalting and fractionation. In general, the binding capacity of C18 resin is about 10 μg peptides per 1 μL of cartridges bed volume. The loading capacity of a C18 LC column is around 3 μg of peptides per 1 μL of column bed volume [42].

5. Byproducts in TMT reaction (side reaction products with H_2O and quenching reaction product with hydroxylamine) also bind to C18 resins and affect subsequent LC/LC-MS/MS analysis. These byproducts can be removed by extensively washing the desalting C18 cartridges with wash buffer (5% ACN and 0.1% TFA, 10× bed volumes).

6. Use different offline basic pH LC columns for the whole proteome and the ubiquitinome analysis based on the loading amount of peptides [42].

7. To reduce the variations caused by heterogeneity of brains tissues, the collected tissue samples can be pre-pulverized and mixed thoroughly before tissue lysis. Meanwhile, it is essential to include at least two replicates for each biological condition in the TMT design for statistical analysis.

8. Do not centrifuge the lysate at high speed to remove the insoluble proteins. Digestion of these insoluble proteins with trypsin can increase the number of identified proteins [49].

9. Quality control (QC) steps (i.e., Western blot of positive and negative control proteins) before TMT labeling are necessary to avoid wasting of reagents and time on low-quality samples. QC results can also be used to identify potential sample mislabeling during sample processing.

10. The enrichment of ubiquitinated peptides should be performed before TMT labeling because the di-GG peptides are not recognized by ubiquitin remnant motif K-ε-GG antibodies after TMT labeling. For other PTM analysis, such as phosphorylation and methylation, the enrichment is often performed after TMT labeling to eliminate experimental variation during the enrichment.

11. Desalted peptides should be dried completely to remove trace amount of TFA. Residual TFA may reduce pH lower than 7 when the peptides are dissolved in the IAP buffer, which may affect the interaction between di-GG peptides and antibody. It is important to confirm the pH value before the enrichment.

12. It is critical to evaluate the efficiency of antibody-based enrichment of di-GG peptides. K48 and K63 chains consist of a substantial portion of total ubiquitinome. Comparison of K48 and K63 chain intensity between input and flow-through by MS can help estimate the enrichment efficiency of ubiquitinome.

13. Among the two commercially available K-GG monoclonal antibodies, the antibody from Lucerna was generated against GG-modified histone; the one from Cell Signaling Technology was produced against the sequence CXXXXXXKGGXXXXXX

(X = any amino acid except Cys and Trp). Both antibodies gave similar and partially overlapping results in a comparative study, but the antibody from Lucerna led to ~30% less coverage of the ubiquitinome. Thus, we describe the use of the antibody from Cell Signaling Technology in this chapter. The antibody can distinguish K-ε-GG peptides from M1-GG peptides (linear peptide modified on the N-terminal amine group) [47]. In addition to the antibody-based enrichment, ubiquitinome may be analyzed through ubiquitinated protein purification using a range of affinity reagents [50, 51], but the purification is typically complicated by co-purified unmodified proteins [52].

14. The K-GG antibody is non-covalently attached to protein A beads. Thus, the antibody is co-eluted with di-GG peptides and may affect the LC-MS/MS analysis. During TMT labeling process, the antibody is labeled with TMT reagent and becomes highly hydrophobic. The TMT-labeled antibody is removed by desalting, as it binds to C18 resin and is not eluted. The alternative way to limit the antibody's impact on LC-MS/MS is to cross-link the antibody to beads with dimethyl pimelimidate (DMP) prior to enrichment.

15. When performing the labeling efficiency test, TMT-labeled samples should be analyzed before unlabeled samples to reduce the impact of LC carryover on the calculation of labeling efficiency.

16. The MS parameters are optimized for TMT16-labeled peptides based on our previous settings for TMT11-labeled peptides [8]. Compared to TMT11, TMT16 has an increased mass, and therefore MS1 scan range starts from a higher m/z (450 instead of 410). The MS2 ion time (110 ms) matches the time required to obtain MS scans at the 60,000 resolution on a Q Exactive HF MS [53]. The NCE is optimized to balance peptide fragment ion intensity and TMT reporter ion intensity. A lower collision energy (32%) is used in TMT16 experiments and 35% in TMT11 studies [26]. In addition, we use a relatively short dynamic exclusion time (10 s) because for extensively fractionated samples (40 fractions), the MS time is sufficient to scan all detected precursor ions. During the LC gradient, the peak width is around 20 s, so 10 s exclusion allows at least two MS2 scans for each precursor ion: one at the beginning of the peak and the other at the center point of the peak to improve the MS2 signal.

Acknowledgments

This work was partially supported by the National Institutes of Health (R01GM114260, R01AG047928, R01AG053987, and RF1AG064909) and ALSAC (American Lebanese Syrian Associated Charities). The MS analysis was performed in the Center of Proteomics and Metabolomics at St. Jude Children's Research Hospital, partially supported by NIH Cancer Center Support Grant (P30CA021765).

References

1. Zhang Y, Fonslow BR, Shan B et al (2013) Protein analysis by shotgun/bottom-up proteomics. Chem Rev 113(4):2343–2394

2. Aebersold R, Mann M (2016) Mass-spectrometric exploration of proteome structure and function. Nature 537(7620):347–355

3. Huttlin EL, Bruckner RJ, Paulo JA et al (2017) Architecture of the human interactome defines protein communities and disease networks. Nature 545(7655):505–509

4. Yu J, Peng J, Chi H (2019) Systems immunology: integrating multi-omics data to infer regulatory networks and hidden drivers of immunity. Curr Opin Sys Biol 15:19–29

5. Toby TK, Fornelli L, Kelleher NL (2016) Progress in top-down proteomics and the analysis of proteoforms. Annu Rev Anal Chem (Palo Alto, Calif) 9(1):499–519

6. Peng J, Gygi SP (2001) Proteomics: the move to mixtures. J Mass Spectrom 36 (10):1083–1091

7. Wang H, Yang Y, Li Y et al (2015) Systematic optimization of long gradient chromatography mass spectrometry for deep analysis of brain proteome. J Proteome Res 14(2):829–838

8. Bai B, Tan H, Pagala VR et al (2017) Deep profiling of proteome and phosphoproteome by isobaric labeling, extensive liquid chromatography, and mass spectrometry. Methods Enzymol 585:377–395

9. Huttlin EL, Jedrychowski MP, Elias JE et al (2010) A tissue-specific atlas of mouse protein phosphorylation and expression. Cell 143 (7):1174–1189

10. Kim W, Bennett EJ, Huttlin EL et al (2011) Systematic and quantitative assessment of the ubiquitin-modified proteome. Mol Cell 44 (2):325–340

11. Mertins P, Mani DR, Ruggles KV et al (2016) Proteogenomics connects somatic mutations to signalling in breast cancer. Nature 534 (7605):55–62

12. Vasaikar S, Huang C, Wang X et al (2019) Proteogenomic analysis of human colon cancer reveals new therapeutic opportunities. Cell 177 (4):1035–1049. e1019

13. Stewart E, McEvoy J, Wang H et al (2018) Identification of therapeutic targets in rhabdomyosarcoma through integrated genomic, pigenomic, and proteomic analyses. Cancer Cell 34(3):411–426. e419

14. Wang H, Diaz AK, Shaw TI et al (2019) Deep multiomics profiling of brain tumors identifies signaling networks downstream of cancer driver genes. Nat Commun 10(1):3718

15. Bai B, Hales CM, Chen PC et al (2013) U1 small nuclear ribonucleoprotein complex and RNA splicing alterations in Alzheimer's disease. Proc Natl Acad Sci U S A 110 (41):16562–16567

16. Bai B, Wang X, Li Y et al (2020) Deep multilayer brain proteomics identifies molecular networks in Alzheimer's disease progression. Neuron 105:975–991.e7. [Epub ahead of print]:online 8 January 2020.

17. Liu H, Sadygov RG, Yates JR 3rd (2004) A model for random sampling and estimation of relative protein abundance in shotgun proteomics. Anal Chem 76(14):4193–4201

18. Cox J, Hein MY, Luber CA et al (2014) Accurate proteome-wide label-free quantification by delayed normalization and maximal peptide ratio extraction, termed MaxLFQ. Mol Cell Proteomics 13(9):2513–2526

19. Venable JD, Dong MQ, Wohlschlegel J et al (2004) Automated approach for quantitative analysis of complex peptide mixtures from tandem mass spectra. Nat Methods 1(1):39–45

20. Ludwig C, Gillet L, Rosenberger G et al (2018) Data-independent acquisition-based SWATH-MS for quantitative proteomics: a tutorial. Mol Syst Biol 14(8):e8126

21. Bache N, Geyer PE, Bekker-Jensen DB et al (2018) A novel LC system embeds analytes in

pre-formed gradients for rapid, ultra-robust proteomics. Mol Cell Proteomics 17 (11):2284–2296

22. Thompson A, Schafer J, Kuhn K et al (2003) Tandem mass tags: a novel quantification strategy for comparative analysis of complex protein mixtures by MS/MS. Anal Chem 75 (8):1895–1904

23. Ross PL, Huang YN, Marchese JN et al (2004) Multiplexed protein quantitation in Saccharomyces cerevisiae using amine-reactive isobaric tagging reagents. Mol Cell Proteomics 3 (12):1154–1169

24. Frost DC, Greer T, Li L (2015) High-resolution enabled 12-plex DiLeu isobaric tags for quantitative proteomics. Anal Chem 87(3):1646–1654

25. Rauniyar N, Yates JR 3rd (2014) Isobaric labeling-based relative quantification in shotgun proteomics. J Proteome Res 13 (12):5293–5309

26. Thompson A, Wolmer N, Koncarevic S et al (2019) TMTpro: design, synthesis, and initial evaluation of a proline-based isobaric 16-plex tandem mass tag reagent set. Anal Chem 91 (24):15,941–15,950

27. Hogrebe A, von Stechow L, Bekker-Jensen DB et al (2018) Benchmarking common quantification strategies for large-scale phosphoproteomics. Nat Commun 9(1):1045

28. Ting L, Rad R, Gygi SP et al (2011) MS3 eliminates ratio distortion in isobaric multiplexed quantitative proteomics. Nat Methods 8(11):937–940

29. Niu M, Cho JH, Kodali K et al (2017) Extensive peptide fractionation and y1 ion-based interference detection method for enabling accurate quantification by isobaric labeling and mass spectrometry. Anal Chem 89 (5):2956–2963

30. Wenger CD, Lee MV, Hebert AS et al (2011) Gas-phase purification enables accurate, multiplexed proteome quantification with isobaric tagging. Nat Methods 8(11):933–935

31. Savitski MM, Mathieson T, Zinn N et al (2013) Measuring and managing ratio compression for accurate iTRAQ/TMT quantification. J Proteome Res 12(8):3586–3598

32. Wuhr M, Haas W, McAlister GC et al (2012) Accurate multiplexed proteomics at the MS2 level using the complement reporter ion cluster. Anal Chem 84(21):9214–9221

33. Peng J, Schwartz D, Elias JE et al (2003) A proteomics approach to understanding protein ubiquitination. Nat Biotechnol 21 (8):921–926

34. Udeshi ND, Mani DR, Eisenhaure T et al (2012) Methods for quantification of in vivo changes in protein ubiquitination following proteasome and deubiquitinase inhibition. Mol Cell Proteomics 11(5):148–159

35. Udeshi ND, Svinkina T, Mertins P et al (2013) Refined preparation and use of anti-diglycine remnant (K-epsilon-GG) antibody enables routine quantification of 10,000s of ubiquitination sites in single proteomics experiments. Mol Cell Proteomics 12(3):825–831

36. Rose CM, Isasa M, Ordureau A et al (2016) Highly multiplexed quantitative mass spectrometry analysis of ubiquitylomes. Cell Syst 3 (4):395–403. e394

37. Wang X, Li Y, Wu Z et al (2014) JUMP: a tag-based database search tool for peptide identification with high sensitivity and accuracy. Mol Cell Proteomics 13(12):3663–3673

38. Li Y, Wang X, Cho JH et al (2016) JUMPg: an integrative proteogenomics pipeline identifying unannotated proteins in human brain and cancer cells. J Proteome Res 15(7):2309–2320

39. Xu P, Duong DM, Peng JM (2009) Systematical optimization of reverse-phase chromatography for shotgun proteomics. J Proteome Res 8(8):3944–3950

40. Peng J, Cheng D (2005) Proteomic analysis of ubiquitin conjugates in yeast. Methods Enzymol 399:367–381

41. Na CH, Jones DR, Yang Y et al (2012) Synaptic protein ubiquitination in rat brain revealed by antibody-based ubiquitome analysis. J Proteome Res 11(9):4722–4732

42. Pagala VR, High AA, Wang X et al (2015) Quantitative protein analysis by mass spectrometry. Methods Mol Biol 1278:281–305

43. Peng J, Elias JE, Thoreen CC et al (2003) Evaluation of multidimensional chromatography coupled with tandem mass spectrometry (LC/LC-MS/MS) for large-scale protein analysis: the yeast proteome. J Proteome Res 2 (1):43–50

44. Elias JE, Gygi SP (2007) Target-decoy search strategy for increased confidence in large-scale protein identifications by mass spectrometry. Nat Methods 4(3):207–214

45. Nielsen ML, Vermeulen M, Bonaldi T et al (2008) Iodoacetamide-induced artifact mimics ubiquitination in mass spectrometry. Nat Methods 5(6):459–460

46. Xu P, Duong DM, Seyfried NT et al (2009) Quantitative proteomics reveals the function of unconventional ubiquitin chains in proteasomal degradation. Cell 137(1):133–145

47. Bustos D, Bakalarski CE, Yang Y et al (2012) Characterizing ubiquitination sites by peptide

based immunoaffinity enrichment. Mol Cell Proteomics 11(12):1529–1540

48. Chen PC, Na CH, Peng J (2012) Quantitative proteomics to decipher ubiquitin signaling. Amino Acids 43(3):1049–1060

49. Pirmoradian M, Budamgunta H, Chingin K et al (2013) Rapid and deep human proteome analysis by single-dimension shotgun proteomics. Mol Cell Proteomics 12(11):3330–3338

50. Xu P, Peng J (2006) Dissecting the ubiquitin pathway by mass spectrometry. Biochim Biophys Acta 1764(12):1940–1947

51. Gao Y, Li Y, Zhang C et al (2016) Enhanced purification of ubiquitinated proteins by engineered tandem hybrid ubiquitin-binding domains (ThUBDs). Mol Cell Proteomics 15 (4):1381–1396

52. Seyfried NT, Xu P, Duong DM et al (2008) Systematic approach for validating the ubiquitinated proteome. Anal Chem 80 (11):4161–4169

53. Kelstrup CD, Jersie-Christensen RR, Batth TS et al (2014) Rapid and deep proteomes by faster sequencing on a benchtop quadrupole ultra-high-field orbitrap mass spectrometer. J Proteome Res 13(12):6187–6195

Chapter 16

Quantification and Identification of Post-Translational Modifications Using Modern Proteomics Approaches

Anja Holtz, Nathan Basisty, and Birgit Schilling

Abstract

Post-translational modifications (PTMs) occur dynamically, allowing cells to quickly respond to changes in the environment. Lysine residues can be targeted by several modifications including acylations (acetylation, succinylation, malonylation, glutarylation, and others), methylation, ubiquitination, and other modifications. One of the most efficient methods for the identification of post-translational modifications is utilizing immunoaffinity enrichment followed by high-resolution mass spectrometry. This workflow can be coupled with comprehensive data-independent acquisition (DIA) mass spectrometry to be a high-throughput, label-free PTM quantification approach. Below we describe a detailed protocol to process tissue by homogenization and proteolytically digest proteins, followed by immunoaffinity enrichment of lysine-acetylated peptides to identify and quantify relative changes of acetylation comparing different conditions.

Key words Acetylation, Post-translational modifications, Mass spectrometry, Data-independent acquisition, Quantification

1 Introduction

Proteomics techniques utilizing mass spectrometry (MS) have quickly become a preferred method to measure relative changes in protein abundance as well as significant changes in post-translational modifications (PTMs) (*see* also Chaps. 8, 13–15). Protein profiles consist of many unique proteoforms carrying different PTMs, which can exist simultaneously and change dynamically in response to environmental and other stimuli. One of the more common PTMs is lysine acetylation, which can regulate a multitude of physiological processes via changed protein-protein interactions, gene expression, and cellular location.

Post-translational modifications have been studied for decades; however, recent studies utilizing mass spectrometry have revolutionized the analysis of PTM profiles and identified thousands of novel PTM sites. Lysine acetylation has been reported to regulate many cellular pathways, specifically of mitochondrial proteins

Katrin Marcus et al. (eds.), *Quantitative Methods in Proteomics*, Methods in Molecular Biology, vol. 2228,
https://doi.org/10.1007/978-1-0716-1024-4_16, © The Author(s) 2021

[1, 2]. However, many PTMs including lysine acetylation occur at a relatively low stoichiometry. Thus, enriching acetylated peptides via antibody-based affinity enrichment protocols have been established as a key methodology and workflow to measure the dynamics of the acetylome both in mammalian [3–5] and bacterial model systems [6–8]. Here, we present a detailed protocol to allow for a routine workflow of robust identification and quantification of lysine acetylation sites by affinity enrichment of PTM-containing peptides followed by mass spectrometric analysis. This will provide a standardized protocol for the study of lysine acetylation sites to streamline the identification and quantification of PTMs. This method provides a strategy to study PTMs with relatively low starting material, even with capabilities for multiplexing and enriching for multiple PTMs simultaneously [9] in an unbiased approach using label-free DIA-MS [10–12] (*see* also Chaps. 8, 22–24) in combination with multiple different software programs, such as Skyline and Spectronaut [13, 14] or others (*see* Chap. 31).

2   Materials

2.1 Tissue Lysis and Tryptic Protein Digestion

1. Tissue sample to be digested.

2. 2 mL Safe-Lock tubes.

3. Lysis buffer: 8 M (w/v) urea, 100 mM (w/v) triethylammonium bicarbonate (TEAB) pH 8.5, 1× protease inhibitor cocktail (Pierce), 5 μM (w/v) trichostatin A (TSA), 5 mM (w/v) nicotinamide, and 75 mM (w/v) sodium chloride (NaCl).

4. TissueLyser II and 5 mm stainless steel beads.

5. Bioruptor sonicator.

6. Bicinchoninic acid (BCA) protein assay.

7. Reducing reagent: 1 M (w/v) dithiothreitol (DTT), freshly prepared in deionized water.

8. Deionized water (referred to as H_2O).

9. Alkylation reagent: 200 mM (w/v) iodoacetamide (IAA), freshly prepared in H_2O.

10. Dilution buffer: 50 mM (w/v) triethylammonium bicarbonate (TEAB) in H_2O.

11. Digestion enzyme: modified sequencing grade trypsin.

2.2 Desalting of Proteolytic/Tryptic Peptides After Digestion (Oasis/HLB)

1. Oasis HLB (hydrophilic-lipophilic balanced) 1 cc Vac cartridge, 30 mg sorbent per cartridge, 30 μm particle size.

2. Extraction manifold, 20-port vacuum manifold.

3. HPLC-MS grade acetonitrile (ACN) and water (H_2O).

4. HPLC-MS grade formic acid (FA).

5. HLB Solvent A: 0.2% (v/v) FA in HPLC-MS grade H_2O.

6. HLB Solvent B: 80% (v/v) ACN, 20% (v/v) of 0.2% FA in HPLC-MS grade H_2O.

2.3 Anti-Acetyl Immunoaffinity Enrichment

1. PTMScan Acetyl-Lysine Motif [Ac-K] Immunoaffinity Beads.

2. PTMScan Immunoaffinity (IAP) Buffer: 50 mM MOPS, 10 mM Na_3PO_4, 50 mM NaCl in water at pH 7.2.

3. 1× phosphate-buffered saline (PBS): 0.01 M phosphate-buffered saline (0.0027 M KCl, 0.138 M NaCl) pH 7.4 at 25 °C.

4. Wide-bore 200 μL pipet tips.

2.4 Small-Scale Acetyl-Peptide Desalting Prior to MS Analysis

1. Empore Octadecyl (C18) 47 mm Extraction Disks (3 M).

2. 18-Gauge blunt-tipped needle and plunger.

3. VWR 200 μL low-binding pipet tips.

4. Multi SafeSeal Sorenson 0.65 mL microcentrifuge tubes.

5. Snap Cap Low Retention 1.5 mL and 2 mL graduated microcentrifuge tubes.

6. StageTip Solvent A: 0.2% FA in 99.8% HPLC-MS grade H_2O (v/v).

7. StageTip Solvent B: 0.2% FA/50% (v/v) HPLC-MS grade ACN/49.8% HPLC-MS H_2O (v/v/v).

2.5 Chromatography and Mass Spectrometry: Nanoflow HPLC-MS/MS

All HPLC-MS/MS buffers are "HPLC-MS grade."

1. Mobile Phase A: 2% ACN/98% water/0.1% formic acid (v/v/v).

2. Mobile Phase B: 98% ACN/2% water/0.1% formic acid (v/v/v).

3. Nanoflow liquid chromatography: Ultra Plus nano-LC 2D HPLC (Eksigent) connected to a cHiPLC system (Eksigent) with a C18 pre-column chip (200 μm × 0.4 mm ChromXP C18-CL chip, 3 μm, 120 Å, SCIEX) and an analytical C18 column chip (75 μm × 15 cm ChromXP C18-CL chip, 3 μm, 120 Å).

4. Mass spectrometer: orthogonal quadrupole time-of-flight (QqTOF): TripleTOF 6600 system (SCIEX) or any other high-resolution mass spectrometry system.

3   Methods

3.1   Tissue Lysis

1. Harvest tissue of interest (here mouse liver), and take a portion ~50 mg wet weight to process immediately or freeze at −80 °C until ready.

2. Chill TissueLyser adapter sets to −20 °C for 1 h.

3. Prepare and label 2 mL Safe-Lock tubes on dry ice.

4. Add frozen tissue, and then add one stainless steel bead to each of the labeled tubes.

5. Add 500 μL ice-cold lysis buffer containing protease and deacetylase inhibitors (nicotinamide and TSA).

6. Vortex briefly, and spin to cover the whole tissue in lysis buffer. Add more lysis buffer if the tissue is not completely covered.

7. Place and balance tubes on chilled adapter sets. Homogenize with the TissueLyser II at 30 Hz twice for 3 min at 4 °C.

8. Remove bead with tweezer. Clean tweezer with deionized water and then HPLC-grade methanol, and dry between samples to prevent cross-contamination.

9. Spin briefly.

10. Sonicate on Bioruptor sonicator for ten cycles of 30 s on/30 s off at 4 °C.

11. Centrifuge homogenized tissue lysate for 15 min at 14,000 × g, 4 °C.

12. Transfer supernatant to new 1.5 mL tubes while avoiding the lipid layer above the cleared lysate and any pellet at the bottom of the tube.

13. Determine protein concentration using the BCA assay.

3.2   Tryptic Digestion

1. Remove an aliquot of lysate containing ~5 mg of soluble protein according to BCA assay (or more input material if available; see **Note 1**).

2. Add DTT to a final concentration of 4.5 mM to reduce disulfide bonds for 30 min at 37 °C with agitation.

3. Cool reduced lysate to room temperature (RT).

4. Add IAA to a final concentration of 10 mM to alkylate free thiols. Allow reaction to proceed for 30 min at RT in the dark.

5. Dilute reduced and alkylated proteins tenfold with 50 mM TEAB.

6. Add trypsin to initiate protein digestion (enzyme to protein ratio = 1:50, wt/wt) at 37 °C overnight with 1400 agitation.

7. Quench the digestion by adding FA to a final concentration of 1% (v/v) FA.

8. Remove undigested proteins and lipids by centrifugation for 10 min at RT with 1800 × *g*.

9. Desalt the supernatant containing peptides (*see* Subheading 3.3).

3.3 Desalt Tryptic Peptides with Oasis HLB Cartridges

1. Apply vacuum to the Oasis HLB 1 cc cartridges (30 mg sorbent; max. Binding capacity, 5 mg) using the vacuum extraction manifold, and condition the cartridges twice with 800 μL of organic HLB Solvent B.

2. Equilibrate cartridges three times with 800 μL of aqueous HLB Solvent A.

3. Load the acidified tryptic peptides onto the cartridge (here from 5 mg protein digest).

4. Wash the bound peptides with 800 μL of HLB Solvent A three times.

5. Place new, labeled 1.5 mL microcentrifuge tube into the vacuum manifold to collect eluting peptides.

6. Elute the peptides with the addition of 800 μL HLB Solvent B.

7. Elute once more with 400 μL HLB Solvent B.

8. Mix the elution with the vortexer, and then remove 2.4 μL (or 10 μg) for independent and parallel protein-level quantification (also *see* **Note 2**).

9. Concentrate/dry the eluted peptides completely using a SpeedVac (*see* **Note 3**).

3.4 Anti-Acetyl Immunoaffinity Enrichment

1. Resuspend the dried peptides in 1.4 mL cold IAP buffer, and mix by pipetting. Do not vortex.

2. Pipette 2 μL of the resuspended peptide solution onto litmus paper to check that the pH is neutral (between 7 and 8).

3. Centrifuge at 10,000 × *g* for 5 min at 4 °C. A small pellet may appear. Keep peptide solution on ice while preparing the antibody-bead conjugate (*see* **Note 4**).

4. Prepare the PTMScan Acetyl-Lysine antibody beads for peptide affinity enrichment by adding 1 mL cold 1× PBS to one tube of antibody-conjugated beads, and mix by pipetting. The ratio of PTMScan Acetyl-Lysine Motif antibody-conjugated beads to peptide starting material should be ¼ of a tube of antibody beads for 5 mg of peptides (*see* **Note 1**).

5. Transfer the slurry of antibody-conjugated beads to a new 1.5 mL microcentrifuge tube, and centrifuge at 2000 × *g* for 30 s at RT to prevent beads from sticking to the side of the tube.

6. Remove the PBS buffer by aspiration, and leave a small volume in the bottom to avoid disrupting the beads.

7. Wash the antibody beads with 1 mL cold 1× PBS, and centrifuge at 2000 × g for 30 s at RT. Remove the majority of PBS by aspiration.

8. Repeat the PBS wash step twice for a total of four PBS washes.

9. Resuspend the washed beads from one tube of PTMScan Acetyl-Lysine antibody in 440 μL PBS, and mix several times by pipetting with wide-bore 200 μL pipet tips.

10. Place four 100 μL aliquots of bead suspension into 1.5 mL microcentrifuge tubes. To ensure consistent bead quantities in the 100 μL aliquots, about 40 μL of beads will remain in the original tube (*see* **Note 5**).

11. Centrifuge the aliquoted beads at 2000 × g for 30 s at RT. Visually check that each tube has approximately the same quantity of antibody-conjugated beads.

12. Remove all 1× PBS by aspiration using a 0.2 mm flat gel loading pipet tip.

13. Transfer the resuspended peptides from **step 3** in Subheading 3.4 directly onto the prepared PTMScan Acetyl-Lysine Motif antibody-conjugated beads.

14. Incubate the peptides and antibody-conjugated bead mixture at 4 °C overnight on an end-over-end rotator or gentle mixer.

15. Centrifuge the peptide/bead mixtures at 2000 × g at 4 °C for 30 s.

16. Remove the supernatant, which contains unbound peptides, and save for further applications.

17. Wash the peptide-bound beads with 1 mL cold IAP buffer, and mix by inverting the tube five times.

18. Centrifuge at 2000 × g, 4 °C for 30 s. Remove the IAP wash solution by aspiration, and leave a small volume of IAP to avoid disrupting the beads.

19. Repeat the IAP wash once for a total of two washes.

20. Wash the peptide-bound beads with 1 mL ice-cold HPLC-MS water, and mix by inverting five times.

21. Centrifuge at 2000 × g, 4 °C for 30 s.

22. Remove the water wash solution by aspiration, and leave a small volume to avoid disrupting the beads.

23. Repeat the water wash twice for a total of three washes.

24. After the last water wash, centrifuge once more for 30 s at 2000 × g, 4 °C to collect any remaining volume to the bottom.

25. Aspirate the remaining water with 0.2 mm gel loading flat pipet tip while avoiding the beads.

26. Add 55 µL 0.15% TFA in HPLC-MS water to the peptide-bound beads. Incubate at RT for 10 min. Mix by tapping the bottom of the tubes intermittently.

27. Centrifuge the mixture for 30 s at 2000 × *g*, RT.

28. Remove and transfer the eluted peptides with 0.2 mm gel loading flat pipet tip to a new, labeled 0.65 mL microcentrifuge tube.

29. Add 45 µL 0.15% TFA in HPLC-MS water to the peptide-bound beads, not the eluted peptides. Incubate the mixture at RT for 10 min with intermittent agitation by tapping the bottom of tubes.

30. Centrifuge the mixture for 30 s at 2000 × *g*, RT. Remove the second elution with a 0.2 mm gel loading flat pipet tip, and combine with the first elution.

31. Centrifuge the eluted peptides at 12,000 × *g* at RT for 5 min to pellet any beads that may have carried over. Store eluted peptides on ice for immediate desalting.

3.5 Small-Scale Acetyl-Peptide Desalting with C18 StageTips

1. Prepare the C18 StageTips for desalting as described by Rappsilber et al.: assemble a set of three disks (punched out with a 18-gauge needle from an Octadecyl C18 Extraction Disk membrane) in a low-binding 200 µL pipet tip, held together in a 0.65 mL Eppendorf tube with a hole in the bottom so the solvent can flow upon centrifugation into a 2 mL collection tube.

2. Condition the StageTip with 100 µL of 100% ACN by passing the supernatant through the assembly by centrifugation at 3000 × *g* for 1 min.

3. Wash the StageTip with 100 µL of Stage Tip Solvent B by centrifugation at 3000 × *g* for 1 min (*see* **Note 6**).

4. Equilibrate the StageTip with 100 µL of Stage Tip Solvent A by centrifugation at 3000 × *g* for 1.5 min. Repeat this step for a total of two equilibration washes.

5. Load the acidified immunoaffinity peptide elution from **step 31** onto the StageTip, and centrifuge at 3000 × *g* for 1.5 min.

6. Wash the peptides bound to the StageTip with 100 µL of Solvent A by centrifugation at 3000 × *g* for 1.5 min. Repeat this step for a total of two washes.

7. Elute the peptides with 50 µL of Stage Tip Solvent B into a new Eppendorf tube, and centrifuge at 3000 × *g* for 3 min to ensure all elution volume passes through.

8. Dry the peptide eluate completely using a SpeedVac.

9. Resuspend the peptides in an appropriate volume of mobile phase A of your LC-MS system, e.g., 7 µL of 2% ACN/98%

water/0.1% formic acid (v/v/v), and add a retention time standard, such as 0.5 µL of indexed retention time standard (iRT from Biognosys or other standards).

10. Vortex the peptide solution for 10 min at 4 °C, and then centrifuge for 2 min at 12,000 × g and 4 °C.

11. Transfer the supernatant to an autosampler vial for nano LC-MS/MS (*see* **Note 7**).

3.6 Nanoflow LC-MS/MS Analysis

1. Samples are analyzed by reverse-phase HPLC-ESI-MS/MS using the Eksigent Ultra Plus nano-LC 2D HPLC system combined with a cHiPLC System, directly connected to a quadrupole time-of-flight TripleTOF 6600 mass spectrometer (SCIEX). Typically, mass resolution for precursor ion scans is ~45,000 (TripleTOF 6600), and fragment ion resolution was ~15,000 ("high sensitivity" product ion scan mode) (*see* **Note 8**). After injection, peptide mixtures are transferred onto a C18 pre-column chip and washed at 2 µL/min for 10 min with the Mobile Phase A (loading solvent). Subsequently, peptides are transferred to the analytical column ChromXP C18-CL chip and eluted at a flow rate of 300 nL/min typically with a 2–3 h gradient using aqueous and acetonitrile solvent buffers (Mobile Phases A and B).

2. *Data-dependent acquisitions (DDA).* For spectral library building, initial data-dependent acquisitions (DDA) are carried out to obtain MS/MS spectra for the 30 most abundant precursor ions (100 ms accumulation time per MS/MS) following each survey MS1 scan (250 ms), yielding a total cycle time of 3.3 s.

3. *Data-independent acquisitions (DIA).* For label-free relative quantification, all study samples are analyzed by data-independent acquisitions (DIA), using a 64-variable window SWATH acquisition strategy [10, 12]. Briefly, instead of the Q1 quadrupole transmitting a narrow mass range through to the collision cell, windows of variable width (5–90 m/z) are passed in incremental steps over the full mass range (m/z 400–1250). The cycle time of 3.2 s includes a 250 ms precursor ion scan followed by 45 ms accumulation time for each of the 64 DIA-SWATH segments.

3.7 Identification and Quantification of Acetylation Sites Using DDA and DIA

1. Mass spectrometric data from data-dependent acquisitions (DDA) is analyzed with the database search engine ProteinPilot 5.0 (SCIEX) using parameters such as trypsin digestion, cysteine alkylation set to iodoacetamide, and lysine acetylation, and in our case species *Mus musculus*, false discovery rates of 1% are used (*see* **Note 9**).

2. Using the database search engine results generated above, MS/MS spectral libraries are generated in Skyline-daily

v19.1.1.248, an open-source data processing workspace for quantitative proteomics, DIA raw data files are imported into Skyline, and both MS1 precursor ion scans and MS2 fragment ion scans are extracted for all acetylated peptides present in the spectral libraries. In Skyline typically 6–10 MS2 fragment ions are extracted per acetylated peptide based on ranking from the corresponding MS/MS spectra in the spectral libraries, and fragment peak areas are summed per peptide.

3. Relative quantification of acetylation levels and comparisons of different conditions or strains (for example, knockout versus wild-type) can be performed directly in Skyline using integrated statistical algorithms. Statistical assessment of peak selection can be done within Skyline using mProphet, which was adjusted to specifics of DIA data. Alternatively, the corresponding extracted acetylation site peak areas can be exported and subjected to other open-source programs, such as mapDIA which is specialized for processing and statistical analysis of quantitative proteomics data from DIA-MS (*see* **Note 8**).

3.8 Anticipated Results

1. Depending on the experimental design, typically 1000–2000 acetylation sites or more can be identified and quantified from enrichment of 5 mg of protein lysate or 1 mg of isolated liver mitochondria.

2. Typically, 1–20 acetylation sites are detected per protein, but this varies clearly for each protein.

3. Workflow reproducibility can be assessed between replicates by using the coefficients of variations, CV, which we observe typically as <20%.

4. The affinity workflow using the PTMScan Acetyl-Lysine Motif [Ac-K] Immunoaffinity Beads to enrich for acetylated peptides typically yields high acetylation enrichment, with 50–70% of the peptides detected being acetylated.

4 Notes

1. We recommend a minimum of 5 mg whole protein input or 1 mg of protein isolated from liver mitochondria for the acetyl-lysine affinity enrichments. A maximum of 20 mg of protein from whole liver or 4 mg of protein from isolated liver mito-chondria can be used following this protocol. Higher protein amounts are preferred for more robust results. Subsequently, the amount of antibody beads used should be changed according to the amount of protein used to maintain a high percentage of acetyl-peptide enrichment (50–70%). Ideally, an entire

tube of PTMScan Acetyl-Lysine Motif antibody-conjugated beads should be used for 20 mg of peptides, and the amount of beads will change proportionately with the amount of protein used.

2. We normalize acetyl-peptide peak areas by dividing them by their corresponding protein-level areas. This allows us to determine which changes are truly based on the acetylation profile dynamics rather than changes in overall protein abundance. In **step 9** from Subheading 3.3 a small aliquot (100 μg) of digested and desalted protein was set aside to assess the total protein-level abundance changes for normalization.

3. Freeze the peptide resuspension solution quickly after drying, and keep it frozen throughout the drying process. This will allow for easy and consistent resuspension, resulting in a fluffy white/yellow powder after drying. If an oily film forms, the drying process was not performed optimally.

4. All steps during peptide and antibody-bead conjugate preparation should be done on ice. Centrifugation should be performed at 4 °C.

5. When creating aliquots of the antibody-bead conjugate, ensure there are approximately the same number of beads per aliquot by vortexing the beads immediately before removing aliquots.

6. Ensure the StageTips (packed with C18 disks) remain damp throughout all steps until after elution. Check the StageTips frequently during centrifugation and adjust the time accordingly, and remove StageTips which are flowing more quickly than others so they do not dry. Alternatively, C18 ZipTips can be used for desalting.

7. Even though there may be no visible material after centrifugation, handle samples delicately without agitation to avoid resuspending any small particulates.

8. Other high-resolution mass spectrometric systems from other instrument vendors will be able to similarly perform these label-free, high-resolution workflows.

9. To potentially increase the number of identified acetylated peptides, additional database search engines can be used after conversion of the raw data to mzXML format using msConvert from ProteoWizard.

Acknowledgments

This work was supported by the National Institute of Allergy and Infectious Disease (R01 AI108255 to BS) and the National Institute of Diabetes and Digestive and Kidney Diseases (R24

DK085610 to Eric Verdin; R01 DK090242 to Eric Goetzman), and the NIH shared instrumentation grant for the TripleTOF system (1S10 OD016281). N.B. was supported by a postdoctoral fellowship from the Glenn Foundation for Medical Research.

References

1. Rardin MJ, Newman JC, Held JM et al (2013) Label-free quantitative proteomics of the lysine acetylome in mitochondria identifies substrates of SIRT3 in metabolic pathways. Proc Natl Acad Sci U S A 110(16):6601–6606

2. Still AJ, Floyd BJ, Hebert AS et al (2013) Quantification of mitochondrial acetylation dynamics highlights prominent sites of metabolic regulation. J Biol Chem 288 (36):26209–26219

3. Carrico C, Meyer JG, He W et al (2018) The mitochondrial acylome emerges: proteomics, regulation by sirtuins, and metabolic and disease implications. Cell Metab 27(3):497–512

4. Meyer JG, Softic S, Basisty N et al (2018) Temporal dynamics of liver mitochondrial protein acetylation and succinylation and metabolites due to high fat diet and/or excess glucose or fructose. PLoS One 13(12):e0208973

5. Wang G, Meyer JG, Cai W et al (2019) Regulation of UCP1 and mitochondrial metabolism in brown adipose tissue by reversible succinylation. Mol Cell 74(4):844–857. e847

6. Christensen DG, Baumgartner JT, Xie X et al (2019) Mechanisms, detection, and relevance of protein acetylation in prokaryotes. mBio 10 (2):e02708

7. Kuhn ML, Zemaitaitis B, Hu LI et al (2014) Structural, kinetic and proteomic characterization of acetyl phosphate-dependent bacterial protein acetylation. PLoS One 9(4):e94816

8. Zhang J, Sprung R, Pei J et al (2009) Lysine acetylation is a highly abundant and evolutionarily conserved modification in Escherichia coli. Mol Cell Proteomics 8(2):215–225

9. Basisty N, Meyer JG, Wei L et al (2018) Simultaneous quantification of the acetylome and succinylome by 'One-Pot' affinity enrichment. Proteomics 18(17):e1800123

10. Collins BC, Hunter CL, Liu Y et al (2017) Multi-laboratory assessment of reproducibility, qualitative and quantitative performance of SWATH-mass spectrometry. Nat Commun 8 (1):291

11. Gillet LC, Navarro P, Tate S et al (2012) Targeted data extraction of the MS/MS spectra generated by data-independent acquisition: a new concept for consistent and accurate proteome analysis. Mol Cell Proteomics 11(6): O111.016717

12. Schilling B, Gibson BW, Hunter CL (2017) Generation of high-quality SWATH (R) acquisition data for label-free quantitative proteomics studies using TripleTOF(R) mass spectrometers. Methods Mol Biol 1550:223–233

13. Bruderer R, Bernhardt OM, Gandhi T et al (2015) Extending the limits of quantitative proteome profiling with data-independent acquisition and application to acetaminophen-treated three-dimensional liver microtissues. Mol Cell Proteomics 14(5):1400–1410

14. MacLean B, Tomazela DM, Shulman N et al (2010) Skyline: an open source document editor for creating and analyzing targeted proteomics experiments. Bioinformatics 26 (7):966–968

Chapter 17

Affinity Enrichment Chemoproteomics for Target Deconvolution and Selectivity Profiling

Thilo Werner, Michael Steidel, H. Christian Eberl, and Marcus Bantscheff

Abstract

In order to understand the full mechanism of action of candidate drug molecules, it is critical to thoroughly characterize their interactions with endogenously expressed pharmacological targets and potentially undesired off-targets. Here we describe a chemoproteomics approach that is based on functionalized analogs of the compound of interest to affinity enrich target proteins from cell or tissue extracts. Experiments are designed as competition binding assays where free parental compound is spiked at a range of concentrations into the extracts to compete specific binders off the immobilized compound matrix. Quantification of matrix-bound proteins enables generation of dose–response curves and half-binding concentrations. In addition, the influence of the affinity matrix on the equilibrium is determined in rebinding experiments. TMT10 isobaric mass tags enable analyzing repeat binding and dose-dependent competition samples in a single mass spectrometry analysis run, thus enabling the efficient identification of targets, apparent dissociation constants, and selectivity of small molecules in a single experiment. The workflow is exemplified with the kinase inhibitor sunitinib.

Key words Chemoproteomics, Competition binding assay, Affinity enrichment, Tandem mass tag, Selectivity profiling

1 Introduction

Small molecule inhibitors are widely used as either chemical probes to understand the function of their target proteins in relevant model system or as drugs. For both applications, it is important to understand the activities of the molecule, in particular its affinity to the pharmacological target and its selectivity towards off-targets. Over the last decades, multiple chemoproteomics techniques have evolved to specifically address these questions, for example, affinity or activity-based protein profiling (ABPP), cellular thermal shift assay (CETSA) [1] and its proteome-wide extension thermal proteome profiling (TPP) [2], drug affinity responsive target stability (DARTS) [3], and others. In 2007 a quantitative chemoproteomics target profiling technology with kinobeads has been developed

Katrin Marcus et al. (eds.), *Quantitative Methods in Proteomics*, Methods in Molecular Biology, vol. 2228,
https://doi.org/10.1007/978-1-0716-1024-4_17, © Springer Science+Business Media, LLC, part of Springer Nature 2021

which enables the determination of the selectivity of kinase inhibitors against the kinome expressed in cell or tissue extracts. In this technology, multiple pan-kinase inhibitors were immobilized on Sepharose beads and used to enrich endogenously expressed kinases by incubation with cellular extracts [4]. By spiking unmodified kinase inhibitors at various concentrations into the extracts prior to the enrichment and subsequent quantitative mass spectrometric analysis, the reduction of kinase binding to the kinobeads as a function of free inhibitor concentration can be recorded. By fitting a dose-response function, half-binding concentrations (IC_{50}s) can be determined for each kinase. Based on the principles of Cheng-Prusoff relationship, Sharma et al. [5] extended this method to account for kinase segregation to the bead matrix and how this influences IC_{50} values to derive apparent dissociation constants (K_ds). In recent years, the kinome coverage increased substantially due to better affinity matrices in combination with faster and more sensitive mass spectrometers. Together, this enabled the analysis of more than 350 kinases in a single experiment [6]. In a recent landmark study, Klaeger et al. reported kinobeads selectivity profiles for a total of 240 kinase inhibitors [7]. The recent development of higher multiplexing isobaric mass tags (TMT8–11) [8] (*see* also Chaps. 8, 13–15, 28) enabled combining more data points per experiment and is crucial for the approach described in this chapter.

The basic concept of affinity enrichment chemoproteomics is not restricted to kinases, but also other target classes can be covered with the respective matrices ((HDACs) [9], dioxygenase [10]) as well as for target identification efforts, e.g., of molecules active in phenotypic screens, provided that functionalization of the compounds is feasible under preservation of their activity [11, 12].

With the protocol we describe here, it is possible to identify targets of small molecule inhibitors, their binding potencies, and their selectivity profiles in a single 10-plex experiment. At first an investigated compound or a functionalized analog thereof is immobilized on Sepharose beads. For determination of efficacy targets, knowledge on the structure-activity relationships is useful to modify the compound under preservation of its activity. For a full characterization of on- and off-targets, however, it is best to link the molecule at different positions as off-targets may follow different structure-activity relationships. As modifying a small molecule and linking it up on a matrix might alter its protein binding properties, experiments are performed in a competition binding format where increasing concentrations of the underivatized test compound of interest are spiked into aliquots of cell extracts prior to adding the bead matrix (samples 5–10 in Fig. 1). Thus, the free compound competes with the immobilized analog for binding to the (off-)target proteins. The affinity matrix is then separated from unbound proteins. Bound proteins are eluted, digested with

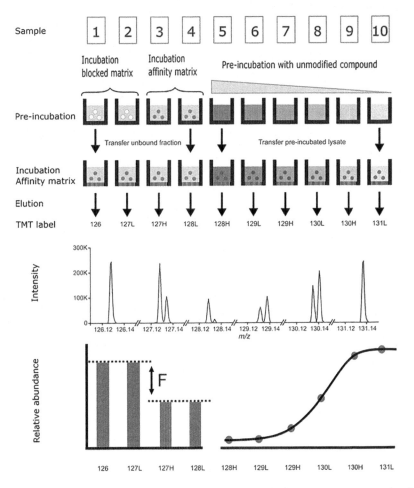

Fig. 1 Schematic representation of the chemoproteomics experiment to assess the correction factor F and dose-dependent binding in single multiplexed mass spectrometric experiment. The first four conditions are used to calculate the correction factor F (average samples 1 and 2/average samples 3 and 4). A dose–response curve is fitted to the relative abundances of the reporter ions matching to samples 5–10

trypsin, and subsequently identified and quantified using isobaric mass tagging of tryptic peptides and LC-MS/MS. Captured amounts of target and off-target proteins of the free compound are dose-dependently reduced, thus enabling the determination of concentrations of half-maximal binding (IC_{50}). Apparent dissociation constants (K_d^{app}) are derived from the IC_{50} values by taking into account the amount of (off-)target sequestered by the affinity matrix and thus influencing the equilibrium between the (off-)target and the unmodified compound, using the Cheng-Prusoff relationship (IC_{50}/K_d^{app} correction factor, Fig. 1 samples 1–4). Sequential binding experiments with only the bead matrix are performed to identify the respective IC_{50}/K_d^{app} correction factor for each captured protein. Duplicate sequential binding experiments and competition binding experiments using six different

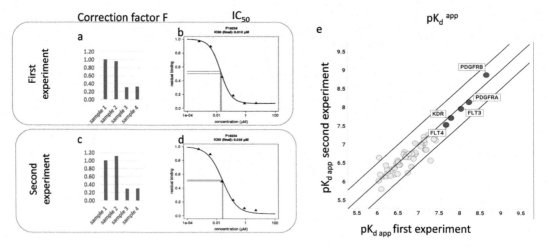

Fig. 2 Determination of apparent K_ds exemplified by profiling of sunitinib with a functionalized analog (LA—sunitinib [13]) in extracts from human placenta and BV-173 cells. **(a)/(b)** The IC_{50} to K_d conversion factor F is calculated as the ratio binding/rebinding (average samples 1 and 2/average samples 3 and 4). **(b)/(d)** Six-point dose–response curve (5 μM–0.64 nM ibrutinib) to calculate IC_{50} value for the sunitinib target PDGFR. **(e)** Replicate analysis of pK_d^{app} derived from two independent experiments. All proteins with a $pK_d^{app} > 7.5$ are marked. Plotted is the equality line and equality line ±0.5. (For uncharacterized affinity matrices, replicate experiments are required to confidently determine specific binders)

inhibitor concentrations are analyzed in a single 10-plexed TMT mass spectrometric experiment. Captured proteins may bind either directly to the affinity matrix or indirectly through protein complex partners. Here we illustrate the approach using sunitinib as an example; *see* Fig. 2.

2   Materials

Reagents and equipment listed in the following paragraphs were primarily sourced from Europe. Unless otherwise stated, the suppliers of laboratory chemicals are not critical, but reagent purity should be at least analytical grade. Likewise, most materials required might be purchased from alternative suppliers.

2.1 Whole Cell Lysis and Tissue Lysis

1. Lysis buffer: 50 mM Tris–HCl, pH 7.5, 5% (w/v) glycerol, 1.5 mM $MgCl_2$, 150 mM NaCl, 25 mM NaF, 1 mM sodium vanadate.

2. Detergent: 20% IGEPAL CA-630 added to the lysis buffer after the bead raptor set to a final concertation of 0.8% (w/v).

3. Bead Ruptor 24 Elite Bead Mill Homogenizer (Omni International).

4. Bead Ruptor 7 mL microtubes pre-filled with 2.8 mm ceramic (Omni International).

5. End-over-end shaker (Roto-Shake Genie, Scientific Industries, Inc.).

6. Ultracentrifugation: any ultracentrifuge and rotor may be used provided it can process the appropriate volumes and delivers the required g-force (e.g., $20,000 \times g$ step, Sorvall SLA600; $100,000 \times g$ step, Ti50.2).

7. Bradford protein assay (Bio-Rad).

2.2 Preparation of Affinity Matrix

1. NHS-activated Sepharose 4 Fast Flow in isopropanol (GE Healthcare Life Sciences).

2. DMSO, isopropanol, methanol.

3. Linkable sunitinib (Pfizer).

4. Triethylamine (TEA, Sigma, 99% pure).

5. End-over-end shaker (Roto-Shake Genie, Scientific Industries, Inc.).

6. 15 mL conical polypropylene tubes (Falcon, BD Biosciences, USA).

7. Laboratory centrifuge with rotor suitable for 15 mL conical tubes (e.g., Multifuge 3S-R Heraeus).

8. 2-Aminoethanol (Sigma-Aldrich).

9. HPLC system with UV detector.

2.3 Rebinding and Competition Binding

1. DP buffer: 50 mM Tris–HCl, pH 7.5, 5% (w/v) glycerol, 1.5 mM $MgCl_2$, 150 mM NaCl, 1 mM sodium vanadate filtered through 0.22 μm filter (Millipore, USA).

2. Ultracentrifuge with matching rotor (e.g., Ti50.2).

3. Sunitinib (Pfizer).

4. Detergent, e.g., Nonidet P-40 or IGEPAL CA-630.

5. Laboratory centrifuge as in Subheading 2.1.

6. Microlute, 2 mL 96-well filter plate with matching top and bottom sealing mats (Porvair Sciences).

7. End-over-end shaker (Roto-Shake Genie, Scientific Industries, Inc.).

8. Enhanced $2\times$ sample buffer (enhanced $2\times$ SB): 200 mM Tris–HCl, 250 mM Tris base, 20% glycerol, 4% SDS, 0.01% bromophenol blue (10 mL of 1% stock).

9. Vacuum concentrator centrifuge UNIVAPO 150 H/ECH (UniEquip).

10. Vacuum manifold for 96-well plates (e.g., MultiScreen HTS Vacuum Manifold, Millipore).

2.4 Protein Digestion

1. Ethanol, water, formic acid (all HPLC grade).

2. Sera-Mag Speed beads, 45152105050250, 1 μm avg. part. size, suspension (GE Healthcare).

3. Sera-Mag Speed beads, 65152105050250, 0.70–1.10 μm avg. part. size, suspension (5% solids) (GE Healthcare).

4. Cleanup solution: 2.64 mL ethanol, 0.88 mL 15%(v/v) formic acid in water, and 176 μL Sera-Mag beads slurry.

5. Digestion solution: 1 vial LysC (20 μg) (Wako) and 1 vial trypsin (20 μg) (Promega) in 0.2 mL 0.1 M HEPES (4-(2-hydroxyethyl)-1-piperazineethanesulfonic acid), pH 8.50.

6. M CAA (chloroacetamide) stock (in 0.1 M HEPES) (store frozen at −20 °C).

7. M TCEP (Tris(2-carboxyethyl)phosphine) stock (in 1 M HEPES) (store frozen at −20 °C).

8. MultiScreen™HTS-GV 0.45 μM 96-well filter plates (Millipore).

9. 96-well plate shaker (e.g., Titramax 1000 plate shaker, Heidolph).

10. Laboratory centrifuge for 96-well plates (e.g., Heraeus Megafuge 40 centrifuge, Thermo Scientific).

2.5 Labeling of Tryptic Digests with Isobaric Mass Tags

1. TMT10plex™ Isobaric Label Reagent (Thermo Fisher Scientific).

2. 96-well plate shaker (e.g., Titramax 1000 plate shaker, Heidolph).

3. Vacuum centrifuge (SpeedVac).

4. Stop solution: 2.5% ethanolamine in 0.1 M HEPES, pH 8.5.

5. Wash solution: 60/40 (v/v) 0.2 M HEPES, pH 8.5, and 0.4 mL acetonitrile.

2.6 Peptide Cleanup

1. TFA, methanol, acetonitrile.

2. Laboratory centrifuge.

3. Empore Cation 47 mm Extraction Disks (3 M).

4. Empore Octadecyl C18 47 mm Extraction Disks (3 M).

5. Vacuum centrifuge (SpeedVac) for 96-well plates.

2.7 Mass Spectrometry

1. Solution A: 0.1% formic acid, 3.5% DMSO in water.

2. Solution B: 70% acetonitrile/26.4% H2O/3.5% DMSO/0.1% formic acid (v/v/v/v).

3. Custom-made 35 cm reversed phase columns (Reprosil), 100 μm.

4. Nano-flow HPLC system, e.g., RLSC nano, pre-column, and separation column (Thermo Fisher Scientific).

5. Mass spectrometer equipped with online nano-spray ion source, e.g., Q Exactive HF or Lumos Fusion (Thermo Fisher Scientific).

2.8   Data Analysis

1. Mascot.

2. isobarQuant (https://github.com/protcode/isob/archive/master.zip).

3. Xcalibur2.2 or Xcalibur3 (part of Thermo Foundation 3 software).

4. Python 2.7, 32-bit installation.

5. R [http://www.r-project.org/].

6. DRC package (http://www.bioassay.dk).

3   Methods

3.1   Cell and Tissue Lysis

This protocol describes a generic cell lysis protocol for frozen cell pellets or tissue using 0.8% (w/v) IGEPAL CA-630 as detergent. In this example human placenta is lysed.

1. Weigh the organ before lysis, and calculate the amount of buffer to be used. Use 4–5 volumes of lysis buffer.

2. Thaw the placenta in a petri dish on ice, and cut it into pieces.

3. Transfer the organ pieces into 7 mL Bead Ruptor tubes (with 2.8 mm ceramic beads). For 13.6 g placenta, use 54.84 mL detergent free lysis buffer and equally distribute into 27 tubes.

4. Operate the Bead Ruptor according to the manufacturer's specifications. For placenta we recommend the following program: 5.5 m/s (speed), 15 s (interval), 3 cycles, 30s dwell.

5. Transfer the content of the Bead Ruptor tubes (lysate + beads) into a beaker.

6. Add IGEPAL CA-630 (stock 20%) to final concentration of 0.8% and transfer to 50 mL tubes (*see* **Note 1**).

7. Incubate for 30 min at 4 °C on an end-over-end shaker at speed 3.

8. Incubate for 30 min rotating in cold room.

9. Centrifuge for 10 min at 20,000 $\times g$ at 4 °C.

10. Transfer supernatant to a UZ-polycarbonate tube.

11. Spin supernatant for 1 h at 140,000 $\times g$ at 4 °C.

12. Save supernatant (remove and discard the lipid layer), and transfer supernatant into fresh 50 mL conical tube.

13. Determine protein concentration by Bradford assay.

14. Typical protein concentrations range from 5 to 20 mg/mL.

15. Freeze aliquots in liquid nitrogen and store at −80 °C.

3.2 Preparation of Affinity Matrix

Sepharose beads are available with different functional groups enabling immobilization of probe compounds with a variety of different chemical procedures. Likewise, availability of suitable residues on the probe compound itself is an important parameter for selection of the appropriate immobilization chemistry and, thus, the required functional group on the Sepharose resin. The generic protocol described below uses NHS-activated Sepharose beads and requires a primary or secondary amine function on the probe compound for immobilization. As example sunitinib and a functionalized analog of sunitinib [13] were chosen.

1. Wash 1 mL (settled volume) of NHS-Sepharose beads three times with 10 mL of DMSO using a 15 mL conical tube. Gently spin down beads after each washing step using a suitable centrifuge, and discard supernatant.

2. After the last washing step, resuspend beads in 1 mL of DMSO.

3. Add 10 μL of a 100 mM (see **Note 2**) probe compound (linkable sunitinib) solution in DMSO to generate an affinity matrix with a 1 mM ligand density (see **Note 3**).

4. Add 15 μL of TEA, shake, and settle the beads using a centrifuge.

5. Pipette 20 μL of supernatant into a HPLC autosampler vial and dilute with 20 μL methanol.

6. Incubate beads at room temperature in darkness on an end-over-end shaker for 16–20 h.

7. Pipette 20 μL of supernatant into a HPLC autosampler vial and dilute with 20 μL methanol.

8. Add 50 μL of 2-aminoethanol to beads, and incubate on an end-over-end shaker overnight to block remaining NHS groups of Sepharose beads.

9. Wash beads once with 10 mL DMSO and three times with isopropanol. Gently spin down beads after each washing step using a suitable centrifuge, and discard supernatant.

10. After the last washing step, resuspend beads in 1 mL isopropanol, and keep beads in the refrigerator (or at −20 °C in a tightly sealed conical tube until further use).

11. For the generation of blocked bead, (needed as control in the rebinding experiment), follow the same protocol as above, but without probe compound.

Table 1
Sample composition

Sample	First incubation step	Second incubation step
1	Lysate + blocked beads + DMSO	1 h NBF[a] of sample 1 + affinity matrix
2	Lysate + blocked beads + DMSO	1 h NBF[a] of sample 2 + affinity matrix
3	Lysate + affinity matrix + DMSO	1 h NBF[a] of sample 3 + fresh affinity matrix
4	Lysate + affinity matrix + DMSO	1 h NBF[a] of sample 4 + fresh affinity matrix
5	Lysate + 5 μM sunitinib	1 h lysate + 5 μM sunitinib + affinity matrix
6	Lysate + 0.83 μM sunitinib	1 h lysate + 0.83 μM sunitinib + affinity matrix
7	Lysate + 0.14 μM sunitinib	1 h lysate + 0.14 μM sunitinib + affinity matrix
8	Lysate + 0.024 μM sunitinib	1 h lysate + 0.024 μM sunitinib + affinity matrix
9	Lysate + 0.0039 μM sunitinib	1 h lysate + 0.0039 μM sunitinib + affinity matrix
10	Lysate + 0.00064 μM sunitinib	1 h lysate + 0.00064 μM sunitinib + affinity matrix

The first incubation step is performed with a 20% increased input of material (600 μL) to ensure that the second incubation step can be carried out with exactly 500 μL pre-incubated lysate
[a]NBF = non-bound fraction

3.3 Rebinding and Competition Binding Experiment

Layout: Each experiment consists of ten samples that are processed in two steps (see Table 1). Both incubation steps are carried out in 96-well filter plates, and multiple experiments can be performed in parallel.

1. Dilute lysate further with DP buffer containing 0.4% (w/v) IGEPAL CA-630 to a final protein concentration of 5 mg/mL.

2. Ultracentrifuge for 20 min at $100,000 \times g$ at 4 °C (see **Note 4**).

3. Discard pellet and keep supernatant on ice.

4. Dissolve test compound (sunitinib) in DMSO, and prepare a 30 mM stock solution.

5. Dilute stock in DMSO to the following concentrations: 1000, 166, 27.8, 4.6, 0.7, and 0.13 μM.

6. Wash 500 μL probe matrix twice with 10 mL DP buffer and once with 10 mL DP buffer containing 0.4% (w/v) IGEPAL CA-630, allow beads to settle after each wash step, and discard supernatant.

7. After the last washing step, resuspend the beads in 1×DP-Buffer/0.4% IGEPAL to get a 5% slurry (e.g., 300 μL beads to be resuspended in 6 mL buffer).

8. For the first incubation step, add 20 μL of the 5% bead slurry (i.e., 21 μL dry beads) to the wells of samples 1–4.

9. Use blocked beads for samples 1 and 2 and affinity matrix beads for samples 3 and 4. In the first incubation step, the wells of samples 5–10 have no beads.

10. Add 300 μL in 1×DP-Buffer/0.4% IGEPAL to pre-wet.

11. Let the buffer run through without letting the beads dry out.

12. Close filter plate tightly with bottom sealing mat.

13. Samples 1 and 2: Add 0.6 mL precleared lysate (3 mg) to the prepared wells in the filter plate (A) with 21 μL washed and equilibrated blocked beads.

14. Samples 3 and 4: Add 0.6 mL precleared lysate (3 mg) to the prepared wells in the filter plate (A) with 21 μL washed and equilibrated capturing beads.

15. Samples 5–10: Add 0.6 mL precleared lysate (3 mg) to wells in filter plate A.

16. Add 3 μL of each 200× stock compound-solution (1:200 dilution) or 3 μL DMSO as control (0.5%) to the respective well of the filter plate (A).

17. Close filter plate A with heated top sealing mat tightly.

18. Incubate the samples for 1 h at 4 °C on an end-over-end shaker.

19. Once the first incubation step is done, lay filter plate A upside down on a bench, and carefully remove the bottom sealing mat.

20. Place filter plate A on top of a fresh 96-well 1.2 mL square storage plate, and spin down briefly to remove sample from the top lid (1 min, 250 × g, 4 °C).

21. Remove the top sealing mat, and centrifuge again (1 min, 250 × g, 4 °C) to collect all the samples in the 1.2 mL storage plate.

22. Transfer 500 μL of the pre-incubated lysate or lysate/compound mixture to filter plate B containing 17.5 μL beads of the affinity per well.

23. Seal the plate, and incubate the samples 1 h at 4 °C on an end-over-end shaker.

24. Wash the beads in the filter plate with 10 mL 1×DP-buffer/0.4% IGEPAL (add 5× 2 mL) using a vacuum manifold (*see* **Note 5**).

25. Wash the beads in the filter plate with 6 mL 1×DP-buffer/0.4% IGEPAL (add 3× 2 mL) using a vacuum manifold.

26. Close filter plate with fresh bottom mat tightly.

27. Add 50 μL enhanced 2× sample buffer (incl. 50 mM DTT) to each well.

28. Close top with adhesive sealing tape at Incubate plate at 50 °C in incubator, shake for 30 min at 140 rpm.

29. Remove adhesive sealing tape and bottom mat without losing eluate.

30. Place filter plate onto fresh 96-well plate, and centrifuge for 2 min at 250 × g at RT.

31. Lyophilize frozen samples for 2 h using the vacuum concentrator centrifuge.

3.4 Protein Digestion

This protocol is a modification of the magnetic bead-based SP3 approach (solid-phase sample preparation) [14] utilizing 96-well filter plates instead of magnets for bead retention. This enables highly parallelized sample processing without the need for special equipment.

1. Prepare beads on the day of the experiment. Homogenize both bead solutions, and transfer 20 μL of each into a 1.5 mL tube. Wash the beads with 200 μL water, and store the final suspension in 200 μL water.

2. Prepare cleanup solution. Volumes are for 80 samples, respectively, 8 TMT10 experiments.

3. Prepare the digestion solution right before the experiment (*see* **Note 6**). For one 96-well plate, use 2.925 mL 0.1 M HEPES, 0.255 mL trypsin solution, 0.225 mL LysC solution, 0.18 mL CAA stock, and 0.045 mL TCEP stock.

4. Dissolve lyophilized samples in water (25 μL).

5. Place the filter plate on top of a waste plate.

6. Homogenize prepared cleanup solution. Add 40 μL cleanup solution per sample to the wells used for digestion. Add 20 μL of samples to 40 μL cleanup solution (*see* **Note 7**).

7. Cover filter plate using a plastic lid, and shake for 15 min at 500 rpm on a Titramax shaker to ensure complete protein binding.

8. Centrifuge for 1 min at 1000 × g (Megafuge). Discard waste plate, and place filter plate on top of a new waste plate.

9. Wash beads four times with 200 μL 70% EtOH by centrifuging for 2 min at 1000 × g. Discard flow-through after each centrifugation step.

10. Add 40 μL of digest solution to each cleaned sample, and place the filter plate on top of a washed collection plate.

11. Seal the plate with a plastic foil, and digest the samples overnight on Titramax orbital shaker (500 rpm) at room temperature.

12. Centrifuge for 1 minute at 1000 × *g*. Add 10 μL water, and repeat centrifugation to ensure complete peptide elution.

13. Lyophilize samples in the collection plate.

3.5 Labeling of Tryptic Digests with Isobaric Mass Tags

1. Resuspend peptides in 10 μL water, and add 10 μL TMT10plex™ reagent (5 mg in 580 μL acetonitrile). Seal plate with a plastic foil. Incubate for 1 h at RT, shaking at 500 rpm on a Titramax shaker.

2. Add 5 μL of freshly prepared stop solution. Incubate for 15 min at RT, shaking at 500 rpm on a Titramax shaker.

3. Pool all samples, corresponding to one experiment, in a 1.5 mL tube by using a 200 μL pipette tip.

4. Add 10 μL wash solution in all wells, and pool residual solution into the corresponding tubes.

5. Dry sample in a SpeedVac.

3.6 Peptide Cleanup

This part of the protocol has been adopted from Rappsilber et al. [15]. Other techniques using reversed phase and strong cation exchange material to clean up peptides can be used.

1. Stage tip generation: Punch three SCX plugs that fit into tip of a 200 μL pipette tip. Place the disks in the pipette tip, and push the 5 mm above the end of the tip. Push firmly but not too hard. Thereafter, punch three plugs out of a C18 extraction disc, and place them on the top.

2. Dissolve dried peptides in 100 μL of 6% TFA. Ensure that the pH is below 4.

3. Place the C18 SCX stage with an adaptor ring into an 1.5 mL tube.

4. To pre-wet the disks, add 100 μL methanol, and centrifuge for 2 min at 2000 × *g*.

5. Equilibrate the disks twice with 100 μL of 0.5% TFA, 2% acetonitrile solution, and centrifuge for 2 min at 2000 × *g*. Discard the flow-through.

6. Apply the sample to the tip, and centrifuge for 2 min at 2000 × *g*.

7. Wash the disks with 100 μL of 0.5% TFA, 2% acetonitrile. Centrifuge for 2.5 min at 2000 × *g*.

8. Elute the desalted peptides to the SCX plugs with 100 μL of 0.5% TFA in 60% acetonitrile (centrifuge for 4 min at 1000 × *g*), followed by 200 μL of 0.5% TFA in 60% acetonitrile (centrifuge for 2.5 min at 2000 × *g*).

9. Place the tip in a new collection tube, and eluate the peptides with 50 μL of 5% NH3 in 80% acetonitrile. Centrifuge at 2000 × g for 1 min. Repeat this step once.

10. Transfer the into a 96-well plate for LC-MS analysis. Dry sample in a SpeedVac and store at −20 °C.

3.7 Mass Spectrometry

Appropriate instrumentation for mass spectrometric analysis of the samples generated in Subheading 3.6 is available at many institutions and core facilities. Many different instrument platforms are well suited for this purpose. For TMT10 it is important to use a mass spectrometer that can resolve the 6 mDa between some of the isobaric mass labels at the m/z range 126–131 dalton. As an example, in the following section, typical instrument settings for the currently very common Q-Exactive mass spectrometers will be described.

1. Resuspend samples in 15 μL 0.1% formic acid/H_2O.

2. Inject 33% of the sample into a nano-LC system coupled online to the mass spectrometer.

3. Peptides are separated on a custom-made 35 cm reversed phase columns (Reprosil) with 100 μm inner diameter using a linear elution gradient from 2% to 40% solution B (see **Note 8**) within 120 min at a flow rate of 300 nL/min.

4. Peptide masses were detected at 60,000 resolution with an ion target value of 3E6 in MS mode and at 30,000 resolution with an ion target of 2E5 in MS/MS mode (see **Note 9**). Ion accumulation time was kept at 250 ms for MS and 64 ms for MS/MS spectra.

5. The ten most intense multiply charged ions per MS spectrum are fragmented. Using the Apex trigger option to "2–10s," dynamic exclusion 40 s and exclude isotopes on (see **Note 10**).

6. Intact peptide ions were isolated with isolation width set to 0.4 Th, and HCD fragmentation was performed using 35% normalized collision energy.

3.8 Data Analysis

1. For protein identification, tandem mass spectra are typically converted to peak lists and submitted to a database search against a non-redundant sequence database of the species of interest (see **Note 11**). A variety of free and commercial search engines such as Mascot [16], Sequest [17], and Andromeda [18] are available for this purpose.

2. For protein quantification, multiple software packages were developed. We use and recommend for TMT samples isobarQuant [19]. Follow the pre- and post-Mascot process as described in the quick start guide of isobarQuant. Set the

label of the vehicle control (sample 1; *see* Subheading 3.3) as reference for the calculation of fold changes.

3. IC_{50} values are calculated by performing a sigmoidal curve fit using R and the DRC package. Measured IC_{50} values can be influenced by the affinity matrix essentially following the Cheng-Prusoff relationship. In order to determine by how much the binding of the matrix to the protein influences the binding equilibrium between free inhibitor and its target and to calculate apparent dissociation constants (K_d^{app}), depletion factors F (defined by $F = 1 + [B]/K_dB$, where $[B]$ is the bead ligand concentration and K_dB is the dissociation constant of the protein from the beads). This factor is determined by the ratio of bound/rebound protein amounts (samples 1, 2/ samples 3, 4; *see* Subheading 3.3).

4. Not all identified proteins have bound specifically to the matrix. To eliminate irreproducible and unspecific results, perform at least two independent experiments.

4 Notes

As some proteins can have very distinct expression patterns, the choice of the cells or tissue can have an impact on the detection of proteins. Here we have chosen a mix of extracts from human placenta and BV-137 cells to achieve comprehensive coverage of the VGFR family members.

1. It is important to add the detergent after the homogenization in the beads ruptor.

2. 100 mM is quite a high concentration for some probe compounds. Alternatively, DMF might be used for increased solubility. Turbid solutions indicate incomplete dissolution of the test compound that might cause problems during subsequent steps in the experiment.

3. The ligand density is a critical parameter for the characteristics of the affinity matrix and depends on the functionalized analogs. Too high ligand densities result in high target depletion, and thus, the determination of correction factors is impaired. Also, it can lead to a high background of unspecific binding proteins. If the ligand density is too low, the amounts of captured target proteins can be reduced, and in extreme cases, these can no longer be detected. For uncharacterized functionalized analogs, a 1 mM ligand density is usually a good starting point.

4. The ultracentrifugation step after thawing of lysate is critical. If it is not performed, it can lead to myosin action aggregation on the affinity matrix resulting in a high protein background.

5. Vacuum Manifold: Release the vacuum before addition of new buffer. It is important to fill the well to the top so that the walls are rinsed completely. The washing steps can also be performed in a centrifuge.

6. We recommend to use HEPES to avoid unwanted side reaction products generated with the triethylammonium bicarbonate (TEAB) [20].

7. Trapping of proteins on the bead surface is most efficient at 50% EtOH (v/v). Therefore, it is important to immediately add the samples to the filter plates prepared with cleanup solution to avoid evaporation of EtOH.

8. We observed the beneficial effect as described by Hahne et al. [21] only with ultra-pure DMSO. When changing to DMSO-containing solvents, it is recommended to flush the system for at least a day and disconnect it from the mass parameter during that time to avoid contamination.

9. For isobaric mass tags with neutron encoded spacing, it is important to limit the MS/MS target ion setting to avoid coalescence effects [22].

10. For the quantification using isobaric mass tags, it is important to select as pure as possible precursor ions for fragmentation. Co-isolation of other peptides leads to co-fragmentation and ratio compression.

11. Typical settings for database search include a mass tolerance of 10 ppm for peptide precursor ions and 20 mDa for fragment masses. Variable modifications include methionine oxidation, N-terminal acetylation of proteins, and TMT modification of peptide N termini. Carbamidomethylation of cysteine residues and TMT modifications of lysine residues are set as fixed modifications.

Acknowledgments

We would like to thank Jürgen Stuhlfauth and Toby Mathieson for contributions to the development of the protocol.

References

1. Molina DM, Jafari R, Ignatushchenko M et al (2013) Monitoring drug target engagement in cells and tissues using the cellular thermal shift assay. Science 341:84–87

2. Savitski MM, Reinhard FBM, Franken H et al (2014) Tracking cancer drugs in living cells by thermal profiling of the proteome. Science 346:1255784

3. Lomenick B, Hao R, Jonai N et al (2009) Target identification using drug affinity responsive target stability (DARTS). Proc Natl Acad Sci 106:21,984–21,989

4. Bantscheff M, Eberhard D, Abraham Y et al (2007) Quantitative chemical proteomics reveals mechanisms of action of clinical ABL kinase inhibitors. Nat Biotechnol 25:1035–1044

5. Sharma K, Weber C, Bairlein M et al (2009) Proteomics strategy for quantitative protein interaction profiling in cell extracts. Nat Methods 6:741–744

6. Eberl HC, Werner T, Reinhard FB et al (2019) Chemical proteomics reveals target selectivity of clinical Jak inhibitors in human primary cells. Sci Rep 9:1–14

7. Klaeger S, Heinzlmeir S, Wilhelm M et al (2017) The target landscape of clinical kinase drugs. Science 358:eaan4368

8. Werner T, Becher I, Sweetman G et al (2012) High-resolution enabled TMT 8-plexing. Anal Chem 84:7188–7194

9. Bantscheff M, Hopf C, Savitski MM et al (2011) Chemoproteomics profiling of HDAC inhibitors reveals selective targeting of HDAC complexes. Nat Biotechnol 29:255–265

10. Joberty G, Boesche M, Brown JA et al (2016) Interrogating the druggability of the 2-oxoglutarate-dependent dioxygenase target class by chemical proteomics. ACS Chem Biol 11:2002–2010

11. Abrahams KA, Chung C, Ghidelli-Disse S et al (2016) Identification of KasA as the cellular target of an anti-tubercular scaffold. Nat Commun 7:1–13

12. Becher I, Werner T, Doce C et al (2016) Thermal profiling reveals phenylalanine hydroxylase as an off-target of panobinostat. Nat Chem Biol 12:908–910

13. Ramsden N, Perrin J, Ren Z et al (2011) Chemoproteomics-based design of potent LRRK2-selective lead compounds that attenuate Parkinson's disease-related toxicity in human neurons. ACS Chem Biol 6:1021–1028

14. Hughes CS, Foehr S, Garfield DA et al (2014) Ultrasensitive proteome analysis using paramagnetic bead technology. Mol Syst Biol 10:757

15. Rappsilber J, Mann M, Ishihama Y (2007) Protocol for micro-purification, enrichment, pre-fractionation and storage of peptides for proteomics using StageTips. Nat Protoc 2:1896–1906

16. Perkins DN, Pappin DJ, Creasy DM et al (1999) Probability-based protein identification by searching sequence databases using mass spectrometry data. Electrophoresis 20:3551–3567

17. Eng JK, McCormack AL, Yates JR (1994) An approach to correlate tandem mass spectral data of peptides with amino acid sequences in a protein database. J Am Soc Mass Spectrom 5:976–989

18. Cox J, Neuhauser N, Michalski A et al (2011) Andromeda: a peptide search engine integrated into the MaxQuant environment. J Proteome Res 10:1794–1805

19. Franken H, Mathieson T, Childs D et al (2015) Thermal proteome profiling for unbiased identification of direct and indirect drug targets using multiplexed quantitative mass spectrometry. Nat Protoc 10:1567–1593

20. Ting L, Rad R, Gygi SP et al (2011) MS3 eliminates ratio distortion in isobaric multiplexed quantitative proteomics. Nat Methods 8:937–940

21. Hahne H, Pachl F, Ruprecht B et al (2013) DMSO enhances electrospray response, boosting sensitivity of proteomic experiments. Nat Methods 10:989–991

22. Werner T, Sweetman G, Savitski MF et al (2014) Ion coalescence of neutron encoded TMT 10-Plex reporter ions. Anal Chem 86:3594–3601

Chapter 18

2nSILAC for Quantitative Proteomics of Prototrophic Baker's Yeast

Stefan Dannenmaier, Silke Oeljeklaus, and Bettina Warscheid

Abstract

Stable isotope labeling by amino acids in cell culture (SILAC) combined with high-resolution mass spectrometry is a quantitative strategy for the comparative analysis of (sub)proteomes. It is based on the metabolic incorporation of stable isotope-coded amino acids during growth of cells or organisms. Here, complete labeling of proteins with the amino acid(s) selected for incorporation needs to be guaranteed to enable accurate quantification on a proteomic scale. Wild-type strains of baker's yeast (*Saccharomyces cerevisiae*), which is a widely accepted and well-studied eukaryotic model organism, are generally able to synthesize all amino acids on their own (i.e., prototrophic). To render them amenable to SILAC, auxotrophies are introduced by genetic manipulations. We addressed this limitation by developing a generic strategy for complete "native" labeling of prototrophic *S. cerevisiae* with isotope-coded arginine and lysine, referred to as "2nSILAC". It allows for directly using and screening several genome-wide yeast mutant collections that are easily accessible to the scientific community for functional proteomic studies but are based on prototrophic variants of *S. cerevisiae*.

Key words SILAC, Native SILAC, 2nSILAC, Quantitative proteome analysis, Mass spectrometry, Yeast, Prototroph, Mitochondria

1 Introduction

Stable isotope labeling by amino acids in cell culture (SILAC) [1] (*see* also Chaps. 8, 13) combined with high-resolution mass spectrometry (MS) is a well-established technique in functional proteomic research that is widely employed to address a variety of biological questions. SILAC relies on the metabolic incorporation of stable isotope-coded "heavy" amino acids (^{2}H, ^{13}C, and/or ^{15}N) into proteins during cell growth, resulting in the introduction of a distinct mass difference into peptides that contain the amino acid(s) used for labeling. Consequently, differentially labeled cells can be mixed directly after harvesting, prior to subsequent sample

---

Silke Oeljeklaus is co-corresponding author.

Katrin Marcus et al. (eds.), *Quantitative Methods in Proteomics*, Methods in Molecular Biology, vol. 2228, https://doi.org/10.1007/978-1-0716-1024-4_18, © The Author(s) 2021

preparation and liquid chromatography-tandem MS (LC-MS/MS) analysis. Thus, experimental variations resulting from separate sample handling are reduced to a minimum, which makes SILAC an accurate technique for quantitative proteome analysis using MS [2]. This is typically highly beneficial for quantitative proteomic studies that build on multistep protocols for, e.g., the preparation of organellar or other subcellular fractions, the purification of protein complexes, or the enrichment of peptides carrying post-translational modifications.

SILAC is usually performed using heavy arginine (e.g., $^{13}C_6{}^{13}N_4$, Arg10) and lysine (e.g., $^{13}C_6{}^{13}N_2$, Lys8) for labeling. When combined with the standard protease trypsin for the proteolytic digest, all peptides contain at least one labeled amino acid (except for the C-terminus of a protein), thereby maximizing the number of SILAC peptide pairs available for accurate protein quantification. However, heavy arginine is often metabolically converted into heavy proline ($^{13}C_5{}^{13}N_1$, Pro6, in case Arg10 was used for labeling) [3–8], which results in satellite peaks originating from peptides containing heavy proline that are generally not considered for quantification. Depending on the extent of arginine-to-proline conversion, this may significantly impair the accuracy of quantitative data. A frequently used strategy to counteract the generation of heavy proline is the addition of unlabeled proline to the culture medium [4, 8, 9]. A further requirement for accurate quantification in generic SILAC experiments is the complete incorporation of heavy arginine and lysine into the proteome, which may be compromised in cells or organisms that are prototrophic for these amino acids.

The baker's yeast *Saccharomyces cerevisiae* is widely accepted as eukaryotic model organism, and since many vital biological processes are conserved during evolution, knowledge gained in research using yeast is often transferable to the human system [10, 11]. Numerous SILAC-based proteomic studies employing *S. cerevisiae* greatly contributed to a better understanding of many fundamental cellular functions in eukaryotes. However, commonly used *S. cerevisiae* strains such as BY4741 [12] and its derivatives are prototrophic. Thus, SILAC studies usually rely on yeast in which arginine and lysine auxotrophies have been introduced through deletion of genes coding for enzymes of the arginine and lysine biosynthetic pathways to render the strains amenable for complete SILAC labeling.

The ease of working with *S. cerevisiae*, which includes a short generation time, easy cultivation in defined media under controlled conditions, and simplicity of genetic manipulation, has fueled the generation of several yeast mutant strain collections such as gene deletion strains [13, 14] and strains expressing genes fused to an epitope tag for tandem affinity purification [15, 16], a library of GFP fusion strains [17], and the SWAp-Tag libraries

[18, 19]. These strain collections are readily available to the scientific community and, thus, provide valuable resources of yeast mutants for global and targeted functional proteomic studies, but they are based on the strain BY4741 lacking the auxotrophies for arginine and lysine. Previous studies, however, showed that prototrophic yeast can efficiently be labeled with heavy lysine [20–23], a concept referred to as "native SILAC" (nSILAC) [21].

In this chapter, we describe a generic nSILAC strategy that allows for metabolic labeling of prototrophic yeast using both heavy lysine and arginine, thus exploiting the full potential of SILAC for accurate proteome quantification. We refer to this strategy as "complete native SILAC" or "2nSILAC" [8]. It allows for the utilization of yeast from the strain collections without further genetic manipulations for the introduction of auxotrophies. 2nSILAC works efficiently for cells grown on different carbon sources and is therefore well-suited to study a large variety of biological questions. It is compatible with protocols for the purification of organelles, subcellular fractions, protein complexes, or peptide fractions, making it a universal tool for the study of protein functions. We exemplarily describe the application of 2nSILAC for the analysis of mitochondrial gene deletion effects on the mitochondrial proteome. Thus, we here provide a protocol for the cultivation of yeast cells under respiratory growth conditions, which is generally used for the study of mitochondria in *S. cerevisiae*, and for the preparation of subcellular fractions enriched in mitochondria (Fig. 1).

2   Materials

Buffers and solutions described in this chapter should be prepared with ultrapure water ($\geq$18.2 MΩ $\times$ cm resistivity, Milli-Q quality); reagents and solvents for LC-MS sample preparation and subsequent LC-MS analysis should be HPLC grade.

2.1 Metabolic Labeling of Prototrophic Yeast Using 2nSILAC

Isogenic *S. cerevisiae* strains that are prototrophic for arginine and lysine, such as BY4741 [12] and a BY4741 deletion strain [13] that lacks the gene encoding the mitochondrial protein of your interest (*see* **Note 1**).

2.1.1 Yeast Strains

2.1.2 Culture Media

1. YPD agar plates: 1% (w/v) yeast extract, 2% (w/v) peptone, 2% (w/v) glucose, 2% (w/v) agar containing appropriate selection markers (if applicable).

2. Amino acid mix (12.5×): 125 mg/L adenine, 1000 mg/L L-aspartic acid, 250 mg/L L-histidine, 625 mg/L L-isoleucine, 1250 mg/L L-leucine, 250 mg/L L-methionine, 625 mg/L L-

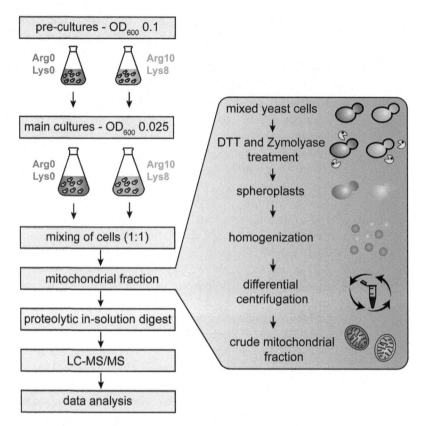

Fig. 1 *Workflow for the proteomic analysis of mitochondrial fractions by 2nSILAC.* Pre-cultures of cells, inoculated at an OD_{600} of 0.1, are cultivated overnight in medium containing light arginine (Arg0) and lysine (Lys0) or the corresponding isotope-coded heavy variants (Arg10, Lys8). Main cultures are inoculated at an OD_{600} of 0.025 and grown until they reach exponential growth phase. Differentially SILAC-labeled cells are mixed in equal ratio and treated with DTT and Zymolyase to digest the cell walls. The resulting spheroplasts are homogenized followed by differential centrifugation to generate a crude mitochondrial fraction. Proteins are digested in solution using LysC and trypsin and analyzed by LC-MS/MS

phenylalanine, 1250 mg/L L-threonine, 625 mg/L L-trypto-phan, 625 mg/L L-tyrosine, 250 mg/L uracil, 1750 mg/L L-L-valine.

3. Amino acids for stable isotope labeling: $^{12}C_6{}^{14}N_4$-L-arginine (Arg0), $^{13}C_6{}^{15}N_4$-L-arginine (Arg10), $^{12}C_6{}^{14}N_2$-L-lysine (Lys0) and $^{12}C_6{}^{14}N_2$-L-lysine (Lys8).

4. Synthetic complete (SC) medium: 0.17% (w/v) YNB without amino acids, 0.5% (w/v) ammonium sulfate, 3% (v/v) glyc-erol/0.02% (w/v) glucose (*see* **Notes 2** and **3**), 1× amino acid mix, pH adjusted to 5.5 with KOH; "light" medium is supple-mented with 50 mg/L Arg0 and Lys0; "heavy" medium is supplemented with 50 mg/L Arg10 and Lys8 (*see* **Notes 4** and **5**).

5. Sterile water and sterile pipette tips.

6. Spectrophotometer to measure optical density at 600 nm (OD_{600}).

2.2 Preparation of Whole Cell Lysates

1. Lysis buffer: 8 M urea, 75 mM NaCl, 50 mM Tris–HCl (pH 8.0), 1 mM EDTA.

2. Screw cap microcentrifuge tubes (2 mL).

3. Glass beads (0.4–0.6 mm in diameter).

4. Bradford reagent, bovine serum albumin (BSA), and a spectrophotometer (or any alternative method) to determine protein concentrations.

5. Minilys® benchtop homogenizer (Peqlab/VWR).

2.3 Small-Scale Preparation of Mitochondria-Enriched Fractions

1. Dithiothreitol (DTT) buffer: 100 mM Tris–H_2SO_4 (pH 9.4), 10 mM DTT.

2. Zymolyase buffer: 20 mM potassium phosphate (pH 7.4), 1.2 M sorbitol.

3. Zymolyase® 20-T (MP Biomedicals Life Sciences).

4. Homogenization buffer: 10 mM Tris-HCl (pH 7.4), 0.6 M sorbitol, 1 mM EDTA, 1 mM PMSF.

5. SEM buffer: 250 mM sucrose, 1 mM EDTA, 10 mM MOPS (pH 7.2).

6. Thermomixer.

7. 3 mL syringe (e.g., Omnifix®, Braun).

8. 0.8 × 22 mm blunt end needle (e.g., Sterican®, Braun).

9. BSA and Bradford reagent to determine protein concentrations.

10. Liquid nitrogen.

2.4 LC–MS Sample Preparation

2.4.1 Acetone Precipitation

1. Acetone (100%, ice-cold).

2. Urea buffer: 8 M urea in 50 mM NH_4HCO_3 (ABC).

2.4.2 Proteolytic In-Solution Digest

1. 50 mM Tris(2-carboxy-ethyl)-phosphine (TCEP) in H_2O.

2. 500 mM iodoacetamide (IAA) or 2-chloroacetamide (CAA) dissolved in 50 mM ABC.

3. 100 mM DTT in H_2O.

4. LysC endoproteinase (Promega), 40 ng/μL dissolved in 50 mM ABC (*see* **Note 6**).

5. Sequencing grade modified trypsin (Promega), 20 ng/μL dissolved in 50 mM ABC.

6. 1% (v/v) trifluoroacetic acid (TFA).

7. 0.1% (v/v) TFA.

2.4.3 Desalting *of Peptides*	1. 100% methanol (MeOH). 2. Buffer A: 0.5% (v/v) acetic acid (AcOH). 3. Buffer B: 80% (v/v) acetonitrile (ACN)/0.5% (v/v) AcOH. 4. 200 μL pipette tips. 5. Luer lock needle (gauge, G16; outer diameter, 1.7 mm). 6. C18 material (3 M Empore or Affinisep). 7. Glass vials with cap (CS-Chromatographie Service GmbH, Germany; article number 300 101), septa (CS-Chromatographie Service GmbH, Germany; article number 300 351), and inserts for LC analysis (Macherey-Nagel GmbH & Co. KG, Germany; article number 702 968). 8. Vacuum concentrator.

2.5 LC-MS/MS
and Data Analysis

1. Solvent A: 0.1% (v/v) formic acid (FA).

2. Solvent B: 86% (v/v) ACN/0.1% (v/v) FA.

3. Nano UHPLC system (e.g., an UltiMate™ 3000 RSLCnano system; Thermo Fisher Scientific, Dreieich, Germany) equipped with a C18 pre-column (e.g., PepMap™ C18 pre-column; length, 5 mm; inner diameter, 0.3 mm; Thermo Scientific) for washing and preconcentration of peptides and an analytic C18 reversed-phase nano LC column (e.g., Acclaim™ PepMap™ RSLC column; length, 50 cm; inner diameter, 75 μm; particle size, 2 μm; Thermo Scientific) for peptide separation.

4. ESI-MS instrument (e.g., Q Exactive Plus mass spectrometer; Thermo Fisher Scientific, Bremen, Germany).

5. Software for protein identification and quantification (e.g., MaxQuant/Andromeda, www.maxquant.org) [24–26].

6. Protein sequence database for *S. cerevisiae* (*see* **Note 7**).

7. Software tools for further data processing (e.g., statistics, filtering, clustering, GO term enrichment, etc.) and data visualization such as Perseus [26], MATLAB (https://www.mathworks.com/products/matlab.html), or R (https://www.r-project.org/).

3 Methods

We here describe a sample preparation and analysis pipeline for the study of mitochondrial gene deletion effects on the mitochondrial proteome using 2nSILAC. Individual protocols include the cultivation of yeast cells under respiratory conditions, which are generally used for the study of mitochondria in *S. cerevisiae* (*see* **Note 8**); the preparation of mitochondria-enriched fractions; LC-MS sample preparation, analysis, and data processing; the assessment of heavy arginine and lysine incorporation as well as arginine-to-proline conversion; and the analysis and visualization of SILAC-MS data obtained in replicate experiments.

3.1 Cultivation and Metabolic Labeling of Yeast Cells

1. Plate yeast strains (BY4741 control and gene deletion strain) on YPD agar plates (with or without selection markers, depending on the requirements of the yeast) and incubate for 2–3 days at 30 °C.

2. Prepare light and heavy SC medium and sterilize by autoclaving.

3. Use a sterile pipette tip to inoculate 1 mL of sterile water with cells from the agar plates and determine the OD_{600} of the cell suspensions.

4. Use cell suspensions to prepare a pre-culture for each strain by inoculating 10 mL of light or heavy SC medium at an initial OD_{600} of 0.1 (see **Note 9**).

5. Incubate pre-cultures overnight at 160 rpm and 30 °C.

6. Determine the OD_{600} of the pre-cultures (see **Note 10**) and dilute them into 150 mL of fresh SC medium at an initial OD_{600} of 0.025.

7. When the cells reach the exponential growth phase (OD_{600} 0.5–1.5), take a 20 mL aliquot of the heavy-labeled cells to analyze the incorporation of Arg10 and Lys8 into the proteins and the extent of heavy arginine-to-proline conversion using whole cell lysates (see Subheading 3.2). Collect the cells by centrifugation for 5 min at $5000 \times g$ and 4 °C. Remove the supernatant and proceed as described in Subheading 3.2.

8. Mix the remaining differentially labeled cells in equal ratio based on their OD_{600}.

9. Harvest cells by centrifugation for 5 min at $5000 \times g$ and 4 °C. Remove the supernatant, store cells at −20 °C (if necessary), and proceed with the preparation of a mitochondrial fraction (Subheading 3.3).

3.2 Preparation of Whole Cell Lysates

When performing SILAC experiments, we routinely check the incorporation of the isotopically labeled amino acids and the extent of arginine-to-proline conversion. To this end, we prepare whole cell lysates of heavy labeled cells and analyze the proteins by LC-MS.

1. Resuspend the cell pellet (see Subheading 3.1, **step** 7) in 500 μL of lysis buffer, transfer the sample to a 2-mL screw cap microcentrifuge tube, and add 300 mg of glass beads.

2. Disrupt and lyse the cells by bead beating using a Minilys® homogenizer (see **Note 11**) for 4 min at 5000 rpm. Perform

this step twice with at least 2 min cooling on ice between the cycles.

3. Remove cell debris and glass beads by centrifugation for 5 min at 15,000 × g and 4 °C.

4. Determine the protein concentration, e.g., by using the Bradford assay [27] with BSA as protein standard according to the standard protocol.

5. Adjust the final protein concentration to 1 μg/μL using lysis buffer for subsequent proteolytic in-solution digest (Subheading 3.4.2).

3.3 Small-Scale Preparation of Mitochondria-Enriched Fractions

We here describe a protocol for a small-scale preparation of mitochondria-enriched fractions [8], which is based on a medium-scale protocol published recently [28]. Briefly, cell walls are removed by treating cells with DTT and Zymolyase, a mixture of lytic enzymes. The resulting spheroplasts are homogenized using a syringe and mitochondria are enriched via differential centrifugation (see Notes 12 and 13).

1. Resuspend cells (see Subheading 3.1, step 9) in 1.5 mL of DTT buffer and transfer them to a 2 mL microcentrifuge tube.

2. Incubate cells for 10 min at 1000 rpm and 30 °C using a thermomixer.

3. Pellet cells by centrifugation for 5 min at 1500 × g, discard supernatant, and wash cells with 1.5 mL of Zymolyase buffer.

4. Centrifuge for 5 min at 1500 × g, discard supernatant, and resuspend the pellet in 1.5 mL of Zymolyase buffer containing 4 mg/mL Zymolyase 20-T.

5. Incubate for 30 min at 1000 rpm and 30 °C in a thermomixer.

6. Harvest spheroplasts by centrifugation for 5 min at 900 × g and 4 °C (see Note 14).

7. Resuspend pellet in 1.5 mL of ice-cold homogenization buffer and homogenize the spheroplasts by drawing the sample 20 times up and down through a 0.8 × 22 mm blunt end needle attached to a 3-mL syringe.

8. Remove cell debris by centrifugation for 10 min at 900 × g and 4 °C.

9. Transfer supernatant to a new microcentrifuge tube.

10. Take a small aliquot (5%) of the supernatant (S0.9) for immunoblot analysis (Fig. 3a).

11. Resuspend the pellet of step 8 in 1.5 mL of ice-cold homogenization buffer and take an equal aliquot of 5% of the pellet fraction (P0.9) for immunoblot analysis (Fig. 3a).

12. Proceed with the supernatant from step 9 and repeat step 8.

13. Transfer the supernatant to a new microcentrifuge tube and pellet mitochondria by centrifugation for 5 min at $16,800 \times g$ and 4 °C. Take an aliquot (5%) of the supernatant (S16.8) for immunoblot blot analysis (Fig. 3a).

14. Wash the mitochondria-containing pellet with 2 mL of ice-cold SEM buffer and centrifuge again for 10 min at $900 \times g$ and 4 °C.

15. Transfer the supernatant to a new microcentrifuge tube and centrifuge for 5 min at $16,800 \times g$ and 4 °C.

16. Resuspend the mitochondrial pellet in 200 µL of ice-cold SEM buffer and determine the protein concentration, e.g., by using the Bradford assay. Take an aliquot (5%) of the mitochondria-enriched fraction (Mito) for immunoblot analysis (Fig. 3a).

17. Mitochondria-enriched fractions can either directly be used for LC-MS analysis or snap-frozen in liquid nitrogen and stored at −80 °C until further use.

3.4 LC-MS Sample Preparation

3.4.1 Acetone Precipitation

1. Precipitate proteins of mitochondria-enriched fractions by adding the fivefold volume of ice-cold acetone.

2. Incubate samples at −20 °C for at least 2 h.

3. Pellet precipitated proteins by centrifugation for 15 min at $15,000 \times g$ and 4 °C.

4. Remove acetone and dry samples for 5 min under the laminar flow hood.

5. Resuspend proteins in urea buffer (1 µg/µL final concentration) for subsequent proteolytic in-solution digest.

3.4.2 Proteolytic In-Solution Digest

The protocol for proteolytic in-solution digestion of whole cell lysates or mitochondria-enriched fractions is described for 10 µg of protein (*see* **Note 15**).

1. Add 1 µL of 50 mM TCEP and incubate for 30 min at room temperature to reduce cysteine residues (*see* **Note 16**).

2. Add 1 µL of 500 mM IAA or CAA and incubate for 30 min at room temperature to alkylate free thiol groups. When IAA is used, incubate samples in the dark.

3. Quench the alkylation reaction by adding 3 µL of 100 mM DTT.

4. For protein digestion with LysC, dilute samples to 4 M urea using 50 mM ABC, add 100 ng of LysC, and incubate for 4 h at 37 °C (*see* **Note 17**).

5. For digestion with trypsin, dilute the samples further to 1.8 M urea using 50 mM ABC, add 200 ng of trypsin, and incubate overnight at 37 °C.

6. To inactivate the proteases, acidify the samples by adding 45 μL of 1% TFA. Use pH paper to make sure that the samples are acidified (pH < 3.0).

7. Remove insoluble material by centrifugation at 13,800 × g for 5 min and transfer supernatant to a new microcentrifuge tube.

3.4.3 Desalting of Peptides

StageTips can be used to remove urea, salts, and other contaminants prior to MS analysis. The method we describe is adapted from Rappsilber and colleagues [29]. In brief, monolithic C18 material is inserted into 200 μL pipette tips and is used to capture and wash proteolytic peptides. Afterward, peptides are eluted, dried in vacuo and are ready for LC-MS/MS analysis.

1. To prepare StageTips, cut out three disks of C18 matrix with a Luer lock needle and insert them into a 200 μL pipette tip (*see* **Note 18**). Gently push the disks toward the end of the tip without applying too much pressure.

2. Insert the StageTips into microcentrifuge tubes with a hole in the lid or use appropriate adapters to fix the StageTips in the microcentrifuge tubes.

3. Activate StageTips by adding 100 μL of MeOH and centrifuge at 800 × g until StageTips are empty.

4. Wash StageTips by adding 100 μL of buffer B and centrifuge at 800 × g until empty.

5. Equilibrate StageTips by adding 100 μL of buffer A and centrifuge at 800 × g until empty.

6. Load the acidified samples onto the StageTips and centrifuge at 800 × g until empty.

7. Add 100 μL of buffer A for washing and centrifuge at 800 × g until empty. Perform this step twice (*see* **Note 19**).

8. Place glass inserts into microcentrifuge tubes and insert Stage-Tips into the glass vials.

9. Elute peptides by adding 30 μL of buffer B and centrifuge at 800 × g until empty.

10. Repeat **step 9**, combine eluates, and dry peptides in vacuo using a vacuum concentrator.

3.5 LC-MS/MS and Data Analysis

Peptide mixtures are analyzed by nanoflow LC-MS/MS using an UltiMate™ 3000 RSLCnano system coupled to a Q Exactive Plus mass spectrometer.

1. Place the glass inserts containing the peptide mixtures into glass vials and seal the vial with septa and caps.

2. Load the peptide mixtures onto the pre-column of the RSLCnano system and wash and preconcentrate peptides with solvent A for 5 min at a flow rate of 30 μL/min.

3. Switch pre-column and analytical column in line and elute peptides by applying the following gradient at a flow rate of 0.3 μL/min: 4% solvent B for 25 min, 4–39% B in 25 min, 39–54% B in 175 min, 54–95% B in 15 min, and 5 min at 95% B (*see* **Note 20**).

4. Use the following parameters to operate the Q Exactive Plus: MS scans, *m/z* 375–1700 at a resolution of 70,000 at *m/z* 200; automatic gain control, 3×10^6; maximum ion time, 60 ms.

5. Apply a TOP15 method for fragmentation of precursor ions ($z \geq +2$) by higher-energy collisional dissociation. Set the normalized collision energy to 28%. Acquire fragment ion spectra at a resolution of 35,000. Set automatic gain control for MS/MS scans to 10^5 ions and the maximum ion time to 120 ms. Use a dynamic exclusion time for previously selected precursor ions of 45 s.

6. Download the latest version of the MaxQuant/Andromeda software package for protein identification and quantification (*see* **Note 21**).

7. To evaluate labeling efficiencies for Arg10 and Lys8 as well as the extent of heavy arginine-to-proline conversion in whole cell lysates of cells grown in heavy SC medium (*see* Subheading 3.2), use the default settings in MaxQuant (refers to version 1.6.5.0) for protein identification and the required settings for SILAC-based quantification with the following exceptions: disable the option "requantify" (*see* **Note 22**) and add Pro6 as variable modification.

The incorporation efficiency of Arg10 and Lys8 is calculated for individual peptides using the following equation [30]:

$$\text{Incorporation } (\%) = \frac{\text{ratio } (H/L)}{\text{ratio } (H/L) + 1} * 100$$

The degree of incorporation needs to be calculated separately for Arg10- and Lys8-containing peptides since the yeast strain(s) used may exhibit differences in incorporation for heavy arginine and lysine. Use the MaxQuant output file "evidence.txt". Remove entries derived from the reverse and contaminant databases. Select only peptides that contain either lysine or arginine and use the non-normalized H/L ratios. For peptides showing no or complete incorporation of Arg10 or Lys8, H/L ratios cannot be computed since the corresponding heavy or light counterpart is missing. In these cases, the incorporation is manually set to 0% for no incorporation or 100% for complete incorporation. To obtain relative quantitative data of high accuracy, the mean incorporation across all peptides should be >95% for both arginine- and lysine-containing peptides. To assess arginine-to-proline conversion, calculate (i) the percentage of Pro6-containing peptides in relation to all identified peptides or (ii) the relative intensity of Pro6-containing peptides (in %) in relation to the summed MS intensity of all peptides per dataset (*see* **Note 23**).

8. To analyze MS data of 2nSILAC experiments, use MaxQuant default settings for protein identification and SILAC-based quantification. Enable the options "requantify" (*see* **Note 22**) and "match between runs" when analyzing replicates.

9. Use the MaxQuant "proteinGroups.txt" output file for the quantitative analysis of 2nSILAC data. Remove the hits derived from contaminant and reverse database and discard entries that were only identified by site. In case the experimental setup included label switches, adjust the protein ratio H/L by inverting the respective values. Logarithmize the protein abundance ratio ($\log_2$) for each replicate to obtain normally distributed data for statistical tests. The reproducibility between individual replicates can be assessed by ratio-versus-ratio plots and calculation of the Pearson correlation. To analyze which protein groups show significant changes in abundance between the two populations of cells/mitochondria, perform a two-sided one-sample t-test with a hypothetical mean value of 0. Data can be visualized by plotting the mean log ratios against the $\log_{10}$ of the p-values (Fig. 3b). In addition, intensity-dependent and independent outlier tests can be performed to identify significant outliers relative to the overall population (e.g., significance A or B [24]).

4   Notes

1. BY4741 and derivatives thereof are available from Euroscarf (http://www.euroscarf.de). When using different strains than BY4741, make sure that the genotype is compatible with arginine and lysine labeling [8].

2. Growth on respiratory carbon sources such as glycerol induces mitochondrial proliferation and results in higher yields of mitochondria. Addition of 0.02% (w/v) glucose allows for reproducible cell cultivation without affecting the yield.

3. Depending on the experiment, alternative carbon sources such as galactose or glucose may be used. Utilization of galactose results in a slightly lower yield of mitochondria. When cells are grown under respiratory growth conditions using glycerol/

glucose or galactose, we recommend to use the larger set of amino acids described in this section to ensure complete incorporation of heavy arginine and lysine. For the growth of cells under fermentative conditions using glucose, a minimal set of essential amino acids is sufficient ($20\times$ amino acid mix: adenine, L-histidine, L-leucine, L-methionine, L-tryptophan, and uracil, 400 mg/L each).

4. Unlabeled proline (Pro0) may need to be added to the medium to prevent arginine-to-proline conversion. The appropriate concentration of Pro0 should be determined experimentally. Add the same amount of Pro0 to both light and heavy SC medium.

5. In case a different growth medium is used, make sure to omit external sources of light arginine and lysine (e.g., yeast extract) from the medium.

6. LysC is proteolytically active in buffer containing high concentrations of urea and is often used in combination with trypsin to increase the efficiency of protein digestion.

7. A sequence database specific for *S. cerevisiae* is included in the MaxQuant software package; alternatively, it can be downloaded from the *Saccharomyces* Genome Database (https:// downloads.yeastgenome.org/sequence/).

8. When adapting the 2nSLAC strategy to different growth conditions, pilot experiments need to be performed, and culture conditions may need to be adjusted to ensure high incorporation efficiencies for heavy arginine and lysine as well as minimum extents of arginine-to-proline conversion.

9. To obtain quantitative data of high reliability, we strongly recommend to perform at least three biological replicates including a light/heavy label switch experiment to prevent artifacts introduced by the labeling scheme.

10. For 2nSILAC experiments, it is crucial that the OD_{600} of pre-cultures does not exceed a value of 4. Higher cell densities may result in insufficient incorporation of the heavy amino acids (Fig. 2) and an increase in arginine-to-proline conversion [8].

11. Alternatively, cells can be disrupted using a vortex mixer.

12. Please note that mitochondria-enriched fractions prepared according to the protocol we describe here still contain small amounts of other organelles such as the ER and Golgi (Fig. 3a). To obtain mitochondrial fractions of higher purity (>90%), mitochondria-enriched fractions can be further purified by sucrose density gradient centrifugation as described elsewhere [28, 31].

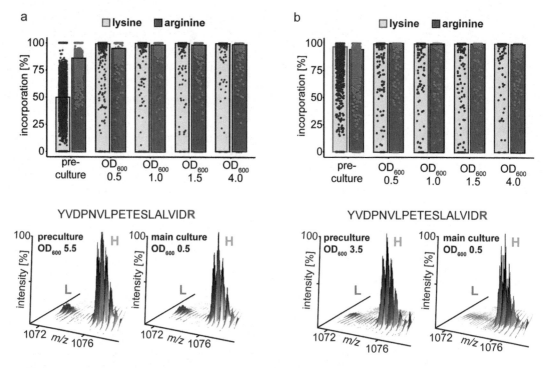

Fig. 2 *Incorporation of isotope-coded arginine and lysine into proteins of prototrophic yeast.* Prototrophic yeast was cultured in medium containing heavy arginine and lysine. Pre-cultures were grown up to an OD_{600} of 5.5 (**a**) or 3.5 (**b**). Corresponding main cultures were cultivated until they reached an OD_{600} of 4.0. As shown for the entirety of arginine- and lysine-containing peptides (**a** and **b**, top) and exemplarily for the arginine-containing peptide YVDPNVLPETESLALVIDR (**a** and **b**, bottom), peptides from cells that were pre-cultured to a high density exhibit incomplete incorporation of heavy amino acids, in particular heavy arginine, at early time points (e.g., at OD_{600} of 0.5). In contrast, peptides from cells that were pre-cultured to a lower density show virtually complete incorporation (>98.5%) of both heavy arginine and lysine. L/H, isotope pattern of the light/ heavy peptide species; m/z, mass-to-charge ratio. Data shown in this figure are taken from Dannenmaier et al. (2018) [8] (https:/pubs.acs.org/doi/10.1021/acs.analchem.8b02557). Permissions related to this material need to be directed to ACS Publications

13. To evaluate the quality of the mitochondrial fraction, we recommend to monitor the enrichment of the mitochondria during individual steps of the differential centrifugation. To this end, take aliquots of samples at critical steps during the preparation (as indicated in the protocol) and analyze them by SDS-PAGE and immunoblotting using antisera against marker proteins for mitochondria and selected other subcellular compartments (Fig. 3a).

14. Suspensions of spheroplasts need to be handled gently; cut off pipette tips to prevent disrupting the organelles.

15. For deep proteome coverage, sample fractionation on protein or peptide level such as SDS-PAGE, strong cation exchange chromatography, high pH reversed-phase chromatography, or other alternatives can be integrated into the workflow.

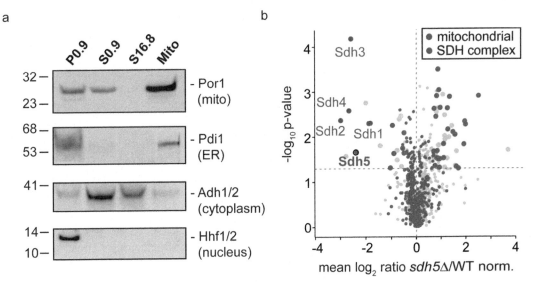

Fig. 3 *Analysis of mitochondrial gene deletion effects on the mitochondrial proteome using 2nSILAC.* (**a**) A mitochondria-enriched fraction (Mito) was generated by differential centrifugation. Equal aliquots of samples were analyzed by SDS-PAGE and immunoblotting using antibodies against marker proteins for different organelles as indicated. The mitochondria-enriched fraction was virtually devoid of Adh1/2 and Hhf1/2, but it still contained a significant amount of the ER protein Pdi1 (*see* **Note 12**). P/S0.9, pellet/supernatant of the second $900 \times g$ centrifugation; S16.8, supernatant of the $16,800 \times g$ centrifugation. (**b**) Mitochondrial proteomes of wild-type (WT) cells and cells lacking the gene *SDH5* (*sdh5Δ*) coding for an assembly factor of the succinate dehydrogenase complex (SDH; complex II of the mitochondrial respiratory chain) are quantitatively compared following the 2nSILAC strategy as described in Fig. 1. The experiment was performed in three independent replicates, and data are visualized by plotting the mean $\log_2$ of normalized (norm.) *sdh5Δ*/WT ratios of proteins quantified in at least two replicates against the negative $\log_{10}$ of their *p*-values determined in a two-sided Student's *t*-test. In addition, an intensity-based outlier *t*-test (Significance B) was performed. Proteins considered to be significantly affected by *SDH5* deletion were required to have a *p*-value of <0.05 in both the Student's *t*-test and the outlier analysis (indicated by large filled circles). The 2nSILAC study shows that deletion of the assembly factor SDH5 leads to a strong reduction of the levels of all four core components of the SDH complex (i.e., Sdh1-4). Data shown in this figure are taken from Dannenmaier et al. (2018) [8] (https:/pubs.acs.org/doi/10.1021/acs.analchem.8b02557). Permissions related to this material need to be directed to ACS Publications

16. Avoid heating the samples since this may lead to carbamylation of lysine residues by the urea present in the buffer in high concentration (i.e., 8 M).

17. Protein digestion with LysC may be omitted. However, we recommend to perform double digestion with both LysC and trypsin, as it improves proteome coverage and accurate relative quantification [8, 32].

18. The estimated binding capacity per disk is 2–4 μg. Make sure not to over- or underpack as this may lead to clogging or loss of sample.

19. After this step, StageTips can be stored at 4 °C for several weeks. In this case, peptides need to be rehydrated by adding 100 µL of buffer A before continuing with the elution.

20. Depending on the complexity of the sample and the analytical column used for peptide separation, slope and duration of the gradient may need to be adjusted to identify maximum numbers of peptides and proteins.

21. Useful information about MaxQuant including computational requirements is provided in literature [24, 25, 33–35] and on the MaxQuant documentation website. We further suggest to subscribe to the MaxQuant Google group for additional information, recommendations, advice for troubleshooting, and for posting questions.

22. The option "requantify" generally allows for the calculation of SILAC ratios in case peptides are only present in the isotope-labeled (i.e., heavy) or unlabeled (light) form by assigning a peptide intensity for the missing counterpart from signals in MS spectra at the m/z value expected for the missing counterpart [33]. This feature is of advantage when high ratios are expected, e.g., in gene deletion studies or in affinity purification MS experiments. However, "requantify" should not be applied to analyze the incorporation of heavy amino acids into peptides/proteins since this would result in incorrect H/L ratios.

23. Please note that the calculation of arginine-to-proline conversion based on identified peptides generally underestimates the occurrence of the conversion since only sequenced Pro6-containing peptides are taken into account.

Acknowledgments

This work was supported by the Deutsche Forschungsgemeinschaft (DFG, German Research Foundation) Project-ID 278002225—RTG 2202, Project-ID 403222702—SFB 1381, Germany's Excellence Strategy (CIBSS—EXC-2189—Project ID 390939984), and the European Research Council (ERC) Consolidator Grant No. 648235.

References

1. Ong S-E, Blagoev B, Kratchmarova I et al (2002) Stable isotope labeling by amino acids in cell culture, SILAC, as a simple and accurate approach to expression proteomics. Mol Cell Proteomics 1:376–386

2. Lau H-T, Suh HW, Golkowski M et al (2014) Comparing SILAC- and stable isotope dimethyl-labeling approaches for quantitative proteomics. J Proteome Res 13:4164–4174

3. Ong S-E, Kratchmarova I, Mann M (2003) Properties of 13C-substituted arginine in stable isotope labeling by amino acids in cell culture (SILAC). J Proteome Res 2:173–181

4. Bendall SC, Hughes C, Stewart MH et al (2008) Prevention of amino acid conversion in SILAC experiments with embryonic stem cells. Mol Cell Proteomics 7:1587–1597

5. Bicho CC, de Lima Alves F, Chen ZA et al (2010) A genetic engineering solution to the "arginine conversion problem" in stable isotope labeling by amino acids in cell culture (SILAC). Mol Cell Proteomics 9:1567–1577

6. Park SK, Liao L, Kim JY et al (2009) A computational approach to correct arginine-to-proline conversion in quantitative proteomics. Nat Methods 6:184–185

7. Lössner C, Warnken U, Pscherer A et al (2011) Preventing arginine-to-proline conversion in a cell-line-independent manner during cell cultivation under stable isotope labeling by amino acids in cell culture (SILAC) conditions. Anal Biochem 412:123–125

8. Dannenmaier S, Stiller SB, Morgenstern M et al (2018) Complete native stable isotope labeling by amino acids of Saccharomyces cerevisiae for global proteomic analysis. Anal Chem 90:10,501–10,509

9. Piechura H, Oeljeklaus S, Warscheid B (2012) SILAC for the study of mammalian cell lines and yeast protein complexes. Methods Mol Biol 893:201–221

10. Kolkman A, Slijper M, Heck AJR (2005) Development and application of proteomics technologies in Saccharomyces cerevisiae. Trends Biotechnol 23:598–604

11. Botstein D, Fink GR (2011) Yeast: an experimental organism for 21st century biology. Genetics 189:695–704

12. Brachmann CB, Davies A, Cost GJ et al (1998) Designer deletion strains derived from Saccharomyces cerevisiae S288C: a useful set of strains and plasmids for PCR-mediated gene disruption and other applications. Yeast 14:115–132

13. Giaever G, Chu AM, Ni L et al (2002) Functional profiling of the Saccharomyces cerevisiae genome. Nature 418:387–391

14. Giaever G, Nislow C (2014) The yeast deletion collection: a decade of functional genomics. Genetics 197:451–465

15. Gavin AC, Bösche M, Krause R et al (2002) Functional organization of the yeast proteome by systematic analysis of protein complexes. Nature 415:141–147

16. Ghaemmaghami S, Huh W-K, Bower K et al (2003) Global analysis of protein expression in yeast. Nature 425:737–741

17. Huh W-K, Falvo JV, Gerke LC et al (2003) Global analysis of protein localization in budding yeast. Nature 425:686–691

18. Yofe I, Weill U, Meurer M et al (2016) One library to make them all: streamlining the creation of yeast libraries via a SWAp-Tag strategy. Nat Methods 13:371–378

19. Weill U, Yofe I, Sass E et al (2018) Genome-wide SWAp-Tag yeast libraries for proteome exploration. Nat Methods 15:617–622

20. Dilworth DJ, Saleem RA, Rogers RS et al (2010) QTIPS: a novel method of unsupervised determination of isotopic amino acid distribution in SILAC experiments. J Am Soc Mass Spectrom 21:1417–1422

21. Fröhlich F, Christiano R, Walther TC (2013) Native SILAC: metabolic labeling of proteins in prototroph microorganisms based on lysine synthesis regulation. Mol Cell Proteomics 12:1995–2005

22. Kaneva IN, Longworth J, Sudbery PE et al (2018) Quantitative proteomic analysis in Candida albicans using SILAC-based mass spectrometry. Proteomics 18:1700278

23. Christiano R, Nagaraj N, Fröhlich F et al (2014) Global proteome turnover analyses of the Yeasts S. cerevisiae and S. pombe. Cell Rep 9:1959–1965

24. Cox J, Mann M (2008) MaxQuant enables high peptide identification rates, individualized p.p.b.-range mass accuracies and proteome-wide protein quantification. Nat Biotechnol 26:1367–1372

25. Cox J, Neuhauser N, Michalski A et al (2011) Andromeda: a peptide search engine integrated into the MaxQuant environment. J Proteome Res 10:1794–1805

26. Tyanova S, Temu T, Sinitcyn P et al (2016) The Perseus computational platform for comprehensive analysis of (prote)omics data. Nat Methods 13:731–740

27. Bradford MM (1976) A rapid and sensitive method for the quantitation of microgram quantities of protein utilizing the principle of protein-dye binding. Anal Biochem 72:248–254

28. Morgenstern M, Stiller SB, Lübbert P et al (2017) Definition of a high-confidence mitochondrial proteome at quantitative scale. Cell Rep 19:2836–2852

29. Rappsilber J, Ishihama Y, Mann M (2003) Stop and go extraction tips for matrix-assisted laser desorption/ionization, nanoelectrospray, and LC/MS sample pretreatment in proteomics. Anal Chem 75:663–670

30. Oeljeklaus S, Schummer A, Suppanz I et al (2014) SILAC labeling of yeast for the study of membrane protein complexes. Methods Mol Biol 1188:23–46

31. Meisinger C, Sommer T, Pfanner N (2000) Purification of Saccharomcyes cerevisiae mitochondria devoid of microsomal and cytosolic contaminations. Anal Biochem 287:339–342

32. Glatter T, Ludwig C, Ahrné E et al (2012) Large-scale quantitative assessment of different in-solution protein digestion protocols reveals superior cleavage efficiency of tandem Lys-C/trypsin proteolysis over trypsin digestion. J Proteome Res 11:5145–5156

33. Tyanova S, Temu T, Cox J (2016) The Max-Quant computational platform for mass spectrometry-based shotgun proteomics. Nat Protoc 11:2301–2319

34. Cox J, Matic I, Hilger M et al (2009) A practical guide to the MaxQuant computational platform for SILAC-based quantitative proteomics. Nat Protoc 4:698–705

35. Tyanova S, Mann M, Cox J (2014) MaxQuant for in-depth analysis of large SILAC datasets. Methods Mol Biol 1188:351–364

Chapter 19

Metabolic Labeling of *Clostridioides difficile* Proteins

Anke Trautwein-Schult, Jürgen Bartel, Sandra Maaß, and Dörte Becher

Abstract

The introduction of stable isotopes in vivo via metabolic labeling approaches (SILAC or ^{15}N-labeling) allows, after combination of differentially treated labeled and unlabeled cells or protein extracts, for correction of protein quantification errors implemented during elaborated sample preparation workflows. The SILAC-based approach uses heavy arginine and lysine to incorporate the label into bacterial strains and cell lines, whereas ^{15}N-metabolic labeling is achieved by cultivation in ^{15}N-salt containing media. In case of *Clostridioides difficile*, the lack in arginine and lysine auxotrophy as well as the Stickland dominated metabolism makes metabolic labeling challenging. Here, a step-by-step guideline for the metabolic labeling of *C. difficile* is described, which combines cultivation in liquid ^{15}N-substituted medium followed by cultivation steps on solid ^{15}N-substituted medium. The described procedure results in a label incorporation rate higher than 97%. Cells prepared by the following method can be used as standard for relative quantification approaches of, e.g., the membrane or surface proteome of *C. difficile*.

Key words ^{15}N-Celtone, Anaerobic condition, *Clostridioides difficile*, Metabolic labeling, Proteomics, Stable isotope label

1 Introduction

The ubiquitous, obligate anaerobic, spore-forming, Gram-positive bacterium *Clostridioides difficile* [1–3], formerly known as *Clostridium difficile*, is one main cause of healthcare-associated infective diarrhea after antibiotic exposure [4] with a mortality rate of approximately 10% [5, 6]. Due to the increasing interest in this pathogenic organism, different proteomics studies were performed to support the identification of new targets for therapeutic strategies or drug development.

Relative quantitative proteome analyses in general are based on the determination of protein amounts in a label-free (*see* Chaps. 8, 16, 20–24) manner or on the introduction of stable isotopes into proteins or peptides [7] (*see* Chaps. 8–10, 13–15, 17–19, 25, 26). For *C. difficile* many relative quantitative proteomics studies were based on chemical labeling of peptides [8–11], label-free quantification approaches [12–16], as well as comparative 2D PAGE-based

Katrin Marcus et al. (eds.), *Quantitative Methods in Proteomics*, Methods in Molecular Biology, vol. 2228,
https://doi.org/10.1007/978-1-0716-1024-4_19, © Springer Science+Business Media, LLC, part of Springer Nature 2021

analysis [17, 18] (*see* Chaps. 4–7). The main advantages of the label-free quantification approaches are a high quantitative proteome coverage and an unrestricted application for a proteomic analysis of each organism [7]. The label-based approaches show in comparison to label-free approaches a higher accuracy as well as reproducibility and thus enable the determination of even small significant changes in protein abundance between different samples. The introduction of stable isotopes into each protein during protein synthesis via the metabolic labeling procedure was first described for bacterial proteomes in 1999 [19]. Usually, the label incorporation rate is sufficient after six to eight generations [7]. Due to the early combination of unlabeled sample ("light") and labeled standard ("heavy") at cellular or protein level, this approach allows for correction of all sources of quantification errors introduced by elaborated sample preparation workflows [7]. The sample preparation of the combined extracts and the co-elution of the "light" and "heavy" peptides in a subsequent LC-MS/MS analysis allows for an abundance comparison of precursor ions in the survey spectrum [19]. Especially, studies of the membrane or surface proteome, which are prone to higher standard deviations due to complex preparation workflows, will be more sensitive by a metabolic labeling strategy compared to a label-free approach.

For a successful metabolic labeling of all tryptic peptides with the SILAC (stable isotope labeling by amino acids in cell culture) approach (*see* Chaps. 13, 18), the organisms should be arginine and lysine auxotroph [20]. *C. difficile* is described as proline, cysteine, leucine, isoleucine, tryptophan, and valine but unfortunately not arginine and lysine auxotroph [21], which makes an application of the often used traditional SILAC labeling strategy impossible. As an alternative for a SILAC metabolic labeling approach, an in vivo ^{15}N-metabolic labeling by cultivation in ^{15}N-salt containing media could be used like described for *Bacillus subtilis* [22]. Unfortunately, due to the Stickland dominated metabolism in *C. difficile* [23, 24], which enables this bacterium to grow with amino acids as sole carbon and/or energy source, the implementation of a metabolic labeling strategy for this anaerobic pathogen was challenging. The pairwise usage of amino acids as electron donor and electron acceptor [24] hampers the incorporation of heavy nitrogen in the proteins of *C. difficile*. In 2018, the here described metabolic labeling procedure was published [25]. The application of the presented protocol resulted in the quantification of 1,110 proteins. Recently, the group of de Koning published a self-prepared ^{15}N-yeastolate medium for an alternative *C. difficile* labeling procedure [26]. A combination of the here in detail explained procedure and the newly described workflow of de Koning could also enhance the proteome coverage for further proteome studies due to the usage of different growth conditions used to label proteins.

In this chapter, a detailed cultivation procedure with ^{15}N-labeled medium as described in Trautwein-Schult et al. 2018 is shown, which resulted in an incorporation rate higher than 97% [25]. The often described *C. difficile* minimal medium [23, 12] supplemented with casamino acids, cysteine, and tryptophan needed to be modified for metabolic labeling due to the very expensive or commercially unavailability of isotopically labeled supplements. Here, ^{15}N-Celtone, a labeled amino acid mixture, was added to the medium instead of casamino acids, cysteine, and tryptophan. The ^{15}N-labeled cells or protein extracts ("heavy") can subsequently be mixed with unlabeled cells or protein extracts ("light"). After digestion of proteins, a relative quantification of their peptide abundances can be performed by comparison of their precursor ion abundance in the survey mass spectrum.

2 Materials

Prepare all solutions in *aqua bidest* and store all reagents at 4 °C (unless indicated otherwise). When working with hazardous chemicals, always follow local bylaws and regulations to protect laboratory staff and the environment. Diligently follow all waste disposal regulations when disposing waste materials.

2.1 Buffer, Media, and Solutions for Preparatory Operations

1. Purified *C. difficile* spores (e.g. Leibniz Institute DSMZ-German Collection of Microorganisms and Cell Cultures) (*see* **Note 1**).

2. 1% (w/v) resazurin: Dissolve 10 mg resazurin sodium salt in 1 mL *aqua bidest* and store the closed tube enclosed with aluminum foil for light protection at 4 °C.

3. 10% (w/v) taurocholate: Dissolve 1 g taurocholic acid sodium salt hydrate in 10 mL *aqua bidest*, sterilize with a 0.2 μm syringe filter, and store 500 μL aliquots at −20 °C.

4. 1× PBS (phosphate buffered saline): Dissolve 8 g NaCl, 0.2 g KCl, 1.78 g $Na_2HPO_4 \times 2 H_2O$, and 270 mg KH_2PO_4 in 800 mL *aqua bidest*. Adjust pH to approximately 7.4 with HCl. Fill up to 1 L with *aqua bidest* and sterilize the buffer with a 0.2 μm syringe filter. Store 1× PBS at room temperature.

5. 5× amino acids stock solution: Dissolve 12.5 g casein hydrolysate (e.g. Carl Roth GmbH & Co. KG), 125 mg L-tryptophan, and 625 mg L-cysteine in 250 mL *aqua bidest*. Prepare this stock solution always fresh and sterilize the stock solution with a 0.2 μm syringe filter directly before use.

6. 10× salts stock solution: Dissolve 50 g $Na_2HPO_4 \times 2 H_2O$, 20 g $NaH_2PO_4 \times H_2O$, 9 g KH_2PO_4, and 9 g NaCl in 1 L *aqua bidest*. Autoclave 10× salts stock solution at 121 °C for 20 min and store at room temperature.

7. $20\times$ ^{15}N-Celtone stock solution: Dissolve 3.75 g Celtone base powder (^{15}N, 98%+, e.g. Cambridge Isotope Laboratories, Inc.) in 75 mL *aqua bidest*. Prepare this stock solution always fresh and sterilize $20\times$ ^{15}N-Celtone stock solution with a 0.2 μm syringe filter directly before use.

8. $100\times$ sugar stock solution: Dissolve 20 g D(+)-glucose anhydrous in 1 L *aqua bidest*. Sterilize $100\times$ sugar stock solution with a 0.2 μm syringe filter and store 10 mL aliquots at 4 °C.

9. $100\times$ ^{15}N-trace salts stock solution: Dissolve 400 mg (^{15}NH$_4$)$_2$SO$_4$ (^{15}N, 99%, e.g. Cambridge Isotope Laboratories, Inc.), 260 mg CaCl$_2$ × 2 H$_2$O, 200 mg MgCl$_2$ × 6 H$_2$O, 100 mg MnCl$_2$ × 4 H$_2$O, 10 mg CoCl$_2$ × 6 H$_2$O, and 1.5 mg NaHSeO$_3$ in 100 mL *aqua bidest*. Sterilize $100\times$ ^{15}N-trace salts stock solution with a 0.2 μm syringe filter and store 10 mL aliquots at 4 °C.

10. $100\times$ trace salts stock solution: Dissolve 400 mg (NH$_4$)$_2$SO$_4$, 260 mg CaCl$_2$ × 2 H$_2$O, 200 mg MgCl$_2$ × 6 H$_2$O, 100 mg MnCl$_2$ × 4 H$_2$O, 10 mg CoCl$_2$ × 6 H$_2$O, and 1.5 mg NaHSeO$_3$ in 100 mL *aqua bidest*. Sterilize $100\times$ trace salts stock solution with a 0.2 μm syringe filter and store 10 mL aliquots at 4 °C.

11. $200\times$ vitamins stock solution: Dissolve 3 mg D(+)-biotin, 10 mg Ca-D(+)-pantothenate, and 10 mg pyridoxine hydrochloride in 50 mL *aqua bidest*. Sterilize $200\times$ vitamins stock solution with a 0.2 μm syringe filter and store 5 mL aliquots at −20 °C.

12. $500\times$ iron stock solution: Dissolve 100 mg FeSO$_4$ × 7 H$_2$O in 50 mL *aqua bidest*. Prepare $500\times$ iron stock solution always fresh and sterilize this stock solution with a 0.2 μm syringe filter directly before use.

13. ^{15}N-CDCM (*C. difficile* ^{15}N-Celtone medium): Mix 100 mL $10\times$ salts stock solution, 50 mL $20\times$ ^{15}N-Celtone stock solution, 10 mL $100\times$ sugar stock solution, 10 mL $100\times$ ^{15}N-trace salts stock solution, 5 mL $200\times$ vitamins stock solution, and 2 mL $500\times$ iron stock solution. Fill up to 1 L with sterilized *aqua bidest*. Afterward, purge medium with oxygen-free gas and reduce medium in the anaerobic chamber overnight (*see* **Note 2**).

14. ^{15}N-CDCM agar plates: Mix solution A containing 100 mL $10\times$ salts stock solution, 50 mL $20\times$ ^{15}N-Celtone stock solution, 10 mL $100\times$ sugar stock solution, 10 mL $100\times$ ^{15}N-trace salts stock solution, 5 mL $200\times$ vitamins stock solution, and 2 mL $500\times$ iron stock solution. Fill up solution A to 500 mL with sterilized *aqua bidest*. Dissolve 20 g agar-agar in 500 mL *aqua bidest* for solution B, add 100 μL 1% (w/v)

resazurin, and autoclave at 121 °C for 20 min to sterilize. Mix solution A and solution B (after cooling to maximal 60 °C) in a 1:1 ratio carefully to prevent blistering. Fill approximately 15 mL ^{15}N-CDCM agar medium in petri dishes and wait until the medium is solid. Afterward transfer the plates inside the anaerobic chamber.

15. BHI medium: Dissolve 37 g brain heart infusion (e.g. Oxoid) in 1 L *aqua bidest*. Add 100 μL 1% (w/v) resazurin and autoclave at 121 °C for 20 min to sterilize. Afterward, purge medium with oxygen-free gas and reduce medium in the anaerobic chamber overnight (*see* **Note 2**).

16. CDMM (*C. difficile* minimal medium): Mix 200 mL 5× amino acids stock solution, 100 mL 10× salts stock solution, 10 mL 100× sugar stock solution, 10 mL 100× trace salts stock solution, 5 mL 200× vitamins stock solution, and 2 mL 500× iron stock solution. Fill up to 1 L with sterilized *aqua bidest*. Afterward, purge medium with oxygen-free gas and reduce medium in the anaerobic chamber overnight (*see* **Note 2**).

2.2 Equipment

1. 20 G × 1 ½" and 21 G × 4 ¾" needles.

2. An anaerobic chamber (37 °C, 5% H$_2$, 95% N$_2$) (*see* **Note 3**).

3. An autoclave.

4. A centrifuge.

5. A photometer.

6. A thermomixer.

7. A vortexer.

8. Afnor plasma bottles in appropriated size (e.g. Zscheile & Klinger GmbH).

9. Aluminum foil.

10. Cell scraper with 25 cm long handle.

11. Chlorobutyl rubber septa type 4434/4106/50 (e.g. Zscheile & Klinger GmbH).

12. Glass beads with diameter of 2.85–3.45 mm.

13. Open-top-style aluminum caps (e.g. Zscheile & Klinger GmbH).

14. Petri dishes 92 × 16 mm with cams.

15. Sterile syringe filters with 0.2 μm cellulose acetate membrane.

3 Methods

Carry out all procedures at room temperature unless otherwise specified.

3.1 Preparatory Operation

Key working steps for the preparatory operations are illustrated in Fig. 1 in a time-resolved scale.

1. Purify *C. difficile* 630Δ*erm* (or *C. difficile* strains of interest) spores as previously described in detail by Edwards and McBride [27], and store purified spores at 4 °C (*see* **Note 1**).

2. Prepare appropriate volumes of liquid BHI medium, autoclave BHI medium in Afnor plasma bottles, and reduce medium before use (*see* **Note 2**).

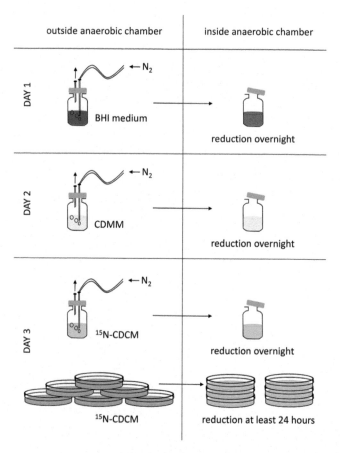

Fig. 1 Simplified working steps with recommended time table for all necessary preparatory operations to have always fresh media at disposal. Preparation of *C. difficile* spores, weighting of ingredients, mixing of stock solutions, as well as autoclaving or sterile filtration steps were omitted in the figure. Oxidized resazurin colorized the medium red, after reduction the medium regain the original color

3. Prepare appropriate volumes of CDMM by mixing sterile stock solutions in an Afnor plasma bottle and reduce medium before use (*see* **Note 2**).

4. Prepare appropriate volumes of ^{15}N-CDCM by mixing sterile stock solutions in an Afnor plasma bottle and reduce medium before use (*see* **Note 2**).

5. Prepare appropriate numbers of ^{15}N-CDCM agar plates. Store prepared petri dishes 24–48 h in the anaerobic chamber until the pink color of oxygen indicator resazurin returns to colorless

3.2 Metabolic Labeling Procedure

Key working steps of the pre-cultures (**steps 1–6**) of the metabolic labeling procedure are illustrated in Fig. 2 in a temporal scale and the metabolic labeling procedure (**steps 7–31**) is schematically presented in Fig. 3.

1. Transfer materials or consumables in the anaerobic chamber at least 1 day before starting the experiment (*see* **Note 4**).

2. For an efficient germination of spores, activate them at 55 °C for 15 min in a thermomixer gently shaking directly before inoculation (*see* **Note 5**).

3. Inoculate 7 mL fresh BHI medium (supplemented with 70 μL 10% (w/v) taurocholate) with 100 μL activated *C. difficile* spores (*see* **Notes 5** and **6**).

4. Incubate germinating spores for 16–24 h at 37 °C (*see* **Notes 7** and **8**).

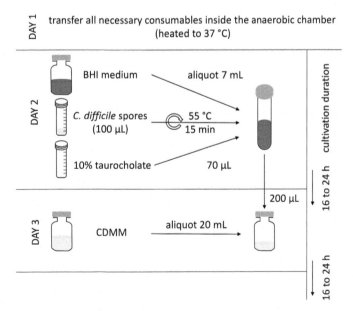

Fig. 2 Simplified scheme for the pre-cultures of *C. difficile* spores for the metabolic labeling procedure over the course of time

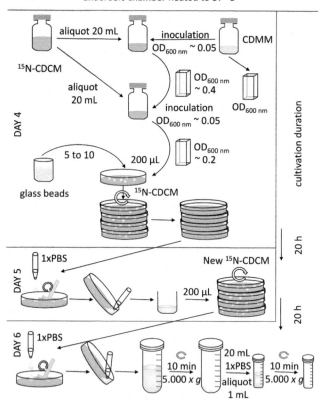

Fig. 3 Simplified working steps for the metabolic labeling procedure of *C. difficile*

5. Use 200 μL of the germinated *C. difficile* spores for an inoculation of 20 mL CDMM (*see* **Note 5**).

6. Incubate *C. difficile* cells for 16–24 h at 37 °C (*see* **Notes 7** and **8**).

7. Determine the optical density of the sample on the next morning at 600 nm (OD_1). Use fresh medium as blank (*see* **Note 9**).

8. Calculate the necessary volume of the cultured *C. difficile* CDMM (V_1) to inoculate 20 mL fresh ^{15}N-CDCM (V_2) to an optical density (OD_2) of approximately 0.05–0.07 with the following formula (*see* **Note 5**).

$$V_1[\text{mL}] = \frac{V_2[\text{mL}] \times OD_2}{OD_1}$$

9. Use the calculated volume of the *C. difficile* CDMM culture to inoculate 20 mL ^{15}N-CDCM (*see* **Note 5**).

10. Determine the optical density of the ^{15}N-CDCM culture at 600 nm. Use ^{15}N-CDCM as blank.

11. Incubate *C. difficile* culture at 37 °C (*see* **Note 7**).

12. Measure the optical density of growing ^{15}N-CDCM culture regularly (every 60–120 min) at 600 nm.

13. After reaching an optical density of approximately 0.4, calculate the necessary volume of the ^{15}N-CDCM culture (V_1) to inoculate 20 mL fresh ^{15}N-CDCM culture (V_2) to an optical density (OD_2) of approximately 0.05–0.07 at 600 nm according the above presented formula (*see* **Notes 5** and **10**).

14. Incubate *C. difficile* culture at 37 °C (*see* **Note 7**).

15. Measure the optical density of growing ^{15}N-CDCM culture regularly (every 60–120 min) at 600 nm.

16. After reaching an optical density of approximately 0.2, pipet 200 μL of ^{15}N-CDCM culture on six fresh, pre-reduced ^{15}N-CDCM agar plates (*see* **Notes 5, 11** and **12**).

17. Spread cells with five to ten sterile glass beads by swivel the petri dishes carefully until the agar surface is dry (*see* **Note 13**).

18. Incubate agar plates for 20 h at 37 °C (*see* **Note 7**).

19. Wet plates with 1 mL 1× PBS, scrap off *C. difficile* cells with cell scraper carefully and admit cell containing buffer with a 1 mL pipet (*see* **Note 14**).

20. Collect the cells of all plates in one bottle (*see* **Note 5**).

21. Repeat **steps 19** and **20** at least once with each plate and collect the cells in the same bottle as before.

22. Collect the cell suspension in a falcon tube and fill up to 12 mL with 1× PBS.

23. Pipet 200 μL of carefully mixed cell suspension on a fresh, pre-reduced ^{15}N-CDCM agar plate. Use 60 plates in total. Repeat **steps 17** and **18** for each plate (*see* **Notes 5, 13,** and **15**).

24. After 20 h, repeat **steps 19–21** with each plate and collect the replaced cells in one bottle (*see* **Note 16**).

25. Transfer cell suspension in centrifugation vessels and centrifuge 10 min at 5,000 × *g* at 4 °C.

26. Discard the supernatant carefully.

27. Suspend cell pellet in 20 mL 1× PBS and mix by vortexing.

28. Aliquot 1 mL cell suspension in 1.7 mL reaction tubes.

29. Centrifuge 10 min at 5,000 × *g* at 4 °C.

30. Discard the supernatant carefully.

31. Store cell pellet at −70 °C until use.

4   Notes

1. The purification of *C. difficile* 630Δ*erm* (or *C. difficile* strains of interest) spores will take several days up to 1 week and should be done independently before the metabolic labeling experiment.

2. Seal Afnor plasma bottles containing medium or other solutions with open-top-style aluminum caps with chlorobutyl rubber septum. Insert a sterile needle (21 G × 4 ¾") through the septum and puncture the septum with a second needle (20 G × 1 ½") to prevent overpressure in the bottle. Supply oxygen-free gas into the liquid with 400–600 mbar (supply at least 10 min for 50 mL, 20 min for 100 mL, and so on). Store reduced medium or solutions with bottles slightly opened in the anaerobic chamber overnight. Subsequently, the reduced solutions should be thoroughly closed and can subsequently be stored at 4 °C.

3. Detailed information about the use and maintenance of an anaerobic chamber are described in Edwards et al. [28].

4. Store an appropriated number of consumables at least overnight in the anaerobic chamber. Try to minimize the number of consumables to save enough space for the following experiment.

5. If you want to prepare different biological replicates, use spores prepared from different single colonies. Additionally, use separate bottles or plates with medium during cultivation procedure and collect cells in different bottles during cell harvesting procedure.

6. Produced number of spores of per ml of *C. difficile* 630Δ*erm* was approximately 7.8×10^7 CFU/mL in the prepared spore stock.

7. Incubate *C. difficile* cells at 37 °C. If the temperature of the anaerobic chamber is not adjustable, use an incubator inside the anaerobic chamber.

8. Cells cultured for 16 h instead of 24 h showed usually a shorter lag phase in fresh medium compared to the longer cultured cells.

9. Start with this step as early as possible on your working day; otherwise the necessary procedure for this day will end very late.

10. *C. difficile* 630Δ*erm* cells cultured in ^{15}N-CDCM as described in this protocol needed approximately 5 h to reach an optical density of 0.4.

11. *C. difficile* 630Δ*erm* cells cultured a second round in ^{15}N-CDCM as described in this protocol needed approximately 3 h to reach an optical density of 0.2 caused by a decreased cell growth ability in this medium.

12. By using 6 fresh, pre-reduced ^{15}N-CDCM agar plates at this step and 60 fresh, pre-reduced ^{15}N-CDCM agar plates on the next cultivation step, the resulting ^{15}N-labeled soluble protein amount will be approximately 1.25–2 mg after cell disruption with ultrasonication as described in Trautwein-Schult et al. [25]. If this labeled material is not enough, use more plates at this procedure step and the following once.

13. Stack up to five petri dishes and swivel the stacked over one another petri dish together until the agar surface is dry. Stacking of plates can save a lot of time.

14. Use the cell scraper carefully to prevent gashing agar inside the petri dishes. Be careful and hamper the transfer of solid medium from the plate into the collecting bottle by carefully pipetting.

15. To reduce storage time of cells in 1× PBS, prepare each fresh ^{15}N-CDCM agar plate with five to ten sterile glass beads before you harvest the cells and/or work together with a colleague.

16. Schedule approximately 1 h for harvesting of the metabolically labeled cells from 60 plates. To reduce time for cells in 1× PBS or on plates, you can work together with a colleague or use petri dishes with bigger size (e.g. petri dishes 150 × 20 mm). Prefer round plates otherwise you will lose cells in the corners of the plates. Additionally, adapt the ^{15}N-CDCM agar volume per plate (e.g. for petri dishes 150 × 20 mm use 40 mL of ^{15}N-CDCM agar).

References

1. Collins MD, Lawson PA, Willems A et al (1994) The phylogeny of the genus Clostridium: proposal of five new genera and eleven new species combinations. Int J Syst Bacteriol 44(4):812–826

2. Yutin N, Galperin MY (2013) A genomic update on clostridial phylogeny: gram-negative spore formers and other misplaced clostridia. Environ Microbiol 15(10):2631–2641

3. Lawson PA, Citron DM, Tyrrell KL et al (2016) Reclassification of *Clostridium difficile* as *Clostridioides difficile* (Hall and O'Toole 1935) Prévot 1938. Anaerobe 40:95–99

4. Freeman J, Bauer MP, Baines SD et al (2010) The changing epidemiology of *Clostridium difficile* infections. Clin Microbiol Rev 23 (3):529–549

5. Planche TD, Davies KA, Coen PG et al (2013) Differences in outcome according to *Clostridium difficile* testing method: a prospective multicentre diagnostic validation study of *C. difficile* infection. Lancet Infect Dis 13 (11):936–945

6. Lessa FC, Mu Y, Bamberg WM et al (2015) Burden of *Clostridium difficile* infection in the United States. N Engl J Med 372(9):825–834

7. Bantscheff M, Schirle M, Sweetman G et al (2007) Quantitative mass spectrometry in proteomics: a critical review. Anal Bioanal Chem 389(4):1017–1031

8. Jain S, Graham C, Graham RL et al (2011) Quantitative proteomic analysis of the heat stress response in *Clostridium difficile* strain 630. J Proteome Res 10(9):3880–3890

9. Chong PM, Lynch T, McCorrister S et al (2014) Proteomic analysis of a NAP1 *Clostridium difficile* clinical isolate resistant to metronidazole. PLoS One 9(1):e82622

10. Chen JW, Scaria J, Mao C et al (2013) Proteomic comparison of historic and recently emerged hypervirulent *Clostridium difficile* strains. J Proteome Res 12(3):1151–1161

11. Sievers S, Dittmann S, Jordt T et al (2018) Comprehensive redox profiling of the thiol proteome of *Clostridium difficile*. Mol Cell Proteomics 17(5):1035–1046

12. Otto A, Maaß S, Lassek C et al (2016) The protein inventory of *Clostridium difficile* grown in complex and minimal medium. Proteomics Clin Appl 10(9-10):1068–1072

13. Chilton CH, Gharbia SE, Fang M et al (2014) Comparative proteomic analysis of *Clostridium difficile* isolates of varying virulence. J Med Microbiol 63(Pt 4):489–503

14. Dresler J, Krutova M, Fucikova A et al (2017) Analysis of proteomes released from in vitro cultured eight *Clostridium difficile* PCR ribotypes revealed specific expression in PCR ribotypes 027 and 176 confirming their genetic relatedness and clinical importance at the proteomic level. Gut Pathogens 9:45

15. Sievers S, Metzendorf NG, Dittmann S et al (2019) Differential view on the bile acid stress response of *Clostridioides difficile*. Front Microbiol 10:258

16. Hofmann JD, Otto A, Berges M et al (2018) Metabolic reprogramming of *Clostridioides difficile* during the stationary phase with the induction of toxin production. Front Microbiol 9:1970

17. Maass S, Otto A, Albrecht D et al (2018) Proteomic signatures of *Clostridium difficile* stressed with metronidazole, vancomycin, or fidaxomicin. Cell 7(11):213

18. Pizarro-Guajardo M, Ravanal MC, Paez MD et al (2018) Identification of *Clostridium difficile* immunoreactive spore proteins of the epidemic strain R20291. Proteomics Clin Appl 12(5):e1700182

19. Oda Y, Huang K, Cross FR et al (1999) Accurate quantitation of protein expression and site-specific phosphorylation. Proc Natl Acad Sci U S A 96(12):6591–6596

20. Ong SE, Blagoev B, Kratchmarova I et al (2002) Stable isotope labeling by amino acids in cell culture, SILAC, as a simple and accurate approach to expression proteomics. Mol Cell Proteomics 1(5):376–386

21. Karasawa T, Ikoma S, Yamakawa K et al (1995) A defined growth medium for *Clostridium difficile*. Microbiology 141(Pt 2):371–375

22. Otto A, Bernhardt J, Meyer H et al (2010) Systems-wide temporal proteomic profiling in glucose-starved *Bacillus subtilis*. Nat Commun 1:137

23. Neumann-Schaal M, Hofmann JD, Will SE et al (2015) Time-resolved amino acid uptake of *Clostridium difficile* 630Delta*erm* and concomitant fermentation product and toxin formation. BMC Microbiol 15:281

24. Bouillaut L, Self WT, Sonenshein AL (2013) Proline-dependent regulation of *Clostridium difficile* Stickland metabolism. J Bacteriol 195(4):844–854

25. Trautwein-Schult A, Maass S, Plate K et al (2018) A metabolic labeling strategy for relative protein quantification in *Clostridioides difficile*. Front Microbiol 9:2371

26. Abhyankar WR, Zheng L, Brul S et al (2019) Vegetative cell and spore proteomes of *Clostridioides difficile* show finite differences and reveal potential protein markers. J Proteome Res 18(11):3967–3976

27. Edwards AN, McBride SM (2016) Isolating and purifying *Clostridium difficile* spores. Methods Mol Biol 1476:117–128

28. Edwards AN, Suarez JM, McBride SM (2013) Culturing and maintaining *Clostridium difficile* in an anaerobic environment. J Vis Exp (79):e50787. https://doi.org/10.3791/50787

Chapter 20

Application of Label-Free Proteomics for Quantitative Analysis of Urothelial Carcinoma and Cystitis Tissue

Kathrin E. Witzke, Frederik Großerueschkamp, Klaus Gerwert, and Barbara Sitek

Abstract

A label-free approach based on a highly reproducible and stable workflow allows for quantitative proteome analysis. Due to advantages compared to labeling methods, the label-free approach has the potential to measure unlimited samples from clinical specimen monitoring and comparing thousands of proteins. The presented label-free workflow includes a new sample preparation technique depending on automatic annotation and tissue isolation via FTIR-guided laser microdissection, in-solution digestion, LC-MS/MS analyses, data evaluation by means of Proteome Discoverer and Progenesis software, and verification of differential proteins. We successfully applied this workflow in a proteomics study analyzing human cystitis and high-grade urothelial carcinoma tissue regarding the identification of a diagnostic tissue biomarker. The differential analysis of only 1 mm^2 of isolated tissue cells led to 74 significantly differentially abundant proteins.

Key words Label-free proteomics, FTIR imaging, Laser microdissection, Urothelial cell carcinoma (UCC), Bladder cancer, AHNAK2

1 Introduction

Label-free mass spectrometry (*see* also Chaps. 8, 16, 21–24), as the name implies, does not use any labeling strategies, making it very cost-effective, but only quantifies by matching identical peptides over several runs. Advantages beside cost-effectiveness and less sample preparation steps are high proteome coverage and high dynamic range. The disadvantages here are, therefore, high measurement times, as each sample or condition has to be measured separately and, furthermore, the separate handling of each sample from acquisition to measurement [1, 2].

In label-free proteomics, there are mainly two approaches of protein quantification, which are spectral counting and ion intensity-based quantification. While in spectral counting, as the name implies, the number of MS/MS fragment ion spectra that were

Katrin Marcus et al. (eds.), *Quantitative Methods in Proteomics*, Methods in Molecular Biology, vol. 2228,
https://doi.org/10.1007/978-1-0716-1024-4_20, © Springer Science+Business Media, LLC, part of Springer Nature 2021

obtained for the peptides of a protein are counted and compared, the second type of label-free quantification measures the chromatographic peak areas of peptide precursor ions. Both strategies are possible, as both the number of MS/MS spectra of a peptide increase with the amount of the corresponding protein [3] and the areas under the curves (AUC) of chromatographic peptide peaks correlate linearly with the corresponding protein abundance [4, 5].

Nowadays, the main approach of protein quantification is ion intensity-based quantification as it relies on measuring physical data and not simply on counting the acquired spectra. In this approach, raw MS data have to be further processed for analysis. This includes, for example, feature detection, retention time alignment, intensity normalization, and peak picking [4].

The sample preparation is one very important step in proteomics for obtaining high-class quantification results. Especially in tissue proteomics, the right sample preparation is crucial due to tissue heterogeneity [6]. Here, a new strategy for automated annotation and isolation of regions of interest (ROI) has been used [7–9]. Conventionally, histological stainings or pathological annotations are necessary to detect ROIs in tissue samples, which are then transferred to unstained adjacent sections for LCM. The transmission of ROIs to adjacent slides, though, implies insurmountable deviations to annotated ROIs. The novel strategy of label-free automated tissue annotation and subsequent isolation via FTIR (Fourier transform infrared)-guided laser capture microdissection was coupled to subsequent label-free LC-MS/MS proteome analysis. By combining these techniques, very homogeneous samples can be obtained that are very accurately annotated, as the same tissue section annotated via FTIR imaging can be used for proteome analysis [10].

In label-free proteomics, the reproducibility and stability of the workflow are of highest importance due to the high sample complexity and separate measurements for each sample. Therefore, all steps of the label-free approach have to be optimized for best results, which includes the before-mentioned sample preparation with protein extraction and digestion, peptide separation by liquid chromatography, and data analysis including identification, quantification, and statistical analysis. One major advantage of label-free proteomics is its compatibility with high-throughput analyses that allow for processing of large numbers of biological samples required for statistically significant quantification.

We describe the application of a label-free approach for the identification of biomarker candidate proteins in the context of urothelial carcinoma diagnosis. We used fresh-frozen human tissue of patients with an inflammation of the bladder (cystitis) in comparison with high-grade urothelial carcinoma and performed label-free tissue annotation via FTIR (Fourier transform infrared) imaging with guided automated laser microdissection for unbiased

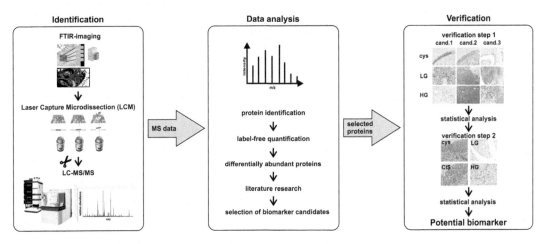

Fig. 1 Workflow of the label-free proteomics approach coupled with FTIR-guided LCM. First the tissue is annotated label-free and regions of interest isolated. After isolation, proteins will be extracted and digested with trypsin. Peptides will be analyzed via LC-MS/MS and generated data evaluated with Proteome Discoverer for identification and Progenesis QI for quantification. For verification of biomarker candidate proteins, immunohistochemistry will be performed

isolation of the tissue sections of interest only and subsequent label-free LC-MS/MS proteome analysis. For that, we used nano-HPLC coupled to an Orbitrap Elite mass spectrometer for the generation of peptide profiles. For quantitative analysis of the data, the software Progenesis QI for proteomics was used. Altogether, 74 proteins were found to show significant differential abundance between the analyzed groups (FDR-adjusted p-value ≤ 0.05 and absolute fold change ≥ 1.5). Verification was performed in two steps with increasing cohort sizes and the addition of more urothelial carcinoma groups (low grade and carcinoma in situ). From three tested candidates in the first step, AHNAK2 was selected for further verification and proposed as a biomarker candidate for the differentiation between cystitis and several subgroups of urothelial carcinoma (Fig. 1).

2 Materials

2.1 FTIR Imaging and Laser Microdissection

1. HM550 cryostat (Thermo Fisher Scientific, Waltham, MA, USA).

2. PET (polyethylene terephthalate) frame slides (Leica, Wetzlar, Germany).

3. Cary 620 IR microscope equipped with a 128×128 pixel liquid nitrogen-cooled mercury cadmium telluride (MCT) focal plane array (FPA) (Agilent Technologies, Santa Clara, CA, USA).

4. Cary 670 spectrometer (Agilent Technologies, Santa Clara, CA, USA).

5. Parker Balston AirDryer Assembly 75-62 (Parker Hannifin Corporation, Lancaster, NY, USA).

6. PALM Microbeam Laser microdissection (LMD) microscope (Carl Zeiss Microscopy GmbH, Jena, Germany).

7. MATLAB (MathWorks, Natick, MA, USA) or equivalent (e.g., R or Python).

2.2 Sample Preparation and Digestion

1. Ultrasonic bath (VWR, Darmstadt, Germany).

2. Centrifuge (Eppendorf, Hamburg, Germany).

3. Lysis buffer: 50 mM ammonium bicarbonate with 0.1% Rapi-Gest SF surfactant (Waters GmbH, Eschborn, Germany).

4. Digestion: 20 mM dithiothreitol (DTT), 100 mM iodoacetamide (IAA), 33 ng/μL trypsin, trifluoroacetic acid (TFA).

2.3 Liquid Chromatography

1. Ultimate 3000 RSLCnano high-performance liquid chromatography system (Dionex, Idstein, Germany).

2. Trap column: Acclaim PepMap100 C18 Nano-Trap column (C18, 100 μm × 2 cm, particle size 5 μm, pore size 100 Å; Thermo Fisher Scientific, Bremen, Germany).

3. Nano column: Acclaim PepMap RSLC Nano Viper C18 analytical column (C18, 75 μm × 50 cm, particle size 2 μm, pore size 100 Å; Thermo Fisher Scientific).

4. Loading solvent: 0.1% (v/v) TFA (MS grade).

5. Gradient solvent A: 0.1% (v/v) Formic acid (FA) (MS grade).

6. Gradient solvent B: 0.1% (v/v) FA (MS grade), 84% (v/v) acetonitrile (ACN) (MS grade).

2.4 Mass Spectrometry

1. LTQ Orbitrap Elite with an online nano-ESI source (Thermo Fisher Scientific).

2. Pico Tip™ emitter Silica Tip™ (New Objective, Woburn, USA).

3. Collision gas: nitrogen.

2.5 Data Analysis

1. Proteome Discoverer v.1.4 (Thermo Fisher Scientific).

2. Mascot v.2.5 (Matrix Science, London, UK).

3. Progenesis QI v.2.0 (Nonlinear Dynamics, Durham, NC, USA).

4. R v.3.4.0 (Free Software Foundation).

3   Methods

3.1   General Practice

Human tissue of patients with an inflammation of the bladder (cystitis) or high-grade UCC was collected during cystectomy surgery according to standard operation procedure. Tissue was washed with isotonic saline solution, slowly frozen on the surface of liquid nitrogen within 8 min and stored at −80 °C. Frozen tissue was sectioned with an HM550 cryostat (Thermo Fisher Scientific, Waltham, MA) at −20 °C, and 10 μm sections were collected on polyethylene terephthalate (PET) frame slides.

3.2   Automatic Annotation Via FTIR Imaging and Laser Microdissection

1. Take the tissue thin section mounted on a PET frame slide and place it under the FTIR imaging microscope (here Cary 620) (*see* also **Note 1**).

2. Thaw the tissue sample under dry air in the FTIR system. At the same time, the system is stabilized with the dry air.

3. Select a clean background position on the PET slide and collect spatially resolved IR spectra in the wave number region 3700–950 cm^{-1} at a spectral resolution of 4 cm^{-1} with co-added 128 scans. Use a 15× objective resulting in a pixel resolution of ~5.5 μm and a field of view (FOV) of 715 μm^2 per FPA field.

4. Select the region of interest (ROI) on the PET slide and collect spatially resolved IR spectra in mapping mode. This allows imaging of larger regions than the FOV by stitching the collected FPA fields afterwards.

5. Stitch the collected spectral hypercubes in MATLAB (or equivalent software) and pre-process the data. The first step is a quality test based on the integral of the amide I band and the signal-to-noise ratio (noise, 2100–2000 cm^{-1}; signal, 1600–1500 cm^{-1}). Then subject all spectra to extended multiplicative scattering correction-based Mie and resonance-Mie scattering correction from 2300 to 950 cm^{-1}.

6. The dataset is now prepared for multivariate data analysis or machine learning algorithms. For bladder, pre-train and use a random forest (RF) classifier to annotate the tissue (*see* also **Note 2**).

7. Select the ROIs from the label-free annotated IR image for isolation via LMD.

8. Select three reference points at the IR imaging system and then transfer the tissue section to the LMD.

9. Find the three selected reference points at the LMD. They are needed for the coordinate transfer.

10. Transfer the coordinates of the ROIs selected from the FTIR imaging results to LMD by two-dimensional Helmert transformation based on three reference points in MATLAB or equivalent software (*see* also **Note 3**).

11. Collect the needed tissue area. For bladder, 10 μm sections were used, and regions of 1 mm^2 were collected in lysis buffer (20 μL/1 mm^2 cells).

3.3 Sample Preparation and Enzymatic Digestion

1. Lyse the cells in lysis buffer (20 μL/1 mm^2 cells) and sonicate the samples upside down on ice for 1 min and finally centrifuge the samples in the upright position for 1 min to transfer them from the lid to the vial itself.

2. Normally, it is necessary to know the concentration of the samples. However, due to LCM isolation of only 1 mm^2 tissue, digest all.

3. Perform a tryptic in-solution digest for the proteolysis of proteins. For reduction, add 3.7 μL DTT (20 mM) to the samples and incubate for 30 min at 60 °C. Afterwards alkylate with 2.2 μL IAA (100 mM) for 30 min at room temperature in the dark. Add trypsin (0.02 μg) to digest the proteins overnight (max. 16 h) at 37 °C. Stop the digestion by adding 1.3 μL 10% TFA to the solution, incubate for 30 min at 37 °C. Afterwards centrifuge the samples (10 min, 16,000 × g) and transfer the supernatant to a glass vial (*see* **Note 4**).

4. Before performing the LC-MS analysis, dry the samples in a vacuum centrifuge and dissolve them in 17 μL 0.1% TFA. Use a sample amount of the whole 1 mm^2 tissue area for one LC-MS/MS measurement (*see* **Note 5**).

3.4 Peptide Separation with Reversed Phase High-Performance Liquid Chromatography

1. For the separation of the digested proteins, perform a reversed phase high-performance liquid chromatography with the Ultimate 3000 RSLCnano high-performance liquid chromatography system (Dionex). Within this, use a system containing a nano-trap column (C18) and a nano-analytical column (C18). The columns need to be heated to 60 °C to allow high flow rates of 400 nL/min at acceptable pressure (*see* **Note 6**).

2. Use a sample volume of 15 μL for injection. First, peptides are pre-concentrated on the trap column for 7 min, while detergents and salts are washed away. Use a flow rate of 30 μL/min for loading.

3. The gradient for peptide separation works as follows: (a) linear gradient from 5 to 40% solvent B over 98 min, followed by (b) 95% B in 2 min, (c) constant 95% B for 7 min, and finally (d) 5 min at 5% B for equilibration. Set the gradient pump flow rate to 400 nL/min.

3.5 Detection of Separated Peptides with Mass Spectrometry

1. Operate the Orbitrap Elite™ mass spectrometer (Thermo Fisher Scientific) in the data-dependent mode to automatically switch between MS and MS/MS acquisition.

2. Set the mass range for survey full scan MS spectra and MS/MS spectra to *m/z* 350–2000.

3. For fragmentation, apply collision-induced dissociation (CID) with nitrogen as collision gas and normalized collision energy of 35. For MS/MS measurements, use a top 20 method based on intensity (*see* **Note 7**). The minimal required signal for precursor ions is 500 counts, and the isolation width is 2 ppm.

4. Reject charge states 1+ and prefer charge states 2+, 3+, and 4+ for precursor ion isolation.

5. Utilize dynamic exclusion with an exclusion duration of 30 s and one repeat count within 30 s. Use exclusion list size of 500 precursor ions with an exclusion mass width of 10 ppm.

6. In the end, export generated data as Thermo .raw file format.

3.6 Identification of Measured Proteins

1. For protein identification, use the software Proteome Discoverer version 1.4 (Thermo Fisher Scientific). Search spectra against the UniProtKB/Swiss-Prot database using the Mascot search engine version 2.5 (Matrix Science).

2. Create a Proteome Discoverer workflow. Use the following parameters: (a) taxonomy setting, homo sapiens; (b) enzyme, trypsin, (c) missed cleavages, allow up to one; (d) dynamic modification, oxidation (methionine); (e) static modification, carbamidomethyl (cysteine); (f) precursor mass tolerance, 5 ppm; (g) fragment mass tolerance, 0.4 Da; (h) false discovery rate (FDR), via p function (identifications with FDR >1% are rejected).

3. Import .raw data files into the Proteome Discoverer Daemon and start the search with the created workflow.

4. Open the results with the Proteome Discoverer Viewer and export results in Excel format.

3.7 Quantitative Proteome Analysis

1. As software for quantitative proteome analysis, use Progenesis QI for proteomics (Nonlinear Dynamics). First, import the LC-MS analysis .raw data files into the program.

2. Select the reference run that all your other runs are aligned to. This can either be done by yourself, or there is the option to let the program select the reference run, either out of all runs or out of a selection you make. An optimal reference run should have the greatest similarity to all other runs (*see* **Note 8**).

3. The alignment step is most important for label-free quantification. Therefore it is necessary to have a very accurate alignment result. First, apply the automatic alignment, but be sure to

check the alignment carefully when it is finished. If the result of the alignment is not good enough, you can often alter the result for the better by manually adding vectors to align specific runs.

4. Progenesis automatically performs feature detection, normalization, and quantification. Always check results and exclude or include features from analysis results. Exclude retention times from washing/equilibration. Include only ion charge states of 2+, 3+, and 4+ with a minimum of three isotope peaks to exclude contaminations from the analysis (*see* **Note 9**).

5. Create the experimental design, in our case cystitis vs. high-grade UCC. Runs which, e.g., did not align properly can be excluded.

6. Identify the quantified proteins. Therefore, import Excel results of priorly obtained identification via Proteome Discoverer. Consider all non-conflicting peptides for protein quantification.

7. Differential analysis is also performed by Progenesis. You can filter based on p- and q-value (≤ 0.05) and fold change (≥ 1.5) and tag them for further analysis.

8. You can perform principal component analysis (PCA) to check if runs cluster based on experimental grouping. Also, check regulation profiles of interesting candidates (*see* **Note 10**).

9. Export results of quantified proteins in Excel format.

3.8 Statistical Analysis

1. Despite the results of the differential analysis by Progenesis, perform a separate statistical analysis via R. Arcsinh-transform normalized protein abundances obtained from Progenesis and use those for t-test calculations. Adjust test p-values for FDR control with the method of Benjamini Hochberg.

2. Use normalized protein abundances obtained from Progenesis for fold change calculation.

3. Consider proteins significantly differentially abundant between experimental groups if they have an absolute fold change ≥ 1.5 and an FDR-corrected p-value ≤ 0.05.

4 Notes

1. You can use other FTIR or IR imaging systems also. The procedure is equivalent. Furthermore, if other options in data collection are more convenient for you, change them. The values given here are only recommendations that have worked well for our work on bladder cancer.

2. For analysis of the pre-processed FTIR imaging datasets, you can use multivariate data analyses like *k*-means clustering or hierarchical cluster analysis. This way, you select your ROIs on spectral similarity. A better and more precise way is to train a supervised cluster algorithm like random forests (RF). Therefore, you need a spectral database for different tissue types that is created by measuring known tissue samples previously. Then the classifier (RF) can be trained, and afterwards it is possible to annotate unknown spectral datasets by the use of this classifier.

3. The coordinate transfer is the same as the transfer of coordinates in geographic sciences. Beside the Helmert transformation, you can use other methods also. It is always helpful to test the accuracy of the transfer with a test target previously.

4. The centrifugation step after stopping the tryptic digest is crucial here to remove excess RapiGest from the samples that could potentially damage your LC system.

5. The workflow can also be applied for body fluids and cell culture experiments. An adapted sample processing could also be carried out successfully for FFPE tissue.

6. On the one hand, heating of columns is needed to reduce the pressure of the system and to get a better peptide separation; however, heated columns need higher flow rates for acceptable pressure. Higher flow rates mean worse sensitivity. Thus, a compromise between separation and sensitivity is necessary.

7. In a top 20 method, the 20 most abundant peptide ions of the full scan are selected for fragmentation and measured for tandem mass spectra in the linear ion trap.

8. Where possible, master mixes of all samples can be used as optimal reference runs for alignment, as they combine features of all runs.

9. Tryptic peptides have charge states between 2+ and 4+, while contaminations mostly have charge states of 1+. If another protease than trypsin is used, other charge states might apply.

10. Checking regulation profiles offers easy insight into the quality of candidate proteins. It can be observed that sometimes only few but high differences can boost statistical significance.

Acknowledgments

This work was supported by the Ministry of Innovation, Science and Research of North-Rhine Westphalia, Germany. The authors would like to thank Lidia Janota, Kristin Fuchs, Stephanie Tautges, and Birgit Zülch for their excellent technical assistance.

References

1. Bantscheff M, Schirle M, Sweetman G et al (2007) Quantitative mass spectrometry in proteomics: a critical review. Anal Bioanal Chem 389(4):1017–1031. https://doi.org/10.1007/s00216-007-1486-6

2. Megger DA, Pott LL, Ahrens M et al (2014) Comparison of label-free and label-based strategies for proteome analysis of hepatoma cell lines. Biochim Biophys Acta 1844 (5):967–976. https://doi.org/10.1016/j.bbapap.2013.07.017

3. Liu H, Sadygov RG, Yates JR 3rd (2004) A model for random sampling and estimation of relative protein abundance in shotgun proteomics. Anal Chem 76(14):4193–4201. https://doi.org/10.1021/ac0498563

4. Megger DA, Bracht T, Meyer HE et al (2013) Label-free quantification in clinical proteomics. Biochim Biophys Acta 1834(8):1581–1590. https://doi.org/10.1016/j.bbapap.2013.04.001

5. Bondarenko PV, Chelius D, Shaler TA (2002) Identification and relative quantitation of protein mixtures by enzymatic digestion followed by capillary reversed-phase liquid chromatography-tandem mass spectrometry. Anal Chem 74(18):4741–4749. https://doi.org/10.1021/ac0256991

6. Mukherjee S, Rodriguez-Canales J, Hanson J et al (2013) Proteomic analysis of frozen tissue samples using laser capture microdissection. Methods Mol Biol 1002:71–83. https://doi.org/10.1007/978-1-62703-360-2_6

7. Grosserueschkamp F, Kallenbach-Thieltges A, Behrens T et al (2015) Marker-free automated histopathological annotation of lung tumour subtypes by FTIR imaging. Analyst 140 (7):2114–2120. https://doi.org/10.1039/c4an01978d

8. Miller LM, Dumas P (2006) Chemical imaging of biological tissue with synchrotron infrared light. Biochim Biophys Acta 1758 (7):846–857. https://doi.org/10.1016/j.bbamem.2006.04.010

9. Ooi GJ, Fox J, Siu K et al (2008) Fourier transform infrared imaging and small angle x-ray scattering as a combined biomolecular approach to diagnosis of breast cancer. Med Phys 35(5):2151–2161. https://doi.org/10.1118/1.2890391

10. Grosserueschkamp F, Bracht T, Diehl HC et al (2017) Spatial and molecular resolution of diffuse malignant mesothelioma heterogeneity by integrating label-free FTIR imaging, laser capture microdissection and proteomics. Sci Rep 7:44829. https://doi.org/10.1038/srep44829

Chapter 21

Quantitative MS Workflow for a High-Quality Secretome Analysis by a Quantitative Secretome-Proteome Comparison

Gereon Poschmann, Nina Prescher, and Kai Stühler

Abstract

Cells secrete proteins to communicate with their environment. Therefore, it is interesting to characterize the proteins which are released from cells under certain experimental conditions the so-called secretome. Here, often proteins from conditioned medium of cultured cells are analyzed, but these additionally might include also contaminating proteins of serum that have not been sufficiently removed or proteins from dying cells. To provide high-quality secretome data and minimize potential contaminants, we describe a quantitative comparison of conditioned medium and the cellular proteome. The described workflow comprises cell cultivation, sample preparation, and final data analysis which is based on the comparison of data from label-free mass spectrometric quantification of proteins from the conditioned medium with corresponding cellular proteomes enabling the detection of bona fide secreted proteins.

Key words Secretome, Contaminants, Quantitative MS

1   Introduction

The main challenge of secretome analysis is the determination of bona fide secreted proteins and to differentiate them from contaminating proteins derived from culture medium or proteins released from dying cells. Therefore, several steps in cell culture, sample preparation, and data analysis have to be considered to obtain high-quality secretome data [1]. Because it has been shown that proteome analysis of supernatant containing fetal calf serum is feasible [2, 3], the standard workflow for secretome analysis facilitates serum-free culture medium conditions. Secreted proteins are usually present in low amounts in the conditioned medium of mammalian cell cultures. For example, cytokines like, e.g., TNF or IL12 exist in the low ng/mL range in the conditioned medium [4, 5]. Therefore contaminations derived from, e.g., serum containing medium or other sources (sample handling, etc.) will interfere with MS-based protein identification of secreted proteins due

Katrin Marcus et al. (eds.), *Quantitative Methods in Proteomics*, Methods in Molecular Biology, vol. 2228,
https://doi.org/10.1007/978-1-0716-1024-4_21, © Springer Science+Business Media, LLC, part of Springer Nature 2021

to masking of lowabundant proteins and will reduce the dynamic range of protein quantification [6]. Thus, quality controls and optimization of cell culture conditions are indispensable prior to secretome analysis to minimize potential contaminants for the collection of high-quality data.

Once the cell culture and sample preparation have been optimized to avoid false-positive identification of secreted proteins data analysis offers an additional step to determine contaminations in secretome studies. One approach to exclude contaminants relies on the assumption that proteins showing a higher abundance in the secretome in comparison with the cellular proteome have a higher likelihood to be truly secreted. Therefore, several groups analyzed both the secretome and proteome and used a data comparison to define bona fide secreted proteins [7–10]. For example, Stiess et al. (2015) and Loei et al. (2012) considered the extra-to-intracellular protein ratio using either dual SILAC or iTRAQ labeling, respectively, to determine bona fide secreted proteins. Luo et al. (2011) published a method that employed a M-A plotting method referring to microarray data analysis to display the proteins which are enriched in the secretome of A549 cells.

Here, we present an experimental approach to analyze protein secretion by comparing data from label-free quantification (*see* also Chaps. 8, 16, 20, 22–24) of the secretome with the proteome [10]. Thus, the significance analysis of microarrays algorithm (SAM) [11] is applied to determine bona fide secreted proteins due to their higher abundance in the secretome compared to the cellular proteome [10]. All these slightly different published approaches considering quantitative data from secretomes as well as cellular proteomes to determine the significant enrichment of secreted proteins and to eliminate potential protein contaminants lead to the identification of bona fide secreted proteins. It is important to note that these approaches rely on a simplified assumption that the comparison of two quite different quantitative data sets allows decreasing the number of false-positive protein identifications. Nevertheless, these experimental approaches allow for the first time to get an idea about the complexity of the secretome.

2   Materials

Prepare all solutions using ultrapure water (prepared by purifying deionized water, to attain a sensitivity of 18 MΩ-cm at 25 °C) and use MS grade reagents especially for MS analysis. Prepare and store all reagents at room temperature (unless indicated otherwise). Diligently follow all waste disposal regulations when disposing waste materials.

For data analysis, the version of the used software and datasets might be of importance. If other software versions as indicated are used, the compatibility with the workflow should be evaluated.

2.1 Cell Culture	1. Phosphate-buffered saline (PBS).

2.1   Cell Culture

1. Phosphate-buffered saline (PBS).

2. Medium (for the A549 cells, we use Dulbecco's Modified Eagle's Medium (DMEM)—high glucose (D5796 Sigma Aldrich) + 10% fetal calf serum (FCS)).

3. Serum-free medium (for the A549 cells, we use DMEM—high glucose (*see* Subheading 2.1, **item 2**)).

4. A549 cells (CCL-185, ATTC/LGC Standards GmbH, Wesel, Germany).

2.2   Collection of Supernatant and Cell Harvest

1. Ice bed.

2. Rubber policeman.

3. Greiner tube (10 mL).

4. Eppendorf cup (2 mL).

5. PBS.

6. 0.2 μm Pall Acrodisc MS syringe filter (Pall, Dreieich, Germany).

7. Refrigerated centrifuge.

2.2.1   Cell Lysis

1. 5 mm steel beads.

2. Ultrasonic bath (VWR, Darmstadt, Germany).

3. TissueLyser (Qiagen, Hilden, Germany).

4. Lysis buffer: 30 mM 2-Amino-2-(hydroxymethyl)-1,3-propanediol (Tris base; 0.36 g/100 mL), 2 M thiourea (15 g/100 mL), 7 M urea (42 g/100 mL), 4% (w/v) 3-[(3-Cholamidopropyl)dimethylammonio]-1-propanesulfonate hydrate (CHAPS; 4 g/100 mL). Weight solid compounds and dissolve in about 95 mL water (*see* **Note 1**). Mix and adjust the buffer to pH 8.5 with hydrochloric acid (HCl). Make up to 100 mL with water and freeze aliquots of 10 mL at −20 °C (*see* **Note 2**).

5. Refrigerated centrifuge.

2.2.2   Precipitation of Proteins from Cell Supernatant

1. 50% (w/v) ice-cold Trichloroacetic acid (TCA) solution.

2. 10% N-Lauroylsarcosine sodium salt solution (*see* **Note 3**).

3. Refrigerated centrifuge.

4. Ice-cold acetone.

5. Eppendorf cup 2 mL.

6. Resolubilization (lysis) buffer: 30 mM Tris base, 2 M thiourea, 7 M urea, 4% (w/v) CHAPS. Adjust pH to 8.5 with HCl (*see* Subheading 2.2.1, **item 4**).

7. Ultrasonic bath.

2.3 Determination of Protein Concentration

1. Pierce 660 nm Protein Assay (Thermo Fisher Scientific, Darmstadt, Germany).

2. DTX Multimode Detector (Beckman Coulter).

3. Non-binding 96-well plate with a flat bottom.

2.4 In-Gel Digestion

1. 4–12% Bis-Tris sodium dodecyl sulfate (SDS)-polyacrylamide gel (Novex NuPAGE, Thermo Fisher Scientific, Darmstadt, Germany).

2. 4× SDS buffer: 600 mM dithiothreitol (DTT) (9.25 g/ 100 mL), 30% glycerin (30 g/100 mL), 12% SDS (12 g/ 100 mL), 150 mM Tris base (1.82 g/100 mL). Weight solid compounds and dissolve in about 95 mL water. Mix and adjust it to pH 7.0 with HCl. Make up to 100 mL with water and freeze aliquots of 10 mL at −20 °C. Before application, thaw 10 mL aliquots, aliquot it to 1 mL, and add some bromophenol blue crystals.

3. 20× 3-(N-morpholino)propanesulfonic acid (MOPS) buffer (500 mL) for BisTris gels: 1 M MOPS (104.63 g/500 mL), 1 M Tris base (60.57 g/500 mL), 2% SDS (10 g/500 mL), 20 mM ethylenediaminetetraacetic acid (EDTA) (20 mL EDTA of a 5 M EDTA stock solution (pH 8.0)). A pH value of 7.7 automatically adjusts itself.

4. Coomassie staining: Staining solution—40% (v/v) ethanol ≥99.8%, p.a. (50 mL/100 mL), 10% acetic acid (10 mL/ 100 mL). Add 25 mg Coomassie and adjust the volume to 100 mL with water.

5. Washing: Washing solution A—10 mM ammonium hydrogen carbonate (NH_4HCO_3) (80 mg/100 mL) in water. Washing solution B: 10 mM NH_4HCO_3 + acetonitrile (1:1) (*see* **Note 4**).

6. Alkylation: 10 mM DTT (0.155 g/100 mL) in 50 mM NH_4HCO_3. Freeze aliquots of 10 mL at −20 °C; 55 mM iodoacetamide (IAM) (1.017 g/100 mL) in 50 mM NH_4HCO_3. Freeze aliquots of 10 mL at −20 °C; 100 mM NH_4HCO_3 (0.8 g/100 mL) in water; 10 mM HCl: 8 µL 37% HCl in 992 µL water.

7. Digestion and extraction: Stock solution trypsin—250 ng/µL (25 µg trypsin/100 µL 10 mM HCl). Prepare 5 µL aliquots in glass tubes. Prepare a working solution of trypsin (0.033 µg/µL) fresh: 5 µL stock solution + 32.5 µL 100 mM 100 mM NH_4HCO_3 (*see* **Note 5**). Extraction solution: 0.1% trifluoroacetic acid (TFA)/acetonitrile (1:1); formic acid.

2.5 LC-MS/MS

1. UltiMate 3000 HPLC system (Dionex/Thermo Scientific, Idstein, Germany).

2. Acclaim PepMap100 trap column (C18, inner diameter, 75 μm; particle size 3 μm; pore size, 100 Å; length, 2 cm; Dionex/Thermo Scientific, Idstein, Germany).

3. Analytical column: Acclaim PepMapRSLC (C18, particle size 2 μm, pore sizes 100 Å, inner diameter 75 μm, length 25 cm; Dionex/Thermo Scientific, Idstein, Germany).

   (a) Loading solvent: 0.1% (v/v) TFA (MS grade).

   (b) Gradient solvent A: 0.1% (v/v) formic acid (MS grade).

   (c) Gradient solvent B: 0.1% (v/v) formic acid (MS grade), 84% (v/v) acetonitrile in water.

4. Q Exactive plus with an online nano-ESI source (Thermo Fisher Scientific, Bremen, Germany).

5. Distally coated SilicaTip™ emitters (New Objective, Woburn, USA).

6. Collision gas: nitrogen.

2.6 Database Search and Quantification

1. Raw files from secretome and proteome analysis.

2. MaxQuant (Max Planck Institute of Biochemistry, Version 1.6.6.0, https://maxquant.org).

3. Homo sapiens dataset UP000005640 downloaded from UniProtKB (in this example on 15 November 2019).

2.7 Processing of Quantitative Data

Perseus (Max Planck Institute of Biochemistry, Version 1.6.6.0, https://maxquant.org/perseus/).

2.8 Supplementary Data Annotation

UniProtKB https://www.uniprot.org/.

---

3 Methods

The described workflow is separated into two parts to generate a list of high-quality secreted proteins. The first part is an optimized sample preparation workflow allowing to obtain high reproducible MS data. The second part allows for the determination of bona fide secreted proteins by an integrated bioinformatic approach considering quantitative secretome and proteome data.

This protocol has been optimized for human cell lines grown under serum-free conditions as the high concentration of proteins from the serum interferes dramatically the secretome analysis. Before starting the experiments, each cell line has to be tested to grow under serum-free conditions without changing the underlying phenotype and extensive stress symptoms. If the viability of the cells is limited under serum-free condition, cultivation time has to be reduced to an optimum of high viability (>90%) and sufficient protein concentration in the supernatant (0.5–1 µg/µL). In this protocol, we analyze the conditioned medium and cellular proteome from A549 cells, each in five replicates. We recommend using replicates prepared from individual culture dishes to consider variances during sample preparation.

3.1 Cell Culture

1. Grow A549 cells in cell culture dishes (10 cm diameter) for adherent cells to 80–90% confluence in appropriate medium at 37 °C and 5% CO_2.

2. Wash the cells thoroughly (at least three times) with PBS and then three times with serum-free medium to reduce the contaminants from serum (*see* **Note 6**).

3. Incubate the cells for 24 h in serum-free medium (*see* **Note 7**).

3.2 Collecting Supernatant and Cell Harvest

1. Carefully collect the supernatant (conditioned medium) after serum-free incubation in appropriate tubes.

2. Centrifuge the supernatant at $1000 \times g$ and 4 °C for 10 min and filter the supernatant with a 0.2 µm Pall Acrodisc filter to remove cell debris and dead cells. (Extracellular vesicles can be removed by extended centrifugation (1 h, $100,000 \times g$.)

3. Freeze conditioned medium at −80 °C or proceed with TCA precipitation.

4. To harvest the cells, place the dishes on ice.

5. Wash the cells twice with ice-cold PBS to remove remnants of the medium.

6. Add 1 mL ice-cold PBS to the cells and harvest the cells carefully with a rubber policeman.

7. Transfer the suspension to a previously weighed 2 mL Eppendorf cup and centrifuge the cell suspension at $1000 \times g$ and 4 °C for 5 min.

8. Discard the supernatant.

9. Determine the pellet weight and store the pellet at −80 °C or continue with cell lysis.

3.2.1 Cell Lysis

1. Resuspend the cell pellet in the threefold amount of weight in lysis buffer (30 mM Tris base, 2 M thiourea, 7 M urea, 4% (w/v) CHAPS (pH 8.5)).

2. Add a 5 mm steel bead to each cup and lyse the cells mechanically for 1 min at 40 Hz in the bead mill.

3. Sonicate the samples six times for 10 s on ice and finally centrifuge the suspension at $16,000 \times g$ and 4 °C for 15 min.

4. Transfer the supernatant (proteome lysate) to a new cup and store it on ice.

5. Repeat the lyses (**steps 1–4**) with the twofold amount of lysis buffer.

6. Combine the supernatants after the second time of centrifugation.

7. Store the lysate at −80 °C or determine the protein concentration in the next step.

3.2.2 Precipitation of Secreted Proteins

1. Add ¼ volume ice-cold 50% TCA (w/v) to the harvested conditioned medium (Subheading 3.2, **step 3**) in order to precipitate proteins.

2. Add 100 μL of a freshly prepared 10% sodium lauroyl sarcosinate (N-lauroyl sarcosine sodium salt in water) solution (final concentration of 0.1%) to the conditioned medium for a better resolubilization of precipitated proteins.

3. Vortex and spin down shortly to remove rests of liquid from the lid.

4. Precipitate the proteins from the conditioned medium on ice for 1 h.

5. Centrifuge at $7100 \times g$ (max.) at 4 °C for 10 min.

6. Carefully remove the supernatant.

7. Add 1 mL ice-cold acetone to wash the pellet.

8. Vortex until the pellet is released from the bottom of the tube (*see* **Note 8**).

9. Centrifuge at $7100 \times g$ at 4 °C for 10 min.

10. Carefully remove and discard the supernatant.

11. Dry the pellet at RT for about 5 min.

12. Redissolve the pellet in 50 μL resolubilization buffer (30 mM Tris base, 2 M thiourea, 7 M urea, 4% (w/v) CHAPS (pH 8.5)) and sonicate the samples six times for 10 s on ice.

13. Store the secretome at −80 °C or determine the protein concentration in the next step.

3.3 Determination of the Protein Concentration

1. To determine the protein concentration of the secretome and proteome samples, dilute the secretome samples 1:6 and the proteome samples 1:11 in water to reach a concentration of about 0.5–1 μg/μL and use the Pierce 660 nm Protein Assay (Fisher Scientific, Schwerte, Germany) (*see* **Note 9**).

2. Measure all protein concentrations in duplicate according to the manufacturer's instructions.

3.4 In-Gel Digestion

1. Dilute 5 μg protein of secretome and proteome samples with 4× SDS buffer and heat the samples (40 °C) for 10 min (*see* **Note 10**).

2. Load the samples onto a 4–12% Bis-Tris sodium dodecyl sulfate (SDS)-polyacrylamide gel (Novex NuPAGE, Thermo Scientific, Darmstadt, Germany) and separate the samples at a voltage of 50 V for 10–15 min until a separation distance of 0.5–1.0 cm has been reached (*see* **Note 11**).

3. The gel is removed from its gel cassette and immediately placed in a fixation solution.

4. Visualize the proteins by Coomassie staining: stain the gel for 2 h with Coomassie staining solution, remove the staining solution, and destain the gel using Coomassie destaining solution until the background staining is removed (*see* **Note 12**). Subsequently, cut out the protein-containing bands.

5. Wash the bands three times alternately with washing solutions A and B to destain the bands (*see* **Notes 13** and **14**).

6. Reduce the proteins with 10 mM DTT at 56 °C for 45 min and afterward alkylate the proteins with 55 mM iodoacetamide in the dark at RT for 30 min.

7. Perform a tryptic in-gel digest. Therefore, add trypsin (1:50, w/w) to digest the protein overnight (max. 16 h) at 37 °C.

8. To extract the tryptic peptides from the gel, cover the band with extraction solution (the band should be completely covered).

9. Sonicate the samples for 15 min on ice and transfer the supernatant to an HPLC vial.

10. Repeat the extraction step (**steps 8** and **9**) and combine the supernatants in the HPLC vial.

11. Remove the acetonitrile by drying the samples in a vacuum centrifuge and resuspended 300 ng of the peptides in 17 μL 0.1% TFA for LC-MS analysis (the volume of 0.1% TFA must be adjusted according to the amount of used protein).

3.5 LC-MS/MS

1. Consider 300 ng digest proteins for MS analysis.

2. First separate peptides by nano liquid chromatography on an Ultimate 3000 rapid separation liquid chromatography system with an Acclaim PepMap100 trap column as a pre-column for 10 min at a flow rate of 6 μL/min using 0.1% TFA as a mobile phase. In the next step, separate the peptides using an Acclaim PepMapRSLC as analytical column with a constant flow rate of 300 nL/min using a 2 h gradient from 4 to 40% of solvent B.

3. Elute the separated peptides into an online coupled Q Exactive plus hybrid quadrupole Orbitrap mass spectrometer via a nano source electrospray interface equipped with distal-coated SilicaTip emitters.

4. The mass spectrometer should operate in positive ion mode with a spray voltage of 1400 V.

5. Use a data-dependent top 10 method and record full scans in profile mode in the Orbitrap analyzer over a scan range from 350 to 2000 m/z with a resolution of 70,000. Set the target value for automatic gain control to 3,000,000 and accumulate for a maximum of 80 ms.

6. Record MS/MS spectra considering an available mass range of 200–2000 m/z and a resolution of 17,500. Isolate up to ten precursors (+2, +3 charge states) within a 2 m/z isolation window and fragment them via higher-energy collisional dissociation. Set the maximum ion time to 60 ms and the value for the automatic gain control to 100,000. Exclude already fragmented precursors from further isolation for the next 100 s.

7. Finally, export LC/MS-MS analysis data as Thermo .raw file format.

3.6 Database Search and Quantification

1. Process the conditioned medium and cellular proteome data in parallel (Fig. 1). Raw files are processed with MaxQuant [12, 13] to extract protein identification and quantification information from the data.

2. Import mass spectrometric .raw files of the measurements from conditioned medium and cell lysates into MaxQuant and assign an individual name in the "experiment" column to each replicate, additionally, name cell lysates with "L_" and conditioned medium samples with "S_". Different parameter groups should not be defined (*see* **Note 15**).

3. Apply standard parameters for the search including carbamidomethylation at cysteines as fixed and methionine oxidation and

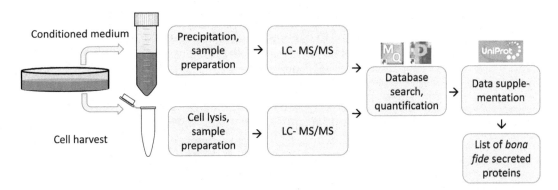

Fig. 1 *Workflow for the determination of high-quality secretome.* In the first part, conditioned medium and cells are harvested separately and prepared for mass spectrometric analysis. Mass spectrometric data was used in the second part for protein identification and quantification and the comparison of protein abundances of proteins from conditioned medium and cellular proteomes. After additional data supplementation, a list of bona fide secreted proteins is generated

N-terminal acetylation as variable modifications. Check the "match between runs" function (Global parameters → Identification → Match between runs), enable label-free quantification (Group-specific parameters → Label-free quantification → LFQ), and specify an appropriate protein dataset. In the current example, we used the UP000005640 Homo sapiens protein sequences from UniProtKB with an appropriate parsing rule (>.*\|(.*)\|). After starting the searches, MaxQuant performs protein identification, quantification of precursor ion intensities, alignment of precursors, and normalization.

3.7 Analysis of Quantitative Protein Data

Perseus [14] is used for first filtering the data and to compare conditioned medium and cell lysates with SAM analysis.

1. Import the "proteinGroups.txt" output file into Perseus. The normalized intensities (LFQ intensities) are specified here as "main" columns.

2. After the import of the data matrix, filter the data. First, remove proteins "identified by site" (Filter rows → Filter rows based on categorical column → Only identified by site → ok) as well as "reverse hits" (Filter rows → Filter rows based on categorical column → Reverse → ok). Before removing contaminants (Filter rows → Filter rows based on categorical column → Potential contaminants → ok), carefully evaluate the amount and intensities especially of potential serum contaminants. Next, remove proteins that have been identified with only one peptide (Filter rows → Filter rows based numerical/main column → Number of columns: 1 × Peptides → Relation 1: $x > 1$ → ok).

3. Perseus (*see* **Note 16**) offers the possibility to supplement the data with functional annotation. Add annotation of different gene ontology (GO) categories, KEGG (curated pathways), and SMART (protein domains) (Annot. Columns → Add annotations → chose the right species (Homo sapiens) and add GOMF slim name, GOBP slim name, GOCC slim name, KEGG name, SMART name → ok).

4. To prepare the data for the statistical analysis, calculate the log2 of the LFQ-intensities (Basic → Transform → log2(x) → ok). Next, define the different groups to be compared. For this, use the "Annot. Rows" (Annot. Rows → categorical annotation rows) function of Perseus. Here, mark conditioned medium samples with S (secretome) and cellular lysates with L. Subsequently, remove proteins showing a lower number as 80% of valid values in at least one group (Filter rows → Filter rows based on valid values → Min. valids: Percentage: 80; Mode: in at least one group → ok). In our example, we need

four valid values either in the conditioned medium or cellular lysate samples. Impute missing values with values from a downshifted normal distribution (Imputation → replace missing values from a normal distribution → ok).

5. Next, calculate a Student's t-test based statistical analysis according to the significance analysis of microarrays method [11] (Tests → Two-sample tests → First group: **S** Second group: **L; S0: 0.8** → ok) and export the newly formed matrix (plain matrix export).

3.8 Data Supplementation

Information about transmembrane helices and signal peptides is provided from UniProtKB.

1. Use Excel or alternative software to extract the first identifier from the "T: Protein IDs" column (*see* **Note 17**).

2. Upload protein identifiers at https://www.uniprot.org/uploadlists/ and check additional annotation including information about signal peptides and transmembrane domains (Fig. 2). After downloading annotation data, map the information to the quantitative/statistical data using Excel or Perseus.

3. Select and evaluate "high-quality" secreted proteins. They should show a "+" in "C: Student's T-test Significant" column, and the "N: Student's T-test Difference" should be >0. Additionally, no transmembrane domains should be annotated in UniProtKB. It is now possible to, e.g., use volcano plots for data evaluation (Fig. 3).

▼ Subcellular location	▼ PTM / Processsing
☑ Intramembrane	☐ Chain
☑ Subcellular location [CC]^i	☐ Cross-link
☑ Topological domain	☐ Disulfide bond
☑ Transmembrane	☑ Glycosylation
	☐ Initiator methionine
	☑ Lipidation
	☐ Modified residue
	☐ Peptide
	☐ Post-translational modification
	☑ Propeptide
	☑ Signal peptide
	☑ Transit peptide

Fig. 2 Annotation selection at UniProtKB to further annotate proteome/secretome data with information of most importantly predicted signal peptides and transmembrane domains

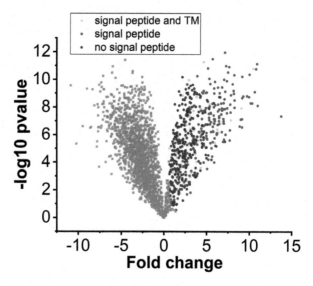

Fig. 3 Volcano plot based comparison of proteins quantified in the comparison of the conditioned medium and the cellular proteome from A549 cells (each $n = 5$). Proteins showing a higher abundance in the conditioned medium and therefore are candidates for secreted proteins are marked by circles. Further annotation of proteins considering signal peptides and transmembrane domains (TM) was added by UniProtKB

4   Notes

1. If the weighed solid compounds don't dissolve under room temperature, use a water bath (max. 35 °C).

2. Thaw buffers early as CHAPS take some time to get completely resolved.

3. Prepare this fresh each time.

4. After everything is solved, the destaining and wash solutions should be filled in Nalgene bottles (Thermo Scientific).

5. You can store the trypsin working solution for max. 7 days at 4 °C.

6. The washing steps are very important to reduce the amount of potential contaminating serum proteins because high-abundant serum proteins, e.g., albumin, might mask the presence of lowabundant secreted proteins.

7. Quality control of cell culture is of high importance. These include regular mycoplasma tests since mycoplasma extensively affects the cell physiology and their metabolism which influences your results. Furthermore, you should consider cell line-specific limits of passage numbers, because the cells can change their morphology in high passages. To control potential cross-contamination with other cell lines or genomic changes,

you can check the cell line identity, for example, by short tandem repeat (STR) profiling. In addition, you have to determine the optimal incubation time with serum-free medium in preliminary studies for your cell type, because serum depriving could interfere with cell proliferation, viability, and metabolism which may possibly lead to experimental biases during qualitative and quantitative secretome analysis. Furthermore, cell numbers and the amount of conditioned medium required for secretome analysis largely depend on the analyzed cell type and incubation time. So you have to determine these factors first. In our experience, 5–10 mL conditioned medium includes sufficient protein amounts to perform secretome analysis.

8. To increase the yield, you can transfer the suspension in an Eppendorf cup and centrifuge at a higher speed.

9. Prior to the determination of the protein concentration, check the pH of the secretome samples. Often, the pH is still in the strong acidic range, which may influence the determination of the protein concentration as well as the running behavior in the gel.

10. For quantitative MS analysis, please prepare sufficient replicates to allow for statistical analysis. Routinely, we consider at least five replicates ($n = 5$).

11. It's always recommended to run a regular SDS-PAGE as quality control.

12. It might necessary to exchange the destaining solution several times. Use for each gel ~100 mL of each solution.

13. Please prepare sufficient replicates for quantitative MS analysis to allow statistical analysis. Routinely, we consider at least five replicates ($n = 5$).

14. Before digestion, the gel pieces should be protected from environmental proteins (e.g., keratins) by keeping the tube closed and covering the gels with a plastic sheet.

15. Label-free quantification and normalization normally require that the abundances of most of the analyzed protein are quite stable. That's normally not the case if conditioned medium samples and cell lysates are compared. Nevertheless, we found in several projects that the LFQ normalization algorithm will help to identify proteins that are overrepresented in the conditioned medium or cellular lysate, respectively.

16. Perseus offers different options for further quality control of the data, e.g., by scatter plots, hierarchical cluster analysis, etc. We recommend to intensively make use of these options to evaluate the quality of the samples and acquired data.

17. Alternatively, data annotation can be carried out for all identifiers in one protein group as the annotation of signal peptides
and transmembrane domains might depend on the respective
protein isoform [10].

Acknowledgments

We'd like to thank Thomas Lenz for performing the A549 cell
culture experiments used to exemplify the provided workflow.

References

1. Schira-Heinen J, Grube L, Waldera-Lupa DM
et al (2019) Pitfalls and opportunities in the
characterization of unconventionally secreted
proteins by secretome analysis. Biochim Biophys Acta Proteins Proteom 1867
(12):140,237

2. Eichelbaum K, Krijgsveld J (2014) Combining
pulsed SILAC labeling and click-chemistry for
quantitative secretome analysis. In: Ivanov IA
(ed) Exocytosis and endocytosis. Springer
New York, New York, NY, pp 101–114

3. Eichelbaum K, Winter M, Diaz MB et al
(2012) Selective enrichment of newly synthesized proteins for quantitative secretome analysis. Nat Biotech 30(10):984–990

4. Mukherjee P, Mani S (2013) Methodologies to
decipher the cell secretome. Biochim Biophys
Acta 1834(11):2226–2232

5. Chevallet M, Diemer H, Van Dorssealer A et al
(2007) Toward a better analysis of secreted
proteins: the example of the myeloid cells
secretome. Proteomics 7(11):1757–1770

6. Weigert C, Lehmann R, Hartwig S et al (2014)
The secretome of the working human skeletal
muscle—a promising opportunity to combat
the metabolic disaster? Proteomics 8
(1–2):5–18

7. Luo X, Liu Y, Wang R et al (2011) A high-
quality secretome of A549 cells aided the discovery of C4b-binding protein as a novel
serum biomarker for non-small cell lung cancer. J Proteome 74(4):528–538

8. Loei H, Tan HT, Lim TK et al (2012) Mining
the gastric cancer secretome: identification of
GRN as a potential diagnostic marker for early
gastric cancer. J Proteome Res 11
(3):1759–1772

9. Stiess M, Wegehingel S, Nguyen C et al (2015)
A dual SILAC proteomic labeling strategy for
quantifying constitutive and cell-cell induced
protein secretion. J Proteome Res 14
(8):3229–3238

10. Grube L, Dellen R, Kruse F et al (2018)
Mining the secretome of C2C12 muscle cells:
data dependent experimental approach to analyze protein secretion using label-free quantification and peptide based analysis. J Proteome
Res 17(2):879–890

11. Tusher VG, Tibshirani R, Chu G (2001) Significance analysis of microarrays applied to the
ionizing radiation response. Proc Natl Acad Sci
U S A 98(9):5116–5121

12. Cox J, Hein MY, Luber CA et al (2014) Accurate proteome-wide label-free quantification by
delayed normalization and maximal peptide
ratio extraction, termed MaxLFQ. Mol Cell
Proteomics 13(9):2513–2526

13. Tyanova S, Temu T, Cox J (2016) The Max-
Quant computational platform for mass
spectrometry-based shotgun proteomics. Nat
Protoc 11(12):2301–2319

14. Tyanova S, Temu T, Sinitcyn P et al (2016) The
Perseus computational platform for compre-
hensive analysis of (prote)omics data. Nat
Methods 13(9):731–740

Chapter 22

Establishing a Custom-Fit Data-Independent Acquisition Method for Label-Free Proteomics

Britta Eggers, Martin Eisenacher, Katrin Marcus, and Julian Uszkoreit

Abstract

Data-independent acquisition (DIA) has recently developed as a powerful tool to enhance the quantification of peptides and proteins within a variety of sample types, by overcoming the stochastic nature of classical data-dependent approaches, as well as by enabling the identification of all peptides detected in a mass spectrometric event. Here, we describe a workflow for the establishment of a sample-fitting DIA method using Spectronaut Pulsar X (Biognosys, Switzerland).

Key words Data-independent acquisition, Label-free quantification, Spectral library generation

1 Introduction

A common and standard proteomic setup consists of liquid chromatography, mainly reversed phase high-performance liquid chromatography (HPLC), coupled to a tandem mass spectrometer (MS). The majority of current proteomic workflows uses peptides for the identification of proteins, a so-called bottom-up approach, in which a mixture of proteins is digested into peptides prior MS analysis. A typical MS method is composed of an MS1 scan followed by an MS2 scan. After the MS1 scan, the Top N most intensive precursor ions (most often Top 10–20) are selected for fragmentation and measured in an MS2 event. This method is called data-dependent acquisition (DDA) (*see* Chaps. 8–10, 13–17, 20, 21, 26). Unfortunately, DDA is faced with several issues, which cannot be solved in a classical MS approach. First and foremost, the identification of low-abundant peptides is quite challenging in a classical DDA approach, due to co-eluting higher-abundant peptides, which will preferentially be picked for fragmentation events. Thus, as a result, their proper quantification is not satisfactory. Furthermore, the stochastic nature of DDA measurements challenges sample reproducibility [1]. Data-independent

Katrin Marcus et al. (eds.), *Quantitative Methods in Proteomics*, Methods in Molecular Biology, vol. 2228,
https://doi.org/10.1007/978-1-0716-1024-4_22, © Springer Science+Business Media, LLC, part of Springer Nature 2021

acquisition (DIA) (*see* Chaps. 8, 16, 23, 24) overcomes this problem, by selecting all eluting peptides in a predefined m/z range for fragmentation [2]. In principle, DIA approaches allow to identify all peptides in a given sample. However, in contrast to classical DDA approaches, the information between precursor ion and its fragments is lost, resulting in the need of alternative data analysis approaches as well as complex algorithms. The most common approach for analyzing DIA data is a spectral library-based approach in which DDA measurements serve as template for the identification of acquired spectra. To mention, also spectral library-free approaches are currently developed, e.g., DIA-Umpire [3] and DirectDIA [1], but will not be discussed in in this chapter. The generation of a well-fitting spectral library is of utmost importance. The quality as well as the similarity of acquired DDA spectra is the most important parameter to consider rather than accumulating large data sets to create a spectral library of large size and identification rates. We therefore strongly recommend to repeatedly measure a sample pool of the actual data set for the creation of an optimal spectral library. These DDA measurements can further be used for the DIA method creation by calculating optimal window sizes utilizing parameters acquired in DDA runs, such as spectrum peak widths. Fractionation of samples may increase the spectral library size but could possibly hinder an optimal identification to quantification rate and should therefore be thoroughly tested [4]. Prior to spectral library generation, one should first optimize LC as well as MS parameters to ensure operating with an optimal method for every sample type analyzed. Therefore, this chapter will explain how to set up an optimal chromatographic method for an adequate elution of peptides, followed by an in-depth explanation on how to establish a sample-fitting DIA method and a suitable spectral library. Crucial LC and MS parameters will be shortly explained and their adaption to different samples discussed.

2   Materials

All buffers should be prepared using either HPLC/LC-MS grade or ultrapure water (prepared by purifying deionized water to attain a conductivity of 18 MΩ cm^{-1} at 25 °C) and analytical grade reagents. For LC loading and gradient solvent, use LC-MS grade acids. All buffers can be prepared and stored at room temperature.

2.1   Liquid Chromatography

1. Dionex Ultimate 3000 Nano LC System (Thermo Fisher Scientific).

2. Trap column; Acclaim PepMap100 C18 Nano-Trap Column (C18, particle size 5 μm, pore sizes 100 Angstrom, I.D. 200 μm, length 2 cm; Thermo Fisher Scientific).

3. Nano column; Acclaim PepMap100 C18 Nano-Trap Column (C18, particle size 3 μm, pore sizes 100 Angstrom, I.D. 75 μm, length 50 cm; Thermo Fisher Scientific).

4. Loading solvent (Buffer A): 0.1% (v/v) trifluoroacetic acid (TFA) (MS grade).

5. Gradient solvent A: 0.1% (v/v) formic acid (FA) (MS grade).

6. Gradient solvent B (Buffer B): 0.1% (v/v) formic acid (FA) (MS grade), 84% (v/v) acetonitrile (ACN) (HPLC-S gradient grade).

2.2 Mass Spectrometry

1. QExactive with an online nanoESI source (Thermo Fisher Scientific, Germany).

2. QExactiveHF with an online nanoESI source (Thermo Fisher Scientific, Germany).

3. Orbitrap Fusion Lumos with an online nanoESI source (Thermo Fisher Scientific, Germany) (*see* **Note 1**).

4. Distal Coated Silica Tips (New Objective, Woburn, USA).

5. Collision gas: nitrogen, helium.

6. iRT peptides prepared after manufacturer's instructions (Biognosys, Switzerland).

2.3 Data Analysis

1. KNIME (version 4.1.0 was used during writing this chapter). KNIME analytics platform is an open-source software, which can be used for various kinds of data science applications and services [5].

2. OpenMS (version 2.4.0). OpenMS is an open-source framework for proteomics, which is used for processing mass spectrometric spectra [6].

3. msConvert. An open-source command line tool, which is part of ProteoWizard. With this tool raw files can be easily converted in a suitable data format—in our case—mzML. format.

4. Thermo Fisher Scientific Xcalibur (v. 4.2.47 was used during writing this chapter).

5. Spectronaut Pulsar X (v. 12.0.20491.7.17149 was used during writing this chapter) for spectral library generation (*see* **Note 2**).

6. Spectronaut Pulsar X for the analysis of data-independent acquisition (*see* **Note 3**).

7. FASTA file (Uniprot KB).

3   Methods

3.1   LC Optimization

An optimal LC gradient is essential for good sample separation. The most crucial parameter to consider is the samples peptide composition. Hydrophilic samples require lower percentage of organic solvent than hydrophobic samples. Likewise, the determination of an optimal chromatographic elution time, usually between 60 and 200 min when operating with a nano flow system, is essential for a good DIA workflow. Here the average peak width should be taken as an essential parameter. As a rule of thumb, complex samples, e.g., cell culture lysates, as well as tissue lysates require longer gradient lengths starting from 120 to 200 min. Furthermore, sample types covering a high dynamic range of peptide abundances, e.g., cerebrospinal fluid (CSF) or muscle tissue, benefit from an extensive chromatographic elution. Fractionated samples as well as laser-microdissected samples or single-cell analysis may be analyzed with shorter gradient lengths starting from 60 min, allowing higher-intensity signals; an example gradient is provided in Table 1, and solvent compositions are shown below:

(a)   Trap column.

- Temperature: 60 °C.
- Flow rate: 30 μL/min.
- Running buffer: 0.1% trifluoroacetic acid.

Table 1
Example gradient (120 min)

Trap column				NC column		
Retention time (min)	Flow μm/min	%B	%C	Retention time (min)	Flow μm/min	%B
0.00	30.000	0.0	0.0	0.00	0.400	5.0
0.00	30.000	0.0	0.0	0.00	0.400	5.0
105.00	30.000	0.0	0.0		0.400	5.0
107.00	30.000	0.0	40.0	105.00	0.400	40.0
109.00	30.000	0.0	40.0	107.00	0.400	95.0
110.00	30.000	0.0	95.0	114.00	0.400	95.0
114.00	30.000	0.0	95.0	115.00	0.400	5.0
115.00	30.000	0.0	0.0	120.00	0.400	5.0
120.00	30.000	0.0	0.0			

(b) Analytical C18 reversed-phase column.

- Temperature: 60 °C.
- Flow rate: 30 μL/min.
- Running buffer A: 0.1% trifluoroacetic acid.
- Running buffer B: 84% acetonitrile.
- Gradient: 5–30% running buffer B over 98 min (*see* **Notes 2** and **3**).

To determine optimal chromatographic settings, one should:

1. Prepare a pool of all your samples to determine optimal gradient length and percentage of organic solvent.

2. Dissolve your samples in 0.1% trifluoroacetic acid (TFA), when working with a trap column device in your HPLC, or in 0.1% formic acid (FA) when working with direct injection on the main column of your HPLC.

3. Create a DDA method using the Thermo Xcalibur Instrument Setup (*see* **Note 4**).

4. Add a Full MS/dd MS^2 (TopN) Experiment and fill in the properties (*see* **Note 5**) (Table 2).

Table 2
Properties of full MS/ ddMS2 (TopN)

General		Full MS		dd-MS^2/dd-SIM		dd Settings	
Run time	0–120 min	Resolution	60,000	Resolution	30,000	Minimum AGC target	1.20e04
Polarity	Positive	AGC target	3e6	AGC target	1e6	Intensity threshold	1.0e5
Default charge	2	Maximum IT	80 ms	Maximum IT	120 ms	Apex trigger	–
Inclusion	–	Scan range	400–1400 *m/z*	Loop count	10	Charge exclusion	Unassigned,1, 5–8 > 8
Exclusion	–			Top N	10	Peptide match	Preferred
Tags	–			Isolation window	1.6 *m/z*	Exclude isotopes	On
				Fixed first mass	100 *m/z*	Dynamic exclusion	10
				(N)CE/ stepped nce	28		

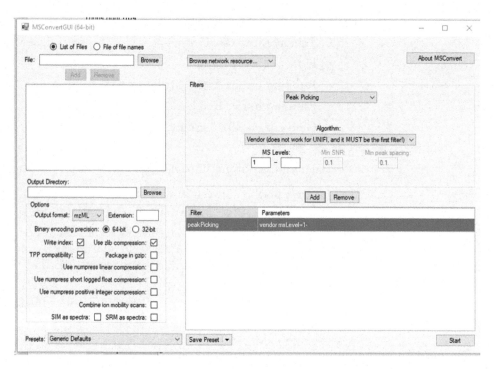

Fig. 1 MSConvertGUI Interface. To convert your raw files into mzML Format, browse your files add an output directory and choose mzML as output format. Filter criteria which shall be set are peak picking for MS1 and MS2 using the Vendor algorithm

5. Run at least five replicates per condition (gradient length, gradient percentage) to determine reproducibility (via calculating the CV, as well as determining peptide overlap) of your chromatographic setting.

6. Convert your raw files into mzml file format, e.g., using msConvert (Fig. 1).

7. Choose 'peak picking' as filter option and the 'vendor algorithm' and enable peak picking on 'MS levels $1 - n$'.

8. Determine average peak widths, e.g., using the 'determine peak width' workflow presented in Fig. 2 (Subheading 3.7) created in KNIME [5].

9. Open the 'Input files' node and add the location where your runs (mzml format) are stored.

10. Run the workflow and open the created histogram in the histogram node to determine average peak widths.

11. Determine peptide identification and quantification rate with a program of your choice, e.g., MaxQuant and Proteome Discoverer 1.4; 2.2; 2.4.

12. Calculate your mean coefficient of variation (CV) as a parameter for method robustness. CVs of technical replicates should be <20%.

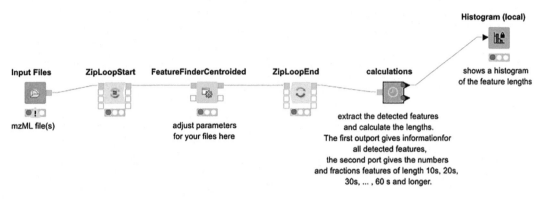

Fig. 2 Overview of the workflow for the determination of peak lengths

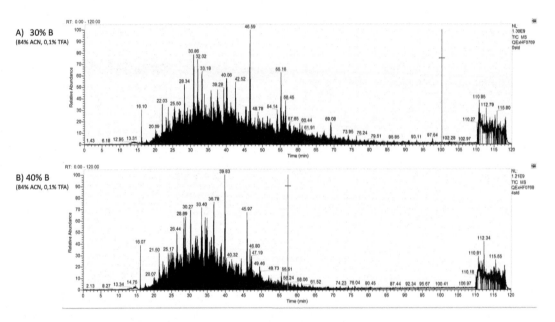

Fig. 3 Elution profiles of two complex cell lysate samples. (**a**) Liquid chromatography gradient with 30% B and with 40% B (**b**). Please refer to Sect. 2.1 for exact buffer composition

13. Elucidate optimal gradient length and organic solvent composition by peptide identification, peptide quantification and their ratios (how many identified peptides could be quantified as well?), as well as by determining elution profiles using XCalibur (Fig. 3a, b, *see* **Note 6**).

3.2 MS Optimization

To establish a robust and optimal DIA method, several MS parameters should be adapted. To ensure that an optimal elution profile of the acquired peaks can be achieved during MS analysis, at least seven to eight data points per peak should be acquired during one scan cycle [7]. To accomplish that, simple calculations have to be carried out taking MS2 resolution and Orbitrap scan time into

Table 3
Orbitrap scan times

Resolution (MS2)	QExactive (scan time Orbitrap)	Resolution (MS2)	QExcativeHF (scan time Orbitrap ms)	Resolution (MS2)	Orbitrap Fusion Lumos
140.000	512	120.000	256	120.000	256
70.000	256	60.000	128	60.000	128
35.000	128	30.000	64	30.000	64
17.500	64	15.000	32	15.000	32

Orbitrap scan times of different Thermo Fisher Scientific mass spectrometry devices depending on MS2 resolution

Table 4
Dependency of peak width, cycle time, and MS2 resolution for a QExactive HF (Thermo Fisher Scientific)

Average peak widths (s)	Resolution (MS2)	QExactive HF (scan time Orbitrap ms)	Cycle Time (s)	Data points per peak	Number of windows
10	30.000	64	~1.5	7	23
20	60.000	128	~3	7	23
30	60.000	128	~4	7	31

account. Depending on the instrument used, scan times can vary drastically. Table 3 is displaying Orbitrap scan times in relation to a certain MS2 resolution for three of the most common Thermo Fisher Orbitrap devices.

As already indicated, if the determined peak widths are <10 s, a maximum cycle time of 1.5 s should be aimed for to acquire a minimum of seven data points per peak (DPPP). Longer average peak widths of >20 s allow longer cycle times up to 3 sec. Table 4 exemplarily provides optimal combinations of peak width, cycle time, and chosen MS2 resolution for a QExactiveHF.

Further parameters to be considered:

Optimizing your normalized collision energy (NCE) is recommended to enhance fragment ion spectra completeness. Please ensure, that DDA and DIA measurements performed match the same LC and MS parameter to ensure optimal comparability.

After optimizing LC and MS parameters, static DIA windows can easily be calculated by using the acquired information described above.

3.3 Calculation of Fixed Window Sizes

1. Determine your mass range (usually 400–1400 m/z) for a standard bottom-up approach.

2. Calculate the optimal number of windows by aiming for a cycle time of 1.5 s for an average of 10 s peak width or a 3 s cycle time for an average peak width of 30 s by using the following formula:

Cycle time/Scan speed of the Orbitrap = number of windows

3. For a resolution of 30,000 on a QExactiveHF instrument, the equation will look as followed:

$$1.5 \ s/0.064 \ s = 23$$

4. Lower your window number by one, because your MS1 scan will make up one scan event.

5. Determine the window size by dividing the total mass range measured with your calculated window number:

Mass range/number of windows = window size

$$1000 \ m/z/22 = 45 \ m/z$$

For a 1000 m/z mass range, window sizes calculated are 45 m/z. To generate a functional MS method, the center of each m/z window has to be calculated and inserted as an inclusion list into the method editor. For our example, the desired inclusion list is listed in Table 5. To ensure an optimal coverage of elution profiles eluting at the edges of our center mass, the center window size is enlarged by 1 m/z to create a window overlap (see **Note 7**).

6. Create your DIA Method using XCalibur.

7. Add an 'Full-MS-SIM' Experiment to your method for an MS1 scan; following parameters may be set accordingly (Table 6) but can be optimized for specific sample types.

8. Add an 'DIA' Experiment to your method for DIA MS2 scans; following parameters shall be set according Table 6.

9. Paste the 'Center window + 1 m/z' values as inclusion list into the XCalibur Method Editor (Table 5); during the measurements every window will repeatedly be scanned every 1.5 s (**Note 7**).

Table 5
Calculation of windows with a window size of 45 m/z and a total number of 22 windows, exemplary displaying the first, second and last window

Nr. of window	Start m/z	End m/z	Center m/z	Center m/z + 1 Da
1	400	445	422.5	423.5
2	445	490	467.5	468.5
22	1355	1400	1377.5	1378,5

Table 6
Properties of full MS-SIM and DIA

Properties of full MS-SIM				Properties of DIA			
General		**Full MS SIM**		**General**		**DIA**	
Runtime	0–120 min	Resolution	60,000	Runtime	0–120 min	Resolution	30,000
Polarity	Positive	AGC target	3e6	Polarity	Positive	AGC target	1e6
Default charge	2	Maximum IT	120 ms	Default charge	2	Maximum IT	Auto
		Scan range	400–1400 *m/z*			Loop count	22
						MSX count	1
						Isolation window	45
						Fixed first mass	100 *m/z*
						NCE	28

Note, that optimal parameters may vary for your sample type and that the parameters presented here were set for an average chromatograpic peak width of 10 s and an Orbitrap scan speed of 64 ms in a QExactive HF device.

3.4 Variable DIA Windows

A more elaborate way to design your DIA experiment is to assign dynamic windows (*see* **Note 8**). It is well known, that peptides do not elute equally over the whole gradient and mass range and by that numbers of features measured can vary greatly along the retention time and *m/z* axes. To ensure that an optimal number of eluting ions is fragmented in one window, variable window sizes can be assigned, by choosing small window sizes for very high numbers of eluting peptides and large window sizes for very small numbers of features in a determined mass range. To do so, we constructed a workflow in KNIME, which automatically choses the optimal dynamic window sizes based on a prior DDA run. Method generation is more time-consuming, because each variable window gets a separate DIA Experiment node. In our case, 22 windows will result in 22 DIA experiment nodes.

1. Set MS1 parameters according to your instrument by adding a 'Full MS—SIM' Experiment. You can orientate yourself on the fixed DIA method creation section, as parameters do not change on MS1 level in a variable window setting.

2. Include your variable DIA windows as an inclusion list.

3. Add an DIA Experiment for each variable window separately.

4. Set DIA parameters as presented in Table 6 but change the 'Loop count' to '1' and the 'Isolation window' parameter to the size of the respective window. If this is the first DIA experiment node, enter the first center window size + 1 Da; if it is the second DIA experiment node, enter the second window size and so on.

3.5 Creation of a Spectral Library

As already indicated in our DIA approach, a spectral library is functioning as template for the assignment of DIA spectra. Spectral library generation is carried out using DDA runs. For an optimal spectra assignment, DDA runs of the measured sample are taken for library generation and are repeatedly measured with DDA. Fractionation of sample by high pH fractionation may increase library size, but may not be optimal for later quantification events. Acquired DDA runs can be used for spectral library generation in Spectronaut Pulsar X. iRT peptides have to be added to every DDA and DIA run acquired to allow retention time alignment.

1. Open Spectronaut Pulsar X.

2. Go to the 'Settings' Icon and open the Pulsar search tab.

3. Open the BGS Factory Setting (default) scheme.

4. Add known modifications of your sample to the 'Applied Modifications' subsection and save the scheme.

5. Go to the 'Library generation tab' and check the BGS Factory Settings for suitability (*see* **Note 9**).

6. Go to the 'Databases' Icon and open the 'Protein Databases' tab.

7. Add a suitable FASTA file (*see* **Note 10**).

8. Add the correct parsing rule.

9. Go to the 'Library' Icon and click 'Generate Library from Pulsar/Search Archives' (*see* **Note 11**).

10. Enter an experiment name and add DDA runs and click next.

11. Assign the correct FASTA file.

12. Determine the Search settings you created in the 'Settings' section.

13. Click skip to last and start creating your spectral library (*see* **Note 12**).

3.6 DIA Method Analysis and Validation

1. Go to the 'Settings' icon and choose the 'DIA analysis' tab.

2. Open the BGS Factory Setting (default) scheme and browse the settings for suitability (*see* **Note 13**).

3. Go the 'Analysis' Icon in Spectronaut Pulsar X and click 'Setup a DIA analysis from File'.

4. Load your acquired DIA data (*see* **Note 13**) and click next.

5. Assign your spectral library and click load.

6. Select the setting scheme for DIA analysis.

7. Assign the FASTA file (*see* **Note 14**) and click next.

8. Choose any run as reference run (*see* **Note 15**), skip to last, and finish.

9. After your analysis is finished, open the 'Post analysis' icon to obtain statistical parameters.

10. Open the 'Report' Icon to download your data as excel format (*see* **Note 16**).

11. Determine your optimal DIA setting as well as a suitable library (fixed, stepped NCE; fixed or dynamic windows) by elucidating critical parameters like identification/quantification ratios, calculation of CVs, and good spectra assignment.

12. Spectra assignment can be reviewed in the 'Analysis' icon. Open a raw file in the drop-down menu and afterwards an m/z slot. Click on one accession to open the 'spectra assignment view'. In Fig. 4 as an example, good and bad spectra assignments are presented.

3.7 Workflows for Data Analysis, Visualization, and Interpretation

In the following paragraphs, the two essential workflows for the creation of a DIA experiment are thoroughly explained. First the determination of peak widths is described and second, a workflow to determine variable windows. Both workflows were creating using KNIME (**Note 17**).

3.7.1 Workflow to Determine Peak Lengths

The peak lengths, or also called feature lengths, which represents the time of a peptide to elute from the HPLC, are a critical number for DIA analyses. As for each peptide enough MS2 spectra should be acquired to perform a plausible quantification, enough cycle times should be planned for the expected average peak lengths. In the following paragraphs, we will briefly describe a workflow for this estimation.

While there are many environments or script languages which could be used for the calculation, KNIME [5] is the environment of choice in our lab for workflows. The *KNIME Analytics Platform* gives an easy-to-use environment to run proteomics workflows on local desktop computers, as we have shown, e.g., in [4, 8–10] (*see* **Notes 17** and **18**).

The proposed workflow is deposited in the KNIME Hub at https://hub.knime.com/julianu/spaces/Public/latest/Determine_peak_lengths.

1. Go to the given address and download the workflow (*see* **Note 19**).

2. If not yet done, also download the *KNIME Analytics Platform* (hereafter referred to as KNIME) from https://www.knime.

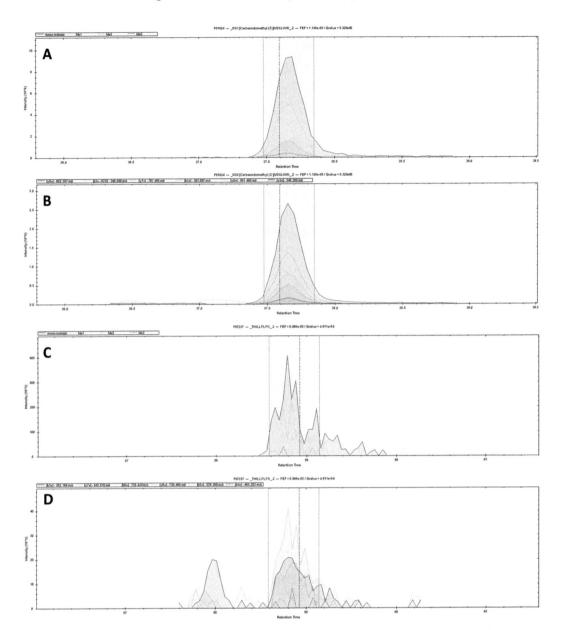

Fig. 4 XIC of P01024 peptide DSC (carbamidomethyl (C)VGSLVVK and P07237 peptide THILLFLPK (**a**) MS1 isotope envelope extracted ion chromatogram (XIC), displaying the monoisotopic peak in red. Following isotopes are visualized in yellow, green, and blue. (**b**) MS2 XIC displaying fragment ions: y6 in red, b3 in yellow, y7 in light green, b3 in green, y8 in turquoise, and y3 in purple. (**c**) MS1 isotope envelope extracted ion chromatogram (XIC), displaying the monoisotopic peak in red. Following isotopes are visualized in yellow, green, and blue. (**d**) MS2 XIC displaying fragment ions: b3 in red, y7 in yellow, b6 in light green, y6 in green, b5 in turquoise, and b4 in purple. XIC of (**a**) and (**b**) show representative MS1 and MS2 spectra. Peak shape and sizes of (**a**) MS1 isoptopes and (**b**) MS2 fragment ions correlate well with each other. XIC of (**c**) and (**d**) are poorly visualised as MS1 isotope and frament ion patterns do not display similar shapes

com/. It is available for Windows, Linux, and MacOS and runs on most current desktop systems.

3. After the installation of KNIME, start the program. The only necessary plugin is OpenMS, which is needed for some file conversion tasks.

4. To install the plugin, refer to (*see* **Note 20**).

5. Import the workflow by clicking onto 'File' -> 'Import KNIME workflow…' and browse under 'select file' to the downloaded file. As soon as the workflow is completely imported, it will appear in the KNIME Explorer to the left.

6. Double-click the workflow to open it. It should load without any warnings. If not, you might still need to install a plugin, for which you will be asked in a pop-up window.

7. Double-click the 'Input Files' node and add your mzML files to be considered (*see* **Note 21**).

8. Open the FeatureDetection Node of OpenMS (*see* **Note 22**).

9. Set up the feature finder and then run the complete workflow by clicking the fast-forward button or pressing 'Shift+F7'.

10. Inspect the lengths of the detected features in two ways: (1) by a histogram over the peak lengths and (2) by a table.

11. Right click on it and select 'View: Histogram View'. Here, you will be presented by a histogram showing how many of the detected features fell into which peak length bin between 10 s, 20 s, …, 60 s or longer.

12. Right click on the 'calculations' node and select the bottom view, to view a table showing the same data. Additionally, you will have a column showing the fraction of features which had a length shorter than the lengths of the given row.

13. Determine which peak length you should take into consideration for the calculation of your cycle time, respectively, the number of DIA bins you can measure on your mass spectrometer.

3.7.2 Workflow for Variable DIA Window Calculation

As discussed before, the m/z range is usually segmented into a fixed number of windows of equal width. This leads to the fact that in some windows, almost no peptide eluates over the complete retention time, while in other windows, the co-elution is quite high. To take the variable distribution of peptides over the measured m/z range better into account, it can therefore be beneficial to use variable DIA windows instead of static windows over the m/z range.

In the presented workflow, the intensity of MS1 events measured by DDA runs is utilized, to create m/z windows for DIA measurement, which have equal amount of MS1 intensities,

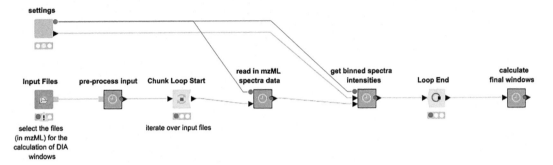

Fig. 5 Overview of the workflow for the calculation of variable *m/z* windows for DIA

instead of equal widths. Under the assumption that most MS1 events correlate to eluting peptides from the column, the co-elution of peptides into DIA windows should be reduced by this.

The proposed workflow is deposited in the KNIME Hub at https://hub.knime.com/julianu/spaces/Public/latest/Calculate_variable_DIA_windows.

1. Download and import the workflow; follow the instructions given above for the peak lengths workflow.

2. Open the workflow (it will look similar to the image shown in Fig. 5).

3. Configure the workflow, by giving some basic configuration settings. For this, double-click on the settings node to be presented by a configuration window. The *binned data minimum* and *maximum* describe the *m/z* range, which will be considered for the calculation of variable windows. All Ms1 events outside of this range will be ignored. The number of final bins is the number of DIA windows you like to measure, as calculated above. The fields 'start RT' and 'end RT' represent the retention time, in seconds, you like to consider (*see* **Note 23**). The intensity threshold can be used to filter out low-abundant noise in the data (*see* **Note 24**).

4. Select the files considered for the calculation in the 'Input Files' node to the left.

5. Double-click the node and browse your files, which must be in mzML format and should be centroided (peak-picked). You can select as many files as you like, to increase the accuracy of your windows, but be aware that this will also increase the time taken for the calculation (*see* **Note 24**).

6. Start the workflow by selecting the 'Execute all executable nodes' (fast-forward button) or simply press *'Shift + F7'*. The files will be processed sequentially. In brief, for each file it is first checked whether it is already centroided. If not, the data will be

peak-picked by the workflow. Afterwards, the file will be filtered using the retention time and m/z parameters, and only MS1 events will be used in all further steps. After this filtering, the intensities of all MS1 events will be binned along the m/z range in the given number of bins. Per default, these will be 1000 bins. The intensities of all measured spectra will be summed up for each m/z bin and normalized for each run. After all runs are preprocessed in this way, all normalized bins will be summed up, and windows containing equal amount of (normalized) intensities/abundances will be calculated. The final result is a table containing the bordering and central points of the windows, together with their widths, like in Table 5.

7. Look at the created table, by clicking right on the far right node after the workflow finished, and select the bottom option with the reading glass and table symbol. This data can be used to set the windows in, e.g., Xcalibur.

8. Select the data in the table and copy it, e.g., into Excel for further processing.

4 Notes

1. DIA method generation in an Orbitrap Fusion Lumos instrument varies from the QExactive Series, as a different version of XCalibur is pre-installed, and will not be discussed in this chapter.

2. MaxQuant and Proteome Discoverer 2.2, 2.3, 2.4 (Thermo Fisher Scientific) may also be used for library creation.

3. Other software tools such as OpenSwath [2] or Skyline [11] may also be used for the analysis of DIA data.

4. You can find the Instrument Setup in the Thermo Xcalibur Roadmap menu.

5. Your DDA MS method has to have certain fixed parameters to enable a DIA analysis. First MS1 and MS2 spectra have to be read out in the Orbitrap; second HCD has to be chosen as fragmentation method. Optimal DDA settings might vary for your sample type and should be therefore adapted prior DIA method development.

6. Elution profiles can directly be interpreted opening the raw file in XCalibur. A good elution profile displays an adequate TIC (usually ranging between 1.0E+09 and 1.0E+10), as well as long elution time. High organic solvent gradients often lead to a fast elution of peptides, in whose no ions are eluting at the end of the gradient (Fig. 3a, b). Low organic solvent percentage though, may cause TIC suppression. Determination of

identification, as well as quantification rates and CV, are necessary for the complete interpretation of data and optimal method establishment. Note that changes in gradient length may be accompanied by adapting the percentage of organic solvent.

7. Overlapping windows are not necessary for a functional DIA method but may improve your analysis outcome.

8. One has to consider that optimal variable window sizes for the same sample might slightly vary over time due to changes in the chromatography system, e.g., after solvent or column exchanges and have therefore be adapted on a regular basis.

9. The BGS factory settings will work perfectly fine, but in some cases adaption of settings is recommended.

10. FASTA files can be found for any species at the UniProt KB website.

11. Library creation can be carried out using MaxQuant and Proteome Discoverer 2.2, 2.3, 2.4 or directly in Spectronaut, by using the Pulsar Search engine. For that, search your DDA runs including the elucidated modifications and use the output for submitting it into a program capable of analyzing DIA data of your choice.

12. The finished library will appear at the 'Library' icon in the 'Spectral library' tab on the left.

13. For determining if a fixed or variable window setup is the optimal solution for you sample, both types should be analyzed in a separated Spectronaut Pulsar analysis, to circumvent the problem of match between runs.

14. Use the same FASTA file as already taken for spectral library generation.

15. The reference run is only taken for later statistical visualization.

16. The export scheme may be adapted to one's personal preferences using the 'Columns' tab.

17. To install a set of nodes, select 'Help' -> 'Install New Software...' in KNIME's main window. Select the community contributions repository under the 'Work with' drop down menu; it should contain an address similar to http://update. knime.org/community-contributions/trusted/4.0. The OpenMS nodes can be found in the 'Bioinformatics & NGS' group or simply by searching for them. Select the OpenMS nodes, click next, accept the license, and restart KNIME finishing the installation. If all went well, you will see the OpenMS logo on the splash screen of KNIME (together with all the other icons) and you will find the OpenMS nodes inside the

'Community Nodes' in the Node Repository (usually left bottom side of screen).

18. If more computational power is required, a KNIME server can be obtained, which allows server side execution of workflows and thus load relief of local computers, as well as access to workflows via a web browser. Recently, the KNIME Hub (https://hub.knime.com) was initiated, which allows easy exchange of workflows in the community.

19. For this, you do not need to register to KNIME. This is only necessary, if you like to upload your own workflows.

20. During the installation of KNIME on Windows, you will be asked how much main memory KNIME is allowed to consume. If you need to change this setting later, or if you work with Linux or Mac OS X, you can simply edit the *knime.ini* file. This file is in the directory into which you installed KNIME, which by default is in 'C:\Program Files\KNIME' on Windows and varies on other operating systems. Open the file with an editor like Notepad, Nano, or Vim and find the line staring with '-Xmx'. Behind this, the amount of used memory is set, like '2048 m' for 2048 megabytes or '4G' for 4 gigabytes. Write in the amount you like to allow KNIME to use for calculations. After saving the file and restarting KNIME, the settings should be activated. Analysis of DIA data can be carried out using several software tools, e.g., open-source-based tools such as OpenSWATH or Skyline. One of the most frequent used software tools is Spectronaut (Biognosys AG, Schlieren, Switzerland), which we chose to analyze our acquired DIA data with.

21. In your workflow, a pre-configured node will be given, but you might want to change some of the settings like the mass tolerances. You could also choose another feature finder algorithm of OpenMS; we currently recommend either the 'FeatureFinderCentroided' or the 'FeatureFinderMultiplex'. For more information on these or recommended settings, please have a look at the OpenMS tutorials.

22. This can be used to filter out common early or late irregularities in the chromatography, like loading time and increase of solvent in the late phase of measurement.

23. Beware that here and in all other settings, not selecting 'change' takes the shown default value, which in the case of the threshold is 100,000. The number of 'calculation bins' represents the resolution for the calculation of windows. Usually, the default of 1000 should be sufficient.

24. It might be a good idea to increase the amount of memory used by KNIME, as described in Note 20.

References

1. Koopmans F, Ho JTC, Smit AB et al (2018) Comparative analyses of data independent acquisition mass spectrometric approaches: DIA, WiSIM-DIA, and untargeted DIA. Proteomics 18(1). https://doi.org/10.1002/pmic.201700304

2. Röst HL, Rosenberger G, Navarro P et al (2014) OpenSWATH enables automated, targeted analysis of data-independent acquisition MS data. Nat Biotechnol 32(3):219–223. https://doi.org/10.1038/nbt.2841

3. Tsou C-C, Avtonomov D, Larsen B et al (2015) DIA-umpire: comprehensive computational framework for data-independent acquisition proteomics. Nat Methods 12 (3):258–264, 7 p following 264. https://doi.org/10.1038/nmeth.3255

4. Barkovits K, Pacharra S, Pfeiffer K et al (2020) Reproducibility, specificity and accuracy of relative quantification using spectral library-based data-independent acquisition. Mol Cell Proteomics 19(1):181–197. https://doi.org/10.1074/mcp.RA119.001714

5. Berthold MR, Cebron N, Dill F et al (2008) KNIME: the konstanz information miner. In: Preisach C, Burkhardt H, Schmidt-Thieme L, Decker R (eds) Data analysis, machine learning and applications. Studies in classification, data analysis, and knowledge organization. Springer, Cham, Switzerland, pp 319–326. https://doi.org/10.1007/978-3-540-78246-9_38

6. Röst HL, Sachsenberg T, Aiche S et al (2016) OpenMS: a flexible open-source software platform for mass spectrometry data analysis. Nat Methods 13(9):741–748. https://doi.org/10.1038/nmeth.3959

7. Bruderer R, Bernhardt OM, Gandhi T et al (2017) Optimization of experimental parameters in data-independent mass spectrometry significantly increases depth and reproducibility of results. Mol Cell Proteomics 16 (12):2296–2309. https://doi.org/10.1074/mcp.RA117.000314

8. Turewicz M, Kohl M, Ahrens M et al (2017) BioInfra.Prot: a comprehensive proteomics workflow including data standardization, protein inference, expression analysis and data publication. J Biotechnol 261:116–125. https://doi.org/10.1016/j.jbiotec.2017.06.005

9. Uszkoreit J, Ahrens M, Barkovits K et al (2017) Creation of reusable bioinformatics workflows for reproducible analysis of LC-MS proteomics data, vol 127. Humana Press, New York, pp 305–324. https://doi.org/10.1007/978-1-4939-7119-0_19

10. Uszkoreit J, Perez-Riverol Y, Eggers B et al (2019) Protein inference using PIA workflows and PSI standard file formats. J Proteome Res 18(2):741–747. https://doi.org/10.1021/acs.jproteome.8b00723

11. MacLean B, Tomazela DM, Shulman N et al (2010) Skyline: an open source document editor for creating and analyzing targeted proteomics experiments. Bioinformatics 26 (7):966–968. https://doi.org/10.1093/bioinformatics/btq054

Chapter 23

Label-Free Proteomics of Quantity-Limited Samples Using Ion Mobility-Assisted Data-Independent Acquisition Mass Spectrometry

Ute Distler, Malte Sielaff, and Stefan Tenzer

Abstract

Over the past two decades, unbiased data-independent acquisition (DIA) approaches have gained increasing popularity in the bottom-up proteomics field. Here, we describe an ion mobility separation enhanced DIA workflow for large-scale label-free quantitative proteomics studies where starting material is limited. We set a special focus on the single pot solid-phase-enhanced sample preparation (SP3) protocol, which is well suited for the processing of quantity-limited samples.

Key words Data-independent acquisition, Label-free quantification, Ion mobility separation, Quantitative proteomics, Bottom-up proteomics, Single pot solid-phase-enhanced sample preparation (SP3)

1 Introduction

Over the past decades, mass spectrometry (MS) has evolved to the analytical technology of choice in proteomics [1]. Currently, "bottom-up" proteomics is most frequently used, where proteins are first proteolytically cleaved into peptides and the resulting peptide mixture is then analyzed by MS acquisition [2]. To obtain comprehensive and accurate proteomic datasets, high proteome coverage, reproducibility, and quantitative precision are critically important. In combination with recent developments in MS instrumentation, current MS acquisition techniques and strategies allow the analysis of complex protein mixtures and to accurately identify and quantify thousands of proteins in a biological sample [3–6]. Over the past years, novel MS acquisition strategies were established that sample in a data-independent manner thus circumventing the problems arising from classical data-dependent acquisition (DDA) (*see* Chaps. 8–10, 13–17, 20, 21, 26), such as stochastic and irreproducible precursor ion selection [7, 8], under-sampling [9], and

Katrin Marcus et al. (eds.), *Quantitative Methods in Proteomics*, Methods in Molecular Biology, vol. 2228,
https://doi.org/10.1007/978-1-0716-1024-4_23, © Springer Science+Business Media, LLC, part of Springer Nature 2021

long instrument cycle times [8]. In contrast to serial peptide fragmentation in DDA, data-independent acquisition (DIA) approaches (*see* Chaps. 8, 16, 20–22, 24) feature parallel fragmentation of all precursor ions, regardless of intensity or other characteristics, resulting in complex but comprehensive product ion data. DIA approaches typically reduce the complexity of the product ion space by dividing it into segments for MS measurement using quadrupole, ion traps, or ion mobility separation (IMS) [10–12] to counterbalance the complexity of biological samples. IMS occurs in the millisecond range and separates ions based on their differential mobility travelling through a buffer gas in an electric field. It can be implemented in between high-resolution chromatographic separation (second time scale) and mass analysis (microsecond timescale) in a loss-less fashion [13]. IMS provides an additional dimension of separation in MS workflows, as ions are not only separated according to their m/z but also to their size and shape. Thus, overall system peak capacity can be significantly increased, while chimeric and composite interferences are concomitantly reduced [14–17]. Especially in highly complex samples, IMS permits the separation of co-eluting features that are indistinguishable by their m/z providing a high overall peak capacity superior to classical LC-MS setups. In the last years, several studies have shown that DIA approaches provide improved quantitative precision, reproducibility as well as proteome coverage [3, 4, 18, 19] when compared with label-free DDA. Consequently, DIA has evolved to the method of choice for label-free quantitative proteomics in many laboratories worldwide. Here, we present a comprehensive protocol for DIA-based label-free quantification of complex proteomes. We provide a detailed description for reproducible sample preparation (i.e., cell lysis and proteolytic digest), high-resolution nanoscale UPLC using reversed phase (C18) columns, IMS enhanced Q-TOF-MS applying drift time-specific collision energies [12], and data processing. Label-free quantification is performed using the Top3 approach, which has already been widely applied in conjunction with MS^E workflows [20] providing reliable and highly reproducible results [12, 21, 22].

2   Materials

All buffers should be prepared using either HPLC/LC-MS grade or ultrapure water (prepared by purifying deionized water to attain a conductivity of 18 MΩ cm^{-1} at 25 °C) and analytical grade reagents. For the preparation of LC mobile phases and wash solutions, use LC-MS grade solvents and acids. Prepare and store all reagents at room temperature (unless indicated otherwise). All waste disposal regulations must be diligently followed when disposing waste materials.

2.1 Common Consumables and Lab Equipment

1. Protein low-binding microfuge tubes (1.5 mL) and pipette tips.

2. Magnetic rack(s) holding up to 24 microfuge tubes (1.5 mL).

2.2 Cell Lysis

1. Lysis buffer: 1% (w/v) sodium dodecyl sulfate (SDS), 1× protease inhibitor cocktail, 50 mM HEPES, pH 8.0 (*see* **Note** 1). Weigh 500 mg of SDS and 596 mg of HEPES and transfer to a fresh 50 mL-polypropylene tube. Add 45 mL of water. Add one tablet of cOmplete protease inhibitor cocktail (Roche, Mannheim, Germany). Mix and adjust pH using 2 N of sodium hydroxide prepared in water. Make up to 50 mL with water. Store 1 mL-aliquots at −20 °C.

2. Benzonase nuclease (Sigma-Aldrich, Saint Louis, MO, USA); optional (*see* **Note** 2).

2.3 Single Pot Solid-Phase-Enhanced Sample Preparation

1. SP3 beads (20 μg solids/μL). Combine 20 μL of Sera-Mag carboxylate-modified magnetic particles (hydrophobic; GE Healthcare, Chicago, IL, USA) and 20 μL of Sera-Mag Speed-Bead carboxylate-modified magnetic particles (hydrophilic; GE Healthcare) in a 1.5 mL-tube. Add 160 μL of water and mix. Place tube with beads on a magnetic rack and remove supernatant after beads have settled. Off the rack, add 200 μL of water and mix. Repeat wash steps two further times. Recover beads in 100 μL of water. Store at 4 °C for up to 1 month (Fig. 1).

2. Dithiothreitol (DTT) stock solution. Weigh 3–5 mg of DTT directly into a fresh 1.5 mL polypropylene tube. Dissolve in lysis buffer to a final concentration of 200 mM. Prepare freshly before use (*see* **Note** 3).

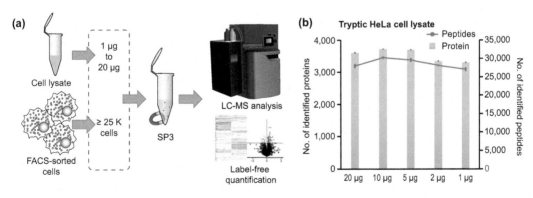

Fig. 1 (**a**) Overview of the presented proteomic approach for samples with limited starting material. After cell lysis and protein extraction, proteins are digested with trypsin applying single pot solid-phase-enhanced sample preparation (SP3) [24]. Tryptic peptides are analyzed by LC-MS, followed by data evaluation and label-free quantification analysis. (**b**) Numbers of identified peptides and proteins after SP3 processing using 1–20 μg of HeLa cell lysate as starting material

3. Iodoacetamide (IAA) stock solution. Weigh 10–20 mg of IAA directly into a fresh 1.5 mL polypropylene tube. Dissolve in lysis buffer to a final concentration of 400 mM. Prepare freshly before use (*see* **Note** 4).

4. 70% (v/v) Ethanol (EtOH). Mix 70 mL of EtOH with 30 mL of water. Store at room temperature for up to 1 month.

5. Ammonium bicarbonate buffer. Weigh 40 mg of ammonium bicarbonate and transfer to 15 mL-polypropylene tube. Dissolve in 10 mL of water to a final concentration of 50 mM. Prepare freshly before use.

6. Trypsin stock solution. Dissolve lyophilized trypsin (Trypsin Gold, MS Grade; Promega, Madison, WI, USA) in 50 mM acetic acid prepared in water to a final concentration of 1 μg/μL. Store 5 μL-aliquots at −80 °C.

7. Digestion buffer. Dilute trypsin stock solution in ammonium bicarbonate buffer to achieve a trypsin-to-protein ratio of 1:25 (w/w) in a final volume of 5 μL (*see* **Note** 5). Prepare freshly before use.

8. Peptide elution solution. Mix 98 μL of water and 2 μL of dimethyl sulfoxide (DMSO) in a 1.5 mL-tube. Prepare freshly before use.

2.4 LC-MS Analysis   For the preparation of LC-MS solvents, we recommend to use headspace grade DMSO.

1. NanoACQUITY UPLC system (Waters Corporation, Milford, MA, USA).

2. SYNAPT G2-S or G2-Si high definition mass spectrometer equipped with a NanoLockSpray dual electrospray ion source (Waters Corporation).

3. Precut PicoTip emitters for nanoelectrospray, outer diameter, 360 μm; inner diameter, 20 μm; inner diameter at tip, 10 μm (New Objective, Woburn, MA, USA; cat. no. FS360-20-10-N-20-C 6.35CT) or 20 μm × 360 μm fused silica capillary (BGB Analytik, Böckten, Switzerland, cat. no. TSP-020375) for open tubular etched emitters (*see* **Note** 6).

4. LC-MS instrument control software: MassLynx (version 4.1 or higher; Waters Corporation).

5. Analytical column: HSS-T3 C18 1.8 μm, 75 μm × 250 mm (Waters Corporation).

6. Autosampler vials: total recovery glass vials, with non-slit poly-tetrafluoroethylene (PTFE)/silicone septa (Waters Corporation, cat. no. 600000750cv).

7. Aqueous mobile phase A: 0.1% (v/v) formic acid (FA) and 3% (v/v) DMSO in water.

8. Organic mobile phase B: 0.1% (v/v) FA and 3% (v/v) DMSO in acetonitrile (ACN).

9. Needle wash: 0.1% (v/v) FA in water.

10. Rear seal wash: 50% (v/v) methanol in water.

11. Lock mass solution: 100 fmol/μL [Glu^1]-fibrinopeptide B in 40% (v/v) ACN, 0.1% (v/v) FA in water.

2.5 Data Analysis

1. ProteinLynx Global SERVER (PLGS, version 2.5.2 or later; Waters Corporation).

2. Freely available ISOQuant software package for label-free quantification analysis (http://www.immunologie.uni-mainz.de/isoquant/).

3 Methods

Unless specified otherwise, all procedures should be carried out at room temperature.

3.1 Cell Lysis

In this section the lysis of cell pellets for proteomic sample preparation by SP3 (*see* Subheading 3.2) is described. The protocol was optimized for cell lines and primary cells isolated by fluorescence-activated cell sorting (FACS) that were subsequently pelleted by centrifugation (*see* **Note 7**).

1. Reconstitute the cell pellet in lysis buffer. Do not exceed a volume of 40 μL. Incubate samples at 95 °C for 5 min to promote cell disruption and protein solubilization (*see* **Note 1**).

2. Sonicate samples using a Bioruptor device (Diagenode, Liège, Belgium) at maximum intensity cycling in 30 s on/off intervals at 4 °C for 15 min in order to shear chromatin (*see* **Note 2**).

3.2 Single Pot Solid-Phase-Enhanced Sample Preparation

1. Reduce disulfide bonds by adding 1 μL of DTT solution per 10 μL of lysate. Incubate samples at 56 °C for 30 min (*see* **Note 1**).

2. Add 2 μL of IAA solution per 10 μL of lysate and incubate the samples in the dark for 30 min in order to promote alkylation of free cysteine residues.

3. Quench the alkylation reaction by the addition of 2 μL of DTT solution per 10 μL of lysate.

4. Resuspend the SP3 beads and add 2 μL of beads to each sample. Vortex-mix the suspension. Avoid excessive pipetting to reduce sample loss.

5. Promote aggregation of proteins and beads by adding ACN to a final concentration of 70% (v/v). Vortex-mix the samples immediately.

6. Incubate the samples for 20 min. During the first 10 min, briefly vortex-mix the samples as soon as sedimentation of protein-bead aggregates is observed. Afterwards, allow the beads to settle.

7. Capture the beads at the tube wall by placing the samples into a magnetic rack. Wait at least 2 min until all beads are trapped. Remove and discard the supernatant (*see* **Note 8**).

8. Add 200 µL of 70% (v/v) EtOH and rinse the pellet on the magnetic rack by carefully pipetting up and down. Remove and discard the supernatant. Repeat this step a second time with 70% (v/v) EtOH and once with ACN. Transfer the tubes with opened lid from the magnetic rack and wait until the residual ACN is evaporated.

9. Resuspend the beads in 5 µL of digestion buffer (*see* **Note 5**). Incubate the samples at 37 °C overnight to promote tryptic digestion of proteins.

10. Resuspend the beads after proteolytic digestion by vortexing the tubes. To promote peptide binding on the beads, add ACN to a final concentration of 95% (v/v). Immediately vortex-mix the samples.

11. Incubate the samples for 20 min. During the first 10 min, briefly vortex-mix the samples as soon as sedimentation of peptide-bead aggregates is observed. Afterwards, allow the beads to settle.

12. Capture the beads at the tube wall by placing the samples into a magnetic rack. Wait at least 2 min until all beads are trapped. Remove and discard the supernatant.

13. Add 200 µL of ACN and rinse the pellet on the magnetic rack by carefully pipetting up and down. Remove and discard the supernatant. Transfer the tubes with opened lid from the magnetic rack and wait until the residual ACN is evaporated.

14. Elute peptides by reconstituting the beads in 10 µL of 2% (v/v) DMSO. Incubate the samples in an ultrasonic bath for 5 min to improve peptide recovery.

15. Briefly centrifuge the tubes and place them onto the magnetic rack. Recover the supernatant containing tryptic peptides after all beads have settled. Make sure not to recover any beads.

16. For LC-MS analysis, mix samples with diluted FA to a final concentration of 0.1% (v/v) FA. For example, if a final sample volume of 20 µL is desired, add 10 µL of 0.2% (v/v) FA.

3.3 LC-MS Analysis

1. Transfer samples for LC-MS analysis into glass vials (*see* equipment setup) before placing them into the autosampler (*see* **Note 9**).

2. Peptides are separated by reversed phase high-performance liquid chromatography using the NanoACQUITY UPLC system (Waters Corporation). After installing a system comprising nano-column (C18), configure the fluidics to direct injection mode (*see* **Note 10**). Use partial loop injection with 0.1% (v/v) FA as a loading solvent. Maximum sample loading amount is 2.6 μL (*see* **Note 11**). After placing the sample into the sample loop, peptides are directly transferred onto the analytical column, which should be heated to 55 °C (*see* **Note 12**).

3. The following gradient is applied for peptide separation: (a) keep at 1% (v/v) B during sample loading at a flow rate of 400 nL/min, (b) increase of B to 5% (v/v) B in 3 min, and (c) elute peptides running a gradient up to 40% solvent B at a flow rate of 300 nL/min over 90 min or 180 min, respectively (*see* **Note 13**). (d) After washing the column with 90% (v/v) mobile phase B at a flow rate of 600 nL/min, (e) re-equilibrate the column at initial conditions for 16 min at a flow rate of 400 nL/min.

4. The SYNAPT G2-S/Si mass spectrometer (Waters Corporation) is operated in data-independent mode using an ion mobility-enhanced MS^E method switching automatically between a high- and low-energy scan to collect precursor and fragment ion information. In the elevated energy scan, we apply drift time-specific collision energies for efficient fragmentation of the precursor ions (*see* **Note 14**). Drift-time specific collision energies can be programmed using the editor for look-up table generation provided by MassLynx (on the tune page controlling the instrument). Set a constant collision energy of 17 eV for the first 20 mobility bins in your look-up file. Afterwards, program a first collision energy ramp starting from ion mobility bin 20 (17 eV) to 110 (45 eV) followed by a second ramp from ion mobility bins 110 (45 eV) to 200 (60 eV).

5. Set up an ion mobility-enhanced MS^E workflow by selecting the MS^E continuum method in the MS method editor of MassLynx. Set the mass range for MS^E data acquisition from m/z 50–2000 with a spectral acquisition time of 0.6 s for both the high- and low-energy scans. Use a collision energy of 4 eV for the low energy scan. Upload the look-up table file containing the drift time-based collision energy information (previous point) into the MS method for precursor ion fragmentation during the high-energy cycles.

6. [Glu$_1$]-fibrinopeptide B is used for post-acquisition lock mass correction of the data. Use the auxiliary solvent manager of the NanoACQUITY UPLC system to pump [Glu$_1$]-fibrinopeptide

B at a flow rate of 1.5 μL/min. Sample [Glu₁]-fibrinopeptide B via the reference sprayer of the NanoLockSpray source with a frequency of 30 s into the mass spectrometer.

7. After data acquisition, export LC-MS data raw data (Waters . raw file format) for data storage and further processing.

3.4 Data Analysis and Label-Free Quantification

1. Initial raw data processing and database search are performed in ProteinLynx Global SERVER (PLGS). Use the following parameters for raw data processing: 300.0 counts for the low-energy threshold, 30.0 counts for the elevated energy threshold, and 750 counts for the intensity threshold (*see* **Note 15**).

2. For database search, download the reviewed proteome (Swiss-Prot) of the analyzed species from http://www.uniprot.org in FASTA format. Add the protein sequences of porcine trypsin and known contaminant proteins such as human keratins using a text editor. A list of common contaminants in FASTA format can be downloaded from the common Repository of Adventitious Proteins (cRAP) at https://thegpm.org/cRAP/. Import the resulting FASTA file into PLGS and use the built-in protein database management tool of PLGS to generate a target-decoy database by adding the reversed protein sequences of the target list. Calculation of false discovery rates (FDR) will be disabled in subsequent analyses if this step is omitted. Use the newly generated target-decoy database to search processed mass spectra for peptide and protein identifications. The following search parameters are used: (a) trypsin as digestion enzyme, (b) a maximum of two missed cleavages for peptide identification, (c) carbamidomethylation of cysteine residues as fixed modification, (d) methionine oxidation as variable modification, (e) a minimum of three identified fragment ions per peptide, and (f) an FDR of 1%.

3. Data post-processing and label-free quantification analysis are performed in ISOQuant [12, 23]. Design a new ISOQuant project from PLGS-pre-processed data using the built-in configuration editor.

4. Retention time alignment, clustering of similar peaks across runs (i.e., based on m/z, retention and drift time information), cross-annotation of peaks within each cluster, normalization of peak intensities, and TOP3-based label-free quantification [21] are automatically performed in ISOQuant. A detailed description of all algorithms can be found in [12, 23] as well as in the "User Manual" provided on http://www.immunologie.uni-mainz.de/isoquant/. All these processes depend on several user-defined parameters, which can be modified using the built-in configuration editor in ISOQuant. We recommend to use the "ISOQuant settings for label-free quantification" as

Table 1
Settings for ISOQuant post-processing and label-free quantification analysis

Parameter name	Value
process.annotation.peptide.maxSequencesPerEMRTCluster	1
process.emrt.clustering.dbscan.minNeighborCount	2
process.emrt.clustering.distance.unit.drift.bin	2
process.emrt.clustering.distance.unit.mass.ppm	6
process.emrt.clustering.distance.unit.time.min	0.2
process.identification.peptide.minSequenceLength	6
process.quantification.maxProteinFDR	0.01
process.quantification.minPeptidesPerProtein	2
process.quantification.peptide.acceptType.IN_SOURCE	False
process.quantification.peptide.acceptType.MISSING_CLEAVAGE	False
process.quantification.peptide.acceptType.NEUTRAL_LOSS_H20	False
process.quantification.peptide.acceptType.NEUTRAL_LOSS_NH3	False
process.quantification.peptide.acceptType.PEP_FRAG_2	False
process.quantification.peptide.acceptType.PEP_FRAG_1	True
process.quantification.peptide.acceptType.PTM	False
process.quantification.peptide.acceptType.VAR_MOD	False
process.quantification.peptide.minMaxScorePerCluster	6
process.quantification.topx.allowDifferentPeptides	True
process.quantification.topx.degree	3

listed in [23]. Some of the most important parameters for the analysis and post-processing of complex proteomic samples (i.e., cell lysates, tissues, etc.) are also summarized in Table 1 (*see* **Note 16**).

5. Run the ISOQuant analysis and save the protein (.xlsx) and peptide (.csv) report files containing the final results of the label-free quantification analysis.

4 Notes

1. The SP3 protocol is compatible with various lysis buffer components [24]. However, acidic conditions can impede protein capture by the carboxylate-coated beads [25]. It is therefore recommended to use lysis buffers with neutral pH. If a urea-based lysis buffer is used, avoid incubation at temperatures above 37 °C to prevent protein carbamylation.

2. Intact chromatin can bind to the magnetic particles and interfere with the SP3 protocol. Chromatin shearing by a suitable ultrasonication device is recommended as sonication further aids cell disruption and protein extraction. Alternatively, enzymatic chromatin degradation can be performed by benzonase treatment of the lysates.

3. DTT is oxygen-sensitive. Always prepare DTT-containing buffers freshly before use and keep on ice.

4. IAA is light-sensitive. Always prepare IAA-containing buffers freshly before use and keep in the dark.

5. Total protein content of the lysates can be determined by colorimetric protein assays, such as the Pierce 660 nm protein assay (Thermo Fisher Scientific, Waltham, MA, USA). However, when the amount of starting material is limited, performing a protein assay likely consumes significant sample amounts. Alternatively, the protein concentration can be estimated, for example, by multiplying the number of analyzed cells with the approximate protein content of the cells. It is recommended to perform preliminary experiments with similar but less rare sample types (e.g., cell lines) in order to determine the approximate protein content of the actual samples of interest.

6. Hydrofluoric acid (HF)-etched open tubular emitters [26] provide higher sensitivity and spray stability as compared to commercial emitters. Moreover, to our experience, they also exhibit longer lifetimes and show less clogging.

7. It is recommended to use protein low-binding tubes and pipette tips for sample processing to minimize sample loss. Ideally, samples (i.e., FACS-sorted cells) are directly collected into the tubes used for further processing. In addition, a swinging bucket rotor might reduce sample loss when pelleting cells by centrifugation.

8. Supernatants of the SP3 protein cleanup procedure can be analyzed by SDS-polyacrylamide gel electrophoresis and Coomassie staining to monitor protein binding efficiency.

9. The temperature of the sample manager should be kept constant at around 6 °C while samples are stored therein. This will guarantee stable conditions during a study avoiding solvent evaporation.

10. The NanoACQUITY UPLC system is equipped with two high-pressure six-port valves. This allows to set up the fluidics either for the use of a precolumn or for direct injection. We recommend to load analytes directly onto the analytical column (i.e., use direct injection mode) to avoid any sample loss during LC separation. However, make sure that no residual salts or other impurities are left in the sample, as they can markedly impair the analysis when the precolumn is omitted.

11. The vendor provides stainless steel sample loops for sample injection. However, we recommend to use a 2.6-μL PEEKsil-sample loop with a 100-μL sample syringe to draw and load your peptide samples. The use of a PEEKsil injection loop minimizes carryover as compared to stainless steel injection loops.

12. It is essential to heat the LC columns for peptide separation. This ensures stable conditions within a single as well as between LC-runs. Moreover, column heating reduces overall system pressure and improves peptide separation. Flow rates might need to be adjusted as the LC system requires a minimum of pressure to work. However, increasing the flow rate typically leads to a loss of sensitivity, and a compromise between separation and sensitivity has to be made. In our hands, a temperature of 55 °C provided the best results using HSS-T3 C18 columns (*see* Subheading 2.4). In case other chromatographic materials are used, we recommend to adjust the temperature to optimize the performance.

13. Using multistep (concave) gradients for nanoLC separation provides more constant peak widths for eluting peptides throughout the whole run, which is especially noticeable at the start and the end of the run and results in slightly higher overall peak capacity.

14. Ramp the transfer cell collision energy from 25 to 55 eV in the elevated energy scan if you program a classical HDMS^E method. However, we highly recommend to ramp the collision energy according to the ion mobility within each IMS cycle, as the ion mobility of a precursor ion correlates with its m/z. Hence, quasi-m/z-specific collision energies can be applied that significantly improve the fragmentation efficiency as compared to the "classical" HDMS^E approach [12]. You should be aware that your ion mobility tuning impacts methods that apply drift time-specific collision energies. Hence, collision energies and ion mobility settings should be adjusted to each other. A detailed description of how to tune the IMS for proteomic samples has been described in detail in [23].

15. For optimal performance, processing parameters may require some adjustment. For each instrument, we recommend to optimize both low-energy threshold (range 100–500, step size 50) and high-energy threshold (range 20–50, step size 5) as processing parameters may vary depending on the experimental and instrument conditions, as, for example, on-column load, detector age, and version of the acquisition software (MassLynx). Lower thresholds typically result in a higher number of reported ions, which may increase the number of identifiable peptides. However, if the thresholds are too low, the number of signals derived from noise will increase while concomitantly reducing the number of identified peptides and substantially increasing data processing times.

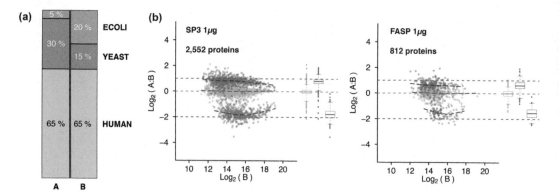

Fig. 2 Hybrid proteome samples were used to evaluate the precision and accuracy of label-free quantification for proteomic samples with low starting material (1 μg of total protein) that were processed by SP3 and FASP. (**a**) Prior to proteolytic digest, two samples were generated mixing yeast, human, and *E. coli* proteins in defined ratios to mimic upregulated (yeast) and downregulated (*E. coli*), as well as nonregulated proteins (human). Sample A contained 65% (w/w) human, 30% (w/w) yeast, and 5% (w/w) *E. coli* proteins. Sample B contained 65% (w/w) human, 15% (w/w) yeast, and 20% (w/w) *E. coli* proteins. (**b**) Log-transformed ratios of average abundances (log$_2$(A:B)) were calculated for each protein from the quantification output of ISOQuant for hybrid proteome samples digested with SP3 (left panel) and FASP (right panel). Calculated log$_2$(A:B) values were then plotted over the log-transformed average abundance in sample B. Only proteins identified in both samples A and B were plotted. The expected log$_2$(A:B) values for human (0), yeast (1), and *E. coli* (−2) proteins are indicated by dashed lines; box plots summarize corresponding log$_2$(A:B) distributions for each species. Lines in the boxes (bounded above and below by the 25th and 75th percentiles) show the medians and whiskers the 2.5% and 97.5% quantiles. Digesting 1 μg of mixed proteome samples using SP3 resulted in over 2550 proteins that could be quantified. Processing the same samples with FASP digest, only 812 proteins were quantified with markedly lower precision and accuracy most likely due to unspecific binding of proteins/ peptides to the filter membranes

16. Mixed proteome samples (*see* Fig. 2a) proved to be a valuable tool for the evaluation of label-free proteomic workflows [22, 23]. These mixtures can be prepared either on the peptide level prior to LC-MS analysis to evaluate different software tools or processing parameters or on the protein level prior to proteolytic digest to evaluate the whole workflow including sample preparation (*see* Fig. 2b).

Acknowledgments

This work was supported by the Deutsche Forschungsgemeinschaft (DFG), grant numbers SFB 1292-Z1 and TE599/3-1 to S.T., DI 2471/1-1 to U.D., as well as by the Forschungszentrum für Immuntherapie (FZI) of the Johannes Gutenberg University Mainz.

References

1. Aebersold R, Mann M (2016) Mass-spectrometric exploration of proteome structure and function. Nature 537:347–355
2. Aebersold R, Mann M (2003) Mass spectrometry-based proteomics. Nature 422:198–207
3. Bruderer R, Bernhardt OM, Gandhi T et al (2017) Optimization of experimental parameters in data-independent mass spectrometry significantly increases depth and reproducibility of results. Mol Cell Proteomics 16:2296–2309
4. Vowinckel J, Zelezniak A, Bruderer R et al (2018) Cost-effective generation of precise label-free quantitative proteomes in high-throughput by microLC and data-independent acquisition. Sci Rep 8:4346
5. Paulo JA, O'Connell JD, Everley RA et al (2016) Quantitative mass spectrometry-based multiplexing compares the abundance of 5000 S. cerevisiae proteins across 10 carbon sources. J Proteome 148:85–93
6. Williams EG, Wu Y, Jha P et al (2016) Systems proteomics of liver mitochondria function. Science 352:aad0189
7. Liu H, Sadygov RG, Yates JR (2004) A model for random sampling and estimation of relative protein abundance in shotgun proteomics. Anal Chem 76:4193–4201
8. Geromanos SJ, Vissers JPC, Silva JC et al (2009) The detection, correlation, and comparison of peptide precursor and product ions from data independent LC-MS with data dependent LC-MS/MS. Proteomics 9:1683–1695
9. Michalski A, Cox J, Mann M (2011) More than 100,000 detectable peptide species elute in single shotgun proteomics runs but the majority is inaccessible to data-dependent LC-MS/MS. J Proteome Res 10:1785–1793
10. Venable JD, Dong M-Q, Wohlschlegel J et al (2004) Automated approach for quantitative analysis of complex peptide mixtures from tandem mass spectra. Nat Methods 1:39–45
11. Purvine S, Eppel J-T, Yi EC et al (2003) Shotgun collision-induced dissociation of peptides using a time of flight mass analyzer. Proteomics 3:847–850
12. Distler U, Kuharev J, Navarro P et al (2014) Drift time-specific collision energies enable deep-coverage data-independent acquisition proteomics. Nat Methods 11:167–170
13. Valentine SJ, Liu X, Plasencia MD et al (2005) Developing liquid chromatography ion mobility mass spectrometry techniques. Expert Rev Proteomics 2:553–565
14. Zhong Y, Hyung S-J, Ruotolo BT (2012) Ion mobility-mass spectrometry for structural proteomics. Expert Rev Proteomics 9:47–58
15. Angel TE, Aryal UK, Hengel SM et al (2012) Mass spectrometry-based proteomics: existing capabilities and future directions. Chem Soc Rev 41:3912–3928
16. Lee S, Li Z, Valentine SJ et al (2012) Extracted fragment ion mobility distributions: a new method for complex mixture analysis. Int J Mass Spectrom 309:154–160
17. Valentine SJ, Ewing MA, Dilger JM et al (2011) Using ion mobility data to improve peptide identification: intrinsic amino acid size parameters. J Proteome Res 10:2318–2329
18. Bruderer R, Bernhardt OM, Gandhi T et al (2016) High-precision iRT prediction in the targeted analysis of data-independent acquisition and its impact on identification and quantitation. Proteomics 16:2246–2256
19. Vowinckel J, Capuano F, Campbell K et al (2014) The beauty of being (label)-free: sample preparation methods for SWATH-MS and next-generation targeted proteomics. F1000Res 2:272
20. Distler U, Kuharev J, Tenzer S (2014) Biomedical applications of ion mobility-enhanced data-independent acquisition-based label-free quantitative proteomics. Expert Rev Proteomics 11:1–10
21. Silva JC, Gorenstein MV, Li G-Z et al (2006) Absolute quantification of proteins by LCMSE: a virtue of parallel MS acquisition. Mol Cell Proteomics 5:144–156
22. Kuharev J, Navarro P, Distler U et al (2015) In-depth evaluation of software tools for data-independent acquisition based label-free quantification. Proteomics 15:3140–3151
23. Distler U, Kuharev J, Navarro P et al (2016) Label-free quantification in ion mobility-enhanced data-independent acquisition proteomics. Nat Protoc 11:795–812
24. Hughes CS, Foehr S, Garfield DA et al (2014) Ultrasensitive proteome analysis using paramagnetic bead technology. Mol Syst Biol 10:757–757
25. Sielaff M, Kuharev J, Bohn T et al (2017) Evaluation of FASP, SP3, and iST protocols for proteomic sample preparation in the low microgram range. J Proteome Res 16:4060–4072
26. Kelly RT, Page JS, Luo Q et al (2006) Chemically etched open tubular and monolithic emitters for nanoelectrospray ionization mass spectrometry. Anal Chem 78:7796–7801

Chapter 24

DIA-MS^E to Study Microglial Function in Schizophrenia

Guilherme Reis-de-Oliveira, Victor Corasolla Carregari, and Daniel Martins-de-Souza

Abstract

Here, we describe a proteomic pipeline to use a human microglial cell line as a biological model to study schizophrenia. In order to maximize the proteome coverage, we apply two-dimensional liquid chromatography coupled with ultra-definition MS^E mass spectrometry (LC-UDMS^E) using a data-independent acquisition (DIA) approach, with an optimization of drift time collision energy.

Key words DIA, Mass spectrometry, Schizophrenia, Microglia, UDMS^E, MS^E

1 Introduction

Schizophrenia is a debilitating psychiatric disorder that affects 21 million people around the world [1]. It is characterized by the presence of positive, negative, and cognitive symptoms [2, 3]. Positive symptoms are related to a loss of connection to reality, of which hallucinations and delusions are the main components used in clinical diagnostics. Negative symptoms include disorganized speech, flattening of affection, and social withdrawal; and cognitive symptoms are associated with deficits in short-term memory and attention span [2, 3].

One current hypothesis used to explain schizophrenia is that imbalances in neurodevelopment starting at the neonatal period affect the onset of the disorder towards the end of adolescence and the beginning of adult life [4]. This is supported by the fact that maternal immune activation (MIA) with poly I:C or LPS during the gestational course revealed shifts in behavior and deficits in prepulse inhibition in adult offspring, which were alleviated after treatment with antipsychotics [5–7]. Additionally, microglia also present a profound change in phagocytosis properties in the MIA model, which suggests a role of these cells in schizophrenia development [8].

Katrin Marcus et al. (eds.), *Quantitative Methods in Proteomics*, Methods in Molecular Biology, vol. 2228, https://doi.org/10.1007/978-1-0716-1024-4_24, © Springer Science+Business Media, LLC, part of Springer Nature 2021

Microglia, the major immune cells in the CNS, play a pivotal role in brain development. Alterations in microglial homeostasis could impair neural processes such as synaptic formation and myelination [9]. According to brain imaging studies, patients with schizophrenia present an overactivation of microglia, and levels of neuroinflammation are related to the severity of psychotic symptoms [10]. These findings may be associated with several genetic studies that have shown impairments in the immune systems of patients with schizophrenia, mainly regarding the major histocompatibility complex (MHC) and in the complement system, which have pivotal roles in microglial homeostasis [11–14]. Proteomic studies have also revealed dysregulations in proteins associated to the immune system in postmortem brain tissue [15, 16] and blood [17, 18] from patients with schizophrenia; however, the mechanisms underlying these processes are still poorly understood.

Mass spectrometry (MS)-based shotgun proteomics have improved the comprehension of psychiatric disorders [19, 20]. The complexity of the proteome has driven advances in MS methods that can be applied to overcome analytical limitations while studying biological samples such as dynamic range and ion suppression [21]. Both issues can be avoided by employing separation techniques (e.g., two-dimensional liquid chromatography) prior to MS analysis, which reduces ion suppression and improves the detection limits of the equipment.

MS acquisition methods have a crucial role in the quality of proteome coverage in biological systems. Data-dependent acquisition (DDA) (*see* Chaps. 8–10, 13–17, 20, 21, 26) is the most employed acquisition mode while performing bottom-up proteomics studies. A DDA method is set to initially perform a MS survey scan, acquiring spectra in a given time of the chromatographic run. Based on the precursor ion intensities, the instrument selects the most intense peptides for a subsequent fragmentation and records the resulting fragment ions. Despite being largely used, the fact that in DDA the selection of ion for fragmentation is based on ion intensity leads to a biased result towards high-abundant peptides, generally resulting from high-abundant proteins [22].

Data-independent acquisition (DIA) (*see* Chaps. 8, 16, 22, 23) is an alternative to overcome the inherent problems of DDA. One type of DIA is MS^E, in which a quadrupole is used as an ion guide and a collision cell applies cycles of low- and high-energy to acquire precursors and fragment ions, respectively [23]. As both parent and daughter ions are acquired in parallel, the identification and quantification of proteins are strictly related to retention time [24, 25]. Moreover, acquisition in MS^E mode can be improved by performing ion mobility separation, which uses electric fields or a gas chamber to separate ions based on their physical properties such as charge or cross section. The latter approach using gas-phase separation, called high-resolution MS^E ($HDMS^E$), allows for the separation of precursor ions with the same m/z but different structures, adding another dimension to peptide identification and quantification [26]. Finally, ultra-definition MS^E ($UDMS^E$) optimizes drift time-specific collision energies of precursor ions, which increases fragmentation efficiency and, therefore, peptide and protein identifications [27, 28]. The MS^E-related approaches have the limitation of generating data profoundly complex, since all precursor ions are fragmented and their daughter ions are simultaneously collected, requiring one of only a few available algorithm options to perform the deconvolution of spectra.

Here we describe a complete pipeline to study the role of microglia in neuropsychiatric disorders using two-dimensional liquid chromatography coupled to a mass spectrometer, acquiring in $UDMS^E$ mode. We use a human microglial cell line as the biological system, describing steps to perform cellular lysis, protein digestion, LC-MS/MS analysis, and data processing (Fig. 1).

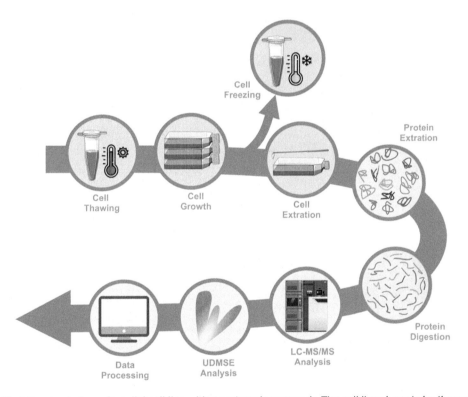

Fig. 1 Workflow to study a microglial cell line with a proteomic approach. The cell lines have to be thawed and cultured in (pink background), followed by protein extraction and digestion (blue background). The peptides are loaded into the LC-MS/MS system, $UDMS^E$ is performed, and the data is processed (yellow background)

2 Materials

2.1 Cellular Processes

1. Immortalized Human Microglia—SV40 (Abm, T0251).
2. Dulbecco's Modified Eagle's Medium, high glucose—DMEM (Merck, D5648).
3. Fetal Bovine Serum—FBS (Gibco®, 12,657–029).
4. DMEM Complete Medium (DMEM-C): 90% DMEM, 10% FBS, 1% Penicillin-Streptomycin.
5. Dulbecco's Phosphate Buffered Saline—DPBS (Merck, D8537).
6. Trypsin-EDTA (0.5%), no phenol red (Gibco®, 15400054).
7. Conical Centrifuge Tube, 15 mL.
8. Water Bath.
9. Cryogenic Vials (Corning®).
10. Cell Culture Flasks (Corning®).
11. Centrifuge with Swinging-Bucket Rotor.
12. Dimethyl sulfoxide—DMSO (Merck, D8418).
13. Cell Scraper.

2.2 In-Gel Digestions

1. Lysis Buffer: 3% sodium dodecyl sulfate (SDS), 160 mM Tris–HCl (pH 6.8), Protease Inhibitor Cocktail (cOmplete™, Mini, EDTA-free).
2. Sample Buffer (6×): 375 mM Tris–HCl (pH 6.8), 9% sodium dodecyl sulfate (SDS), 50% glycerol, 9% 2-mercaptoethanol, 0.03% bromophenol blue.
3. Pestle for microcentrifuge tube.
4. Sonication Probe.
5. Sonication Bath.
6. Centrifuge with Fixed Angle Rotor.
7. Dry bath.
8. Electrophoresis kit.
9. Acetonitrile, HPLC grade (>99.93%).
10. Milli-Q water.
11. AmBic: 50 mM ammonium bicarbonate in Milli-Q water.
12. DTT Solution: 10 mM dithiothreitol in AmBic.
13. IAA Solution: 55 mM iodoacetamide in AmBic.
14. Vacuum concentrator.
15. Trypsin from porcine pancreas (Sigma-Aldrich, T6567).
16. Extraction solution: acetonitrile 50%/formic acid 5%.

2.3 Mass Spectrometry Analysis (LC-MS/MS)	1. Ammonium formate, 20 mM.

2.3 Mass Spectrometry Analysis (LC-MS/MS)

1. Ammonium formate, 20 mM.

2. SYNAPT G2-Si High Definition Mass Spectrometer (Waters Corporation).

3. ACQUITY UPLC System, M Class (Waters Corporation).

4. ACQUITY UPLC M-Class HSS T3 Column, 1.8μm, 75μm × 150 mm, 1/pkg. (Waters Corporation, P/N 186007473).

5. ACQUITY UPLC M-Class Peptide BEH C18 Trap Column, 130 Å, 5μm, 300μm × 50 mm, 1/pkg. (Waters Corporation, P/N 186007471).

6. First Dimensional aqueous phase (A_{1stD}): ammonium formate 20 mM, pH 10.

7. First Dimensional organic phase (B_{1stD}): acetonitrile 100%.

8. Second Dimensional aqueous phase (A_{2ndD}): Milli-Q Water 100%/formic acid 0.1%.

9. Second Dimensional organic phase (B_{2ndD}): acetonitrile 100%/formic acid 0.1%.

10. [Glu1]-fibrinopeptide B solution (Waters Corporation, P/N 700004729).

11. MassLynx v4.1 Software.

2.4 UDMS^E Analysis

1. Driftscope v2.9.

2.5 Data Processing

1. Progenesis® QI for proteomics v4.0.

3 Methods

3.1 Cell Thawing

1. Take the cryogenic vial out of liquid nitrogen and thaw the cells in a water bath at 37 °C (*see* **Note 1**).

2. Quickly transfer all the contents from the vial into a centrifuge tube already containing 10 mL of DMEM-C.

3. Centrifuge the tube at 200 × *g* for 5 min.

4. Discard the supernatant and resuspend the pellet in 1 mL of DMEM-C.

5. Add the solution containing the cells to a T25 cell culture flask containing 4 mL DMEM-C.

6. Incubate the cells at 37 °C with 5% CO_2.

3.2 Cell Growth

1. Change the culture medium 1 day after thawing/trypsinization and every 2 days thereafter (*see* **Note 2**).

2. Carry out the trypsinization procedure when the cellular confluency reaches 90%.

3. Discard the media from cell culture flasks.

4. Wash the cells with 5 mL D-PBS to reduce the FBS concentration.

5. Add 3 mL of trypsin solution ($1\times$) and incubate for 5 min at 37 °C with 5% CO_2.

6. Collect the cellular solution and transfer it to a centrifuge tube already containing 10 mL DMEM-C.

7. Centrifuge the cells at $200 \times g$ for 5 min.

8. Discard the supernatant and resuspend the cells in 4 mL.

9. The cells can be replated with a ratio of 1:3 or 1:4 (cell solution/fresh medium) to keep growing or can be frozen.

3.3 Freezing

1. Set aside the number of cryogenic tubes that will be used depending on the total amount of cells (*see* **Note 3**).

2. Dilute the cells to obtain an approximate concentration of 1×10^6 cells/mL.

3. Add 1.5 mL of cells in solution to each cryogenic vial.

4. Add 150µL of DMSO and homogenize the solution.

5. Freeze the cells by gradually reducing the temperature (*see* **Note 4**).

6. Store the cells at −80 °C or in liquid nitrogen.

3.4 Collecting the Cells

1. Discard the media from cell culture flasks.

2. Wash the cells with 5 mL D-PBS to reduce the FBS concentration.

3. Add 2 mL of D-PBS and gently use a cell scraper to detach the cells from the flask, allowing them to be collected with a pipette.

4. Transfer the cell solution to a 15 mL centrifuge tube.

5. Centrifuge the cells at $200 \times g$ for 5 min.

6. Discard the supernatant.

7. Freeze the pellet at −80 °C until the cell lysis and protein extraction steps.

3.5 Cell Lysis and Protein Extraction (See Note 5)

1. Add 150µL of Lysis Buffer to tubes containing the cell pellets and thoroughly mix with the pipette, performing up-and-down movements.

2. Macerate the cells with a microtube pestle (*see* **Note 6**).

3. Perform an ultrasonication of each sample for 30 s with 10% duty cycle and an output level of 2.

4. Centrifuge the sample at $10,000 \times g$ for 15 min.

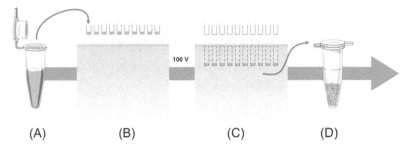

(A)   (B)   (C)   (D)

Fig. 2 In-gel digestion protocol. Proteins extracted from microglial cell lines (**a**) are loaded into a polyacrylamide gel, and electrophoresis is carried out (**b**). When all spots have entered into the resolving gel, each spot must be excised and diced (**c**), followed by further steps of in-gel digestion (**d**)

5. Save the supernatant with proteins and discard the pellet with cellular debris.

6. Quantify the protein present in the sample (*see* **Note 7**).

3.6 In-Gel Digestion (See Fig. 2)

1. Transfer 50μg of protein to a new microtube and add 5μL of Sample Buffer.

2. Heat the sample for 5 min at 95 °C in a dry bath.

3. Centrifuge the sample at $10,000 \times g$ for 5 min.

4. Transfer 30μL of the supernatant from each sample into the gel wells (Fig. 2a).

5. Run the 12% SDS-PAGE gel at 100 V until the entire band has entered the resolving (bottom) gel (Fig. 2b).

6. Cut the gel spots from the beginning of resolving gel down to the upper edge of the bromophenol blue band (Fig. 2c).

7. Dice each sample spot into small pieces and transfer each sample to an individual microtube.

8. Cover the gel in the tubes with 50% ACN in Milli-Q water and incubate for 5 min (Fig. 2d).

9. Discard the solution, taking care to not remove any of the gel.

10. Reduce the disulfide bonds by adding 200μL of DTT Solution (10 mM) and incubate at 80 °C for 30 min.

11. Alkylate the cysteine residues by adding 200μL of IAA Solution (55 mM) and incubate in the dark for 20 min, at room temperature (RT).

12. Dehydrate the gel spots by adding 200μL of 100% ACN and incubating for 10 min at RT, discarding the solution. Perform this step twice.

13. Remove all the solvents from the gel with a vacuum concentrator.

14. Add trypsin solution to the gel to obtain a proportion of 1:50 (trypsin: sample protein) and incubate overnight (16–18 h) at 37 °C in a dry bath (*see* **Note 8**).

15. Spin down any evaporation on the tube lid and walls in a MiniSpin for 1 min.

16. Cover the gel with AmBic solution and incubate at RT for 20 min (*see* **Note 9**).

17. Transfer all the liquid contents to a new microtube (collection tube).

18. Cover the gel with extraction solution, incubate for 10 min at RT, and sonicate for 10 min in a sonication bath and then transfer the solution to the gel-free tube. Perform this step twice.

19. Cover the gel with 100% ACN, incubate for 5 min, and transfer the solution to the collection tube.

20. Evaporate the volume in the collection tubes in a vacuum concentrator and store them at −20 °C until LC-MS/MS analysis.

3.7 Mass Spectrometry Analysis (LC-MS/MS) (See Note 10)

1. Add 50μL of ammonium formate (20 mM) to the sample and quantify the peptide concentration in a spectrophotometer.

2. Load a volume containing 1μg of peptides for each fraction onto a BEH130 C18 column.

3. Use an LC method that fractionates the sample into three elutions before separation by the analytical column.

4. The first dimension of the chromatographic method can be carried out using 13.7%, 18.4%, and 50% B_{1stD} at a flow rate of 2.0μL/min (optional).

5. Perform a binary gradient pumping to the analytical column (HSS T3) for each fraction described above, using 3–40% B_{2ndD} for 29 min, followed by a binary gradient from 40 to 85% B_{2ndD} for 4 min, at a flow rate of 300 nL/min.

6. Use the 100 fmol/μL [Glu1]-fibrinopeptide B as a lock-mass compound.

7. The ionization should use ESI in positive mode. The DIA analysis is performed in high-definition MS^E ($HDMS^E$), using ion mobility separation and CID fragmentation (energy ramp from 19 to 53 eV), with a mass range from 50 to 2000 Da.

3.8 UDMS^E Analysis (Fig. 3)

1. Perform one run to create a representation of the samples (*see* **Note 11**).

2. Use this sample—or pool of samples—to create a drift diagram for each fraction, running in $HDMS^E$ mode.

3. Open DriftScope v2.9 (Fig. 3a).

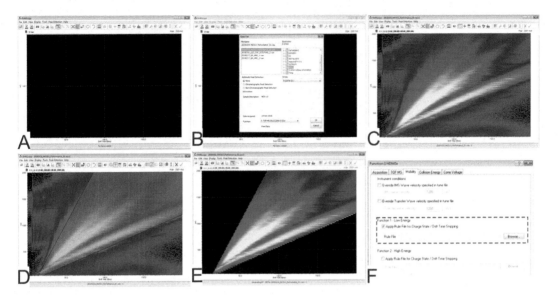

Fig. 3 Step-by-step visualization of using DriftScope® to perform the drift time collision energy optimization (for UDMS^E)

4. Import the raw data acquired from MS/MS analyses with File > Open.

5. Select Function 1: TOF MS (50.0:2000.0) ES+ to optimize the collision energy of precursor ions (Fig. 3b).

6. Use the select tool and enable the band selections (Fig. 3c).

7. Select the up- and downside of the red curve to exclude the background masses marked in blue against a black background (Fig. 3d).

8. Export the Rule file (.rul) in the File > Export Selection Rule.

9. Click on "Accept current selection" (Fig. 3e).

10. Save the image in File > Save Selected View.

11. Import the Rule file into MassLynx in MS File > HDMS^E > Mobility > Function 1: Low Energy (Fig. 3f)

12. Perform the LC-MS/MS of all samples as described in Subheading 3.7.

3.9 Data Processing (See Note 12)

1. Run Progenesis® QI for Proteomics.

2. Create a New Experiment, selecting the type of raw data and mass spectrometer.

3. Import the raw data (.raw), confirming the lock mass calibration with [Glu1]-fibrinopeptide B (m/z 785.8426).

4. Run the Ion Accounting workflow.

5. Specify the peak intensity threshold calculating the optimal threshold with the Homo Sapiens FASTA file or set it manually, leaving it at the default parameters.

6. Following file importing, let the runs align automatically.

7. Set parameters for peak picking: leave at automatic sensitivity with default noise estimation algorithm and +5 maximum ion charge to be detected.

8. Set the parameters for peptide and protein identification: select the Homo sapiens databank, trypsin as the digestion enzyme and up to one missed cleavage.

9. Add carbamidomethyl cysteine as fixed modification and methionine oxidation as a variable modification (*see* **Note 13**).

10. Set the false discovery rate (FDR) to 1%, two or more fragments/peptide, five or more fragments/protein, and two or more peptides/protein as the minimum identification parameters.

11. Choose the relative quantitation method (*see* **Note 14**).

12. Exclude peptides that have an absolute mass error greater than 20 ppm.

13. Export the peptide ion/peptide/protein tables.

4   Notes

1. Perform figure-eight movements in the bath just until the solution has thawed to avoid prolonged exposure of the cells to DMSO.

2. The growth rate can vary with culture conditions; occasionally, cells may be confluent before the medium must be changed.

3. Cell counting can be performed with a Neubauer Chamber, for example.

4. The cells can be stored at −4 °C for 3 h to overnight, followed by overnight freezing at −80 °C. Cells can then be transferred to liquid nitrogen for long-term storage (optional).

5. All steps described below must be performed on ice. The centrifuge should be pre-cooled to 4 °C.

6. The number of grinding motions must be the same for every sample (e.g., 20×).

7. Due to the high concentration of SDS in the lysis buffer, ensure compatibility with the method used. The micro BCA protocol (Thermo Scientific) has been confirmed to be compatible.

8. Ensure that the gel is full covered by the solution at this step, as in all above steps.

9. The reaction can be stopped storing the samples at $-20°$, if necessary.

10. Although the LC-MS/MS analysis described below is specific for an ACQUITY UPLC M-Class coupled to a Synapt G2-Si mass spectrometer, it can easily be adjusted for other systems.

11. Running a pooled group of samples is often beneficial to obtain an overall view of the samples to be run.

12. Using Progenesis, each sample fraction must be processed separately, with the same search parameters, and then later merged.

13. Other protein modifications can be added to study post-translational regulation of biological systems.

14. Progenesis allows three options for relative quantitation: Hi-N, non-conflicting peptides, and all peptides. We recommend using Hi-N with three peptides/protein to quantify proteins.

Acknowledgments

GRO and DMS would like to thank FAPESP for the Funding (under grant numbers 18/01410-1, 18/03673-0, 17/25588-1, 19/00098-7).

Conflicts of Interest: The authors declare no conflicts of interest.

References

1. WHO (2014) WHO |(2014) WHO | Schizophrenia

2. Freedman R (2003) Schizophrenia. N Engl J Med 349:1738–1749

3. Kahn RS, Sommer IE, Murray RM et al (2015) Schizophrenia. Nat Rev Dis Primers 1:15,067

4. Owen MJ, O'Donovan MC, Thapar A et al (2011) Neurodevelopmental hypothesis of schizophrenia. Br J Psychiatry 198:173–175

5. Borrell J (2002) Prenatal immune challenge disrupts sensorimotor gating in adult rats implications for the etiopathogenesis of schizophrenia. Neuropsychopharmacology 26:204–215

6. Zuckerman L, Weiner I (2005) Maternal immune activation leads to behavioral and pharmacological changes in the adult offspring. J Psychiatr Res 39:311–323

7. Estes ML, McAllister AK (2016) Maternal immune activation: implications for neuropsychiatric disorders. Science 353:772–777

8. Mattei D, Ivanov A, Ferrai C et al (2017) Maternal immune activation results in complex microglial transcriptome signature in the adult offspring that is reversed by minocycline treatment. Transl Psychiatry 7:e1120

9. Prinz M, Jung S, Priller J (2019) Microglia biology: one century of evolving concepts. Cell 179:292–311

10. Bloomfield PS, Selvaraj S, Veronese M et al (2016) Microglial activity in people at ultra high risk of psychosis and in schizophrenia: an [(11)C]PBR28 PET brain imaging study. Am J Psychiatry 173:44–52

11. Schizophrenia Working Group of the Psychiatric Genomics Consortium (2014) Biological insights from 108 schizophrenia-associated genetic loci. Nature 511:421–427

12. Network and Pathway Analysis Subgroup of Psychiatric Genomics Consortium (2015) Psychiatric genome-wide association study analyses implicate neuronal, immune and histone pathways. Nat Neurosci 18:199–209

13. Sekar A, Bialas AR, de Rivera H et al (2016) Schizophrenia risk from complex variation of complement component 4. Nature 530:177–183

14. Xu J, Sun J, Chen J et al (2012) RNA-Seq analysis implicates dysregulation of the immune system in schizophrenia. BMC Genomics 13(Suppl 8):S2

15. Velásquez E, Martins-de-Souza D, Velásquez I et al (2019) Quantitative subcellular proteomics of the orbitofrontal cortex of schizophrenia patients. J Proteome Res 18:4240

16. Martins-de-Souza D, Gattaz WF, Schmitt A et al (2009) Prefrontal cortex shotgun proteome analysis reveals altered calcium homeostasis and immune system imbalance in schizophrenia. Eur Arch Psychiatry Clin Neurosci 259:151–163

17. de Witte L, Tomasik J, Schwarz E et al (2014) Cytokine alterations in first-episode schizophrenia patients before and after antipsychotic treatment. Schizophr Res 154:23–29

18. Li Y, Zhou K, Zhang Z et al (2012) Label-free quantitative proteomic analysis reveals dysfunction of complement pathway in peripheral blood of schizophrenia patients: evidence for the immune hypothesis of schizophrenia. Mol BioSyst 8:2664–2671

19. Zuccoli GS, Saia-Cereda VM, Nascimento JM et al (2017) The energy metabolism dysfunction in psychiatric disorders postmortem brains: focus on proteomic evidence. Front Neurosci 11:493

20. Saia-Cereda VM, Cassoli JS, Martins-de-Souza D et al (2017) Psychiatric disorders biochemical pathways unraveled by human brain proteomics. Eur Arch Psychiatry Clin Neurosci 267:3–17

21. Yates JR 3rd (2013) The revolution and evolution of shotgun proteomics for large-scale proteome analysis. J Am Chem Soc 135:1629–1640

22. Ludwig C, Gillet L, Rosenberger G et al (2018) Data-independent acquisition-based SWATH-MS for quantitative proteomics: a tutorial. Mol Syst Biol 14:e8126

23. Silva JC, Denny R, Dorschel CA et al (2005) Quantitative proteomic analysis by accurate mass retention time pairs. Anal Chem 77:2187–2200

24. Silva JC, Gorenstein MV, Li G-Z et al (2006) Absolute quantification of proteins by LCMSE: a virtue of parallel MS acquisition. Mol Cell Proteomics 5:144–156

25. Silva JC, Denny R, Dorschel C et al (2006) Simultaneous qualitative and quantitative analysis of the Escherichia coli proteome: a sweet tale. Mol Cell Proteomics 5:589–607

26. Geromanos SJ, Hughes C, Ciavarini S et al (2012) Using ion purity scores for enhancing quantitative accuracy and precision in complex proteomics samples. Anal Bioanal Chem 404:1127–1139

27. Distler U, Kuharev J, Navarro P et al (2014) Drift time-specific collision energies enable deep-coverage data-independent acquisition proteomics. Nat Methods 11:167–170

28. Distler U, Kuharev J, Navarro P et al (2016) Label-free quantification in ion mobility-enhanced data-independent acquisition proteomics. Nat Protoc 11:795–812

Chapter 25

Detailed Method for Performing the ExSTA Approach in Quantitative Bottom-Up Plasma Proteomics

Andrew J. Percy and Christoph H. Borchers

Abstract

The use of stable isotope-labeled standards (SIS) is an analytically valid means of quantifying proteins in biological samples. The nature of the labeled standards and their point of insertion in a bottom-up proteomic workflow can vary, with quantification methods utilizing curves in analytically sound practices. A promising quantification strategy for low sample amounts is external standard addition (ExSTA). In ExSTA, multipoint calibration curves are generated in buffer using serially diluted natural (NAT) peptides and a fixed concentration of SIS peptides. Equal concentrations of SIS peptides are spiked into experimental sample digests, with all digests (control and experimental) subjected to solid-phase extraction prior to liquid chromatography tandem mass spectrometry (LC-MS/MS) analysis. Endogenous peptide concentrations are then determined using the regression equation of the standard curves. Given the benefits of ExSTA in large-scale analysis, a detailed protocol is provided herein for quantifying a multiplexed panel of 125 high-to-moderate abundance proteins in undepleted and non-enriched human plasma samples. The procedural details and recommendations for successfully executing all phases of this quantification approach are described. As the proteins have been putatively correlated with various noncommunicable diseases, quantifying these by ExSTA in large-scale studies should help rapidly and precisely assess their true biomarker efficacy.

Key words Forward curve, Human plasma, Protein, Proteomics, Stable isotope-labeled standard, Standard curve, Quantification

1 Introduction

The quantitative proteomics field has advanced tremendously during the past decade. This includes the method workflows, automation schemes, analysis regimens, and scope of applications. Improvements have been made in sample throughput and analytical metrics (e.g., precision, sensitivity) in MS-based quantitation [1–3]. The reproducibility of the methodologies and techniques has also been enhanced, as has its utility in interlaboratory studies [2, 4, 5]. These merits are collectively needed for the widespread assessment of protein disease biomarkers toward clinical diagnostic implementation [6].

Katrin Marcus et al. (eds.), *Quantitative Methods in Proteomics*, Methods in Molecular Biology, vol. 2228,
https://doi.org/10.1007/978-1-0716-1024-4_25, © The Author(s) 2021

While MS-based proteomic measurements are increasingly performed in a label-free manner [7, 8] (*see* Chaps. 8, 20–24), higher identification confidence and improved analytics can be obtained through the use of stable isotope-labeled protein or peptide standards (*see* Chaps. 8, 11). These standards can be inserted at a number of points of the analytical workflow and can be constructed in various formats with different isotope-labeling patterns (commonly $^{13}C/^{15}N$, with a minimum mass shift of 6 Da from its unlabeled counterpart) [9]. The classical workflow is "bottom-up," where the enzymatically cleaved peptides serve as molecular surrogates of the proteins of interest. The types of standard(s) used have included recombinant proteins [10, 11], tryptic peptides [12–14], winged (referred to also as extended, flanked, or cleavable) peptides [15, 16], concatenated peptides (abbreviated QconCATs for quantification concatemers) [17, 18], and protein epitope signature tags (PrESTs) [19–21]. In all cases, the labeled standards are designed to resemble the chemical structure of its endogenous (natural or NAT) analogue. Their quantitative effectiveness has been evaluated and compared/contrasted in different sample matrices (e.g., plasma, cerebrospinal fluid, urine), using both protein [22] and peptide [23] standards. These studies have demonstrated the utility of labeled standards, with the selection ultimately guided by experimental design and quantitative application.

Protein quantification with standards can be accomplished in a number of ways. While single-point measurements (using area ratios of unlabeled peptides, from a sample, to labeled peptides, from a spiked-in standard) are the easiest to perform, this method is not analytically accurate for large-scale analyses (due, for example, to plate-to-plate variation) and over wide dynamic ranges. Better quantitative methods utilize standard curves prepared in buffer or in a pooled control sample. These can be generated in a forward or reverse manner [24–27]. Forward curves involve a dilution series of NAT peptide concentrations (derived synthetically or from an endogenous sample) and constant SIS peptide concentrations, while reverse curves are the converse involving a dilution series of SIS peptide concentrations with constant NAT across the calibrant levels. There are advantages and disadvantages of curve-based quantitative strategies, as described and demonstrated previously [28]. One promising and applicable approach uses external standard addition (ExSTA), where forward standard curves are prepared in buffer and constant SIS peptide concentrations are spiked into the experimental samples. The peptide concentrations in the endogenous sample can then be determined by applying the experimental response ratios (i.e., NAT vs. spiked-in SIS) to the regression equation of their peptide-specific standard curves (plot of NAT/SIS peptide response vs. NAT peptide concentration). This strategy has been demonstrated to be a robust means for precisely quantifying endogenous proteins in human plasma samples. Moreover, it is well suited for routine LC-MS/MS processing, large-scale proteomic applications, and sample-limited analyses.

Detailed here is a procedure for preparing samples (both controls and experimental), via the ExSTA approach, for robust, MS-based quantification of human plasma proteins without the use of upfront depletion, enrichment, or LC fractionation. The controls are generated in a BSA/PBS (bovine serum album/phosphate-buffered saline) surrogate matrix. These surrogate controls encompass standard-curve samples comprising eight concentration-level calibrants (spanning a 1000-fold range) and three quality-control (QC) samples (at low, medium, and high levels). The mixtures utilized in these samples are NAT and SIS tryptic peptides, with the SIS mixes also implemented in the plasma sample analysis ($n = 20$). The peptides correspond to a panel of medium-to-high abundance human plasma proteins (125 in total, with one peptide per protein) and were initially carefully selected according to a series of peptide/protein selection rules [29, 30]. The peptide mixtures and the described methods have been validated according to the Clinical Proteomic Tumor Analysis Consortium (CPTAC) guidelines [31, 32] and are verified to be fit-for-purpose for Tier 2, research-based analysis [33]. The assays utilize reversed-phase liquid chromatography (RPLC) in conjunction with tandem MS (MS/MS), performed in the selected/multiple reaction monitoring (SRM/MRM) or parallel reaction monitoring (PRM) acquisition mode. Performing the procedures described herein should produce precise quantitative results for putative biomarker analysis (at the discovery or verification stage) while opening the door for expanded target panels and sample sizes.

2   Materials

2.1 Sample and Solution Preparation Supplies

1. Analytical balance.
2. Adjustable pipettes and pipette tips.
3. Polypropylene Falcon tubes (15 and 50 mL) and screw caps (all from Fisher Scientific; Suwanee, GA, USA).
4. LoBind (i.e., low-binding) microcentrifuge tubes (1.5 mL; Eppendorf; Hauppauge, NY, USA).
5. Glass beaker (250 mL).
6. Glass graduated cylinder (100 mL).
7. Ultralow temperature freezer (capable of −80 °C).
8. Vortex mixer.
9. Minicentrifuge (capable of $\geq 1000 \times g$).
10. Benchtop centrifuge (compatible with 96-well plates).

11. Incubator.

12. Aluminum foil.

13. 96-well deep well plates (1.1 mL) and accessories (e.g., silicone sealing mat, Axygen; Union City, CA, USA; AluminaSeal II™ sealing film; EXCEL Scientific; Victorville, CA, USA).

14. 96-well Oasis HLB (i.e., hydrophilic-lipophilic balance) μElution plates (2 mg sorbent, 30 μm particles; part number 186001828BA; Waters; Milford, MA, USA).

15. Skirted 96-well (150 μL) PCR collection plates (Eppendorf).

16. Positive pressure vacuum manifold (compatible with 96-well plates, e.g., part number 186006961; Waters).

17. Lyophilizer (or SpeedVac) and accessories (e.g., wide-mouth borosilicate glass flasks).

18. Autosampler vials and screw caps.

2.2 Control and Experimental Sample Preparations

1. 10 mg/mL BSA in PBS: In a 1.5 mL LoBind microcentrifuge tube, dissolve 10 mg BSA in 1 mL of PBS solution (*see* **Note 1**). The pH of this solution should be approximately 7.5.

2. Reagent A: 1 M Tris (pH 8.0). In a 250 mL glass beaker, dissolve 12.12 g of Tris in 90 mL of LC-MS grade water and adjust the pH to 8.0 with dropwise addition of 12 M HCl. Transfer the solution to a 100 mL glass graduated cylinder and bring the volume to 100 mL with LC-MS grade water (*see* **Note 1**). Vortex briefly.

3. Reagent B: 9 M urea in 300 mM Tris (pH 8.0). In a 15 mL screw cap polypropylene tube, dissolve 5.4 g of urea in 3 mL of Reagent A and 3 mL of LC-MS grade water (*see* **Note 1**). Vortex until fully solubilized (may take up to 5 min; *see* **Note 2**).

4. Reagent C: 9 M urea and 20 mM DL-dithiothreitol (DTT) in 300 mM Tris (pH 8.0). In a 1.5 mL LoBind microcentrifuge tube, dissolve 15.4 mg of DTT in 200 μL of Reagent B, to prepare a 500 mM DTT solution. Using the 500 mM DTT solution, dilute this 25× by transferring 40 μL to 960 μL of Reagent B (*see* **Note 1**). Vortex briefly.

5. Reagent D: 100 mM iodoacetamide (IAA). In a 1.5 mL LoBind microcentrifuge tube, dissolve 18.5 mg of IAA in 1 mL of LC-MS grade water. Vortex until completely solubilized, and then wrap the tube in aluminum foil to prevent deactivation by light (*see* **Note 1**).

6. Reagent E: 100 mM Tris (pH 8.0). In a 15 mL screw cap polypropylene tube, add 1 mL of Reagent A to 9 mL of LC-MS grade water (*see* **Note 1**). Vortex briefly.

7. 1 mg/mL trypsin. In a 1.5 mL LoBind microcentrifuge tube, dissolve 1 mg of trypsin in 1 mL of Reagent E (*see* **Note 1**).

8. 2% aqueous formic acid (FA) (v/v). In a 15-mL screw cap polypropylene tube, add 200 µL FA to 9.8 mL of LC-MS grade water.

9. 0.1% aqueous FA (v/v). In a 50-mL screw cap polypropylene tube, add 50 µL of FA to 49.95 mL of LC-MS grade water. Vortex briefly.

10. 0.1% FA in 30% LC-MS grade acetonitrile (ACN)/water (v/v/v). In a 1.5 mL LoBind microcentrifuge tube, add 3 µL of FA to 899 µL of LC-MS grade ACN and 2.098 mL of LC-MS grade water. Vortex briefly.

11. 0.1% FA in 50% LC-MS grade ACN/water (v/v/v). In a 15 mL screw cap polypropylene tube, add 5 µL of FA to 2.497 mL of LC-MS grade ACN and 2.497 mL of LC-MS grade water. Vortex briefly.

12. NAT peptide mixture. Rehydrate a lyophilized NAT mix aliquot (*see* **Note 3**) in 60 µL of 0.1% FA/30% ACN (v/v). Vortex and centrifuge briefly, and then store on ice. This stock solution represents level H of the 8-point standard curve and is to be later diluted serially to produce the calibrant levels and three curve QC samples.

13. SIS peptide mixture. Rehydrate a lyophilized SIS mix aliquot (*see* **Note 3**) in 450 µL of 0.1% FA/30% ACN (v/v). Vortex and centrifuge briefly, and then store on ice. This creates a concentration-balanced solution (i.e., MS response signals have been tested to be within an order of magnitude of their unlabeled peptide analogues) for direct addition to experimental and control samples. For alternative SIS peptide configurations, refer to **Note 4**.

2.3 LC-MS Equipment

1. LC system: 1290 Infinity UHPLC (Agilent Technologies; Santa Clara, CA, USA).

2. Analytical column: Zorbax Eclipse Plus RRHD C_{18} RP-UHPLC (150 × 2.1 mm i.d., 1.8 µm particles; Agilent Technologies).

3. Eluent A: 0.1% aqueous FA in LC-MS grade water.

4. Eluent B: 0.1% FA in LC-MS grade ACN.

5. 6495 QqQ (Agilent Technologies) with standard-flow ESI source.

6. Q Exactive™ Plus (Thermo Fisher Scientific) with standard-flow ESI source.

7. Mass spectrometer calibrant: ESI-tuning mix.

2.4 Data Collection and Analysis Software	1. MassHunter Quantitative Analysis (Agilent Technologies).
	2. Xcalibur™ (Thermo Fisher Scientific).
	3. Skyline-daily.
	4. Microsoft Excel.

3   Methods

3.1   Digest Preparations (for Control and Experimental Samples)

1. Using a 96-well deep well plate, add 20 µL of Reagent C to the highlighted wells in Fig. 1a (red wells, A1-H1, A2-H2, and A3-D3 for the experimental samples; green wells, A11 and A12 for the control samples).

2. Add 10 µL of undiluted human plasma to the red wells (i.e., A1-H1, A2-H2, and A3-D3).

3. Add 10 µL of the BSA/PBS solution (at 10 mg/mL) to the green wells (i.e., A11 and A12).

4. Cover the plate with a silicone sealing mat, and then mix, centrifuge briefly, and incubate for 30 min at 37 °C. The concentrations of urea and DTT during the denaturation/reduction are 4.5 mM and 10 mM, respectively.

5. Following this incubation, alkylate the free sulfhydryl groups by adding 20 µL of Reagent D to each of the aforementioned wells (22 in total). The reaction concentration of IAA is 33 mM.

6. Cover the plate with a silicone sealing mat, and then mix, centrifuge briefly, and incubate for 30 min at ambient

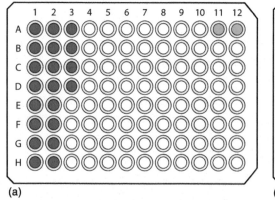
(a)

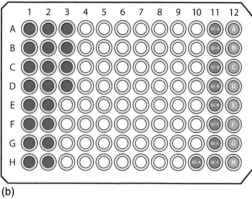
(b)

Fig. 1 Map of 96-well plate illustrating the suggested placement of the control and experimental samples. Part (**a**) is for the digest preparations and (**b**) for the standard spiking to the digests. In both cases, the green/blue wells are for the control samples and red wells for the experimental samples (*n* = 20 human plasma). Regarding the controls in (**b**), the green wells are for the calibration curve sample digests (A, or low NAT peptide concentration, to H, or high), and the blue wells are for the curve QC digests (low, QC-A; medium, QC-B; and high, QC-C)

temperature (i.e., 20–25 °C) in the dark. During this incubation, prepare the trypsin solution as described in **item** 7 of Subheading 2.2.

7. After alkylation, prepare the protein solutions for digestion by adding 272 μL of Reagent E and 35 μL of trypsin (at 1 mg/mL for a 20:1 w/w substrate/enzyme ratio; *see* **Note 5**) to the 22 aforementioned wells. The reagent concentrations at digestion are 0.5 M urea (>1 M inhibits trypsin activity), 1 mM DTT, and 6 mM IAA in 100 mM Tris (pH 8.0).

8. Cover the plate with a silicone sealing mat, and then mix, centrifuge briefly, spin-down, and incubate for 18 h at 37 °C. Prior to the completion of this incubation period (e.g., 2 h beforehand), we recommend to prepare the NAT/SIS peptide standard solutions outlined in Subheading 2.2.

9. After the 18-h incubation, quench the proteolytic digestion by adding 343 μL of 2% FA. This brings the total volume of each well to 700 μL.

10. Cover the plate with a silicone sealing mat and then mix and centrifuge briefly.

11. Combine the two BSA/PBS digests into a 1.5 mL LoBind microcentrifuge tube. This solution will be used in the control sample preparations (for the calibration curve and the QC samples; Subheading 3.3).

12. Store the plate (containing the plasma samples) and microcentrifuge tube (containing the surrogate matrix) on ice until standard spiking (Subheading 3.3).

3.2 Standard Solution Preparations (for the Control Samples)

1. Label seven 1.5 mL LoBind microcentrifuge tubes from Standard G to A. These form a 7-concentration dilution series for the 8-point standard curve, with the highest concentration level being Standard H. This standard reflects the undiluted, concentrated NAT stock that was prepared earlier (in **item 12** of Subheading 2.2).

2. To the Standard G microcentrifuge tube, add 14 μL of Standard H to 21 μL of 0.1% FA/30% ACN. Vortex and centrifuge briefly, and then store on ice. Subsequent concentration levels (i.e., Standards F to A) are to be prepared as outlined in Table 1.

3. For the curve-based QC samples, label five LoBind microcentrifuge tubes as follows: QC-C, ID-1, QC-B, ID-2, and QC-A. The IDs reflect intermediate dilutions and are used only in these QC-sample preparations.

4. Prepare QC-C by adding 18 μL of Standard H to 42 μL of 0.1% FA/30% ACN. Vortex and centrifuge briefly, and then store on ice.

Table 1
Serial dilution of the NAT peptide mix toward an 8-point standard curve (1000-fold range)

Standard	H	G	F	E	D	C	B	A
Level-to-level dilution	1	2.5	2.5	4	2	4	2	2.5
Necessary additions	N/A	14 μL of H	12 μL of G	10 μL of F	15 μL of E	10 μL of D	15 μL of C	10 μL of B
	N/A	21 μL solvent	18 μL solvent	30 μL solvent	15 μL solvent	30 μL solvent	15 μL solvent	15 μL solvent
Total volume (μL)	46	36	40	40	30	40	30	25

Dilutions involve combining an aliquot of a more concentrated standard with a volume of solvent (0.1% FA/30% ACN). Each mixture is to be vortexed and centrifuged before proceeding, with final storage on ice until use.

Table 2
Preparation of three curve QC samples (high, QC-C; medium, QC-B; low, QC-A)

QC sample	QC-C	ID-1	QC-B	ID-2	QC-A
Dilution from standard H	3.33×	6.66×	33.3×	83.3×	250×
Necessary additions	18 μL of H 42 μL of solvent	15 μL of QC-C 15 μL of solvent	15 μL of ID-1 60 μL of solvent	16 μL of QC-B 24 μL of solvent	20 μL of ID-2 40 μL of solvent
Total volume (μL)	60	30	75	40	60

The dilutions involve combining an aliquot of a more concentrated standard with a volume of solvent (0.1% FA/30% ACN). Each mixture is to be vortexed and centrifuged before proceeding, with final storage on ice. The intermediate dilutions (ID-1 and ID-2) can be discarded following the QC-B and QC-A preparations.

5. The subsequent ID (ID-1 and ID-2) and QC (QC-B and QC-A) samples are to be prepared as outlined in Table 2. The ID samples can be discarded following the QC-B and QC-A preparations.

3.3 Standard Spiking to Control/ Experimental Digests

The steps outlined in this section are to be conducted in a fresh 96-well deep well plate.

1. To prepare the curve samples (comprises eight concentration levels, 1:2.5:2.5:4:2:4:2:2.5 dilutions from standard H to A), add the following solutions in succession to wells A12–H12 (*see* green wells in Fig. 1b):

   (a) 40 μL of the BSA/PBS digest (from **step 12** of Subheading 3.1)

   (b) 10 μL of the SIS peptide mixture (rehydrated earlier in **item 13** of Subheading 2.2)

    (c) 10 μL of a NAT standard mix (i.e., add standard A to well A12, down to standard H to well H12; prepared in **step 2** of Subheading 3.2)

    (d) 540 μL of 0.1% aqueous FA.

2. To prepare the curve QC samples, add the following solutions in succession to wells H10 and A11 to H11 (*see* blue wells in Fig. 1b):

    (a) 40 μL of the BSA/PBS digest (from **step 12** of Subheading 3.1)

    (b) 10 μL of the rehydrated SIS peptide mixture

    (c) 10 μL of a NAT QC mix (QC-A to wells H10, A11, B11; QC-B to C11, D11, E11; QC-C to F11, G11, H11; prepared in **step 4** of Subheading 3.2)

    (d) 540 μL of 0.1% aqueous FA.

3. To prepare the experimental plasma samples ($n = 20$), add the following solutions in succession to wells A1–H1, A2–H2, and A3–D3 (*see* red wells in Fig. 1b and **Note 6**):

    (a) 40 μL of a human plasma digest (from **step 12** of Subheading 3.1)

    (b) 10 μL of the rehydrated SIS peptide mixture

    (c) 10 μL of 0.1% FA/30% aqueous ACN

    (d) 540 μL of 0.1% aqueous FA.

4. Cover the plate with a sealing film (AlumaSeal II™) and then mix and centrifuge briefly.

5. Store on ice until performing sample cleanup by solid-phase extraction.

3.4 Extraction and Reconstitution for LC-MS/MS

1. Desalt and concentrate the peptide solutions by solid-phase extraction using a positive pressure vacuum manifold on a 96-well Oasis HLB μElution plate. The extractions are to take place in 37 wells (*see* Fig. 1b for example locations) and involve the following sequential steps (*see* **Note 7**):

    (a) Wash with 600 μL of LC-MS grade methanol.

    (b) Condition with 600 μL of 0.1% aqueous FA.

    (c) Load with 510 μL of each sample (i.e., standard curve samples A–H, QC samples A–C, plasma samples 1–20).

    (d) Wash with 600 μL of LC-MS grade water or 0.1% FA.

    (e) Elute with 75 μL of 0.1% FA/50% ACN into a skirted 96-well microplate.

2. Cover the plate of eluate with a sealing film (AluminaSeal II™) and centrifuge briefly.

3. Puncture the sealing film with a hole (using an 18-gauge needle) at the designated well positions (37 in total), and then cover with parafilm.

4. Freeze the plate and lyophilizer container at −80 °C.

5. Once frozen, remove the parafilm, and then dry down (with lyophilizer) to completion overnight.

6. Rehydrate the 37 samples in 34 µL of 0.1% FA (this will give a final concentration of 1 µg/µL based on an initial plasma protein concentration of 70 mg/mL).

7. Cover the plate with a sealing film, mix, and centrifuge briefly.

8. Remove the sealing film and replace with a silicone sealing mat.

9. Place the sample plate and an autosampler vial containing ~200 µL eluent A (for blank injections) into the LC autosampler (*see* **Note 8**).

10. Prepare for the LC-MS/MS analysis (Subheading 3.5).

3.5 LC System Setup

1. Thermostat the column and autosampler compartments to 50 and 4 °C, respectively.

2. Set up the autosampler for acquisition at a 20 µL/min draw, 40 µL/min eject, and vial/well bottom sensing activated.

3. To facilitate the RPLC peptide separations, use mobile phase compositions of 0.1% aqueous FA in water for A and 0.1% aqueous FA in ACN for B.

4. Use the following ACN gradient (time in min, eluent B composition in %): 0, 2; 2, 7; 50, 30; 53, 45; 53.5, 80; 55.5, 80; 56, 2.

5. To help preserve instrument cleanliness, divert the solvent stream (2% eluent A) to waste at the conclusion of the ACN gradient (at 56 min), and then hold for 4 min to allow column equilibration.

6. Use a 0.4 mL/min flow rate.

7. Use 10 µL injection volumes for all sample types.

3.6 MRM/MS System Setup

1. Operate the ESI source in the positive ion mode with capillary voltage at 3.5 kV.

2. Use the following general parameters: 300 V nozzle voltage, 11 L/min sheath gas at 250 °C, 15 L/min drying gas at 150 °C, 30 psi nebulizer, and iFunnel RF pressures of 200 V (high) and 110 V (low). Ultrahigh-purity nitrogen serves as the carrier gas in all settings.

3. Using the MRM mode (for the 6495 QqQ, *see* **Note 9**), enter the specific acquisition parameters. This encompasses the precursor/product ion *m/z* values, retention times (RTs), and collision energies (*see* Table 3).

Table 3
Specific acquisition parameters for the bottom-up LC-MRM/MS analysis

Protein	Peptide	Product Ion/Charge	NAT Precursor m/z	NAT Product m/z	SIS Precursor m/z	SIS Product m/z	RT (min)	CE (V)
78 kDa glucose-regulated protein	ITPSYVAFTPEGER	y_{12}^{2+}	783.8936	676.8277	788.8977	681.8318	24.1	17
Adipocyte plasma membrane-associated protein	LLEYDTVTR	y_7^{+}	555.2955	883.4156	560.2996	893.4239	24.21	13
Adiponectin	IFYNQQNHYDGSTGK	y_{13}^{2+}	591.2727	756.3291	593.9441	760.3362	8.91	13
Afamin	DADPDTFFAK	y_7^{+}	563.7562	825.4141	567.7633	833.4283	20.36	9
α-1-acid glycoprotein 1	NWGLSVYADKPETTK	y_{13}^{2+}	570.2895	704.8696	572.9609	708.8767	21.15	13
α-1-Antichymotrypsin	EIGELYLPK	y_2^{+}	531.2975	244.1656	535.3046	252.1798	24.8	9
α-1-antitrypsin	LSITGTYDLK	y_7^{+}	555.8057	797.4040	559.8128	805.4182	21.98	17
α-1B-glycoprotein	LQSLFDSPDFSK	b_2^{+}	619.3268	243.1339	623.3339	243.1339	29.68	21
α-2-Antiplasmin	LGNQEPGGQTALK	b_5^{+}	656.8570	542.2575	660.8464	542.2575	16.65	17
α-2-HS-glycoprotein	FSVVYAK	y_5^{+}	407.2289	579.3501	411.2360	587.3643	14.29	9
α-2-macroglobulin	AIGYLNTGYQR	y_7^{+}	628.3251	851.4370	633.3293	861.4453	16.75	21
Antithrombin-III	DDLYVSDAFHK	y_8^{2+}	437.2068	483.7376	439.8782	487.7447	18.34	9
Apolipoprotein A-I	ATEHLSTLSEK	y_{10}^{2+}	405.8787	572.7959	408.5501	576.8030	5.68	9
Apolipoprotein A-II	EQLTPLIK	y_4^{+}	471.2869	470.3337	475.2940	478.3479	19.36	9
Apolipoprotein A-IV	LGEVNTYAGDLQK	b_3^{+}	704.3594	300.1554	708.3665	300.1554	18.95	25
Apolipoprotein B-100	FPEVDVLTK	y_8^{2+}	524.2897	450.7555	528.2968	454.7626	25.8	13
Apolipoprotein C-I	EWFSETFQK	y_7^{+}	601.2798	886.4305	605.2869	894.4447	25.16	13
Apolipoprotein C-II	TYLPAVDEK	b_2^{+}	518.2715	265.1183	522.2786	265.1183	14.64	13

(continued)

Table 3
(continued)

Protein	Peptide	Product Ion/Charge	NAT Precursor m/z	NAT Product m/z	SIS Precursor m/z	SIS Product m/z	RT (min)	CE (V)
Apolipoprotein C-III	GWVTDGFSSLK	b_2^+	598.8009	244.1081	602.8080	244.1081	28.36	17
Apolipoprotein C-IV	ELLETVVNR	y_4^+	536.8082	487.3075	541.8082	497.3075	19.52	17
Apolipoprotein D	VLNQELR	y_5^+	436.2534	659.3471	441.2576	669.3554	7.43	13
Apolipoprotein E	LGPLVEQGR	y_7^{2+}	484.7798	399.7271	489.7840	404.7312	7.23	13
Apolipoprotein L1	VAQELEEK	y_6^+	473.2480	775.3832	477.2551	783.3974	4.26	17
Apolipoprotein M	AFLLTPR	y_5^+	409.2502	599.3875	414.2543	609.3958	21.77	9
Apolipoprotein(a)	TPAYYPNAGLIK	y_7^+	654.3533	712.4352	658.3604	720.4494	22.46	25
Attractin	SVNNVVVR	y_6^+	443.7589	700.4100	448.7630	710.4183	6.69	13
β-2-glycoprotein 1	ATVVYQGER	y_6^+	511.7669	751.3733	516.7710	761.3816	5.84	17
β-Ala-his dipeptidase	ALEQDLPVNIK	y_5^+	620.3508	570.3610	624.3579	578.3752	24.17	17
Biotinidase	SHLIIAQVAK	b_4^+	360.5572	451.2663	363.2286	451.2663	13.82	9
Cadherin-13	INENTGSVSVTR	b_2^+	638.8282	228.1343	643.8324	228.1343	7.18	21
Carbonic anhydrase 1	VLDALQAIK	y_7^+	485.8002	758.4407	489.8073	766.4549	23.62	9
Carboxypeptidase B2	IAWHVIR	y_4^{2+}	298.8518	262.6729	302.1851	267.6729	16.37	9
Cathelicidin antimicrobial peptide	AIDGINQR	y_6^+	443.7407	702.3529	448.7448	712.3612	5.82	17
Cation-independent mannose-6-phosphate receptor	GHQAFDVGQPR	y_4^+	404.5354	457.2518	407.8715	467.2600	7.79	13
CD5 antigen-like	LVGGLHR	y_5^+	376.2323	539.3049	381.2364	549.3131	4.87	13
Ceruloplasmin	IYHSHIDAPK	y_8^{2+}	394.2085	452.7354	396.8799	456.7425	4.29	9

Cholinesterase	YLTLNTESTR	y_8^+	599.3091	921.4636	604.3133	931.4719	13.96	25
Clusterin	ELDESLQVAER	y_3^+	644.8226	375.1987	649.8267	385.2069	15.69	17
Coagulation factor IX	SALVLQYLR	y_5^+	531.8189	692.4090	536.8231	702.4173	30.11	17
Coagulation factor V	SEAYNTFSER	y_8^+	602.2675	987.4530	607.2716	997.4613	9.33	17
Coagulation factor VIII	LHPTHYSIR	y_7^{2+}	375.2051	437.2325	378.5412	442.2366	12.5	9
Coagulation factor X	TGIVSGFGR	y_5^+	447.2456	523.2623	452.2497	533.2706	15.37	13
Coagulation factor XII	EQPPSLTR	y_6^{2+}	464.2483	335.6978	469.2525	340.7019	6.74	9
Complement C1q subcomponent subunit B	IAFSATR	y_5^+	383.2163	581.3042	388.2205	591.3125	9.49	9
Complement C1q subcomponent subunit C	FNAVLTNPQGDYDTSTGK	y_{11}^{2+}	643.3059	584.7595	645.9773	588.7666	19.86	9
Complement C1r subcomponent	GLTLHLK	y_5^{2+}	261.1692	306.1974	263.8406	310.2045	13.81	5
Complement C1r subcomponent-like protein	VVVHPDYR	y_7^{2+}	328.8469	443.2325	332.1830	448.2366	5.75	5
Complement C1s subcomponent	TNFDNDIALVR	b_2^+	639.3279	216.0979	644.3320	216.0979	23.87	17
Complement C2	HAFILQDTK	b_4^+	358.1977	469.2558	360.8691	469.2558	11.94	5
Complement C3	TGLQEVEVK	y_6^+	501.7769	731.3934	505.7840	739.4076	11.71	17
Complement C5	VFQFLEK	y_5^+	455.7553	664.3665	459.7624	672.3807	24.42	9
Complement component C7	LIDQYGTHYLQSGSLGGEYR	b_2^+	753.0343	227.1754	756.3704	227.1754	4.94	25
Complement component C9	LSPIYNLVPVK	y_9^{2+}	621.8765	521.8184	625.8836	525.8255	33.22	17
Complement factor B	EELLPAQDIK	y_6^+	578.3164	671.3723	582.3235	679.3865	19.39	13
Complement factor H	SSQESYAHGTK	y_9^{2+}	398.8510	510.7409	401.5224	514.7480	2.30	5
Complement factor I	VFSLQWGEVK	y_8^+	596.8217	946.4993	600.8288	954.5135	30.84	13
Corticosteroid-binding globulin	WSAGLTSSQVDLYIPK	y_2^+	882.9620	244.1656	886.9691	252.1798	34.46	29

(continued)

Table 3
(continued)

Protein	Peptide	Product Ion/Charge	NAT Precursor m/z	NAT Product m/z	SIS Precursor m/z	SIS Product m/z	RT (min)	CE (V)
Cystatin-C	ALDFAVGEYNK	y_6^+	613.8062	709.3515	617.8133	717.3657	22.72	13
Endothelial protein C receptor	TLAFPLTIR	y_7^+	516.3160	817.4931	521.3202	827.5013	34.09	13
Fetuin-B	LVVLPFPK	y_6^+	456.7995	700.4392	460.8066	708.4534	32.26	9
Fibrinogen α chain	ESSSHHPGIAEFPSR	y_4^+	546.5937	506.2722	549.9298	516.2804	9.76	21
Fibrinogen β chain	HQLYIDETVNSNIPTNLR	b_{13}^{2+}	709.6992	764.3755	713.0352	764.3755	25.7	9
Fibrinogen γ chain	YEASILTHDSSIR	y_{11}^{2+}	497.9195	600.3226	501.2555	605.3267	16.33	13
Fibronectin	HTSVQTTSSGSGPFTDVR	y_6^+	621.9659	734.3832	625.3020	744.3914	12.13	13
Fibulin-1	TGYYFDGISR	y_6^+	589.7775	694.3519	594.7816	704.3601	20.64	17
Galectin-3-binding protein	SDLAVPSELALLK	y_8^+	678.3927	870.5295	682.3998	878.5437	35.83	13
Gelsolin	AGALNSNDAFVLK	b_3^+	660.3513	200.1030	664.3584	200.1030	22.97	25
Glutathione peroxidase 3	QEPGENSEILPTLK	y_{12}^{2+}	777.9041	649.3535	781.9112	653.3606	23.17	25
Haptoglobin	DIAPTLTLYVGK	y_9^{2+}	645.8688	496.2948	649.8759	500.3019	31.67	9
Hemoglobin subunit α	VGAHAGEYGAEALER	y_4^+	510.5829	488.2827	513.9190	498.2910	10.94	9
Hemopexin	NFPSPVDAAFR	y_9^{2+}	610.8066	480.2509	615.8107	485.2550	27.04	13
Heparin cofactor 2	TLEAQLTPR	y_7^+	514.7904	814.4417	519.7945	824.4500	13.89	17
Hepatocyte growth factor-like protein	SPLNDFQVLR	y_7^+	594.8222	891.4683	599.8263	901.4766	28.23	17
Hyaluronan-binding protein 2	VVLGDQDLK	y_7^+	493.7795	788.4149	497.7866	796.4291	13.42	17

Protein	Peptide	Fragment					
Ig mu chain C region	GFPSVLR	y_5^{2+}	388.2267	286.1817	393.2308	291.1859	20.33	9
Insulin-like growth factor I	GFYFNKPTGYGSSSR	y_{13}^{2+}	556.5986	732.3493	559.9347	737.3535	16.87	13
Inter-α-trypsin inhibitor heavy chain H2	SLAPTAAAK	y_6^{+}	415.2425	558.3246	419.2496	566.3388	5.12	9
Intercellular adhesion molecule 1	LLGIETPLPK	y_8^{+}	540.8368	854.4982	544.8439	862.5124	15.88	17
Interleukin-10	AHVNSLGENLK	y_5^{+}	394.5473	560.3039	397.2187	568.3181	7.12	9
Kallistatin	VGSALFLSHNLK	y_{11}^{2+}	429.2470	593.8326	431.9183	597.8397	20.93	9
Keratin-type I cytoskeletal 10	SLLEGEGSSGGGGR	b_2^{+}	631.8022	201.1234	636.8063	201.1234	7.92	25
Keratin-type II cytoskeletal 2 epidermal	YEELQVTVGR	b_2^{+}	597.3117	293.1132	602.3158	293.1132	17.33	17
Kininogen-1	TVGSDTFYSFK	y_9^{+}	626.2982	1051.4731	630.3053	1059.4873	22.22	13
Leucine-rich α-2-glycoprotein	DLLLPQPDLR	y_6^{+}	590.3402	725.3941	595.3444	735.4023	30.62	9
Lipopolysaccharide-binding protein	ITLPDFTGDLR	y_8^{+}	624.3352	920.4472	629.3393	930.4555	34.27	17
L-selectin	AEIEYLEK	y_6^{+}	497.7582	794.4294	501.7653	802.4436	14.16	9
Lysozyme C	AWVAWR	y_4^{+}	394.7137	531.3038	399.7179	541.3121	21.00	9
Mannan-binding lectin serine protease 1	TGVITSPDFPNPYPK	b_3^{+}	816.9170	258.1448	820.9241	258.1448	26.88	25
Mannan-binding lectin serine protease 2	WPEPVFGR	y_5^{+}	494.2560	575.3300	499.2601	585.3383	25.84	17
Metalloproteinase inhibitor 2	EYLIAGK	y_3^{+}	397.2264	275.1714	401.2335	283.1856	11.14	9
Myeloblastin	LVNVVLGAHNVR	y_{10}^{2+}	430.9262	539.8094	434.2623	544.8136	17.45	5
Peroxiredoxin-2	GLFIIDGK	y_6^{+}	431.7553	692.3978	435.7624	700.4120	27.03	9
Phosphatidylinositol-glycan-specific phospholipase D	FGSSLITVR	y_4^{+}	490.2822	488.3191	495.2863	498.3274	21.52	13
Phospholipid transfer protein	AVEPQLQEEER	y_8^{2+}	664.3281	514.7540	669.3322	519.7581	9.36	17
Pigment epithelium-derived factor	LQSLFDSPDFSK	y_{10}^{+}	692.3432	1142.5364	696.3503	1150.5506	28.44	17
Plasma protease C1 inhibitor	FQPTLLTLPR	y_8^{+}	593.3531	910.5720	598.3573	920.5803	32.61	17
Plasma serine protease inhibitor	AVVEVDESGTR	y_9^{2+}	581.2909	496.2382	586.2951	501.2423	7.64	13

(continued)

Table 3
(continued)

Protein	Peptide	Product Ion/ Charge	NAT Precursor m/z	NAT Product m/z	SIS Precursor m/z	SIS Product m/z	RT (min)	CE (V)
Plasminogen	LFLEPTR	b_2^+	438.2529	261.1598	443.2570	261.1598	34.25	9
Plasminogen activator inhibitor 1	VFQQVAQASK	y_5^+	553.3037	504.2776	557.3108	512.2918	7.53	13
Pregnancy zone protein	ISEITNIVSK	y_6^+	552.3190	661.3879	556.3261	669.4021	18.45	17
Protein AMBP	HHGPTITAK	y_2^+	321.1786	218.1499	323.8500	226.1641	2.62	13
Protein S100-A9	DLQNFLK	y_5^+	439.2425	649.3668	443.2496	657.3810	23.70	9
Protein Z-dependent protease inhibitor	ETSNFGFSLLR	y_6^+	635.8250	692.4090	640.8291	702.4173	33.06	17
Prothrombin	ELLESYIDGR	y_6^+	597.8037	710.3468	602.8078	720.3550	24.47	21
Retinol-binding protein 4	YWGVASFLQK	y_8^+	599.8164	849.4829	603.8235	857.4971	34.68	13
Serotransferrin	DGAGDVAFVK	y_7^+	489.7482	735.4036	493.7553	743.4178	14.06	13
Serum albumin	LVNEVTEFAK	y_2^+	575.3111	218.1499	579.3182	226.1641	19.45	37
Serum paraoxonase/arylesterase 1	IFFYDSENPPASEVLR	y_8^+	942.4623	868.4887	947.4665	878.4970	34.14	37
Serum paraoxonase/lactonase 3	IQNVLSEKPR	y_8^{2+}	395.2313	471.7720	398.5674	476.7761	9.01	9
SPARC	LEAGDHPVELLAR	y_{11}^{2+}	473.9245	589.3198	477.2606	594.3240	19.17	9
Tenascin	FTTDLDSPR	y_7^+	526.2564	803.3894	531.2605	813.3976	12.24	17
Tenascin-X	ILISGLEPSTPYR	b_2^+	723.4036	227.1754	728.4077	227.1754	28.20	21
Thrombospondin-1	GTLLALER	y_4^+	436.7636	488.2827	441.7678	498.2910	19.99	13
Thrombospondin-4	KPQDFLEELK	y_4^+	416.2274	518.2821	418.8988	526.2963	23.95	9
Thyroxine-binding globulin	AVLHIGEK	y_6^{2+}	289.5080	348.7056	292.1794	352.7127	7.95	5

Protein	Peptide	Ion						
Tissue factor pathway inhibitor	FYYNSVIGK	y_7^+	545.7820	780.4250	549.7891	788.4392	20.09	9
Tissue-type plasminogen activator	VVPGEEEQK	y_7^+	507.7587	816.3734	511.7658	824.3876	3.49	9
Transferrin receptor protein 1	GFVEPDHYVVVGAQR	y_{13}^{2+}	558.2862	734.8808	561.6223	739.8849	21.50	13
Transthyretin	GSPAINVAVHVFR	y_{11}^{2+}	456.2578	611.8564	459.5939	616.8605	25.73	5
Vascular cell adhesion protein 1	NTVISVNPSTK	y_8^+	580.3195	845.4727	584.3266	853.4869	11.48	17
Vasorin	ESHVTLASPEETR	y_5^+	485.9073	631.3046	489.2434	641.3128	8.14	9
Vitamin K-dependent protein S	SFQTGLFTAAR	y_8^+	599.8144	836.4625	604.8185	846.4707	24.59	21
Vitamin K-dependent protein Z	GLISGWAR	y_5^+	430.2429	576.2889	435.2470	586.2971	24.20	9
Vitronectin	FEDGVLDPDYPR	y_5^+	711.8304	647.3148	716.8346	657.3230	23.14	33
Zinc-α-2-glycoprotein	EIPAWVPFDPAAQITK	y_{10}^+	891.9749	1087.5782	895.9820	1095.5924	41.70	25

For simplicity, shown are the NAT and SIS transition details for the quantifier ions. The PRM acquisitions target the precursor ions, with alternative collision energies stemming from empirical optimizations on the Q Exactive Plus.

4. Use the following general MRM parameters for all transitions in the final method: unit mass resolution (for first and third quadrupole), 1.5 min RT windows, 380 V fragmentor voltage, 5 V cell accelerator potential, and 10 ms dwell times. The total cycle time for this acquisition method run on this LC-MS platform is 700 ms.

3.7 PRM/MS System Setup

1. Operate the ESI source in the positive ion mode.

2. Use the following general parameters: 350 °C capillary temperature, 50 L/min sheath gas, 20 L/min auxiliary gas at 350 °C, 0 sweep gas, 3 kV spray voltage, and S-lens RF level at 50. Ultrahigh-purity nitrogen serves as the carrier gas in all settings.

3. Additional PRM parameters include 17.5 k resolution, 2e5 automatic gain control (AGC) target, 60 ms maximum injection time (IT), and 30 loop count.

4. Using the PRM mode, enter the specific acquisition parameters. This comprises an inclusion list consisting of peptide precursor m/z values and RTs (inserted as "Start" and "End" times over 1.5 min window; *see* Table 3).

3.8 LC-MS/MS Platform Performance Test

1. Purge the pumps with 50% mobile phase B for 5 min at 10 mL/min.

2. Once the UHPLC system is re-equilibrated to the starting conditions (i.e., 2% eluent B), run a solvent blank (at 0.1% FA) to flush the 20 μL loop and connecting tubing.

3. Using the ESI tuning mix, run the automated tuning program to tune or realign a set of m/z values for positive ESI-MS/MS analysis.

4. Following a successful instrument tune, confirm the peptide RTs by injecting 10 μL of a calibrant standard (*see* **Note 10**, for scheduling, and **Note 11**, for data analysis).

5. Input the updated RTs into the acquisition method and use 1.5 min in all target acquisition windows.

3.9 Sample Injection

1. After the performance tests (for LC RTs and mass analyzer tuning, *see* Subheading 3.7) have been successful, a sample batch can be queued for analysis. Worklists vary, but in our practice, the first sample to be injected (all at 10 μL) is the first replicate of the curve QCs (order: A in well H10, B in C11, then C in F11).

2. Inject two solvent blanks after the three QC samples (replicate 1), and then inject a single replicate of the standard curve samples from low concentration (Standard A in well A12) to high concentration (Standard H in H12).

3. Inject two solvent blanks after the calibration-curve samples; then inject the second replicate of the curve QC samples (in the order: A in A11, B in D11, then C in G11).

4. Inject two solvent blanks after the three QC samples (replicate 2), and then inject the experimental samples ($n = 20$), individually, from well A1 to D3.

5. Inject two solvent blanks after the experimental samples, and then inject the final replicate of the curve QCs (order: A in B11, B in E11, C in H11).

6. After the completion of the sample batch, set the LC to an isocratic flow of 0.02 mL/min with 50% eluent B. Divert the solvent stream to waste to help preserve the MS ion source until the next worklist is initiated.

3.10 Data Analysis   Although the analysis of the MRM or PRM data can be performed with several vendors' software programs, for the purposes here, a guide for executing this in Skyline-daily is outlined (*see* [34] for login and installation instructions).

1. Starting with an empty Skyline document set in the "Proteomics Interface," build the MRM or PRM target list. This can be inserted or copied directly from Excel, provided that the column names and order are the same as in the "Skyline Transition List" window. After pasting, click on "Check for Errors," and then save the populated Skyline Window.

2. In the main screen, import the raw data by clicking on "File," "Import," and then "Results." In the "Import Results" window, select "Add single-injection replicates in files" and enable "Show chromatograms during import." After selecting "OK," locate and highlight the appropriate data files (i.e., .d for Agilent, .raw for Thermo) to be imported.

3. In the document grid, select "Views" and "Replicates." In that tab, input the "Analyte Concentration" values for the curve calibrants and QC samples, as illustrated in Fig. 2.

4. In the main screen, select "View," "Document Grid," and then "Peptide Quantification." Enter the peptide-specific concentration multipliers, as outlined in Table 4 (*see* **Note 12**).

5. In the "Quantification" tab, adjust the peptide settings to allow the quantification of the experimental samples by linear regression using the NAT/SIS peptide ratios with a regression weighting of $1/x^2$ ($x =$ concentration).

6. Manually inspect the extracted ion chromatograms (XICs) and calibration curves for the control samples to ensure correct peak selection and accurate integration (*see* Fig. 3 and **Note 13** for projected observations).

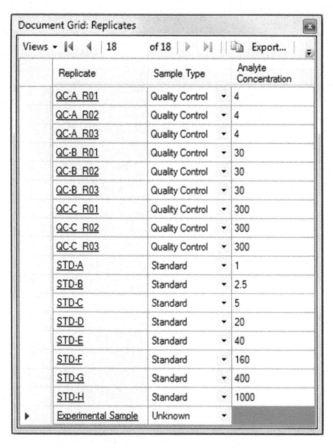

Fig. 2 Peptide concentrations to be entered in Skyline's "Document Grid: Replicates." These fmol/μL values reflect the dilution series of this methods protocol, which comprises an 8-point standard curve (1000-fold range) and 3-point QCs (labeled -A to -C). No concentrations are to be entered for the experimental samples

7. Review the accuracy of the control-based samples, as calculated automatically by Skyline, against the batch acceptance criteria (*see* **Note 14**).

8. Using the control XICs as reference, manually inspect the experimental sample XICs (e.g., for peak symmetry, retention time, interference) and its relation to the curve (*see* Fig. 4).

9. Export the collective Skyline results for further review and analysis. Data extraction pertains to the peaks (e.g., SIS and NAT responses, retention times) and quantified peptides (*see* **Note 15**).

Table 4
Target details for the quantification calculations

Protein	UniProt acc. no.	Peptide	MW (Da)	Concentration multiplier
78 kDa glucose-regulated protein	P11021	ITPSYVAFTPEGER	72,333	15
Adipocyte plasma membrane-associated protein	Q9HDC9	LLEYDTVTR	46,480	4.4
Adiponectin	Q15848	IFYNQQNHYDGSTGK	26,414	3.7
Afamin	P43652	DADPDTFFAK	69,069	32.1
α-1-acid glycoprotein 1	P02763	NWGLSVYADKPETTK	23,512	168.5
α-1-Antichymotrypsin	P01011	EIGELYLPK	47,651	37.5
α-1-antitrypsin	P01009	LSITGTYDLK	46,737	49.4
α-1B-glycoprotein	P04217	LETPDFQLFK	54,254	85.2
α-2-Antiplasmin	P08697	LGNQEPGGQTALK	54,566	7.6
α-2-HS-glycoprotein	P02765	FSVVYAK	39,325	27.6
α-2-macroglobulin	P01023	AIGYLNTGYQR	16,3291	41.3
Antithrombin-III	P01008	DDLYVSDAFHK	52,602	3.3
Apolipoprotein A-I	P02647	ATEHLSTLSEK	30,778	132.3
Apolipoprotein A-II	P02652	EQLTPLIK	11,175	133.2
Apolipoprotein A-IV	P06727	LGEVNTYAGDLQK	45,399	49.7
Apolipoprotein B-100	P04114	FPEVDVLTK	515,605	17.8
Apolipoprotein C-I	P02654	EWFSETFQK	9332	33.3
Apolipoprotein C-II	P02655	TYLPAVDEK	11,284	9.9
Apolipoprotein C-III	P02656	GWVTDGFSSLK	10,852	70.6
Apolipoprotein C-IV	P55056	ELLETVVNR	14,553	9.5
Apolipoprotein D	P05090	VLNQELR	21,276	5.9
Apolipoprotein E	P02649	LGPLVEQGR	36,154	14.9
Apolipoprotein L1	O14791	VAQELEEK	43,974	7.2
Apolipoprotein M	O95445	AFLLTPR	21,253	9.8
Apolipoprotein(a)	P08519	TPAYYPNAGLIK	501,319	4.8
Attractin	O75882	SVNNVVR	158,537	4.2
β-2-glycoprotein 1	P02749	ATVVYQGER	38,298	38.8
β-Ala-his dipeptidase	Q96KN2	ALEQDLPVNIK	56,706	15.4
Biotinidase	P43251	SHLIIAQVAK	61,133	2.9

(continued)

Table 4
(continued)

Protein	UniProt acc. no.	Peptide	MW (Da)	Concentration multiplier
Cadherin-13	P55290	INENTGSVSVTR	78,287	60.5
Carbonic anhydrase 1	P00915	VLDALQAIK	28,870	6.1
Carboxypeptidase B2	Q96IY4	IAWHVIR	48,424	37.9
Cathelicidin antimicrobial peptide	P49913	AIDGINQR	19,301	15.7
Cation-independent mannose-6-phosphate receptor	P11717	GHQAFDVGQPR	274,375	1.9
CD5 antigen-like	O43866	LVGGLHR	38,088	89.8
Ceruloplasmin	P00450	IYHSHIDAPK	122,205	51.4
Cholinesterase	P06276	YLTLNTESTR	68,418	5.3
Clusterin	P10909	ELDESLQVAER	52,495	4.1
Coagulation factor IX	P00740	SALVLQYLR	51,778	70.6
Coagulation factor V	P12259	SEAYNTFSER	251,703	52.3
Coagulation factor VIII	P00451	LHPTHYSIR	267,009	97
Coagulation factor X	P00742	TGIVSGFGR	54,732	6.4
Coagulation factor XII	P00748	EQPPSLTR	67,792	7.8
Complement C1q subcomponent subunit B	P02746	IAFSATR	26,722	11.7
Complement C1q subcomponent subunit C	P02747	FNAVLTNPQGDYD TSTGK	25,774	42.8
Complement C1r subcomponent	P00736	GLTLHLK	80,119	3.7
Complement C1r subcomponent-like protein	Q9NZP8	VVVHPDYR	53,498	3.9
Complement C1s subcomponent	P09871	TNFDNDIALVR	76,684	51.2
Complement C2	P06681	HAFILQDTK	83,268	5.3
Complement C3	P01024	TGLQEVEVK	187,148	161.5
Complement C5	P01031	VFQFLEK	188,305	12.1
Complement component C7	P10643	LIDQYGTHYLQSG SLGGEYR	93,518	293
Complement component C9	P02748	LSPIYNLVPVK	63,173	33.6
Complement factor B	P00751	EELLPAQDIK	85,533	13.8
Complement factor H	P08603	SSQESYAHGTK	139,096	19.1
Complement factor I	P05156	VFSLQWGEVK	65,750	75.8

(continued)

Table 4
(continued)

Protein	UniProt acc. no.	Peptide	MW (Da)	Concentration multiplier
Corticosteroid-binding globulin	P08185	WSAGLTSSQVDL YIPK	45,141	2570.3
Cystatin-C	P01034	ALDFAVGEYNK	15,799	13.8
Endothelial protein C receptor	Q9UNN8	TLAFPLTIR	26,671	12.6
Fetuin-B	Q9UGM5	LVVLPFPK	42,055	81.9
Fibrinogen α chain	P02671	ESSSHHPGIAEFPSR	94,973	45.9
Fibrinogen β chain	P02675	HQLYIDETVNSNIP TNLR	55,928	35.4
Fibrinogen γ chain	P02679	YEASILTHDSSIR	51,512	256.4
Fibronectin	P02751	HTSVQTTSSGSGPF TDVR	262,625	35.1
Fibulin-1	P23142	TGYYFDGISR	77,214	10.1
Galectin-3-binding protein	Q08380	SDLAVPSELALLK	65,331	62.3
Gelsolin	P06396	AGALNSNDAFVLK	85,698	52.7
Glutathione peroxidase 3	P22352	QEPGENSEILPTLK	25,552	6.7
Haptoglobin	P00738	DIAPTLTLYVGK	45,205	515.8
Hemoglobin subunit α	P69905	VGAHAGEYGAE ALER	15,258	11.2
Hemopexin	P02790	NFPSPVDAAFR	51,676	73.2
Heparin cofactor 2	P05546	TLEAQLTPR	57,071	14
Hepatocyte growth factor-like protein	P26927	SPLNDFQVLR	80,320	51.1
Hyaluronan-binding protein 2	Q14520	VVLGDQDLK	62,672	5.7
Ig mu chain C region	P01871	GFPSVLR	49,440	39.6
Insulin-like growth factor I	P05019	GFYFNKPTGYGSSSR	21,841	36.3
Inter-α-trypsin inhibitor heavy chain H2	P19823	SLAPTAAAK	106,463	37.3
Intercellular adhesion molecule 1	P05362	LLGIETPLPK	57,825	10.5
Interleukin-10	P22301	AHVNSLGENLK	20,517	4.8
Kallistatin	P29622	VGSALFLSHNLK	48,542	77.9
Keratin-type I cytoskeletal 10	P13645	SLLEGEGSSGGGGR	58,827	887.6
Keratin-type II cytoskeletal 2 epidermal	P35908	YEELQVTVGR	65,433	326.7

(continued)

Table 4
(continued)

Protein	UniProt acc. no.	Peptide	MW (Da)	Concentration multiplier
Kininogen-1	P01042	TVGSDTFYSFK	71,957	69.9
Leucine-rich α-2-glycoprotein	P02750	DLLLPQPDLR	38,178	5.5
Lipopolysaccharide-binding protein	P18428	ITLPDFTGDLR	53,384	42.8
L-selectin	P14151	AEIEYLEK	42,187	9.3
Lysozyme C	P61626	AWVAWR	16,537	5.3
Mannan-binding lectin serine protease 1	P48740	TGVITSPDFPNPYPK	79,247	70.5
Mannan-binding lectin serine protease 2	O00187	WPEPVFGR	75,702	64.7
Metalloproteinase inhibitor 2	P16035	EYLIAGK	24,399	3.8
Myeloblastin	P24158	LVNVVLGAHNVR	27,807	97.1
Peroxiredoxin-2	P32119	GLFIIDGK	21,892	17
Phosphatidylinositol-glycan-specific phospholipase D	P80108	FGSSLITVR	92,336	4.9
Phospholipid transfer protein	P55058	AVEPQLQEEER	54,739	5.6
Pigment epithelium-derived factor	P36955	LQSLFDSPDFSK	46,312	42.4
Plasma protease C1 inhibitor	P05155	FQPTLLTLPR	55,154	26.1
Plasma serine protease inhibitor	P05154	AVVEVDESGTR	45,675	2.9
Plasminogen	P00747	LFLEPTR	90,569	9.5
Plasminogen activator inhibitor 1	P05121	VFQQVAQASK	45,060	15.3
Pregnancy zone protein	P20742	ISEITNIVSK	163,863	4.1
Protein AMBP	P02760	HHGPTITAK	38,999	175.1
Protein S100-A9	P06702	DLQNFLK	13,242	10.6
Protein Z-dependent protease inhibitor	Q9UK55	ETSNFGFSLLR	50,707	56.5
Prothrombin	P00734	ELLESYIDGR	70,037	33.3
Retinol-binding protein 4	P02753	YWGVASFLQK	23,010	70.4
Serotransferrin	P02787	DGAGDVAFVK	77,064	352.5
Serum albumin	P02768	LVNEVTEFAK	69,367	1396.9
Serum paraoxonase/arylesterase 1	P27169	IFFYDSENPPASEVLR	39,731	46.7
Serum paraoxonase/lactonase 3	Q15166	IQNVLSEKPR	39,607	4.8

(continued)

Table 4
(continued)

Protein	UniProt acc. no.	Peptide	MW (Da)	Concentration multiplier
SPARC	P09486	LEAGDHPVELLAR	34,632	10
Tenascin	P24821	FTTDLDSPR	240,853	5.7
Tenascin-X	P22105	ILISGLEPSTPYR	458,220	10.5
Thrombospondin-1	P07996	GTLLALER	129,383	4.2
Thrombospondin-4	P35443	KPQDFLEELK	105,869	42
Thyroxine-binding globulin	P05543	AVLHIGEK	46,325	6.1
Tissue factor pathway inhibitor	P10646	FYYNSVIGK	35,015	7.8
Tissue-type plasminogen activator	P00750	VVPGEEEQK	62,917	90.4
Transferrin receptor protein 1	P02786	GFVEPDHYVVVGA QR	84,871	25.5
Transthyretin	P02766	GSPAINVAVHVFR	15,887	24.6
Vascular cell adhesion protein 1	P19320	NTVISVNPSTK	81,276	4.4
Vasorin	Q6EMK4	ESHVTLASPEETR	71,713	27.6
Vitamin K-dependent protein S	P07225	SFQTGLFTAAR	75,123	26.7
Vitamin K-dependent protein Z	P22891	GLLSGWAR	44,744	38.7
Vitronectin	P04004	FEDGVLDPDYPR	54,306	75.2
Zinc-α-2-glycoprotein	P25311	EIPAWVPFDPAAQI TK	34,259	30.8

The peptide-specific concentration multipliers are to be inserted in the Skyline method builder, while the resulting fmol/μL concentrations can be converted to ng/mL through aid of the protein molecular weights (MWs)

4 Notes

1. The stability of the prepared reagents (i.e., Reagents A to E, BSA/PBS, and trypsin) is predicated based on their time of preparation and storage conditions. Reagent A (stock Tris solution) can be stored at 4 °C for 1 month. The remaining reagents noted above are to be prepared fresh, on day of use. Reagent C (comprising urea and DTT in Tris), Reagent D (IAA), and trypsin are to be prepared immediately prior to use and stored on ice until dispensed.

2. Do not heat the urea solution in an effort to hasten its solubilization. Elevated temperature (i.e., >37 °C) accelerates urea decomposition, with its product (isocyanic acid) resulting in

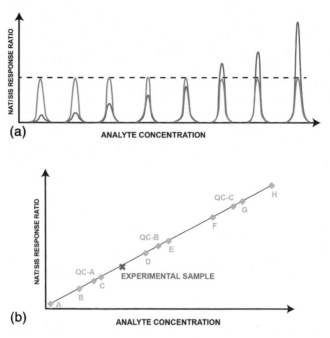

Fig. 3 Theoretical plot of control and experimental sample analysis by bottom-up LC-MS/MS. (**a**) XICs of expected NAT/SIS peptide ratios (NAT in red, SIS in blue) in a buffer-generated, standard curve. (**b**) Placement of 8-point calibrant levels (in green diamonds) and 3-point QCs (in blue diamonds) relative to a plasma sample measurement. Quantitation of this sample is then accomplished through linear regression analysis of the peptide relative response (i.e., NAT/SIS peak area ratio)

carbamylation at ε-amine lysine residues and protein N-termini [35]. To reduce this artifact, dissolve the urea at room temperature with the aid of vortexing only.

3. The NAT and SIS peptides were synthesized (at the University of Victoria-Genome BC Proteomics Centre; Victoria, BC, Canada) on a robotic peptide synthesizer using Fmoc chemistry. The labeled protected amino acids ($^{13}C_6/^{15}N_2$ L-lysine, CNLM-4754-H; $^{13}C_6/^{15}N_4$ L-arginine, CNLM-8474-H), used in the C-terminal labeling of tryptic peptide residues, were obtained from Cambridge Isotope Laboratories, Inc. (Andover, MA, USA). After synthesis, the standard peptides were purified (by HPLC) and characterized (for composition and concentration) and then formulated into mixtures, in a manner similar to that described previously [36]. Lyophilized aliquots of the mixtures are stored at −80 °C, with established stabilities of at least 6 months. Overall, the peptides selected are proteotypic and abide by a series of sequence-specific selection rules (e.g., unique within plasma proteome, devoid of cysteine and methionine, 6–20 residues in length), which makes them well suited to serve as external or internal standards.

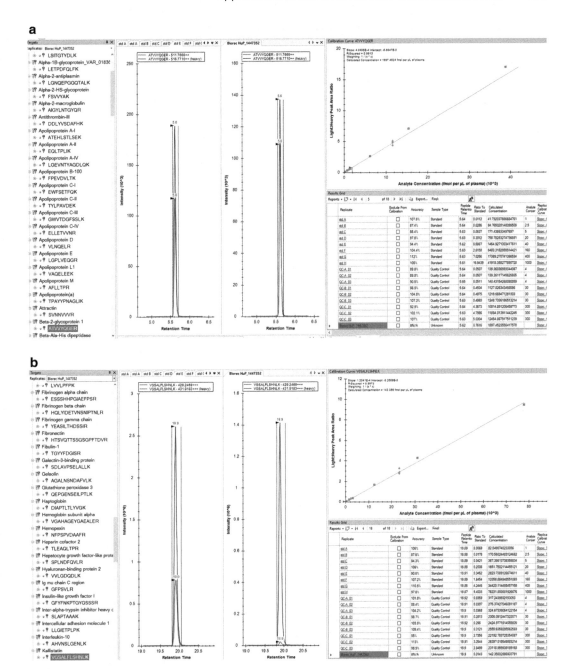

Fig. 4 Bottom-up LC-MRM/MS data for two representative, moderate abundance targets from the 125 protein panel. Illustrated are two Skyline screenshots for peptides ATVVYQGER (from β-2-glycoprotein 1) in (**a**) and VGSALFLSHNLK (from kallistatin) in (**b**). The tabs refer to the monitored panel, two example XICs (for curve calibrant E and an experimental sample), the standard curve (with calibrant levels marked with gray squares, QC samples with green diamonds, experimental sample with red arrow), and the results grid. The concentrations in the human plasma sample (Bioreclamation, lot BRH1447352) shown were calculated to be 72.7 μg/mL for β-2-glycoprotein 1 and 6.9 μg/mL for kallistatin

4. If an alternative SIS peptide labeling scheme is desired, as would be the case with double isotopologue peptide standards (i.e., isotope labeling at a C-terminal amino acid as well as an internal residue), the protocol described herein would need to be amended. The methods article by Eshghi et al. serves as a useful reference guide for the quantitative application of the double SIS peptide approach to different sample types [37]. Regardless of the type of peptide standard, these should be prepared, stored, and handled according to established guidelines for best practices in MS-based proteomics [38].

5. The ideal quantity of trypsin added was previously determined in empirical measurements. The 20:1 (w/w) substrate/enzyme ratio implemented enables efficient tryptic digestion in pooled human plasma samples. Appropriate substrate/enzyme ratios for alternative sample types will need to be determined before application to real experimental samples.

6. The current protocol is amenable to be scaled up to 30 samples. For larger-scale analysis, the procedure will need to be adapted. This affects the quantity of surrogate digest initially prepared (in Subheading 3.1) and the quantity of SIS peptide mixture to be reconstituted (in Subheading 2.3).

7. For all extraction steps, add solvent or sample before applying vacuum. The flow through and washes (extraction steps in Subheading 3.4) are to be collected in an empty basin and can be discarded. In the sample loading step, start at the lowest vacuum setting and then gradually increase the vacuum to load the full sample onto the sorbent bed. Flow rates for sample loading and elution should not exceed 1 mL/min. Also, for improved throughput and precision, all SPE steps should be performed with a multichannel pipette or with a liquid handling robot (e.g., Tecan Freedom EVO 150).

8. Ensure that the sample container of the LC autosampler is configured to read a 96-well plate in one tray (for the digests) and vials in a second tray (with one designated for the solvent blank, e.g., at position A1). This latter tray will need to be loaded first with a vial plate to accommodate the autosampler vial(s).

9. The specific acquisition parameters (i.e., transitions, collision energies, and retention times) will need to be optimized if alternative mass spectrometers are used in place of the 6495 or Q Exactive instrument. It is recommended that these be selected based on empirical measurements using the SIS peptide mixtures as MS infusion solutions.

10. Although we have observed minimal RT shifts in method transfer between laboratories when identical LC-MS conditions/parameters are employed, it is recommended that a

broader detection window of 3 min be initially used for peptide RT verification. This test should require one run only, with standard E being an example standard for injection. After the peptide rescheduling, confirm that the cycle/dwell times enable 10–15 data points across the chromatographic peaks for optimum and accurate peak shapes. It must be noted that although the RT verification is conducted in buffer, in our experience, minimal shifts (maximally 0.5 min) are typically observed when the assay is transferred to human plasma.

11. Analysis of the peptide scheduling data can be accomplished with vendor software, Skyline [39], or other (e.g., Qualis-SIS [40]). If Skyline is used, as advocated here, refer to Skyline's tutorial section for guidance on method refinement for efficient acquisitions. After the retention times have been confirmed for all targets, export the data file to Excel and then update the acquisition method accordingly.

12. The analyte concentration refers to the initial peptide concentration in the standard mixture and is used as a basis for the quantitation calculations. The concentration multipliers are used to automatically convert the calculated values to protein concentrations (expressed in fmol/μL of human plasma).

13. The NAT (synthetic and endogenous) and SIS peptides have identical physicochemical properties and therefore exhibit the same behavior during extraction, separation, and mass analysis (e.g., ionization, fragmentation). The resulting XICs for each peptide pair should therefore co-elute in all sample analyses, with the responses of the NAT standards differing from SIS in the curves and QC samples only (see Fig. 3a). Regarding the ratios, the calibrant levels should approximate a 1:2.5:2.5:4:2:4:2:2.5 dilution from standard H to A (e.g., standard E is 25-fold diluted from standard H), with the QC samples being x-fold from standard H (i.e., 3.33-fold, in QC-C; 33.3-fold, in QC-B; 250-fold, in QC-A; see Fig. 3b). The relative responses and ratios should be confirmed for all sample types.

14. The accuracies of the standard curve and curve QC samples are reported in Skyline as a percent theoretical value. For each peptide, each measured NAT standard in the curve should be within 20% of the theoretical concentration, and at least 6 of the 9 measured NAT concentrations in the QC samples should be within 20% of the theoretical concentrations. In addition, for each peptide, at least 1 of 3 of the QC samples (QC-A, QC-B, or QC-C) should be within 20% of its theoretical concentration. Finally, for each peptide quantified, at least 6 of the 8 measured in the standard curve should be within 20% of the expected concentration.

15. To convert the peptide concentrations (in fmol/μL) to protein concentrations (in ng/mL), multiply the calculated concentrations from Skyline by the molecular weight of a given protein (*see* Table 4) and then divide by 1000. For example, the concentration of ATVVYQGER in a de-identified human plasma sample (from Bioreclamation, lot BRH1447352) was determined to be 1897.5 fmol/μL, which equates to 72.6 μg/mL for β-2-glycoprotein 1 (also referred to as apolipoprotein H or Apo-H).

Acknowledgments

Dr. Borchers is grateful for support from Genome Canada to the Segal Cancer Proteomics Centre through the Genomics Technology Platform (264PRO). Dr. Borchers also appreciates the support from the Segal McGill Chair in Molecular Oncology (McGill University). Dr. Borchers is also grateful for support from the Terry Fox Research Institute, the Warren Y. Soper Charitable Trust, and the Alvin Segal Family Foundation to the Jewish General Hospital (Montreal, QC, Canada).

Competing Interests: The target panel and protocols described in this article are similar to those used in the PeptiQuant™ Plus biomarker assessment kit for protein quantification in human plasma. These kits are commercially available through MRM Proteomics Inc. (Montreal, QC, Canada) where Dr. Borchers is the Chief Scientific Officer, and its partner, Cambridge Isotope Laboratories, Inc., where Dr. Percy is the Senior Applications Chemist for Mass Spectrometry.

References

1. Nie S, Shi T, Fillmore TL et al (2017) Deep-dive targeted quantification for ultrasensitive analysis of proteins in nondepleted human blood plasma/serum and tissues. Anal Chem 89(17):9139–9146

2. Fu Q, Kowalski MP, Mastali M et al (2018) Highly reproducible automated proteomics sample preparation workflow for quantitative mass spectrometry. J Proteome Res 14(2):420–428

3. Chen Y, Vu J, Thompson MG et al (2019) A rapid methods development workflow for high-throughput quantitative proteomic applications. PLoS One 14(2):e0211582

4. Collins BC, Hunter CL, Liu Y et al (2017) Multi-laboratory assessment of reproducibility, qualitative and quantitative performance of SWATH-mass spectrometry. Nat Commun 8(1):291

5. Percy AJ, Tamura-Wells J, Albar JP et al (2015) Inter-laboratory evaluation of instrument platforms and experimental workflows for quantitative accuracy and reproducibility assessment. EuPA Open Proteom 8:6–15

6. Duarte TT, Spencer CT (2016) Personalized proteomics: the future of precision medicine. Proteomes 4(4):29

7. Anand S, Samuel M, Ang CS et al (2017) Label-based and label-free strategies for protein quantitation. In: Methods in molecular biology, vol 1549. Humana Press, New York, pp 31–43

8. Souza GH, Guest PC, Martins-de-Souza D (2017) LC-MSE, multiplex MS/MS, ion mobility, and label-free quantitation in clinical

proteomics. Methods Mol Biol 1546:57–73. Humana Press, New York

9. Percy AJ, Byrns S, Pennington SR et al (2016) Clinical translation of MS-based, quantitative plasma proteomics: status, challenges, requirements, and potential. Expert Rev Proteomics 13(7):673–684

10. Picard G, Lebert D, Louwagie M et al (2012) PSAQ™ standards for accurate MS-based quantification of proteins: from the concept to biomedical applications. J Mass Spectrom 47(10):1353–1363

11. Gilquin B, Louwagie M, Jaquinod M et al (2017) Multiplex and accurate quantification of acute kidney injury biomarker candidates in urine using protein standard absolute quantification (PSAQ) and targeted proteomics. Talanta 164:77–84

12. Bros P, Vialaret J, Barthelemy N et al (2015) Antibody-free quantification of seven tau peptides in human CSF using targeted mass spectrometry. Front Neurosci 9:302

13. Wang Q, Zhang M, Tomita T et al (2017) Selected reaction monitoring approach for validating peptide biomarkers. Proc Natl Acad Sci U S A 114(51):13519–13524

14. Percy AJ, Chambers AG, Yang J et al (2014) Advances in multiplexed MRM-based protein biomarker quantitation toward clinical utility. Biochim Biophys Acta 1844(5):917–926

15. Zhang J, Hong Y, Cai Z et al (2019) Simultaneous determination of major peanut allergens Ara h1 and Ara h2 in baked foodstuffs based on their signature peptides using ultra-performance liquid chromatography coupled to tandem mass spectrometry. Anal Methods 11(12):1689–1696

16. Zhang J, Lai S, Cai Z et al (2014) Determination of bovine lactoferrin in dairy products by ultra-high performance liquid chromatography-tandem mass spectrometry based on tryptic signature peptides employing an isotope-labeled winged peptide as internal standard. Anal Chim Acta 829:33–39

17. Scott KB, Turko IV, Phinney KW (2016) QconCAT: internal standard for protein quantification. Methods Enzymol 566:289–303

18. Cheung CS, Anderson KW, Wang M et al (2015) Natural flanking sequences for peptides included in a quantification concatamer internal standard. Anal Chem 87(2):1097–1102

19. Edfors F, Forsström B, Vunk H et al (2019) Screening a resource of recombinant protein fragments for targeted proteomics. J Proteome Res 18(7):2706–2718

20. Hober A, Edfors F, Ryaboshapkina M et al (2019) Absolute quantification of apolipoproteins following treatment with omega-3 carboxylic acids and fenofibrate using a high precision stable isotope-labeled recombinant protein fragments based SRM assay. Mol Cell Proteomics 18:2433–2446

21. Zeiler M, Straube WL, Lundberg E et al (2012) A Protein Epitope Signature Tag (PrEST) library allows SILAC-based absolute quantification and multiplexed determination of protein copy numbers in cell lines. Mol Cell Proteomics 11(3):O111.009613

22. Oeckl P, Steinacker P, Otto M (2018) Comparison of internal standard approaches for SRM analysis of alpha-synuclein in cerebrospinal fluid. J Proteome Res 17(1):516–523

23. Bronsema KJ, Bischoff R, van de Merbel NC (2012) Internal standards in the quantitative determination of protein biopharmaceuticals using liquid chromatography coupled to mass spectrometry. J Chromatogr B 145:893–894

24. Percy AJ, Michaud SA, Jardim A et al (2017) Multiplexed MRM-based assays for the quantitation of proteins in mouse plasma and heart tissue. Proteomics 17(7). https://doi.org/10.1002/pmic.201600097

25. Thomas SN, Harlan R, Chen J et al (2015) Multiplexed targeted mass spectrometry-based assays for the quantification of N-linked glycosite-containing peptides in serum. Anal Chem 87(21):10830–10838

26. Smit NP, Romijn FP, van den Broek I et al (2014) Metrological traceability in mass spectrometry-based targeted protein quantitation: a proof-of-principle study for serum apolipoproteins A-I and B100. J Prot 109:143–161

27. Razavi M, Johnson LD, Lum JJ et al (2013) Quantification of a proteotypic peptide from protein C inhibitor by liquid chromatography-free SISCAPA-MALDI mass spectrometry: application to identification of recurrence of prostate cancer. Clin Chem 59(10):1514–1522

28. Mohammed Y, Pan J, Zhang S et al (2018) ExSTA: external standard addition method for accurate high-throughput quantitation in targeted proteomics experiments. Proteomics Clin Appl 12(2):1600180

29. Chiva C, Sabidó E (2017) Peptide selection for targeted protein quantitation. J Proteome Res 16(3):1376–1380

30. Mohammed Y, Domański D, Jackson AM et al (2014) PeptidePicker: a scientific workflow with web interface for selecting appropriate

peptides for targeted proteomics experiments. J Prot 106:151–161

31. Whiteaker JR, Halusa GN, Hoofnagle AN et al (2016) Using the CPTAC assay portal to identify and implement highly characterized targeted proteomics assays. Methods Mol Biol 1410:223–236. Humana Press, New York

32. Whiteaker JR, Halusa GN, Hoofnagle AN et al (2014) CPTAC assay portal: a repository of targeted proteomic assays. Nat Methods 11 (7):703–704

33. Carr SA, Abbatiello SE, Ackermann BL et al (2014) Targeted peptide measurements in biology and medicine: best practices for mass spectrometry-based assay development using a fit-for-purpose approach. Mol Cell Proteomics 13(3):907–917

34. MacCoss_laboratory Skyline-daily. https://proteome.gs.washington.edu/software/test/brendanx/Skyline-test/. Accessed Nov 2019

35. Kollipara L, Zahedi RP (2013) Protein carbamylation: in vivo modification or in vitro artefact? Proteomics 13(6):941–944

36. Michaud SA, Sinclair NJ, Pětrošová H et al (2018) Molecular phenotyping of laboratory mouse strains using 500 multiple reaction monitoring mass spectrometry plasma assays. Commun Biol 1:78

37. Eshghi A, Borchers CH (2018) Multiple reaction monitoring using double isotopologue peptide standards for protein quantification. Methods Mol Biol 1788:193–214. Humana, New York

38. Hoofnagle AN, Whiteaker JR, Carr SA et al (2016) Recommendations for the generation, quantification, storage, and handling of peptides used for mass spectrometry-based assays. Clin Chem 62(1):48–69

39. MacLean B, Tomazela DM, Shulman N et al (2010) Skyline: an open source document editor for creating and analyzing targeted proteomics experiments. Bioinformatics 26 (7):966–968

40. Mohammed Y, Percy AJ, Chambers AG et al (2015) Qualis-SIS: automated standard curve generation and quality assessment for multiplexed targeted quantitative proteomic experiments with labeled standards. J Proteome Res 14(2):1137–1146

Chapter 26

Quantitative Cross-Linking of Proteins and Protein Complexes

Marie Barth and Carla Schmidt

Abstract

Cross-linking, in general, involves the covalent linkage of two amino acid residues of proteins or protein complexes in close proximity. Mass spectrometry and computational analysis are then applied to identify the formed linkage and deduce structural information such as distance restraints. Quantitative cross-linking coupled with mass spectrometry is well suited to study protein dynamics and conformations of protein complexes. The quantitative cross-linking workflow described here is based on the application of isotope labelled cross-linkers. Proteins or protein complexes present in different structural states are differentially cross-linked using a "light" and a "heavy" cross-linker. The intensity ratios of cross-links (i.e., light/heavy or heavy/light) indicate structural changes or interactions that are maintained in the different states. These structural insights lead to a better understanding of the function of the proteins or protein complexes investigated. The described workflow is applicable to a wide range of research questions including, for instance, protein dynamics or structural changes upon ligand binding.

Key words Mass spectrometry, Protein structure, Protein interactions, Cross-linking, Quantification, BS3, Ligand binding

1 Introduction

Proteins and protein complexes are key players in the cell. Their function is often modulated by structural changes upon loss or formation of interactions with their ligands, such as proteins, sugars, nucleotides, ions, or lipids. Assessing these structural changes is therefore important for fully understanding their function. Quantitative cross-linking coupled with mass spectrometric analysis allows analyzing these structural changes in solution and is, therefore, well suited to study dynamic proteins and protein complexes in different structural states [1]. In this chapter we describe the workflow of quantitative cross-linking, including the cross-linking reaction, sample preparation, mass spectrometric analysis, as well as data analysis.

Katrin Marcus et al. (eds.), *Quantitative Methods in Proteomics*, Methods in Molecular Biology, vol. 2228, https://doi.org/10.1007/978-1-0716-1024-4_26, © The Author(s) 2021

1.1 Cross-Linking   Cross-linking of proteins or protein complexes, in general, is the covalent linkage of two functional groups of amino acid side chains resulting in inter- or intramolecular linkage. Partial hydrolysis of the cross-linker can further lead to modification of the reactive amino acids. Two strategies are commonly followed: photo-induced or chemical cross-linking. Chemical cross-linking reagents usually contain two identical (homobifunctional) or different (heterobifunctional) reactive groups as well as a linker [2]. Two amino acids are cross-linked when the linker length correlates with the distance between these two amino acids. Zero-length cross-linkers, consequently, do not introduce a linker and only link reactive groups in close proximity [3, 4]. Chemical cross-linkers are, for instance, reactive toward primary amines [5], arginine residues [6], sulfhydryl groups [7–9], or in the case of bifunctional cross-linkers, carboxyl, and amine groups [10, 11]. In addition, there are photo-reactive cross-linkers such as the UV-inducible amino acids photo-methionine or photo-leucine [12]. In some cases, cross-linking reagents contain a third functional group which is used for affinity enrichment (e.g., through a biotin moiety) [13]. The variety of chemical specificity and linker length therefore provides a selection of cross-linking approaches which can be applied to answer diverse research questions [14].

Of the available reagents, N-hydroxysuccinimide (NHS) esters are most commonly used [15]. One prominent, water soluble example is bis(sulfosuccinimidyl)suberate (BS3), which targets primary amines of lysine residues or the proteins' N-termini. Side reactions with hydroxyl groups of serine, threonine, and tyrosine residues also occur when using BS3 or analogue cross-linkers [16]. Due to their polarity, lysine residues are typically located at the solvent-accessible surface of proteins and therefore represent a suitable amino acid target making NHS-esters popular cross-linking reagents.

1.2 Sample Preparation and LC-MS/MS Analysis   Following the cross-linking reaction, proteins or protein complexes are hydrolyzed using a specific endoproteinase such as trypsin. Enzymatic hydrolysis yields linear and cross-linked peptides. The latter include intra-peptide cross-links (so-called loop links) and inter-peptide cross-links [17]. Of these, the inter-peptide cross-links provide information on distance constraints between the cross-linked amino acids of different proteins or within the same protein.

Cross-linked peptides are usually low abundant when compared with their linear counterparts. Therefore, enrichment strategies including affinity chromatography, size exclusion chromatography, or ion-exchange chromatography are employed. Cross-linking reagents containing biotin labels or click-based affinity tags have been developed for affinity chromatography [18–21] while size exclusion chromatography [22] or cation exchange chromatography [23] make use of the increased size or higher charge of cross-linked di-peptides.

Obtained cross-linked peptides are then analyzed by liquid chromatography-coupled tandem mass spectrometry (LC-MS/MS) following standard proteomic procedures. To increase the identification rate of cross-linked peptides, doubly charged precursors, which mostly correspond to linear peptides or intra-cross-linked peptides (*see* above), are often excluded from selection for MS/MS while highly charged cross-linked di-peptides are preferably analyzed [24].

1.3   Data Analysis

Identification of cross-linked peptides is challenging for the following reasons: the masses of the two cross-linked peptides and the linker mass of the cross-linker add up to the precursor mass. Depending on the sample complexity and the required database, a large number of cross-linked peptide combinations are possible, and the analysis can be computationally challenging [25]. In addition, fragmentation of a precursor ideally leads to series of fragment ions of both peptides. However, during collision-induced dissociation or high-energy collision dissociation of cross-linked peptides, the fragmentation of one peptide is typically favored [26]. Moreover, tryptic peptides often yield more y-ions than b-ions [27]. This suboptimal product ion distribution further affects the correct and confident assignment of cross-linked peptides. Therefore, specialized software has been developed; Examples are: xQuest [28], XlinkX [29], Kojak [30], StavroX [31], MeroX [32], XiSearch [33], and pLink [34]. Although the various software tools are constantly developed and improved, the identification of cross-linked peptides often relies on manual validation [15]. Particularly, the analysis of complex samples can therefore be laborious.

Having identified a set of cross-links, these are further analyzed and often visualized in network plots, cartoon representations, or, if available, in high-resolution structures. Available visualization tools are, for instance, xiView/xiNet [35, 36], XLink-DB [37], xVis [38], ProXL [39], or CLMSVault [40]. If crystal structures are available, distance restraints of cross-linked amino acid residues can be extracted using software tools such as Xwalk [41] or Xlink Analyer [42].

1.4   Quantitative Cross-Linking

Applied in the described way, cross-linking contributed considerably to the structural and functional understanding of molecular machines [43]. Recently, the field of cross-linking was expanded, and methods for quantitative cross-links were developed. Note that label-free quantification, which is often employed in proteomic studies, is not fully applicable to cross-linking analyses, mostly due to the difficulties in reproducibility of low abundant peptides such as cross-linked peptides (*see* above). Therefore, targeted approaches such as selected reaction monitoring or parallel reaction monitoring have been explored [44, 45]. Very recently, a quantitative cross-linking workflow including data-independent acquisition was developed [46, 47] enabling large scale quantitative analysis with high reproducibility [46].

On the other hand, a variety of label-based quantification methods is more commonly used. These are based on metabolic labelling (e.g., SILAC; stable isotope labelling by amino acids in cell culture) [48, 49], isobaric labelling using tandem mass tags (TMT [50] or iTRAQ [51]), or isotope labelled cross-linkers. Isotope labelled cross-linking reagents were first introduced to facilitate the identification of cross-links in complex samples [52]. They differ in the number of incorporated heavy stable isotopes leading to a characteristic mass shift in the mass spectrum and resulting in peak pairs of the cross-linked peptides. Nonetheless, differentially labelled cross-linkers (light and heavy) have the same physicochemical properties such as elution time, ionization, and fragmentation. Cross-linking different states of the protein with light and heavy cross-linkers, respectively, therefore allows the quantification of protein interactions in the different states through intensities of the peak pairs [53]. More precisely, by extracting ion chromatograms and comparing the intensity ratios of the two states (light versus heavy), protein interactions that vary or are maintained in the two states are observed. Differences in peak intensities represent structural changes, while the same intensities correspond to interactions that are not affected.

Initially, several strategies have been followed to analyze quantitative cross-linking data, including manual [53] as well as automatic data analysis [54]. To facilitate data analysis, specialized software has been developed and improved for quantitative cross-linking experiments, for instance, xTract [55], XiQ [54], or Spectronaut (Biognosys) [56, 46].

1.5 Quantitative Cross-Linking Workflow and Example Data

Usually, prior to quantitative cross-linking, optimal experimental conditions are determined by varying the cross-linker concentration or modifying the reactions conditions. For this, the protein or protein complex of interest in one state, for instance, in the apo-form without bound ligand, is used. Applying these conditions, the different states of the protein (or protein complex) are then cross-linked with differentially labelled cross-linkers (e.g., BS3-d0 and BS3-d4, respectively). After cross-linking, the two samples are pooled in a 1:1 ratio, the samples are prepared for LC-MS/MS analysis, and cross-link identification is processed. Finally, protein interactions of the different protein states are quantified through their abundances in the mass spectra. The complete workflow is shown in Fig. 1.

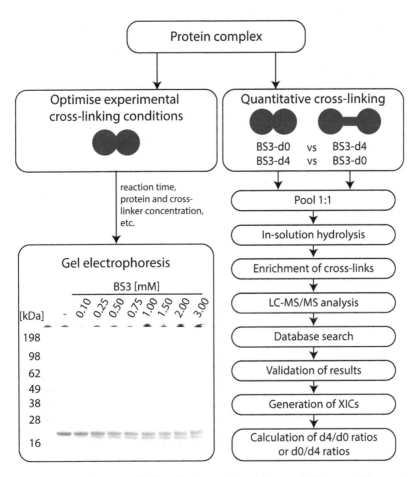

Fig. 1 *Workflow of quantitative cross-linking.* First, experimental cross-linking conditions such as reaction time, protein or cross-linker concentration, etc. are optimized and followed, for instance, by gel electrophoresis (lhs). The protein or protein complex is then cross-linked in different states using light (e.g., BS4-d0) and heavy (e.g., BS3-d4) cross-linkers, respectively (rhs). After cross-linking, the samples are pooled and processed for LC-MS/MS analysis and identification of cross-linked peptides. As a control, the differentially labelled cross-linking reagents are swapped for the different states

Figures 2 and 3 show comparative cross-linking results of Calmodulin (CaM) in the Ca^{2+}-free and Ca^{2+}-bound states which both regulate the function of many proteins [57, 58]. The Ca^{2+}-free form of CaM differs structurally from the Ca^{2+}-bound form (compare high-resolution structures shown in Figs. 2 and 3). Due to structural changes upon Ca^{2+}-binding, lysine residues 76 and 95 reach close proximity (Fig. 2a). Probing the two states, i.e., Ca^{2+}-free and Ca^{2+}-bound, with a light and heavy cross-linker, different intensities in the mass spectrum of the cross-link between these two residues are observed (Fig. 2b). Peak intensities corresponding to the apo-CaM form are lower than those observed for the Ca^{2+}-bound form. In other words, the structural changes occurring due to Ca^{2+}-binding result in a closer proximity of lysine

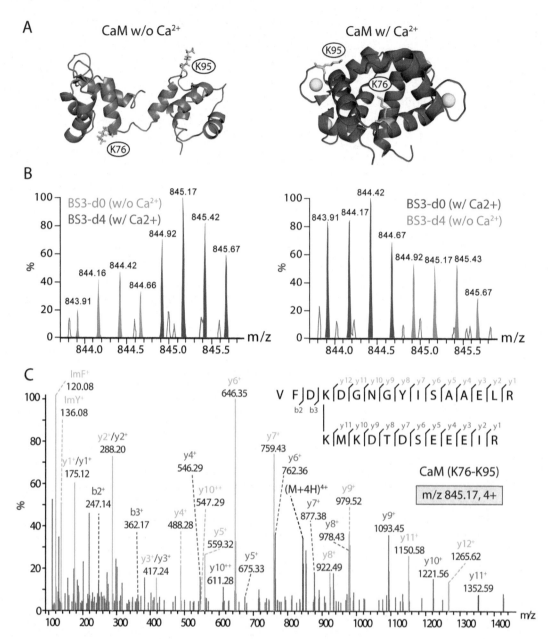

Fig. 2 *Quantitative cross-linking of CaM in two states.* (**a**) Crystal structures of apo-CaM (blue, lhs, PDB: 1DMO, state 1) and CaM with Calcium (red, rhs, PDB: 1PRW, state 2) are shown. Residues 76 and 95 are highlighted (green). (**b**) Mass spectra of the cross-link CaM (K76-K95) obtained from CaM cross-linked in state 1 using BS3-d0 and in state 2 using BS3-d4 (lhs). The light and heavy cross-linkers were also swapped (rhs). The isotope envelope corresponding to the cross-link of CaM in state 1 (blue) and in state 2 (red) are highlighted. (**c**) Tandem mass spectrum of the cross-link CaM (K76-K95). Series of y-ions of the two cross-linked peptides are assigned (green and purple)

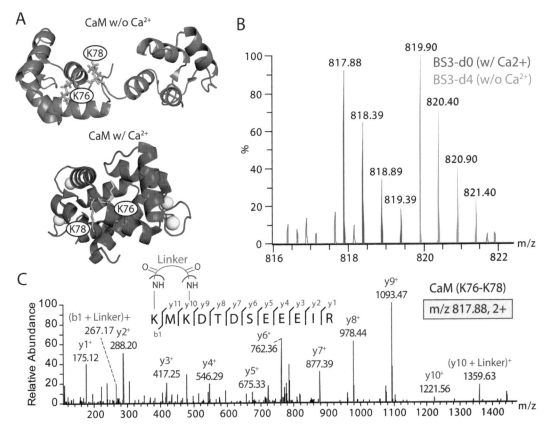

Fig. 3 *Identification and quantification of a loop-link between residues K76 and K78 of CaM.* **(a)** Crystal structures of apo-CaM (blue, top, PDB: 1DMO, state 1) and CaM with Calcium (red, bottom, PDB: 1PRW, state 2) are shown. Residues K76 and K78 are highlighted (green). **(b)** Mass spectra of the cross-link CaM (K76-K78) obtained from CaM cross-linked in state 1 using BS3-d4 and in state 2 using BS3-d0. The isotope envelope corresponding to the cross-link of CaM in state 1 (blue) and in state 2 (red) are highlighted. **(c)** Tandem mass spectrum of the loop-link CaM (K76-K78). Series of y-ions of the peptide are assigned (purple)

76 and 95. On the other hand, the abundance of the loop-link observed between lysine residues 76 and 78 does not differ for the two states (Fig. 3). Ca^{2+}-binding has no effect on this region of the protein structure (Fig. 3a), and equal intensities are observed for the loop-linked peptide cross-linked with the light (BS3-d0) and heavy (BS3-d4) cross-linkers (Fig. 3b). The sequences of the cross-linked peptides and the localization of the cross-linked residues were confirmed by their tandem mass spectra (Figs. 2c and 3c).

2   Materials

Prepare all solutions using ultra-pure water (HPLC grade). All other reagents should have highest available purity.

2.1 Chemical Cross-Linking

1. Protein or protein complex in a suitable buffer (*see* **Note 1**).

2. BS3-d0 stock solution: Dissolve 1.43 mg of BS3-d0 in 100 µL water to obtain 25 mM stock. Dilute 10-times to obtain 2.5 mM stock solution. Dilute with water to obtain lower cross-linker concentrations. Freshly prepare the solution before use (*see* **Note 2**).

3. BS3-d4 stock solution: Dissolve 1.43 mg of BS3-d4 in 100 µL water to obtain 25 mM stock. Dilute 10-times to obtain 2.5 mM stock solution. Dilute with water to obtain lower cross-linker concentrations. Freshly prepare the solution before use (*see* **Note 2**).

4. 1 M Tris–HCl, pH 7.5: Dissolve 6.04 g Tris in 40 mL water. Adjust pH to 7.5 using HCl, and add water to a final volume of 50 mL. Store buffer in a glass bottle at room temperature for up to 3 months.

5. Gel electrophoresis materials (sample buffer, running buffer, polyacrylamide gels, molecular weight marker, coomassie staining solution, gel electrophoresis chamber, power supply, etc.).

2.2 In-Solution Digestion

2.2.1 Ethanol Precipitation

1. 3 M sodium acetate, pH 5.3: Dissolve 24.6 g of sodium acetate in 100 mL of water and adjust the pH to 5.3 using acetic acid. Store the solution in a glass bottle for several months.

2. Ethanol 100% (v/v) can be stored in a glass bottle for several months at −20 °C to provide an ice-cold solution.

3. 80% (v/v) ethanol: Mix 80 mL 100% ethanol and 20 mL water, and store the solution in a glass bottle at −20 °C.

2.2.2 In-Solution Digestion in the Presence of RapiGest

1. 25 mM ammonium bicarbonate (NH_4HCO_3), pH 8.0: Dissolve 19.75 mg NH_4HCO_3 in 10 mL water. Freshly prepare before use (*see* **Note 3**).

2. 1% (m/v) *Rapi*Gest: Dissolve one aliquote (1 mg) *Rapi*Gest (*Rapi*Gest SF Surfactant, Waters Corporation) in 100 µL 25 mM NH_4HCO_3. The solution can be stored at −20 °C for several weeks.

3. 50 mM dithiothreitol (DTT): Dissolve 7.7 mg in 1 mL 25 mM NH_4HCO_3. Freshly prepare before use.

4. 100 mM iodoacetamide (IAA): Dissolve 18.5 mg in 1 mL 25 mM NH_4HCO_3. Freshly prepare the solution before use and protect from light by wrapping with aluminum foil.

5. Trypsin stock solution: 0.1 µg/µL sequencing grade modified trypsin (Promega Corporation) in 50 mM acetic acid. Dissolve 20 µg trypsin in 200 µL 50 mM acetic acid (*see* **Note 4**).

6. 5% Trifluoroacetic acid (TFA): Mix 95 µL water and 5 µL TFA.

2.3 Enrichment of Cross-Links

1. Size-exclusion chromatography system equipped with size-exclusion chromatography column for peptides.

2. Mobile phase: 30% (v/v) acetonitrile (ACN), 0.1% (v/v) TFA.

2.4 LC-MS/MS Analysis

1. Sample solution: 2% (v/v) ACN, 0.1% (v/v) formic acid (FA). Mix 98 mL water with 2 mL ACN and 100 μL FA. Store the solution in a glass bottle for several weeks at −4 °C.

2. Mobile phase A: 0.1% (v/v) FA. Mix 100 mL water and 100 μL FA. Degas the solution for 30 min in an ultrasonic bath.

3. Mobile phase B: 80% (v/v) ACN and 0.1% (v/v) FA. Mix 20 mL water, 80 mL ACN and 100 μL FA. Degas the solution for 30 min in an ultrasonic bath.

4. LC-MS system for proteomic studies (*see* **Notes 5** and **6**).

2.5 Identification of Cross-Linked Peptides

1. Cross-link analysis software (*see* **Note 7**).

2. Data processing tool (*see* **Note 8**).

3. Software tool for visualization of cross-links (*see* **Note 9**).

2.6 Quantification of Identified Cross-Links

1. MS software (*see* **Note 10**).

3 Methods

3.1 Chemical Cross-Linking

3.1.1 Identification of Optimal Cross-Linker Concentration

1. Prepare 5–10 sample tubes containing 10 μM of the protein or protein complex and add increasing amounts of BS3-d0. The BS3-d0 concentration should be ranging from 0 to 5 mM.

2. Incubate for 1 h at 25 °C and 350 rpm.

3. Quench the reaction by addition of 1 M Tris–HCl to a final concentration of 10–20 mM Tris.

4. Prepare samples for gel electrophoresis according to manufacturer's protocols and perform gel electrophoresis. After gel-electrophoresis, wash and stain the gel using coomassie staining solution.

5. Examine the gel and determine optimal protein to cross-linker ratio (*see* **Note 11**).

3.1.2 Quantitative Cross-Linking

1. Prepare 2 reaction mixtures: One containing 10 μM of the protein in state 1 (mixture A) and the other in state 2 (mixture B). Add the optimal BS3-d0 concentration to mixture A (i.e., state 1) and BS3-d4 to mixture B (state 2). For determination of optimal cross-linker concentration, *see* Subheading 3.1.1.

2. Incubate for 1 h at 25 °C and 350 rpm.

3. Quench the reaction by addition of Tris (10–20 mM final concentration; *see* also Subheading 3.1.1, **step 3**).

4. Pool equal amounts of the protein cross-linked with BS3-d0 (mixture A) and BS3-d4 (mixture B).

5. Repeat **steps 1–4** and swap light and heavy cross-linkers (i.e., incubate state 2 with BS3-d0 and state 1 with BS3-d4).

3.2 Sample Preparation for LC-MS/MS Analysis

3.2.1 Ethanol Precipitation and in-Solution Digestion

1. Add water to the cross-linked proteins to reach a final volume of 200 μL.

2. Add 20 μL (1/10 vol.) 3 M sodium acetate (pH 5.3).

3. Add 600 μL (3 vol.) 100% (v/v) ice-cold ethanol and vortex the sample for several seconds.

4. Incubate the sample for at least 2 h at −20 °C (*see* **Note 12**).

5. Centrifuge the sample mixture for 30 min at $16,200 \times g$ and 4 °C.

6. Carefully remove the supernatant and wash the protein pellet by addition of 1 mL 80% (v/v) ice-cold ethanol (*see* **Note 13**).

7. Centrifuge for 30 min at $16,200 \times g$ and 4 °C.

8. Carefully remove the supernatant (*see* **Note 13**).

9. Dry the pellet in a vacuum centrifuge.

10. Dissolve the protein pellet in 10 μL of 1% (m/v) *Rapi*Gest (*see* **Note 14**).

11. Add 10 μL of 50 mM DTT and incubate for 30 min at 60 °C and 500 rpm.

12. Add 10 μL of 100 mM IAA and incubate for 1 h at 37 °C and 500 rpm.

13. Add 70 μL of trypsin solution in 25 mM NH_4HCO_3 at a trypsin-to-protein ratio of 1:20 (w/w).

14. Incubate over night at 37 °C.

15. Add 20 μL of 5% (v/v) TFA and incubate for 2 h at 37 °C.

16. Centrifuge for 30 min at $16,200 \times g$.

17. Transfer the supernatant into a new tube (*see* **Note 15**).

18. Dry the peptides in a vacuum centrifuge.

3.2.2 Enrichment of Cross-Linked Di-Peptides

1. Dissolve the peptides in 60 μL of 30% (v/v) ACN, 0.1% (v/v) TFA and sonicate for 1 min.

2. Centrifuge for 1 min at $16,200 \times g$.

3. Load 50 μL of the peptide solution into the sample loop of the chromatography system.

4. Separate the peptides isocratically at a flow rate of 50 µL/min using a size exclusion column. Collect 50 µL fractions (*see* **Note 16**).

5. Dry peptides in a vacuum centrifuge.

3.3 LC-MS/MS Analysis

1. Dissolve dried peptides in 2% (v/v) ACN, 0.1% FA. Typically, each fraction is dissolved in 8 µL and 5 µL are injected onto the nano-LC system (*see* **Note 17**).

2. Perform LC-MS/MS analysis (*see* **Notes 5 and 6**).

3.4 Identification of Cross-Linked Peptides

The identification of cross-links can be performed using different software. Detailed tutorials are available (*see* **Note 7**). General aspects are listed in the following steps.

1. Generate a protein database containing the proteins of interest in FASTA format.

2. Configure protein database and mass of the cross-linker (for instance, deuterated and non-deuterated BS3).

3. Upload your raw data (*see* **Note 18**).

4. Specify search-parameters, e.g., enzyme used for hydrolysis, number of allowed missed cleavage sites, fixed and variable modifications, instrument parameters.

5. The result file usually contains information on inter- and intra-molecular cross-links as well as mono- and loop-links (*see* **Note 8**).

6. Inspect spectra of cross-linked peptides and manually validate peptide sequences and localization of cross-linked residues (*see* **Notes 19** and **20**). Inspect corresponding MS spectra, and identify the characteristic isotope patterns of light and heavy cross-linked peptides (*see* **Note 21**). Discard false-positive hits, i.e., cross-links without characteristic peak pair or with low-quality MS/MS spectra.

7. Visualize validated cross-links manually or using available software tools (*see* **Note 9**).

3.5 Quantification of Identified Cross-Links

1. Extract ion chromatograms for all validated cross-links. Extract ion chromatograms for both forms, the light and heavy cross-linked peptides (*see* **Note 10**).

2. Calculate BS3-d0-to-BS3-d4 ratios for each cross-linked peptide pair.

3. Analyze the calculated ratios to deduce structural information on your protein or protein complex (*see* **Note 22**).

4   Notes

1. Purification and handling of the protein sample is specific to the protein or protein complex of interest. If possible, use ammonia-free buffers like PBS or HEPES to avoid quenching of the cross-linker.

2. Avoid repeated freeze-thaw cycles and long storage duration in solution.

3. Filter the solution using a syringe filter. The pH should be between 7.5 and 9.

4. The use of modified trypsin reduces proteolytic self-digestion. Depending on the product specification of trypsin, different resuspension buffers are used.

5. Every laboratory uses its own setup. Usually, peptides are first loaded onto a reversed-phase C18 pre-column to concentrate the peptides. A long HPLC column is used to achieve optimal separation of low abundant cross-linked peptides and to enable a high analytical depth during LC-MS/MS analysis.

6. The settings of the mass spectrometer strongly depend on the instrument used. Typically, mass spectra are acquired in data-dependent mode, and tandem-MS spectra are acquired for triply or higher charged peptides. The application of high-resolution mass spectrometers facilitates identification of cross-linked peptides.

7. Following software can be used for identification of cross-links: xQuest [28], XlinkX [29], Kojak [30], StavroX [31], MeroX [32], XiSearch [33], and pLink [34].

8. The Croco [59] software tool can be employed to convert result files from several search engines into other formats.

9. xiView/xiNet [35, 36], XLink-DB [37], xVis [38], ProXL [39], CLMSVault [40], Xwalk [41], or Xlink Analyzer [42] are used for downstream analysis and visualization of cross-links.

10. Extracted ion chromatograms (XIC) can be generated manually using vendor-specific software (e.g., Thermo Xcalibur Qual Browser v4.0, Thermo Fisher scientific) or computationally using xTract [55], XiQ [54], or Spectronaut (Biognosys) [56, 46] software.

11. With increasing amounts of BS3, protein gel bands are usually less resolved (Fig. 1). For protein complexes, intensities of monomeric bands decrease, while protein bands of covalently linked proteins are observed at higher molecular weight. Too high concentrations of the cross-linker might induce protein aggregation, and aggregates might remain in the loading cavities of the gel.

12. Incubation over night at $-20\ ^\circ C$ is recommended. The sample can also be incubated for several days at $-20\ ^\circ C$.

13. Protein pellets can be very small or transparent. If the protein pellet is difficult to recognize, do not remove all liquid to ensure that the pellet is not removed with the supernatant.

14. Avoid formation of foam by gently pipetting the sample up and down. When the protein pellet does not dissolve add another 10 µL of 1% (m/v) *Rapi*Gest and multiply the required volumes of the following steps by two.

15. Avoid transferring decomposed *Rapi*Gest. When *Rapi*Gest is transferred to the fresh sample tube, repeat centrifugation and transfer to a fresh tube (Subheading 3.2.1, **steps 16** and **17**).

16. Due to their size, cross-linked peptides elute in early fractions during size exclusion chromatography. Combine fractions that show low intensity (usually early and late fractions).

17. Adjust the volume for dissolving dried peptides. Note that the required injection volume depends on the sensitivity of the mass spectrometer and needs to be adjusted to obtain sufficient signal.

18. Depending on the used software, different file formats are supported.

19. In most cases, inspection of the top scoring spectra of an identified cross-link is sufficient for manual validation.

20. Cross-links are validated confidentially when series of y- or b-ions of at least 4 successive amino acids of both peptides are observed and the most intense peaks of the spectrum are assigned.

21. The isotope effect of deuterium might lead to earlier elution times for cross-linked peptides containing the deuterated linker. For detection of the peak pair, expand the elution time window toward earlier elution times for deuterated cross-links and toward later elution times for the non-labelled analogue.

22. Ratios of approx. 1 indicate no change in intensity of the individual protein interaction in the two states. Ratios >1 correspond to increased intensity and <1 to decreased intensity of the protein interactions in the different states.

Acknowledgments

We acknowledge funding from the Federal Ministry for Education and Research (BMBF, ZIK programme, 03Z22HN22), the European Regional Development Funds (EFRE, ZS/2016/04/78115), and the MLU Halle-Wittenberg.

References

1. Chen ZA, Rappsilber J (2018) Protein dynamics in solution by quantitative crosslinking/mass spectrometry. Trends Biochem Sci 43(11):908–920

2. Tinnefeld V, Sickmann A, Ahrends R (2014) Catch me if you can: challenges and applications of cross-linking approaches. Eur J Mass Spectrom (Chichester, Eng) 20(1):99–116

3. Staros JV, Wright RW, Swingle DM (1986) Enhancement by N-hydroxysulfosuccinimide of water-soluble carbodiimide-mediated coupling reactions. Anal Biochem 156(1):220–222

4. Bitan G, Teplow DB (2004) Rapid photochemical cross-linking—a new tool for studies of metastable, amyloidogenic protein assemblies. Acc Chem Res 37(6):357–364

5. Bragg PD, Hou C (1975) Subunit composition, function, and spatial arrangement in the Ca2+−and Mg2+−activated adenosine triphosphatases of Escherichia coli and Salmonella typhimurium. Arch Biochem Biophys 167(1):311–321

6. Zhang Q, Crosland E, Fabris D (2008) Nested Arg-specific bifunctional crosslinkers for MS-based structural analysis of proteins and protein assemblies. Anal Chim Acta 627(1):117–128

7. Wu CW, Yarbrough LR (1976) N-(1-pyrene) maleimide: a fluorescent cross-linking reagent. Biochemistry 15(13):2863–2868

8. Kovalenko OV, Yang XH, Hemler ME (2007) A novel cysteine cross-linking method reveals a direct association between claudin-1 and tetraspanin CD9. Mol Cell Proteomics 6(11):1855–1867

9. Kim Y, Ho SO, Gassman NR et al (2008) Efficient site-specific labeling of proteins via cysteines. Bioconjug Chem 19(3):786–791

10. Nagao R, Suzuki T, Okumura A et al (2010) Topological analysis of the extrinsic PsbO, PsbP and PsbQ proteins in a green algal PSII complex by cross-linking with a water-soluble carbodiimide. Plant Cell Physiol 51(5):718–727

11. Rampler E, Stranzl T, Orban-Nemeth Z et al (2015) Comprehensive cross-linking mass spectrometry reveals parallel orientation and flexible conformations of plant HOP2-MND1. J Proteome Res 14(12):5048–5062

12. Suchanek M, Radzikowska A, Thiele C (2005) Photo-leucine and photo-methionine allow identification of protein-protein interactions in living cells. Nat Methods 2(4):261–267

13. Zhang H, Tang X, Munske GR et al (2009) Identification of protein-protein interactions and topologies in living cells with chemical cross-linking and mass spectrometry. Mol Cell Proteomics 8(3):409–420

14. Mattson G, Conklin E, Desai S et al (1993) A practical approach to crosslinking. Mol Biol Rep 17(3):167–183

15. Iacobucci C, Piotrowski C, Aebersold R et al (2019) First community-wide, comparative cross-linking mass spectrometry study. Anal Chem 91(11):6953–6961

16. Kalkhof S, Sinz A (2008) Chances and pitfalls of chemical cross-linking with amine-reactive N-hydroxysuccinimide esters. Anal Bioanal Chem 392(1–2):305–312

17. Schilling B, Row RH, Gibson BW et al (2003) MS2Assign, automated assignment and nomenclature of tandem mass spectra of chemically crosslinked peptides. J Am Soc Mass Spectrom 14(8):834–850

18. Trester-Zedlitz M, Kamada K, Burley SK et al (2003) A modular cross-linking approach for exploring protein interactions. J Am Chem Soc 125(9):2416–2425

19. Fujii N, Jacobsen RB, Wood NL et al (2004) A novel protein crosslinking reagent for the determination of moderate resolution protein structures by mass spectrometry (MS3-D). Bioorg Med Chem Lett 14(2):427–429

20. Hurst GB, Lankford TK, Kennel SJ (2004) Mass spectrometric detection of affinity purified crosslinked peptides. J Am Soc Mass Spectrom 15(6):832–839

21. Chowdhury SM, Du X, Tolic N et al (2009) Identification of cross-linked peptides after click-based enrichment using sequential collision-induced dissociation and electron transfer dissociation tandem mass spectrometry. Anal Chem 81(13):5524–5532

22. Leitner A, Reischl R, Walzthoeni T et al (2012) Expanding the chemical cross-linking toolbox by the use of multiple proteases and enrichment by size exclusion chromatography. Mol Cell Proteomics 11(3):M111.014126

23. Fritzsche R, Ihling CH, Gotze M et al (2012) Optimizing the enrichment of cross-linked products for mass spectrometric protein analysis. Rapid Commun Mass Spectrom 26(6):653–658

24. Leitner A, Walzthoeni T, Aebersold R (2014) Lysine-specific chemical cross-linking of protein complexes and identification of cross-linking sites using LC-MS/MS and the xQuest/xProphet software pipeline. Nat Protoc 9(1):120–137

25. Singh P, Panchaud A, Goodlett DR (2010) Chemical cross-linking and mass spectrometry

as a low-resolution protein structure determination technique. Anal Chem 82 (7):2636–2642

26. Trnka MJ, Baker PR, Robinson PJ et al (2014) Matching cross-linked peptide spectra: only as good as the worse identification. Mol Cell Proteomics 13(2):420–434

27. Trnka MJ, Burlingame AL (2010) Topographic studies of the GroEL-GroES chaperonin complex by chemical cross-linking using diformyl ethynylbenzene: the power of high resolution electron transfer dissociation for determination of both peptide sequences and their attachment sites. Mol Cell Proteomics 9 (10):2306–2317

28. Rinner O, Seebacher J, Walzthoeni T et al (2008) Identification of cross-linked peptides from large sequence databases. Nat Methods 5 (4):315–318

29. Liu F, Rijkers DT, Post H et al (2015) Proteome-wide profiling of protein assemblies by cross-linking mass spectrometry. Nat Methods 12(12):1179–1184

30. Hoopmann MR, Zelter A, Johnson RS et al (2015) Kojak: efficient analysis of chemically cross-linked protein complexes. J Proteome Res 14(5):2190–2198

31. Gotze M, Pettelkau J, Schaks S et al (2012) StavroX—a software for analyzing crosslinked products in protein interaction studies. J Am Soc Mass Spectrom 23(1):76–87

32. Gotze M, Pettelkau J, Fritzsche R et al (2015) Automated assignment of MS/MS cleavable cross-links in protein 3D-structure analysis. J Am Soc Mass Spectrom 26(1):83–97

33. Mendes ML, Fischer L, Chen ZA et al (2019) An integrated workflow for crosslinking mass spectrometry. Mol Syst Biol 15(9):e8994

34. Yang B, Wu YJ, Zhu M et al (2012) Identification of cross-linked peptides from complex samples. Nat Methods 9(9):904–906

35. Graham M, Combe C, Kolbowski L et al (2019) xiView: a common platform for the downstream analysis of crosslinking mass spectrometry data. bioRxiv:561829. https://doi.org/10.1101/561829

36. Combe CW, Fischer L, Rappsilber J (2015) xiNET: cross-link network maps with residue resolution. Mol Cell Proteomics 14 (4):1137–1147

37. Zheng C, Weisbrod CR, Chavez JD et al (2013) XLink-DB: database and software tools for storing and visualizing protein interaction topology data. J Proteome Res 12 (4):1989–1995

38. Grimm M, Zimniak T, Kahraman A et al (2015) xVis: a web server for the schematic visualization and interpretation of crosslink-derived spatial restraints. Nucleic Acids Res 43 (W1):W362–W369

39. Riffle M, Jaschob D, Zelter A et al (2016) ProXL (protein cross-linking database): a platform for analysis, visualization, and sharing of protein cross-linking mass spectrometry data. J Proteome Res 15(8):2863–2870

40. Courcelles M, Coulombe-Huntington J, Cossette E et al (2017) CLMSVault: a software suite for protein cross-linking mass-spectrometry data analysis and visualization. J Proteome Res 16(7):2645–2652

41. Kahraman A, Malmstrom L, Aebersold R (2011) Xwalk: computing and visualizing distances in cross-linking experiments. Bioinformatics 27(15):2163–2164

42. Kosinski J, von Appen A, Ori A, Karius K et al (2015) Xlink analyzer: software for analysis and visualization of cross-linking data in the context of three-dimensional structures. J Struct Biol 189(3):177–183

43. Leitner A, Faini M, Stengel F et al (2016) Crosslinking and mass spectrometry: an integrated technology to understand the structure and function of molecular machines. Trends Biochem Sci 41(1):20–32

44. Barysz H, Kim JH, Chen ZA et al (2015) Three-dimensional topology of the SMC2/SMC4 subcomplex from chicken condensin I revealed by cross-linking and molecular modelling. Open Biol 5(2):150005

45. Chavez JD, Eng JK, Schweppe DK et al (2016) A general method for targeted quantitative cross-linking mass spectrometry. PLoS One 11(12):e0167547

46. Muller F, Kolbowski L, Bernhardt OM et al (2019) Data-independent acquisition improves quantitative cross-linking mass spectrometry. Mol Cell Proteomics 18(4):786–795

47. Muller F, Graziadei A, Rappsilber J (2019) Quantitative photo-crosslinking mass spectrometry revealing protein structure response to environmental changes. Anal Chem 91 (14):9041–9048

48. Chavez JD, Schweppe DK, Eng JK et al (2015) Quantitative interactome analysis reveals a chemoresistant edgotype. Nat Commun 6:7928

49. Chavez JD, Schweppe DK, Eng JK et al (2016) In vivo conformational dynamics of Hsp90 and its interactors. Cell Chem Biol 23 (6):716–726

50. Yu C, Huszagh A, Viner R et al (2016) Developing a multiplexed quantitative cross-linking mass spectrometry platform for comparative structural analysis of protein complexes. Anal Chem 88(20):10,301–10,308

51. Ross PL, Huang YN, Marchese JN et al (2004) Multiplexed protein quantitation in Saccharomyces cerevisiae using amine-reactive isobaric tagging reagents. Mol Cell Proteomics 3 (12):1154–1169

52. Muller DR, Schindler P, Towbin H et al (2001) Isotope-tagged cross-linking reagents. A new tool in mass spectrometric protein interaction analysis. Anal Chem 73(9):1927–1934

53. Schmidt C, Robinson CV (2014) A comparative cross-linking strategy to probe conformational changes in protein complexes. Nat Protoc 9(9):2224–2236

54. Fischer L, Chen ZA, Rappsilber J (2013) Quantitative cross-linking/mass spectrometry using isotope-labelled cross-linkers. J Proteome 88:120–128

55. Walzthoeni T, Joachimiak LA, Rosenberger G et al (2015) xTract: software for characterizing conformational changes of protein complexes by quantitative cross-linking mass spectrometry. Nat Methods 12(12):1185–1190

56. Bruderer R, Bernhardt OM, Gandhi T et al (2015) Extending the limits of quantitative proteome profiling with data-independent acquisition and application to acetaminophen-treated three-dimensional liver microtissues. Mol Cell Proteomics 14(5):1400–1410

57. Jurado LA, Chockalingam PS, Jarrett HW (1999) Apocalmodulin. Physiol Rev 79 (3):661–682

58. Bahler M, Rhoads A (2002) Calmodulin signaling via the IQ motif. FEBS Lett 513 (1):107–113

59. Bender J, Schmidt C (2020) The CroCo crosslink converter: a user-centred tool to convert results from crosslinking mass spectrometry experiments. Bioinformatics 36:1296–1297

Chapter 27

Missing Value Monitoring to Address Missing Values in Quantitative Proteomics

Vittoria Matafora and Angela Bachi

Abstract

Many classes of key functional proteins such as transcription factors or cell cycle proteins are present in the proteome at a very low concentration. These low-abundance proteins are almost entirely invisible to systematic quantitative analysis by classical data dependent proteomics methods (DDA). Moreover, DDA runs in shotgun proteomics experiments are plenty of missing values among the replicates due to the stochastic nature of the acquisition method, thus hampering the robustness of the quantitative analysis. Here, we have overcome these obstacles designing a robust workflow named missing value monitoring (MvM) in order to follow low abundance proteins dynamics.

Key words Proteomics, DDA, DIA, Missing values, Protein quantitation

1 Introduction

Quantitative proteomics is the method of choice to measure changes in global protein levels in biological samples [1–3] and, recently, has also been optimized to reveal the comprehensive proteome of a single cell type [4]. In the last decade, quantitative proteomics analysis was mainly performed on high-resolution mass spectrometers based on data-dependent acquisition (DDA). With this method, the most abundant precursors, usually the top 10–20 most intense precursors per cycle, are selected for fragmentation, and the acquired MS/MS spectra are successively matched against an appropriate database. Although extremely powerful, this method shows some weakness in reproducibly quantifying low-abundance peptides. This is due to the stochastic nature of the fragmentation as the selection is based on the strongest signals [5]. As a result, different subsets of low-abundance peptides are selected in each run causing high variation across replicates. The number of missing values can be substantial among replicates (>30%) and reduces the number of quantifiable proteins [6] (*see* also Chap. 1). Recently it has been introduced a novel method

Katrin Marcus et al. (eds.), *Quantitative Methods in Proteomics*, Methods in Molecular Biology, vol. 2228,
https://doi.org/10.1007/978-1-0716-1024-4_27, © Springer Science+Business Media, LLC, part of Springer Nature 2021

named "Parallel Accumulation - SErial Fragmentation" (PASEF) [7] that improves protein coverage enabling hundreds of MS/MS events per second at full sensitivity. However, it requires latest technology for its application.

In the last years, data-independent acquisition (DIA, or SWATH) (Chaps. 16, 22–24) has become a good alternative to DDA [8, 9] (Chaps. 9, 10, 13–17, 20, 21, 26). With this method, all peptides, within a predefined m/z selection window, are simultaneously fragmented. The acquisition is repeated sequentially in stepped selection windows, usually in the 300–1100 m/z range. It has been shown that DIA outperforms DDA in terms of fewer missing values, a higher number of quantified proteins in shorter analysis time, and lower coefficients of variation across replicates [10]. Nevertheless, DIA presents several limitations: i.e., the increased spectral complexity, manual inspection of spectra, and the requirement of a DDA derived spectral library for peptide identification.

Here, we present a novel workflow named missing value monitoring (MvM) (Fig. 1) which overcomes these limitations. It combines the high accuracy of peptides identification obtained by DDA with the high reproducibility of peptides quantitation obtained by DIA [11].

2   Materials

2.1   Stage Tip Preparation

1. Empore reversed-phase extraction disks from 3 M (C18 (ODS or Octadecyl) reversed- phase material, 3 M product number 2215).

2.2   Protein Digestion

Prepare all solutions using UHQ (Ultra High Quality) water obtained from a Milli-Q system and analytical grade reagents.

1. Urea buffer: 8 M Urea, 100 mM Tris-HCl.

2. Reduction buffer: dissolve dithiothreitol (DTT) (10 mM final concentration) in the digestion buffer.

3. Alkylation buffer: dissolve iodoacetamide (IAA) (50 mM final concentration) in the digestion buffer.

4. Trypsin stock solution: 0.1 μg/μL Trypsin in 1 mM HCl.

5. Digestion buffer: 40 mM NH_4HCO_3.

6. Spin ultrafiltration units of nominal molecular weight cut off of 10 kDa (Millipore).

2.3   Materials for LC-MS/MS.

1. Elution solution: 80% acetonitrile (ACN), HPLC-MS grade, 5% formic acid (FA).

2. Solvent A: 2% ACN, HPLC-MS grade, 0.1% FA.

3. Solvent B: 80% ACN, 0.1% FA.

MvM workflow

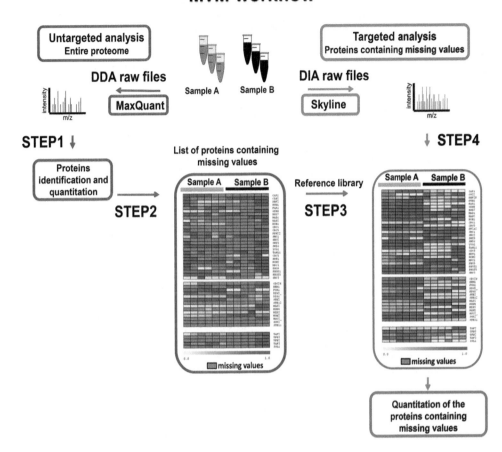

Fig. 1 *MvM workflow.* For each biological sample, perform LC-MS/MS acquisitions alternating DDA and DIA methods. (Step 1) Analyze .raw files generated by DDA method by MaxQuant software. For label free analysis, quantify all proteins with LFQ intensity. (Step 2) Derive the list for the proteins carrying missing values among the replicates and use Skyline to interrogate both DDA and DIA files for all these proteins. (Step 3) For each peptide belonging to the targeted protein, search in DDA raw file the precursor corresponding to the best ID event in MaxQuant and perform the "identity propagation" from DDA to DIA by manually picking all the precursors and their related transitions for each run. (Step 4) Integrate the peak area for each precursor and derive protein quantitation

2.4	*Software*	1. MaxQuant (https://www.maxquant.org/) [12].
		2. Skyline (https://skyline.ms/project/home/software/Skyline/begin.view) [13].

3   Methods

3.1	*Protein Digestion*	For both DDA and DIA the same protocol for protein digestion is applied (FASP Protocol), [14].

1. 50 μg of extracted proteins are dissolved in urea buffer.

2. Protein disulfide bonds reduction is carried out with 0.01 M DTT in urea buffer, 30 min at room temperature.

3. Reduced cysteine alkylation is performed with 0.05 M iodoacetamide in urea buffer at room temperature for 15 min in complete darkness.

4. Protein digestion is performed using 1 μg of trypsin added in 95 μL of 40 mM NH_4HCO_3. Samples are incubated overnight at 37 °C, and then 1 μg of trypsin is added for further 3 h of incubation.

5. Peptides are collected and desalted by C18 Stage Tip (Proxeon Biosystems, Denmark). The Stage Tip workflow is from the method described by Rappsilber et al. [15].

3.2 Mass Spectrometry Analysis

1. For the MS analysis, four technical replicates are performed, two in DDA and two in DIA mode. For each replicate 0.5–1 μg of sample is injected into a quadrupole Orbitrap Q-exactive HF mass spectrometer (Thermo Scientific).

2. Peptide separation is achieved on a linear gradient from 95% solvent A to 55% solvent B over 222 min and from 55 to 100% solvent B in 3 min (see **Note 1**) at a constant flow rate of 0.25 μL/min on UHPLC Easy-nLC 1000 (Thermo Scientific) where the LC system is connected to a 23 cm fused-silica emitter of 75 μm inner diameter (New Objective, Inc. Woburn, MA, USA), packed in-house with ReproSil-Pur C18-AQ 1.9 μm beads (Dr Maisch Gmbh, Ammerbuch, Germany) using a high-pressure bomb loader (Proxeon, Odense, Denmark).

3. For DDA mode, the mass spectrometer is operated with dynamic exclusion enabled (exclusion duration of 15 s), MS1 resolution of 70,000 at m/z 200, MS^1 automatic gain control target of 3×10^6, MS^1 maximum fill time of 60 ms, MS^2 resolution of 17,500, MS^2 automatic gain control target of 1×10^5, MS^2 maximum fill time of 60 ms, and MS^2 normalized collision energy of 25. For each cycle, one full MS^1 scan range of 300–1650 m/z was followed by 12 MS^2 scans using an isolation window of 2.0 m/z.

4. For the DIA mode, the mass spectrometer is operated with a MS^1 scan at resolution of 35,000 at m/z 200, automatic gain control target of 1×10^6, and scan range of 490–910 m/z, followed by a DIA scan with a loop count of 10. DIA settings were as follows: window size of 20 m/z, resolution of 17,500, automatic gain control target of 1×10^6, and normalized collision energy of 30.

3.3 MvM Workflow

1. For DDA data analysis, process DDA .raw files with MaxQuant software making use of the Andromeda search engine [12]. Upload the proper complete UniProtKB/Swiss-Prot protein sequence database (*see* **Note 2**). Specify trypsin as enzyme specificity (*see* **Note 3**) and allow at least two missing cleavages per peptide. Select alkylation of cysteine by carbamidomethylation as fixed modification, and oxidation of methionine and N-terminal acetylation as variable modifications. Set mass tolerance to 5 and 10 ppm for parent and fragment ions, respectively. Set the false discovery rate (FDR) <0.01 for peptide spectrum matches (PSMs) and proteins (a reverse decoy database is generated within Andromeda). Require at least two peptides identifications per protein, of which at least one peptide has to be unique to the protein group. For label-free proteins quantitation, consider a minimum ratio count of two and use the "LFQ intensities." Include the "match between run" option.

2. For DIA data analysis, process DIA .raw files using Skyline [13] and generate spectral libraries by MaxQuant search results. From the list of proteins quantified by DDA label free analysis, choose the proteins with missing values and target them for Skyline analysis. For each protein, select the peptides identified in DDA analysis. Upload the MS/MS library in Skyline by importing msms.txt file generated by MaxQuant. For each peptide, manual check the presence of the precursor and co-eluting transitions (*see* **Note 4**). Exactly, manually select all MS^1 and MS^2 peaks based on retention times and presence of the same transitions identified in shotgun analysis (remind that all the peptides have been previously identified in at least one of the runs as we are just filling missing values with this procedure). Use the following criteria: a window of 5 ppm tolerance for precursors and 10 ppm for transitions (the same used for DDA analysis) (*see* **Note 5**), minimum three unique transitions for each peptide, matching co-elution of both precursor and product ion transitions, and minimum two peptides for each protein with iDOTP (Isotope Dot Product) values close to 1 (*see* **Note 6**). Extract ion currents on MS^1 or MS^2 peaks for peptide quantitation. Peak areas for all selected precursors and transitions in Skyline and use for quantitation (*see* **Note 7**). Use the option normalized area included in skyline for each peptide of a given protein. With this option, all the peak areas related to each peptide are normalized against the most abundant specie across the samples analyzed. To get protein abundance, perform the average of the areas of all peptides belonging to the same protein (*see* **Note 8**).

3. For the normalization of the areas across all the samples, choose a housekeeping protein. Check, among the proteins identified by DDA, for a protein whose abundance resembles

the input of each sample by looking for the correlation of the LFQ intensity of the chosen protein with the summed intensity of each sample (*see* **Note 9**). Analyze the housekeeping protein with DIA by targeting at least three peptides (for details see the previous paragraph). Use the area of the housekeeping protein to normalize the area of all the other proteins analyzed by DIA.

4. Combine DIA and DDA proteins quantitation and use it for statistical analysis. The proteins quantified by DIA will present less zero values and consequently less coefficient of variation among the replicates when compared to DDA quantitation. This increased robustness of the data generates more significant proteins when statistical tests are applied (*see* **Note 10**). MvM method enables successful quantitation for high as well as low abundant proteome, attesting the potential of MvM workflow for the measurement of any set of proteomic analysis.

4   Notes

1. LC gradient slope and length must be adjusted according to sample complexity. For less complex samples (as sub-organelle related proteome or interactome), a shorter gradient/acquisition method is preferred while for complex samples (as entire proteome), longer gradient are preferred (we suggest at least 200 min).

2. The choice of the protein sequence database strictly depends on the source of the sample to be analyzed. More complete is the chosen database, more the chances to cover the entire proteome are high.

3. Trypsin is the preferred enzyme for its high specificity, but also other proteolytic enzymes are suitable for the analysis.

4. Even though Skyline is able to do retention time alignments, we suggest to do it manually, as we observed that the manual alignment allows a retention time drift among the replicates lower than the one obtained automatically by Skyline.

5. We empirically observed that increasing error tolerance above 15 ppm generates ambiguous assignments, while almost all of the peaks were selected with an error lower than 2 ppm both for precursors and transitions.

6. For transitions refinement, check if all transitions are of good quality and reproducible over the samples; check that the relative transition intensity is constant in all runs. If certain transitions/precursors/peptides are of low quality (low intensity, not co-eluting with the other transitions, shouldered, etc.) or irreproducible over runs, remove them from the analysis.

7. Label free quantitation in DIA analysis can be done based on the peak areas of the precursors or of the transitions. Generally, when possible, use the precursors areas as the criteria of choice as it is the same used in shotgun analysis; however when the peak of the precursor is disturbed by any interference, use the summed area of the transitions for quantitation. We have estimated that both measurements correlate [11] therefore are suitable for the analysis.

8. We used the normalized area option to get protein quantitation from DIA analysis. We have adopted these criteria as the values of the area are very different across the different peptides belonging to the same protein (it is known that peptides belonging to the same proteins ionize differently depending on their chemical sequence); therefore, the average is affected by the most abundant species.

9. In order to check the quality of the proteins used as housekeeping, we suggest to use one protein quantitated both in DDA (LFQ intensity) and in DIA (area normalized on the housekeeping protein), and to do correlation analysis, a correct value should be $R^2 > 0.98$.

10. We estimate that the amount of missing values in our datasets is about 25% of the identified proteome. The coefficient of variation between high abundant proteins (copy numbers>1000) measured with DIA and DDA is comparable, while for low abundant proteins DIA outperforms DDA. As results by using MvM we doubled the number of the low abundant proteins statistically regulated in our dataset [11].

References

1. Lundberg E, Borner GHH (2019) Spatial proteomics: a powerful discovery tool for cell biology. Nat Rev Mol Cell Biol 20(5):285–302

2. Pappireddi N, Martin L, Wuhr M (2019) A review on quantitative multiplexed proteomics. Chembiochem 20(10):1210–1224

3. Sinitcyn P, Rudolph JD, Cox J (2018) Computational methods for understanding mass spectrometry–based shotgun proteomics data. Annu Rev Biomed Data Sci 1(1):207–234

4. Schoof EM, Rapin N, Savickas S et al (2019) A quantitative single-cell proteomics approach to characterize an acute myeloid leukemia hierarchy. bioRxiv:745679. https://doi.org/10.1101/745679

5. Michalski A, Cox J, Mann M (2011) More than 100,000 detectable peptide species elute in single shotgun proteomics runs but the majority is inaccessible to data-dependent LC-MS/MS. J Proteome Res 10(4):1785–1793

6. O'Brien JJ, Gunawardena HP, Paulo JA et al (2018) The effects of nonignorable missing data on label-free mass spectrometry proteomics experiments. Ann Appl Stat 12(4):2075–2095

7. Meier F, Beck S, Grassl N et al (2015) Parallel accumulation-serial fragmentation (PASEF): multiplying sequencing speed and sensitivity by synchronized scans in a trapped ion mobility device. J Proteome Res 14(12):5378–5387

8. Ludwig C, Gillet L, Rosenberger G et al (2018) Data-independent acquisition-based SWATH-MS for quantitative proteomics: a tutorial. Mol Syst Biol 14(8):e8126

9. Doerr A (2014) DIA mass spectrometry. Nat Methods 12:35

10. Hu A, Noble WS, Wolf-Yadlin A (2016) Technical advances in proteomics: new developments in data-independent acquisition. F1000Res 5:10.12688/f1000research.7042.1

11. Matafora V, Corno A, Ciliberto A et al (2017) Missing value monitoring enhances the robustness in proteomics quantitation. J Proteome Res 16(4):1719–1727

12. Cox J, Mann M (2008) MaxQuant enables high peptide identification rates, individualized p.p.b.-range mass accuracies and proteome-wide protein quantification. Nat Biotechnol 26(12):1367–1372

13. Pino LK, Searle BC, Bollinger JG et al (2020) The skyline ecosystem: informatics for quantitative mass spectrometry proteomics. Mass Spectrom Rev 39:229–244

14. Wisniewski JR, Zougman A, Nagaraj N et al (2009) Universal sample preparation method for proteome analysis. Nat Methods 6(5):359–362

15. Rappsilber J, Ishihama Y, Mann M (2003) Stop and go extraction tips for matrix-assisted laser desorption/ionization, nanoelectrospray, and LC/MS sample pretreatment in proteomics. Anal Chem 75(3):663–670

Chapter 28

Quantitative Proteome Data Analysis of Tandem Mass Tags Labeled Samples

Oliver Pagel, Laxmikanth Kollipara, and Albert Sickmann

Abstract

In mass spectrometry-based proteomics, relative quantitative approaches enable differential protein abundance analysis. Isobaric labeling strategies, such as tandem mass tags (TMT), provide simultaneous quantification of several samples (e.g., up to 16 using 16plex TMTpro) owing to its multiplexing capability. This technology improves sample throughput and thereby minimizes both measurement time and overall experimental variation. However, TMT-based MS data processing and statistical analysis are probably the crucial parts of this pipeline to obtain reliable, plausible, and significantly quantified results. Here, we provide a step-by-step guide to the analysis and evaluation of TMT quantitative proteomics data.

Key words TMT, Relative quantitative proteomics, Data analysis

Abbreviations

FDR   False discovery rate
HPLC   High-performance liquid chromatography
MS   Mass spectrometer/mass spectrometry
MS/MS   Tandem mass spectrometry
SD   Standard deviation
TMT   Tandem mass tags

1 Introduction

In the last decade, isobaric labeling of peptides with tandem mass tags (TMT) [1, 2] has proven its potential in LC-MS/MS-based proteomics as it allows robust relative quantification of proteins [3, 4] (*see* Chaps. 8, 9, 13–15, 17). Furthermore, labeling of peptides using TMT reagents increases analysis throughput and typically allows multiplexing of up to 16 different samples (i.e., 16plex TMTpro) [5] (*see* also Chap. 15), thus reducing the

Katrin Marcus et al. (eds.), *Quantitative Methods in Proteomics*, Methods in Molecular Biology, vol. 2228,
https://doi.org/10.1007/978-1-0716-1024-4_28, © Springer Science+Business Media, LLC, part of Springer Nature 2021

LC-MS measurement time and inter-sample variation. Whereas, label-free approaches (*see* Chaps. 8, 16, 20–24) require individual sample analysis and stable instrumentation, usually over longer time periods. The critical part, however, is the analysis of the generated TMT-based MS data that can be relatively complex. There are several data processing platforms currently available, e.g., MaxQuant [6], PEAKS studio, [7] and Proteome Discoverer (Thermo Scientific), which can perform protein identification, relative quantification (typically reported as TMT ratios), and data normalization based on the user-defined settings. Despite the ease and straightforwardness of these software applications, data analysis and evaluation steps—especially normalization and setting up significance thresholds—require careful attention to achieve reliable results. For this, the raw protein ratios list can be exported from the respective software, and further statistical analysis can be done manually using Microsoft Excel or by script-based software such as R. In this chapter, we elucidate the TMT 10-plex data evaluation process by using an already published data [8] as an example.

2 Materials

2.1 Raw MS and MS/MS Data

1. Raw LC-MS/MS data of TMT10Plex multiplexed samples acquired at high resolution ($\geq$60,000) MS settings for both MS and MS/MS scans, e.g., by a Q Exactive high field (HF) mass spectrometer (Thermo Scientific).

2.2 Database Analysis Software

1. Data analysis software: Proteome Discoverer (PD, version 1.4 or higher, Thermo Scientific). The following nodes must be included in the workflow.
2. Spectrum file importer node.
3. Spectrum selector node.
4. Mascot [9] node (version 2.6, Matrix Science).
5. Percolator node [10] for false discovery rate (FDR) estimation.
6. Reporter ions quantifier node.

2.3 Data Processing

1. Microsoft Excel (2013 or higher) or script-based software such as R.

2.4 Data Interpretation and Visualization

1. Microsoft Excel (2013 or higher).
2. Stand-alone data visualization software such as InstantClue [11], Perseus [12] or Bioconductor implemented in R [13].

3   Methods

The following section describes TMT10Plex data analysis based on a published study [8] (Data availability: http://proteomecentral. proteomexchange.org/cgi/GetDataset?ID=PXD006006).   The multiplexed sample was pre-fractionated using high-pH (8.0) reversed-phase chromatography prior to LC-MS/MS analysis.

3.1   Database Search

1. Data analysis software: PD version 1.4 (Thermo Scientific) and the following node settings must be included in the workflow (*see* Fig. 1).

2. Spectrum file importer (*see* **Note 1**).

3. Use the default settings of the spectrum selector node as provided by the PD.

4. Use Mascot [9] (version 2.6, Matrix Science) as search algorithm and set the following parameters: precursor and fragment ion tolerances of 10 ppm and 0.02 Da for MS and MS/MS, respectively; trypsin as enzyme with a maximum of two missed cleavages; carbamidomethylation of Cys and TMT on Lys and and N-termini as fixed modification; oxidation of Met as variable modification.

5. Use a human Uniprot "target" database (*see* **Note 2**) as PD generates random decoy hits *on the fly*.

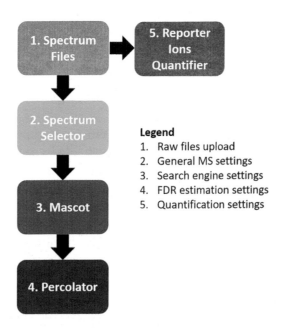

Fig. 1 Flow diagram representation of different nodes and their interconnections for TMT-based MS data processing using Proteome Discoverer software (Thermo Scientific)

6. The FDR estimation should be performed by Percolator [10]. Use a maximum delta Cn of 0.05, strict/relaxed FDR targets of 0.01/0.05 and use q-value based validation.

7. Use the "Reporter Ions quantifier" node and select TMT10Plex as the quantification method. Open the "Quantification" tab in the PD toolbar and add/remove TMT channels according to pre-defined experimental design. Add all possible TMT channel ratios (start with 126/126). Open the "Quantification Method Editor" window, and check "Show the Raw Quan Values" under "Ratio Calculation."

8. In the "Quan Channels" tab, insert the correct values for "Reporter Ion Isotopic Distribution" (*see* **Note 3**).

3.2 Data Visualization, Filtering and Export from PD

1. Open the .msf file of the study in PD. Apply data reduction filters on the peptide-spectrum match (PSM) level, such as search engine rank 1, high confidence corresponding to ≤1% FDR.

2. In the proteins tab, right click in the column header section and select "Enable row filtering."

3. Right click into the protein table and disable "Protein Grouping."

4. Set "1" in the "# proteins" column and "≥2" in the "# unique peptides" column to pre-filter the data.

5. Export the data to MS Excel by right clicking into the protein table and selecting "Export to Excel workbook." Choose "Proteins" in the appearing menu.

3.3 Data Normalization and Scaling Using MS Excel

1. Open the MS Excel file from Subheading 3.2, **step 5** and sort the data by accessions (*see* Fig. 2). This Excel work sheet tab represents the original raw data exported from the PD.

2. Copy the accessions and all TMT ratios into a new Excel work sheet tab and for each protein calculate the log2 value of the ratios for each TMT channel.

3. Calculate the median over all log2 ratio-medians, and subtract it from the single ratio medians to yield factors for ratio normalization.

4. Normalize the original ratio values by subtraction (log-scale) of the respective normalization factors and calculate the row-median over all ratios.

5. For each row, subtract the median of all normalized log2 ratios to obtain normalized, scaled log2 abundance values. These reflect the corrected abundance of each TMT channel (i.e., differentially labeled samples).

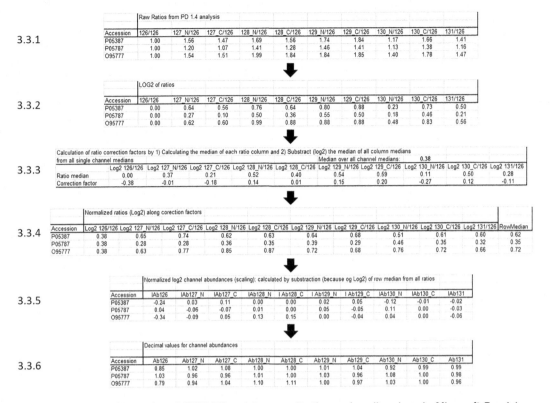

| 3.3.1 | Raw Ratios from PD 1.4 analysis |

Fig. 2 Step-by-step illustration of TMT10Plex data normalization and scaling done in Microsoft Excel (*see* Subheading 3.3)

6. Calculate the decimal normalized channel abundances from the log2 channel abundances by using latter as exponent for the binary radix.

7. Clean up all data by removal of values of "0" or "formula errors" caused by missing values or divisons by zero, since they will interfere with subsequent data analysis using Excel.

3.4 Data Consolidation and Ratio Calculation

1. Copy the decimal, normalized abundance factors of different TMT channels (e.g., TMT126, TMT127N,…, TMT131) to a new Excel work sheet tab and merge them with the columns required for further analysis (at least protein accession) from the original table in Subheading 3.3, **step 1**.

2. If replicates are present, calculate the respective mean values (e.g., MV (R1, R2, R3) AD80 10 nM vs. MV (R1, R2, R3) Control) in new columns, remove values of "0" and "formula errors."

3. Calculate the log2 values for each ratio in a new column and remove "formula errors" (*see* **Note 4**).

3.5.5

Initial channels (unique, quan="used")

126	127_N	127_C	128_N	128_C	129_N	129_C	130_N	130_C	131
4.94E+04	4.46E+04	4.87E+04	4.80E+04	5.52E+04	4.24E+04	7.21E+04	2.96E+04	6.98E+04	3.98E+04
3.14E+04	3.46E+04	2.61E+04	3.80E+04	3.18E+04	3.52E+04	4.07E+04	3.15E+04	4.21E+04	3.79E+04
3.83E+04	4.58E+04	4.59E+04	5.82E+04	4.77E+04	4.99E+04	5.30E+04	3.19E+04	5.04E+04	4.62E+04

⬇

3.5.6

Normalization of abundance: The sum of each channel is compared to the median of all channel sums, resulting in a normalization factor

	126	127_N	127_C	128_N	128_C	129_N	129_C	130_N	130_C	131
Intensity sum	4.07E+10	5.18E+10	4.38E+10	5.62E+10	5.23E+10	5.67E+10	5.88E+10	4.38E+10	5.48E+10	4.72E+10
Median all	5.20E+10									
Factor	7.83E-01	9.95E-01	8.42E-01	1.08E+00	1.00E+00	1.09E+00	1.13E+00	8.41E-01	1.05E+00	9.08E-01

⬇

3.5.7

Corrected channels

126	127_N	127_C	128_N	128_C	129_N	129_C	130_N	130_C	131
6.30E+04	4.48E+04	5.78E+04	4.44E+04	5.49E+04	3.89E+04	6.38E+04	3.52E+04	6.62E+04	4.39E+04
4.01E+04	3.47E+04	3.11E+04	3.52E+04	3.17E+04	3.23E+04	3.60E+04	3.75E+04	3.99E+04	4.18E+04
4.89E+04	4.61E+04	5.45E+04	5.38E+04	4.75E+04	4.58E+04	4.69E+04	3.80E+04	4.78E+04	5.09E+04

⬇

3.5.8

Ratios of corrected intensity sums (per condition)			LOG2 of ratios		
10nM/ctrl	100nM/ctrl	10nM/100nM	log2 10nM/ctrl	log2 100nM/ctrl	log2 10nM/100nM
0.83	1.00	1.19	-0.26	0.00	0.26
0.94	1.07	1.14	-0.10	0.10	0.19
0.98	0.89	0.90	-0.02	-0.17	-0.15

⬇

3.5.9

	log2 10nM/ctrl	log2 100nM/ctrl	log2 10nM/100nM
sigma	0.21	0.21	0.17
3sigma	0.63	0.62	0.50
median	0.01	0.01	-0.01
Down	-0.62	-0.61	-0.51
Up	0.64	0.62	0.49

Fig. 3 Schematic representation of all table calculation steps to define threshold for regulations performed in Excel (*see* Subheading 3.5)

3.5 Threshold Definition for Regulations

1. Go to the "Peptides" tab in PD, right-click and deselect "Peptide grouping" to view PSMs (*see* Fig. 3).

2. Again, right click on the column headers and choose "Enable row filters."

3. Set "1" to the "# proteins" column and "Used" to the "Quan." column to choose only unique, quantified PSMs.

4. Export the PSMs by right clicking into the table, choose "Export to Excel workbook" and open the file.

5. Copy the TMT channel intensities into a new Excel work sheet tab.

6. Calculate the sums of all TMT channel intensities and the median over the sums. Divide the channel sums by the median to obtain correction factors for the PSM intensities.

7. Apply the correction factors to the quan. Channels by dividing each channel value by its respective factor.

8. Form the ratios of choice (the same as in Subheading 3.4, **step 2**) by adding all replicate's corrected intensities per condition and dividing them by, e.g., the control replicate's sum (e.g., sum(10 nM)/sum(control)). Form the log2 values of the obtained ratios.

9. Calculate the median and standard deviation (*see* **Note 5**) of each log2 ratio column, calculate 3*SD. The regulation thresholds are defined per log2 ratio by median + 3*SD (Thr_{up}) and median − 3*SD (Thr_{down}).

10. To check the data distributions, open the "data" tab in Excel, activate the "data analysis toolpack" in Data- > Options- > Add-ins and plot each ratio column as a histogram. The data distribution should be of quasi-Gaussian shape and centered on "0" (*see* **Note 6**).

3.6 Data Evaluation According to Ratios and p-Value

For an illustration of the most important processing steps of this section, *see* Fig. 4.

1. Go back to the table from Subheading 3.4. If there are at least 3 data points of a certain protein (use the if-function in Excel) for each condition, perform a Student's t-test (*see* **Note 7**) of the normalized abundances for both conditions to be compared in new columns (e.g., treatment 1/2/3 vs. control 1/2/3).

2. Copy (as values) regulation thresholds obtained in Subheading 3.5, **step 9** and prepare two columns (named "regulation" and "ratio(decimal)") per ratio you want to evaluate.

3. In new columns, use the "if" function to compare each average log2 ratio with their corresponding thresholds (Fig. 4). Name the columns with, e.g., "10 nM vs. Ctrl."

4. Fill the decimal average ratios of all proteins into the "ratio (decimal)" column.

3.6.1

Accession	Log2 values of mean value ratios			ttest for both value populations (biol. triplicates) adressed per ratio		
	Log2 10nM vs Ctrl	Log2 100nM vs Ctrl	Log2 10nM vs 100nM	P 10nM vs Ctrl	P 100nM vs Ctrl	P 10nM vs 100nM
P05387	0.03	0.00	-0.03	0.78	0.99	0.58
P05787	0.05	0.05	0.00	0.26	0.45	0.99
O95777	0.21	0.11	-0.10	0.14	0.36	0.13

3.6.3 $=if(log2ratio_{avr.}>=Thr_{up};"UP";if(log2ratio_{avr.}<=Thr_{down};"DOWN";"\#"))$

3.6.4

Evaluation of regulated features: 3xsigma of ratio distribution in all PSMs of global dataset:LOG2 values					
-0.62	Down	-0.61	Down	-0.51	Down
0.64	Up	0.62	Up	0.49	Up
10nM vs Ctrl		100nM vs Ctrl		10nM vs 100nM	
Regulation	Ratio (decimal)	Regulation	Ratio (decimal)	Regulation	Ratio (decimal)
#	1.02	#	1.00	#	0.98
#	1.03	#	1.04	#	1.00
#	1.16	#	1.08	#	0.93

Fig. 4 Overview of calculations (in Excel) for data evaluation according to ratios and *p*-value (*see* Subheading 3.6)

5. Copy the evaluation columns from Subheading 3.4, **step 5**, the protein accession, the *p*-values and the average log2 ratios into a new Excel work sheet tab (as values!). This sheet can now be used to sort, filter, and visualize the data.

6. Calculate the −log10 *p*-values for each condition pair, and insert the numbers as values in a new column (*see* **Note 8**).

4   Notes

1. In case of a multiplexed sample subjected to pre-fractionation (e.g., high-pH) prior to LC-MS/MS analysis, mark all respectiveraw files together and press OK when importing via the file importer node.

2. The latest, curated protein database in FASTA format of the organism of choice (e.g., human) can be downloaded from SwissProt accessible via www.uniprot.org [14].

3. Include the isotopic correction factors, which were shipped together with your TMT reagents. Each batch of TMT has its own factors, be sure to use the correct ones (*see* batch-No.)

4. Transformation to the log-scale turns data into a symmetrical entity, which is needed to perform downstream statistical evaluation.

5. If the normalization of sample amounts was performed on the experimental and on the data levels correctly, the median log2 ratio should converge to 0.

6. This validates the usage of the standard deviation as criterion to define regulation thresholds. If the data shape shows several maxima or is asymmetrical/shifted, either sampling, normalization, or calculations are wrong. Check the data and repeat the experiment if the error is not comprehensible.

7. Use the appropriate one depending on the nature and background of samples. A *t*-test with wrong assumption of the null hypothesis is of no value at all.

8. Data or subsets of the data can be uploaded, e.g., to online tools such as DAVID [15], Panther [16] or STRING [17] to perform gene ontology or network analyses or use the Bioconductor package [13] in *R* to statistically evaluate the data in more detail. Stand-alone visualization software such as Instant-Clue [11] and Perseus [12] will help to plot the results in a meaningful way. The −log10-*P*-values and log2 ratios can be used to perform a crude Volcano-plot in Excel to get a first overview of regulated features.

Acknowledgments

OP, LK, and AS acknowledge the support by the Ministerium für Kultur und Wissenschaft des Landes Nordrhein-Westfalen, the Regierende Bürgermeister von Berlin—inkl. Wissenschaft und Forschung, and the Bundesministerium für Bildung und Forschung.

References

1. Thompson A, Schäfer J, Kuhn K et al (2003) Tandem mass tags: a novel quantification strategy for comparative analysis of complex protein mixtures by MS/MS. Anal Chem 75 (8):1895–1904. https://doi.org/10.1021/ac0262560

2. Thompson A, Schäfer J, Kuhn K et al (2003) Tandem mass tags: a novel quantification strategy for comparative analysis of complex protein mixtures by MS/MS. Anal Chem 75 (18):4942–4942. https://doi.org/10.1021/ac030267r

3. Bantscheff M, Lemeer S, Savitski MM et al (2012) Quantitative mass spectrometry in proteomics: critical review update from 2007 to the present. Anal Bioanal Chem 404 (4):939–965. https://doi.org/10.1007/s00216-012-6203-4

4. Bantscheff M, Schirle M, Sweetman G et al (2007) Quantitative mass spectrometry in proteomics: a critical review. Anal Bioanal Chem 389(4):1017–1031. https://doi.org/10.1007/s00216-007-1486-6

5. Thompson A, Wölmer N, Koncarevic S et al (2019) TMTpro: design, synthesis, and initial evaluation of a proline-based isobaric 16-Plex tandem mass tag reagent set. Anal Chem 91 (24):15,941–15,950. https://doi.org/10.1021/acs.analchem.9b04474

6. Cox J, Mann M (2008) MaxQuant enables high peptide identification rates, individualized p.p.b.-range mass accuracies and proteome-wide protein quantification. Nat Biotechnol 26(12):1367–1372. https://doi.org/10.1038/nbt.1511

7. Zhang J, Xin L, Shan B et al (2012) PEAKS DB: de novo sequencing assisted database search for sensitive and accurate peptide identification. Mol Cell Proteomics 11(4): M111.010587. https://doi.org/10.1074/mcp.M111.010587

8. Plenker D, Riedel M, Bragelmann J et al (2017) Drugging the catalytically inactive state of RET kinase in RET-rearranged tumors. Sci Transl Med 9(394):eaah6144. https://doi.org/10.1126/scitranslmed.aah6144

9. Perkins DN, Pappin DJ, Creasy DM et al (1999) Probability-based protein identification by searching sequence databases using mass spectrometry data. Electrophoresis 20 (18):3551–3567. https://doi.org/10.1002/(sici)1522-2683(19991201)20:18<3551::aid-elps3551>3.0.co;2-2

10. Kall L, Canterbury JD, Weston J et al (2007) Semi-supervised learning for peptide identification from shotgun proteomics datasets. Nat Methods 4(11):923–925. https://doi.org/10.1038/nmeth1113

11. Nolte H, MacVicar TD, Tellkamp F et al (2018) Instant clue: a software suite for interactive data visualization and analysis. Sci Rep 8 (1):12,648–12,648. https://doi.org/10.1038/s41598-018-31154-6

12. Tyanova S, Temu T, Sinitcyn P et al (2016) The Perseus computational platform for comprehensive analysis of (prote)omics data. Nat Methods 13(9):731–740. https://doi.org/10.1038/nmeth.3901

13. Huber W, Carey VJ, Gentleman R et al (2015) Orchestrating high-throughput genomic analysis with bioconductor. Nat Methods 12 (2):115–121. https://doi.org/10.1038/nmeth.3252

14. Consortium TU (2015) UniProt: a hub for protein information. Nucleic Acids Res 43 (D1):D204–D212. https://doi.org/10.1093/nar/gku989

15. Huang DW, Sherman BT, Lempicki RA (2009) Systematic and integrative analysis of large gene lists using DAVID bioinformatics resources. Nat Protoc 4(1):44–57. https://doi.org/10.1038/nprot.2008.211

16. Mi H, Muruganujan A, Ebert D et al (2018) PANTHER version 14: more genomes, a new PANTHER GO-slim and improvements in enrichment analysis tools. Nucleic Acids Res 47(D1):D419–D426. https://doi.org/10.1093/nar/gky1038

17. Szklarczyk D, Gable AL, Lyon D et al (2019) STRING v11: protein-protein association networks with increased coverage, supporting functional discovery in genome-wide experimental datasets. Nucleic Acids Res 47(D1): D607–d613. https://doi.org/10.1093/nar/gky1131

Chapter 29

Mining Protein Expression Databases Using Network Meta-Analysis

Christine Winter and Klaus Jung

Abstract

Public databases featuring original, raw data from "Omics" experiments enable researchers to perform meta-analyses by combining either the raw data or the summarized results of several independent studies. In proteomics, high-throughput protein expression data is measured by diverse techniques such as mass spectrometry, 2-D gel electrophoresis or protein arrays yielding data of different scales. Therefore, direct data merging can be problematic, and combining the summarized data of the individual studies can be advantageous. A special form of meta-analysis is network meta-analysis, where studies with different settings of experimental groups can be combined. However, all studies must be linked by one experimental group that has to appear in each study. Usually that is the control group. Then, a study network is formed and indirect statistical inferences can also be made between study groups that appear not in each of the studies.

In this chapter, we describe the working principle of and available software for network meta-analysis. The applicability to high-throughput protein expression data is demonstrated in an example from breast cancer research. We also describe the special challenges when applying this method.

Key words Batch effects, Biological databases, Data merging, Data mining, Network meta-analysis, Protein expression data, Publication guidelines, Reproducibility, Research synthesis

1 Introduction

1.1 Public Data and Reproducibility of Research

For a long period of time, researchers from all areas of science were used to publish results of their experiments and studies only in the form of statistical analyses, which means data were presented in a summarized form. The raw, original data were often left locked in the researchers filing cabinet and not provided to the public. It can be regarded as a little irony of history, that first with an increasing emergence of large high-throughput data sets from molecular biology, the research community and scientific journals started to request for the public availability of raw, original research data.

Now, such data can often be retrieved from public biological databases such as Gene Expression Omnibus (GEO) [1], ArrayExpress (AE) [2], or the PRIDE Archive [3]. These databases enable

Katrin Marcus et al. (eds.), *Quantitative Methods in Proteomics*, Methods in Molecular Biology, vol. 2228, https://doi.org/10.1007/978-1-0716-1024-4_29, © Springer Science+Business Media, LLC, part of Springer Nature 2021

researchers on the one hand to reproduce the analysis of others in order to get a better understanding of an article and on the other hand to perform research synthesis by means of meta-analyses, data fusion, and data mining. Ideally, for the replication of an analysis, not only the raw data but also the analysis code is provided by the authors of a scientific publication [4]. Original study data is also of importance in the context of the so called "reproducibility crisis," a term that has been used a lot in the last 5 years to describe the problem that published research findings could often not be reproduced by other groups [5]. By bringing together the information of multiple studies, meta-analyses are based on larger sample sizes and can thus increase the statistical power to uncover experimental effects. A further advantage is that contradictory findings between studies can be detected. Consider, for example, that a protein was found to be down regulated by a certain factor in one study but showed an upregulation by the same factor in a different study.

1.2 Research Synthesis and Meta-Analyses of High-Throughput Omics Data

The most widely known form of meta-analysis is a "two-stage" meta-analysis [6], which means to take the summarized data of several individual studies (e.g., in the form of p-values of effect estimates from each study) and to combine them to a new result. Thus the first stage refers to the output of the individual studies and the second to the combination of these. One of the first ideas of such combination statistics is Fisher's combination p-value. Consider k p-values for a particular protein from k independent experiments on the same research question (i.e., same setting of experimental groups). According to Fisher's method, these p-values can be combined to a new test statistic that follows a χ^2-distribution:

$$X^2 \doteq 2 \sum_{i=1}^{k} \ln\left(p_i\right)$$

from which a new combined p-value can be calculated. This technique and derivatives have also been proposed to combine the results of high-throughput transcriptome expression experiments, either measured by DNA microarrays or by RNA-sequencing [7, 8], and can in principle also be used for high-throughput protein expression data. Besides p-value combination, the combination of fold changes from the individual studies is frequently used. The fold change describes by how many folds a protein is up- or downregulated in an experimental group compared to a reference group. Combination of p-values and fold changes is of course done for each feature (i.e., gene or protein) separately.

While combination methods are based on summary statistics from the individual studies, the availability of public, high-throughput expression data enables researchers also to perform "single-stage" meta-analyses, meaning the data from the individual studies are merged and then are jointly analyzed in a single step.

There are several advantages and disadvantages of single- and two-stage meta-analyses. Combing the raw data from independent studies can be complicated when batch effects between the studies are assumed. In proteomics, batch effects can especially be the case when data were measured by different technologies such as mass spectrometry, 2-D gel electrophoresis or protein microarrays. In many cases, steps for batch effect removal can be successful [9]. Combining raw data can, however, be also advantageous because preprocessing steps such as normalization [10] or variance stabilization [11] can be harmonized between the independent studies.

While there is a very high availability of gene expression data sets (several thousand), only 187 data sets flagged by the keyword "proteomic profiling by array" were listed by a query result on ArrayExpress in August 2019.

1.3 Network Meta-Analysis

Classical meta-analysis, regardless of single- or two-stage approaches, is based on the concept of equal study groups in the independent studies. In most cases, two-group designs with a treatment and a control group are considered for meta-analyses. In contrast, a network meta-analysis allows for different experimental groups in the independent studies, with the only restriction that at least one group must be the same over all studies. The most simple network meta-analysis would comprise two independent studies. For example, study 1 compares treatment A versus control, and study 2 compares treatment B versus control. Let's assume treatment A and B are two different drugs for the same disease. The control groups (labelled by C) of both studies would constitute the link between the two studies. This simple study network would allow for an indirect comparison between the study groups A and B, i.e., a comparison that was not made in the two original studies.

More complex study networks can be considered as long as treatment effects are consistent within a network. Assume, in the context of protein expression data, the log fold change between A and C is equal to 5 and that between B and C is also 5. If in an independent third study an upregulation from A to B by a log fold change of also 5 would be reported, the results were inconsistent.

Generally, a study network can be represented by a graph were each node represents a study group and edges that connect the nodes represent independent studies. When visualizing a graph, direct comparisons, i.e., comparisons made in the original studies, are usually illustrated by solid lines and indirect comparisons by dashed lines. It should also be remarked that a network meta-analysis can incorporate a classical meta-analysis in that way that a particular comparison between two study groups can be supported by several independent studies. Then, multiple lines would be drawn between the specific two groups. The performance of network meta-analysis in high-dimensional expression data when a comparison is supported by multiple independent studies was studied in Winter et al. (2019) [12].

2   Materials

To illustrate the working principle of network meta-analysis on protein expression data, we have selected two public available sets of high-throughput protein expression data related to breast cancer research (*see* **Note 1**). The first data set was provided within the R-package "mixOmics" [13] and contains expression levels for 142 proteins measured in tissue samples of $n = 150$ women from three different breast cancer subtypes: $n = 45$ with subtype "Basal," $n = 30$ with subtype "Her2," and $n = 75$ with subtype "LumA." Originally the data were measured using mass spectrometry as part of the Cancer Genome Atlas Project [14]. The second data set was retrieved from ArrayExpress under the accession number GSE68114, containing protein expression levels observed by means of protein arrays (*see* **Note 2**). This data set involves 2352 proteins measured in $n = 45$ samples from women with basal-like breast cancer and $n = 45$ healthy controls. After manually matching the protein names (*see* **Note 3**) of both studies, 13 features remained for network meta-analysis: BAX, GATA3, PCNA, PDCD4, PDK1, PTEN, PEA15, SMAD3, EIF4E, P53, STAT3, GSK3A, STAT5A.

Aiming for a network meta-analysis of the two studies, a study network can be build (*see* **Note 4**), consisting of four nodes representing the four groups Control, Basal, Her2, and LumA (Fig. 1), where both studies contribute with samples to the Basal group. Thus, this group can be used to link both independent studies. Furthermore, there are four direct edges representing the

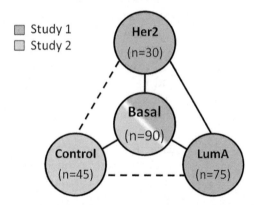

Fig. 1 Study network based on two independent sets of protein expression data from breast cancer research. Study 1, labelled in green, contributed with data for the Basal subtype and a control group. Study 2, labelled in orange, contributed with data for the three subtypes Basal, Her2 and LumA but no control group. Solid edges represent comparisons that were made in the individual studies. Dashed edges represent indirect comparisons enabled by network meta-analysis

comparisons already made in the individual studies, and two indirect edges (Her2 versus Control and LumA versus Control), that were not yet performed in the individual studies.

3   Methods

3.1   Normalization and Differential Expression Analysis

Both data sets were first normalized using the quantile method [15]. Differential expression analysis was performed for both data sets individually and also for the merged data set. Batch effects after data merging were removed according to Johnson et al. (2007) [16] (cf. following subsection). The linear models for microarray data, implemented in the R-package "limma" were employed to perform comparison between each pair of groups. Raw p-values were adjusted to control for a false discovery rate (FDR) of 5%.

3.2   Analysis of Merged Data

After merging the two data sets, using the intersect of 13 proteins, a batch effect between the two studies can be observed (Fig. 2). We

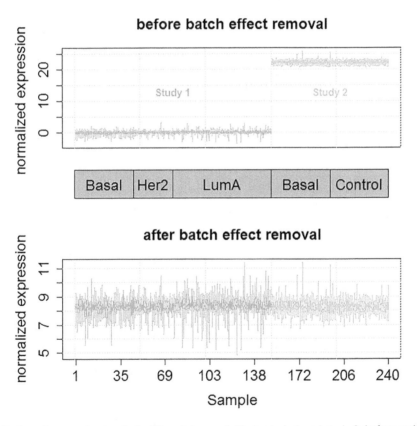

Fig. 2 Distribution of expression levels for 13 proteins available in study 1 and study 2, before and after batch effect removal. Each sample from the four different study groups is represented by one boxplot. Still after batch effect removal samples from study 1 show a higher variance then those of study 2, indicated by longer whiskers

used a batch effect model proposed by Johnson et al. (2007) [16] that includes an additive and a multiplicative batch effect:

$$\Upsilon_{ijp} = \alpha_p + X\beta_p + \gamma_{ip} + \delta_{ip}\varepsilon_{ijp}.$$

In this model, Υ_{ijp} represents the expression level of protein p in sample i of study j. The overall expression level is given by α_p, X is a model matrix indicating the study groups and β_p is a vector of additional group specific expression (i.e., in terms of proteomics, the fold change). Additive and multiplicative batch effects are represented by γ_{ip} and δ_{ip}, respectively, and ε_{ijp} is a normally distributed error term. Estimation and removal of batch effects is implemented in the function "ComBat" of the R-package "sva." The results of applying this function to our merged data set can also be seen in Fig. 2.

An extended model was considered by Hornung et al. (2016) [17], who include an additional term with random latent factors, Z_{ijl}, to allow for different correlation structures between the batches.

$$\Upsilon_{ijp} = \alpha_p + a_{ij}^T\beta_p + \gamma_{ip} + \sum_{l=1}^{m_j} b_{jpl}Z_{ijl} + \delta_{ip}\varepsilon_{ijp}.$$

Their method is implemented in the R-package "Bapred."

3.3 Network Meta-Analysis Based on Individual Study Results

It has been shown that in some cases, steps for batch removal can still leave a batch related bias in the merged data and the subsequent analysis [18]. Especially in proteomics, when expression data was measured by different types of technologies (mass spectrometry, 2-D gel electrophoresis, protein arrays), data merging and batch effect removal can become a difficult task. Therefore, specialized methods that rely on the results of the individual studies can be helpful. Different methods for such kind of network meta-analysis have been proposed and are either based on electrical networks [19], Bayesian [20, 21], or multivariate models [22]. Implementations of these methods are available in the R-packages "netmeta," "gemtc," and "pcnetmeta." In general, these tools need as input the fold changes and their related standard errors from the differential expression analysis of the individual studies. Using the breast cancer example described in the Materials section, input data would be represented separately for each protein in the input matrices as follows:

$$
F = \begin{array}{c} Control \\ Basal \\ Her2 \\ LumA \end{array}
\begin{array}{cccc} C. & B. & H. & L. \end{array}
\left(\begin{array}{cccc}
0 & f_{CB} & f_{CH} & f_{CL} \\
f_{BC} & 0 & f_{BH} & f_{BL} \\
f_{HC} & f_{HB} & 0 & f_{HL} \\
f_{LC} & f_{LB} & f_{LH} & 0
\end{array}\right)
$$

and

$$SE = \begin{array}{c} \\ Control \\ \\ Basal \\ \\ Her2 \\ \\ LumA \end{array} \begin{array}{cccc} C. & B. & H. & L. \\ \begin{pmatrix} 0 & s_{CB} & s_{CH} & s_{CL} \\ s_{BC} & 0 & s_{BH} & s_{BL} \\ s_{HC} & s_{HB} & 0 & s_{HL} \\ s_{LC} & s_{LB} & s_{LH} & 0 \end{pmatrix} \end{array},$$

with F representing the log fold changes for each group comparison and SE the related standard errors (*see* **Note 5**). Of course, the log fold changes and standard errors on the diagonal—comparison of each group with itself—are zero. Please remark also that $f_{ij} = -f_{ji}$, i.e., if the log fold change for comparing group i versus group j is positive, the log fold change for the reverse comparison would be negative. The log fold changes and standard errors would only be available as input data for those comparisons that had been performed in the original studies. In the breast cancer example, the indirect comparisons Her2 versus Control and LumA versus Control would first be indirectly estimated from the input data (cf. Fig. 1).

4 Results and Discussion

4.1 Network Meta-Analysis Versus Analysis of Merged Data

Differential expression analysis was performed in three different ways. First, each study was analyzed by itself. Second, data from the two studies were merged and after batch effect removal jointly analyzed. Finally, the results from the individual studies were used to perform network meta-analysis as described in Subheading 3.2. Analysis of merged data and network meta-analysis allowed to make indirect inferences between Her2 or LumA versus control. Only the 13 proteins described above which could be matched between the two studies were subjected to merged data and network meta-analysis. Figure 3 shows the log fold changes +/− standard error for each group comparison based on the three types of analysis. Figure 4 additionally lists the log fold change and FDR-adjusted p-values of the significant findings.

We were particularly interested in the comparison between the analysis of the merged data and the analysis by network meta-analysis. Regarding the direction of fold changes, i.e., up- or down-regulation, it can be observed that this was mainly the same for both types of analysis. However, in most cases there was a difference in the size of the fold change. When looking at the results obtained for the direct group comparisons (Basal versus Control, Her2 versus Basal, LumA versus Basal and LumA versus Her2), the fold changes calculated by network meta-analysis were overall closer to

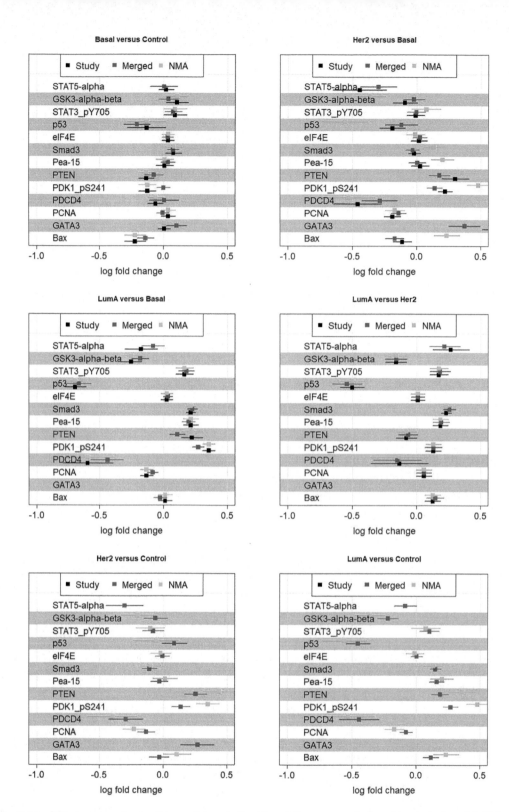

Fig. 3 Log fold changes +/− confidence intervals for 13 proteins commonly found in study 1 and 2. Values were calculated for each group comparison in the study network. The above four plots show results for study-specific (black, bottom indicator per row) as well as merged data (red, middle indicator per row) and network meta-analysis (green, top indicator per row). The two plots at the bottom only show indirectly calculated log fold changes by using either merged data or network meta-analysis

A) Basal vs. LumA

	Study		merged		NMA	
	p_{FDR}	logFC	p_{FDR}	logFC	p_{FDR}	logFC
GATA3	< 0.01	1.96	< 0.01	1.62	< 0.01	1.96
GSK3	0.01	-0.25	0.01	-0.18	0.01	-0.25
PDK1	< 0.01	0.35	< 0.01	0.27	< 0.01	0.35
PTEN	0.02	0.22			0.03	0.22
TP53	< 0.01	-0.69	< 0.01	-0.66	< 0.01	-0.69
PCNA	0.02	-0.13			0.03	-0.13
PDCD4	0.01	-0.59	< 0.01	-0.44	0.01	-0.59
PEA15	< 0.01	0.21	< 0.01	0.20	< 0.01	0.21
SMAD3	< 0.01	0.22	< 0.01	0.23	< 0.01	0.22
STAT3	0.03	0.17	0.01	0.18	0.04	0.17

B) Control vs. LumA

	Study		merged		NMA	
	p_{FDR}	logFC	p_{FDR}	logFC	p_{FDR}	logFC
GATA3			< 0.01	1.52	< 0.01	1.96
PDK1			< 0.01	0.27	< 0.01	0.48
PTEN			0.01	0.19	< 0.01	0.36
GSK3			0.01	-0.22	0.02	-0.36
p53			< 0.01	-0.45	0.01	-0.56
PDCD4			0.01	-0.45		
PEA15			0.01	0.16		
SMAD3			< 0.01	0.15		

C) Her2 vs.LumA

	Study		merged		NMA	
	p_{FDR}	logFC	p_{FDR}	logFC	p_{FDR}	logFC
GATA3	< 0,01	1.29	< 0,01	1.25	< 0,01	1.29
STAT3	< 0.05	0.18				
p53	< 0,01	-0.50	< 0,01	-0.54	< 0,01	-0.50
PEA15	0.01	0.19	0.01	0.19	0.02	0.19
SMAD3	< 0,01	0.23	< 0,01	0.26	< 0.01	0.23
Bax			0.08	0.15		

D) Basal vs. Her2

	Study		merged		NMA	
	p_{FDR}	logFC	p_{FDR}	logFC	p_{FDR}	logFC
GATA3	< 0.01	0.67	0.03	0.37	< 0.01	1.96
GSK3					0.02	-0.36
PDK1	< 0,01	0.22			< 0.01	0.48
PTEN	0.04	0.30			< 0.01	0.36
TP53					0.01	-0.56
PCNA	0.04	-0.19				

E) Control vs. Her2

	Study		merged		NMA	
	p_{FDR}	logFC	p_{FDR}	logFC	p_{FDR}	logFC
GATA3					< 0.01	0.66
PDK1					< 0.01	0.35
PTEN					< 0.01	0.44

Fig. 4 Tables A to E show significantly different expressed proteins with their log fold changes and FDR-adjusted p-values found in direct and indirect comparisons, using original study results, or analysis of merged data or results from network meta-analysis (NMA)

the fold changes determined in the individual study data, while the fold changes calculated in the merged data were different from the original data. Based on these results, we assume also for the indirect comparisons that fold changes from network meta-analysis are more accurate than those from merged data analysis. Standard errors, however, appeared to be of similar size for both types of analysis.

4.2 Biological Interpretation

4.2.1 Differences Regarding the Method of Analysis

Both studies evaluated the expression of numerous proteins, but we found a small intersection containing 13 proteins (BAX, EIF4E, GATA3, GSK3A, P53, PCNA, PDCD4, PDK1, PEA15, PTEN, SMAD3, STAT3, STAT5A). Just regarding this overlap of the two studies, the biggest difference in the proteome can be found between the cancer subtypes Basal and LumA, regardless of the type of data analysis. In contrast, the comparison of Basal and Control group revealed no significant differences in protein expression.

There are also differences considering the analysis method. Comparing, e.g., the protein expression in cells of the Basal and HER2 cancer subtypes, there is only one protein (GATA3) upregulated after data merging. But when using NMA, there can be found five differentially expressed proteins (GATA3, GSK3A, PDK1, PTEN, P53). The comparison of Basal or LumA proteomes on the other hand revealed very similar results with up to ten differentially expressed proteins.

A closer look at the individual proteins reveals that some proteins are found as differentially expressed independently from the analysis method, e.g., PDCD4. This nuclear protein has been found downregulated in the original study after data merging as well as when using network meta-analysis. Whereas BAX expression alters just between Her2 and LumA cells after data merging, GATA3 levels differ in almost all comparisons.

4.2.2 Role of Significant Proteins in Breast Cancer

All examined proteins are described in relation with diverse forms of cancerous diseases. The reason can be found in the specific function of each of these proteins in regulation of cell cycle and apoptosis. Instead of a comprehensive presentation and discussion of all proteins, in this chapter, we only discuss exemplarily a few results to give an impression of the possible biological interpretation.

The transcription factor GATA3 (GATA binding protein 3) is known for its role in t-cell differentiation and maturation [23, 24] but is also important during mammary gland development, especially for luminal epithelial cell differentiation [25, 26]. Furthermore, GATA3 is suggested as a reliable prognostic marker for breast cancer. The protein can be detected in most of the estrogen receptor positive tumors [27]. Comparing different cancer subtypes, LumA cells showed the highest GATA3 expression. Mehra et al. (2005) analyzed cDNA data sets of over 300 patient and found low GATA3 levels associated with shorter overall survival in contrast to cancer cells that showed higher expression of this transcription factor [28]. GATA3 seems to prevent the metastatic spread by regulating cadherin expression. In absence of E-cadherin, tumor cells are incapable of attach to cells in, e.g., the lung and thus creating metastases [29, 30].

5   Discussion

In this chapter we have demonstrated the applicability of network meta-analysis in proteomics. Due to batch effects and due to different laboratory tools to measure high-throughput protein expression data, network meta-analysis provides a practical alternative to the merging of multiple independent data sets and subsequent joint analysis of the merged data.

Currently, the public availability of high-throughput protein expression data is not as extensive as gene expression data sets. The authors therefore recommend that researchers in that field make their data publicly available when they publish their research result.

While in our data example we had only a small overlap of 13 proteins between the two breast cancer studies, larger overlaps can be reached when authors provide their full, unfiltered expression data. In addition, we recommend publishing authors to follow the guidelines of the PRISMA statement to provide relevant information of their study (e.g., about the experimental design and experimental factors) to facilitate meta-analysis by other researchers in the field.

6 Notes

1. When selecting appropriate study data for network meta-analysis, the analyst should follow the general guidelines for meta-analysis listed by the PRISMA statement (http://www.prisma-statement.org/). The PRISMA statement provides a checklist for selecting studies starting with a list of query keywords in a database and continues with criteria for removing studies that don't fit to the primary question of the meta-analysis.

2. To date, nearly 400 data sets from proteomic profiling experiments are available in the ArrayExpress database. Data sets which are also linked to the GEO database can be easily imported into the R analysis environment using the function getGEO in the "GEOquery" package by specifying the accession number of the data set.

3. Not only when using merged data analysis but also when using network meta-analysis, the analyst has to match protein names between the different independent studies. A helpful tool to obtain alias names of different protein is the UniProt database (www.uniprot.org). In this database, alias names and protein identifiers for a large number of species is available. The R-functions "match," "intersect," or "grep" are also helpful to obtain an intersect list of proteins between the individual studies.

4. The R-package "igraph" provides useful functions for drawing network diagrams.

5. Analyst should be careful to annotate log fold changes with the correct sign $(+/-)$ for up- and downregulation. False annotation can lead to inconsistencies in the study network. Furthermore, the function in the "netmeta" package can produce warnings for inconsistencies in the variances of the log fold changes or due to negative variance estimates. In that case, the parameter tol.multiarm can be used to compensate for these difficulties.

References

1. Edgar R, Domrachev M, Lash AE (2002) Gene expression omnibus: NCBI gene expression and hybridization array data repository. Nucleic Acids Res 30(1):207–210

2. Brazma A, Parkinson H, Sarkans U et al (2003) ArrayExpress—a public repository for microarray gene expression data at the EBI. Nucleic Acids Res 31(1):68–71

3. Vizcaíno JA, Csordas A, Del-Toro N et al (2015) 2016 update of the PRIDE database and its related tools. Nucleic Acids Res 44 (D1):D447–D456

4. Hofner B, Schmid M, Edler L (2016) Reproducible research in statistics: a review and guidelines for the Biometrical Journal. Biom J 58(2):416–427

5. Baker M (2016) 1500 scientists lift the lid on reproducibility. Nature 533(7604):452

6. Burke DL, Ensor J, Riley RD (2017) Meta-analysis using individual participant data: one-stage and two-stage approaches, and why they may differ. Stat Med 36(5):855–875

7. Marot G, Foulley JL, Mayer CD et al (2009) Moderated effect size and P-value combinations for microarray meta-analyses. Bioinformatics 25(20):2692–2699

8. Rau A, Marot G, Jaffrézic F (2014) Differential meta-analysis of RNA-seq data from multiple studies. BMC Bioinformatics 15(1):91

9. Lazar C, Meganck S, Taminau J et al (2012) Batch effect removal methods for microarray gene expression data integration: a survey. Briefings Bioinform 14(4):469–490

10. Callister SJ, Barry RC, Adkins JN et al (2006) Normalization approaches for removing systematic biases associated with mass spectrometry and label-free proteomics. J Proteome Res 5 (2):277–286

11. Kreil DP, Karp NA, Lilley KS (2004) DNA microarray normalization methods can remove bias from differential protein expression analysis of 2D difference gel electrophoresis results. Bioinformatics 20(13):2026–2034

12. Winter C, Kosch R, Ludlow M et al (2019) Network meta-analysis correlates with analysis of merged independent transcriptome expression data. BMC Bioinformatics 20(1):144

13. Rohart F, Gautier B, Singh A et al (2017) mixOmics: an R package for 'omics feature selection and multiple data integration. PLoS Comput Biol 13(11):e1005752

14. Weinstein JN, Collisson EA, Mills GB et al (2013) The cancer genome atlas pan-cancer analysis project. Nature Gen 45(10):1113

15. Bolstad BM, Irizarry RA, Åstrand M et al (2003) A comparison of normalization methods for high density oligonucleotide array data based on variance and bias. Bioinformatics 19 (2):185–193

16. Johnson WE, Li C, Rabinovic A (2007) Adjusting batch effects in microarray expression data using empirical Bayes methods. Biostatistics 8 (1):118–127

17. Hornung R, Boulesteix AL, Causeur D (2016) Combining location-and-scale batch effect adjustment with data cleaning by latent factor adjustment. BMC Bioinformatics 17(1):27

18. Nygaard V, Rødland EA, Hovig E (2016) Methods that remove batch effects while retaining group differences may lead to exaggerated confidence in downstream analyses. Biostatistics 17(1):29–39

19. Rücker G (2012) Network meta-analysis, electrical networks and graph theory. Res Synth Methods 3(4):312–324

20. van Valkenhoef G, Lu G, de Brock B et al (2012) Automating network meta-analysis. Res Synth Methods 3(4):285–299

21. Dias S, Sutton AJ, Ades AE, Welton NJ (2013) Evidence synthesis for decision making 2: a generalized linear modeling framework for pairwise and network meta-analysis of randomized controlled trials. Med Decis Making 33 (5):607–617

22. Zhang J, Carlin BP, Neaton JD et al (2014) Network meta-analysis of randomized clinical trials: reporting the proper summaries. Clin Trials 11(2):246–262

23. Skapenko A, Leipe J, Niesner U et al (2004) GATA-3 in human T cell helper type 2 development. J Exp Med 199(3):423–428

24. Ho IC, Tai TS, Pai SY (2009) GATA3 and the T-cell lineage: essential functions before and after T-helper-2-cell differentiation. Nat Rev Immunol 9(2):125

25. Asselin-Laba ML, Sutherland KD, Barker H et al (2007) Gata-3 is an essential regulator of mammary-gland morphogenesis and luminal-cell differentiation. Nature Cell Biol 9(2):201

26. Kouros-Mehr H, Slorach EM, Sternlicht MD et al (2006) GATA-3 maintains the differentiation of the luminal cell fate in the mammary gland. Cell 127(5):1041–1055

27. Fararjeh AFS, Tu SH, Chen LC et al (2018) The impact of the effectiveness of GATA3 as a prognostic factor in breast cancer. Hum Pathol 80:219–230

28. Mehra R, Varambally S, Ding L et al (2005) Identification of GATA3 as a breast cancer prognostic marker by global gene expression meta-analysis. Cancer Res 65 (24):11,259–11,264

29. Dydensborg AB, Rose AAN, Wilson BJ et al (2009) GATA3 inhibits breast cancer growth and pulmonary breast cancer metastasis. Oncogene 28(29):2634

30. Yan W, Cao Q, Arenas RB et al (2010) GATA3 inhibits breast cancer metastasis through the reversal of epithelial-mesenchymal transition. J Biol Chem 285(18):14,042–14,051

Chapter 30

A Tutorial for Variance-Sensitive Clustering and the Quantitative Analysis of Protein Complexes

Veit Schwämmle and Christina E. Hagensen

Abstract

Data clustering facilitates the identification of biologically relevant molecular features in quantitative proteomics experiments with thousands of measurements over multiple conditions. It finds groups of proteins or peptides with similar quantitative behavior across multiple experimental conditions. This co-regulatory behavior suggests that the proteins of such a group share their functional behavior and thus often can be mapped to the same biological processes and molecular subnetworks.

   While usual clustering approaches dismiss the variance of the measured proteins, VSClust combines statistical testing with pattern recognition into a common algorithm. Here, we show how to use the VSClust web service on a large proteomics data set and present further tools to assess the quantitative behavior of protein complexes.

   Key words Proteomics, Bioinformatics, Protein complexes, Multivariate analysis, Cluster analysis, Differential analysis, Pattern recognition, Biological pathways

1 Introduction

The acquisition of large-scale omics data is becoming a standard procedure in biological experiments. With the rise of vast amounts of data coming from proteomics, transcriptomics, and metabolomics experiments, there is an increasing need for user-friendly and powerful tools for their analysis. Moreover, we do not only see an increase in the amount of experiments but also in their complexity where thousands of molecules, often called features, are quantified over multiple different experimental conditions such as different time points or multiple disease states, and furthermore often contain multiple samples (replicates) per condition. Hence, the resulting multidimensional quantitative data poses challenges to its analysis by standard statistical tests where in general only pairs of conditions are compared or a score for an overall change is calculated.

Katrin Marcus et al. (eds.), *Quantitative Methods in Proteomics*, Methods in Molecular Biology, vol. 2228, https://doi.org/10.1007/978-1-0716-1024-4_30, © Springer Science+Business Media, LLC, part of Springer Nature 2021

In order to support biological interpretation in these multi-dimensional data sets, cluster analysis looks for feature groups behaving quantitatively the same way. The similar behavior often can be mapped to affected biological pathways and functional feature groups. *K*-means clustering is a widely used approach to determine these groups but suffers from being highly sensitive to outliers and additionally does not allow discarding noisy features that are, e.g., located between clusters. In contrast, fuzzy *c*-means clustering [1] is a soft clustering extension of *k*-means that circumvents these issues by assigning a membership of each feature to each cluster. The method can be adjusted to the potential noise in the system and thus be adapted to only provide the feature groups that show enhanced co-expression [2].

Current clustering approaches do not apply any statistical testing and are mostly applied to averaged data, where the means or medians over all values of a feature in a condition were taken. This creates challenges to the data interpretation by the necessity to combine the results from the clustering with the result from statistical tests like ANOVA or LIMMA [3]. Herein, it is, for example, unclear whether to apply the clustering method on filtered (only significantly changing features) or unfiltered data.

Variance-sensitive clustering combines both approaches by adapting the fuzzy *c*-means algorithm to take into account feature variance [4] (*see* Fig. 1 for a comparison with other multivariate analysis approaches). We here present a tutorial to use the VSClust web service where the users can carry out cluster analysis using variance-sensitive and standard fuzzy *c*-means clustering. We will exemplify the workflow on a large breast cancer proteomics data set which we will analyze further using tools to investigate the quantitative behavior of protein complexes.

For biological interpretation, protein complexes play an important role as they are involved in most functional control processes in the cell [6]. It is furthermore becoming clear that many complexes are tightly controlled by degradation of protein subunits that do not form the complex [7]. This active control of protein abundance is one of the major reasons for the low correlation between transcriptome and proteome. We will assess the quantitative behavior of protein complexes using the two web services ComplexBrowser [8] and CoExpresso [9].

2 Materials

All tools and all data resources are freely available through the given web pages.

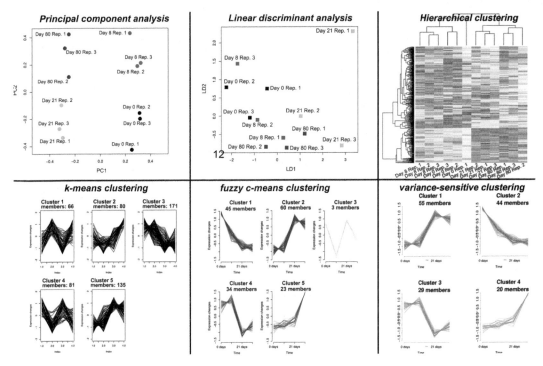

Fig. 1 Comparison of different approaches for multivariate analysis applied to a data set measuring phosphorylated, acetylated and glycosylated peptides along mouse brain development [5]. *Principal component analysis* is a powerful approach to simplify data visualization by mapping the different samples to a lower dimensional space. Here, the different developmental states can be distinguished. *Linear discriminant analysis* is a similar but supervised method for dimensionality reduction to further distinguish different feature groups. The method fails here, mostly due to non-linear changes of the peptides. *Hierarchical clustering* is the most used approach for cluster analysis and does not require to pre-determine the number of clusters in the data. The unclear interpretation of grouping severely complicates the functional analysis in large data sets consisting of thousands of features. *K-means clustering* forces each feature into one of the clusters and is thus highly error-prone in noisy data. *Fuzzy c-means* corrects for noise but does not take into account feature variance. *VSClust* provides a combined way for variance-sensitive clustering. Given that the data consists of a time course, we can assume that the most common quantitative profiles (clusters) will not show changes as the ones found in cluster 1 and 4 of the *k*-means clustering. Fuzzy *c*-means corrects for this effect but still fails by obtaining cluster 3 which is unlikely to represent a cellular process in mouse brain development. VSClust corrects for feature variance and only shows clusters likely to correspond to time-dependent behavior

2.1   Resources

The cluster analysis with VSClust will take place on basis of proteomics data containing nearly 10,000 proteins measured across five breast cancer subtypes, each coming in nine replicates.

Download of quantitative breast cancer data

Download the file *PMC6453966_Apps_in.csv* from the VSClust source code repository: https://bitbucket.org/veitveit/vsclust/src/master/

For that,

1. Click on the file name to open the page of the file.

2. Click on "…" next to the *Edit* button and select *Open raw*.

3. Save the file on your computer (*see* **Note 1**).

2.2 Software

The here presented data analysis is based on web-based software tools which are implemented in Shiny. For further testing and data preprocessing, we recommend installing the R software environment.

We present three different ways to run the web services VSClust, ComplexBrowser, and CoExpresso:

(a) *Docker virtual machines:* For running the Shiny apps locally and without restricted usage, you only need an installation of the Docker container platform. For instructions, *see* https://www.docker.com. When opening the docker command line interface (PowerShell on Windows or a Bash shell on a Linux computer), carry out the following command to download the software container containing VSClust:

   docker pull veitveit/vsclust

   Press Return afterwards. This downloads the most recent version of VSClust in a fully operable Linux environment to your computer. In order to make your analysis reproducible, we recommend using the latest version (release-0.7 when writing this manuscript). The other tools ComplexBrowser and CoExpresso can be pulled by substituting *vsclust* with *complexbrowser* and *coexpresso* (*see* **Note 2**).

   For running VSClust, write *docker run -it -p3838:3838 veitveit/vsclust* and press again Return (*see* **Note 3**). Now you should be able to access the app via *http://localhost:3838/VSClust* in your browser. We recommend using Chrome, Firefox or Safari (*see* **Note 4**). The program can be stopped using *ctrl-c* on the command line interface where you started the Docker image.

(b) *Web service at author's institution:* This is the easiest way to run the software. However, usage might be limited when having too many other users. Use the following URL in your browser:

   http://computproteomics.bmb.sdu.dk/Apps/VSClust

   The other tools ComplexBrowser and CoExpresso can be accessed by substituting VSClust with the respective software name (*see* **Note 5**).

(c) *Run the Shiny apps in Rstudio and R software environment* (www.r-project.org): R provides a versatile scripting language for statistical data analysis. We furthermore recommend installing RStudio (www.rstudio.com) for improved editing and running of the scripts. This also requires installation of all necessary R libraries and associated software. Follow the instructions on https://bitbucket.org/veitveit/vsclust (*see also* **Note 6**), https://bitbucket.org/michalakw/complexbrowser and https://bitbucket.org/veitveit/coexpresso

3 Methods

You will analyze data from a study investigating the deep proteome in breast cancer subtypes [10].

3.1 Data Preparation

We use quantitative profiles of almost 10,000 proteins found in all samples. The original file from the publication (Supplementary Table 1 in ref. [10]) has been formatted to only keep the column with the protein names and the columns with the quantitative values over all samples. We further converted the gene names into UniProt accession numbers. For file input, we recommend using simple csv files that are limited to necessary information (*see* **Note 7**).

3.2 Statistics and Clustering Using VSClust

You will now analyze the quantitative data in VSClust. VSClust is based on fuzzy c-means clustering which is a more noise-permitting version of the well-known k-means clustering. The major difference consists in not assigning a fixed cluster identity to each protein.

We will exemplify the analysis on the basis of the given example data set. This description however is generally applicable. For that, the user only needs to prepare their quantitative data in a simple csv-file and set the respective parameters like number of replicates and number of experimental conditions (*see* **Note 8**).

Description of parameters:

(a) Input file (here: *PMC6453966_Apps_in.csv*)

Important: csv-files as input need to fulfill a certain standard (*see* **Note 9**).

We will start by loading the data and visualizing its projection on principal components, also to find out whether we can proceed with the analysis.

(b) Column names? (here: checked)

(c) Gene/protein identifiers in second column? (here: unchecked) (*see* **Note 10**).

(d) Estimate variance levels from replicated quantifications? (here: checked) (*see* **Note 11**).

(e) Paired tests (here: unchecked) (*see* **Note 12**).

(f) Replicates are grouped (here: unchecked) (*see* **Note 13**).

(g) Number of replicates (here: 9)

(h) Number of conditions (here: 5)

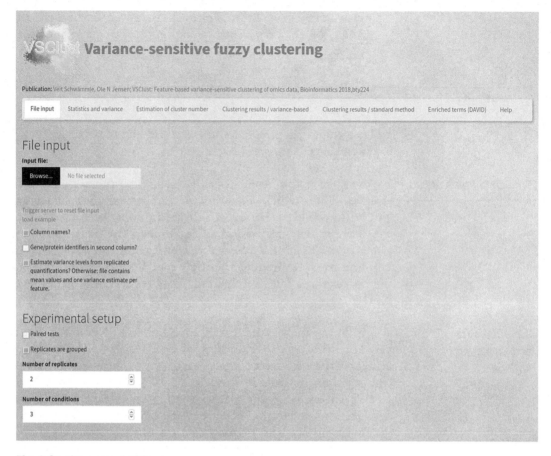

Fig. 2 Starting page of VSClust

1. Upload the file *PMC6453966_Apps_in.csv* to VSClust (*see* also Fig. 2). Set the input parameters to the values given above. Then, the software should automatically open the tab *Statistics and variance*

2. Inspect your data upload and check whether the replicated samples show up as separable groups in the second PCA plot (Fig. 3, right).

   At this stage, VSClust performs statistical tests (LIMMA also known as moderated *t*-test [3]) for differential expression between the different conditions with respect to the first condition in the uploaded file.

   In addition, the different samples and proteins are shown after projection of the full data to the first two principal components and after scaling each feature to *z*-values (*see* **Note 14**).

   Figure 2 shows the distribution of the proteins and their individual variance. A variance estimate is calculated per protein using the LIMMA package and shown by the circle size. The right figure gives an idea about the similarity between the

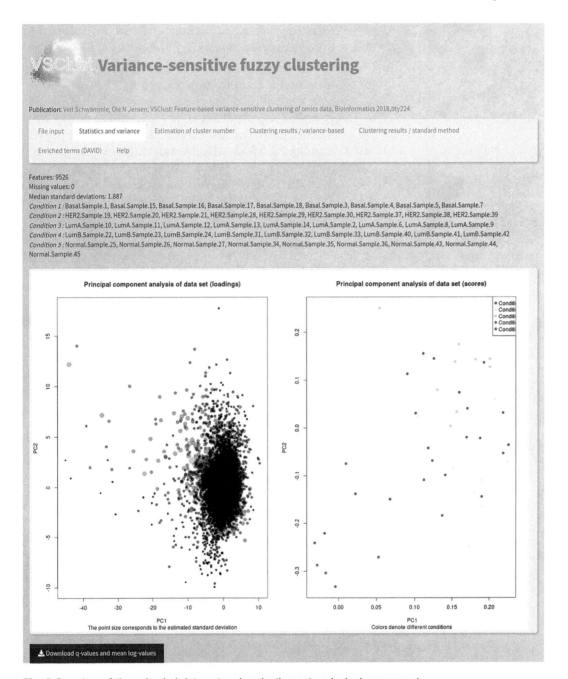

Fig. 3 Overview of the uploaded data set and projection onto principal components

different samples. Here, samples/replicates from the same experimental condition should roughly arrange in groups to ensure that the data is not too noisy for further analysis. Completely mixed samples from different conditions are unlikely to display regulatory differences of statistical significance.

You can download an intermediate file containing the results from the statistical tests.

3. Determine the most suitable number of clusters by moving to the tab *Estimation of cluster number.*

Run the estimation of the cluster number moving the slider to 15 (given the large number of proteins, this will take up to 1h!). This runs both VSClust and standard fcm clustering with cluster numbers 3–15 to calculate validation indices that indicate the number of clusters in the data set (*see* **Note 15**).

Look at the plots of the two validation indices (Minimum centroid distance and Xie-Beni index) and the number of proteins that still get assigned to clusters (Fig. 4) (*see* **Note 16**).

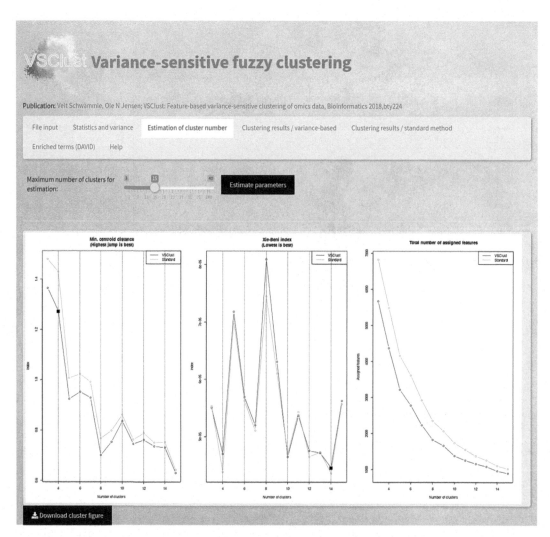

Fig. 4 Estimation of cluster number from data structure (left panel: Minimum centroid distance versus number of clusters, middle panel: Xie-Beni index versus number of clusters). Observe that the number of clustered features decreases drastically with the number of clusters (right panel)

Given the mostly hierarchical structure of biological data, one can choose alternative numbers according to the criteria described in **Note 15**

The variance-sensitive version of the fuzzy *c*-means accounts for the uncertainty of a measurement and thus is more stringent in filtering out proteins with unclear cluster assignments.

4. The validation indices indicate the preferred cluster numbers by a black square (*see* **Note 17**). Use these numbers when running the clustering in the tabs *Clustering results/variance-based* and *Clustering results/standard method* (*see* **Note 18**). In the case of the example data set, the best fitting cluster number according to the minimum centroid distance is four. Run the results also for a cluster number of seven afterwards.

Look at the resulting figures (*see* also Fig. 5). Each cluster corresponds to a group of proteins with similar expression profiles between the 5 breast cancer subtypes (Basal, HER2, LumA, LumB, Normal from left to right). Which cancer types/conditions show more similar behavior? Are there groups of co-regulated proteins that distinguish one or multiple cancer type from the other ones? For instance, clusters 1 and 2 show very distinctive behavior in breast cancer type Basal, and cluster 3 and 4 indicate that breast cancer types 1 (Basal) and 4 (LumB) are highly altered when assessing the common behavior of co-changing protein groups.

5. Download the clustering results (button *Download results*) and open the file in Excel or Libreoffice (*see* **Note 19**).

The different columns describe how the proteins are distributed over the different clusters:

Cluster: cluster to which the protein belongs most

Mean of log A – Mean of log F: averaged and standardized value (z-value, *see* previous **Note 14**) of each protein in each experimental condition.

isClusterNumber: TRUE means that the protein was assigned to a cluster. Otherwise the protein is not shown in the figure and should not be used for the further analysis

maxMembership: highest degree of membership to the best fitting cluster. Only values larger than 0.5 are considered (otherwise *isClusterNumber* is set to FALSE)

membership.of.cluster.n: degree to which the protein belongs to each cluster. The sum of all membership values is 1. For running another data set *see* **Note 20**.

6. The last tab of VSClust allows interrogating the DAVID web service (*see* **Note 21**) for enriched GO terms and biological pathways. We use this feature as a fast check for interesting biological content but recommend using the original interface or other tools for more thorough biological interpretation. For

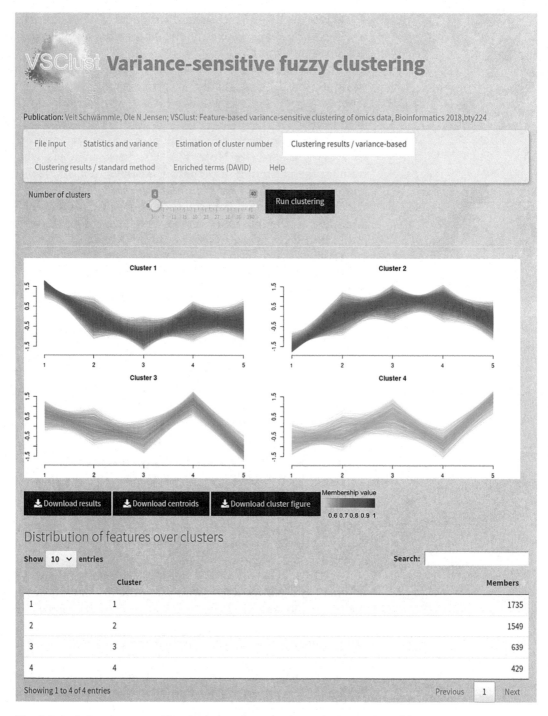

Fig. 5 Output of variance-sensitive clustering after selecting a cluster number of four

the clustering results with 4 or 7 clusters, select *KEGG pathway* in the *Information source (DAVID)* menu and *UNIPROT_ACCESSION* in the *ID type (DAVID)* menu. Figure 6 shows the output for a cluster number of seven. VSClust shows enriched

Fig 6 Summary of enrichment analysis for each cluster

gene annotations both for the variance-sensitive and the standard fuzzy *c*-means clustering (*see* **Note 22**). We observe *Ribosome* (cluster 2) and *Endocytosis* (cluster 1) as the most enriched pathways. They are strongly related to protein complexes, and we will look deeper into their behavior in what follows.

3.3 Quantitative Behavior of Protein Complexes

There is increasing consensus that many proteins being subunits of complexes are tightly regulated by selective degradation [7]. This leads to

(a) common quantitative changes of protein expression profiles and

(b) most visible differences between the transcriptome and proteome

In the following, we shortly describe how to look into quantitative protein complex behavior. More specifically, we will carry out supervised analysis with respect to protein complexes from manual annotation defined in CORUM [11] or Complex Portal [12].

We will look into the behavior of two protein complexes within the different breast cancer subtypes. The following two Shiny apps allow interactively investigating protein complexes, and the co-regulatory behavior of any protein group (*see* **Note 23**).

ComplexBrowser takes a csv-file and runs some basic quality control on the quantitative data. The second stage of the analysis includes summarization of protein changes to describe the quantitative behavior of entire protein complexes. For that, it uses factor analysis for weighted averaging, *see* also [13] and [14].

1. Take the same input file *PMC6453966_Apps_in.csv* and upload it to *ComplexBrowser.*

   Change the number of replicates (9) and number of conditions (5) accordingly, tick *Is data log-transformed* and *Replicates are grouped,* and untick *Are q-values included.* Then press the button *Run QC.* You can now check the data for correlation within replicates, missing values and extract results from statistical testing (*see* **Note 24**). For instance, confirm again whether the replicates of the same breast cancer subtype form separable groups (tab PCA on the lower left panel).

2. We now proceed further to the protein complex analysis by pressing *Complex analysis* on the top of the sidebar menu. Leave all parameters at their default and select *CORUM* and the correct species of the data (Human) (*see* **Note 25**).

3. Select *Ribosome, cytoplasmic* in the table. For that, click on the field below *Complex_name* and write *Ribosome* and then select the correct entry. The visualization will be triggered when you select the complex in the now shortened table (Fig. 7). The figure *Subunits expression profiles* in the upper right panel shows that most proteins follow very similar changes with first breast cancer subtype (basal) being the most abundant. The strong co-regulation of the proteins is confirmed when looking at the correlation between all protein profiles (*Protein correlation map*) and the correlation between the quantitative protein profiles (*Protein expression map*) in the lower left panel. Despite

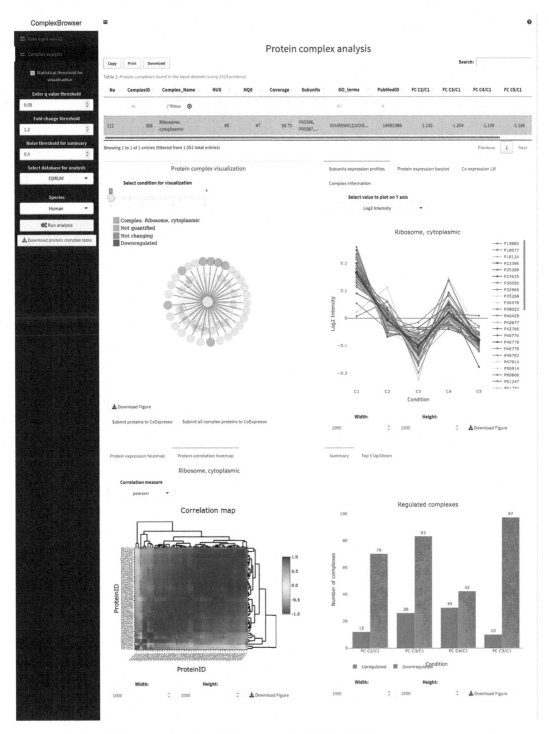

Fig. 7 Visualization of the behavior of the ribosome complex in ComplexBrowser

the high correlation, the overall changes in protein complex abundance, given by the complex fold change *FC C2/C1, FC C3/C1, ...*, are rather low with a maximum of a 1.2-fold change between subtype 3 and subtype 1 (*see* **Note 26**).

Now repeat the procedure for the complexes related to *Endocytosis* (*see* **Note 27**). For that, we enter the GO term Endocytosis (*GO:0006897*) into the Search on the top right of the table. This yields nine complex entries. In the following we will compare the behavior of the complex with the lowest noise value (AP3 adapter complex) in breast cancer cells to its general behavior in human cells. In ComplexBrowser, this complex shows (figure not shown) only slight co-regulation with a peak in subtype four. However, the *Subunits expression profiles* and the *Protein correlation map* show that protein Q13367 (gene name *AP3B2*) exhibits much larger fold changes suggesting a specific role or a different stoichiometry by, e.g., building a multimer in the complex.

We will now submit the observed protein group of this complex to the CoExpresso application (press the button *submit proteins to CoExpresso*). This will open a new browser tab with the CoExpresso application (*see* **Note 28**).

CoExpresso is a light-weight program that is based on the data in ProteomicsDB [15] and with that provides protein coverage over up to 150 different human cell types and tissues (*see* **Note 29**). It calculates evidence for statistically significant co-regulation within a chosen protein group across all cell types where these proteins are co-expressed. With this, we can assess the general behavior of a protein group such as protein subunits of a protein complex.

4. CoExpresso shows several visualizations for the co-regulatory behavior of an arbitrary protein group in human cells. In addition, it provides values for the statistical significance for the co-regulation by comparison of the observed patterns to null distributions. The *Correlation map* shows that there is slight similarity across the available 11 tissues, further confirmed by many high *p*-values from the statistics. We also see that *AP3B2* again deviates the most from the common behavior (*see*, e.g., *Weighted abundance profiles*, right panel), thus indicating its peculiar role (Fig. 8).

For an example of a highly co-regulated protein group, select *Exosome* in the lowest field of the sidebar on the left.

4   Notes

1. If the file opens in the web browser, then right click anywhere in the browser and choose "Save as."

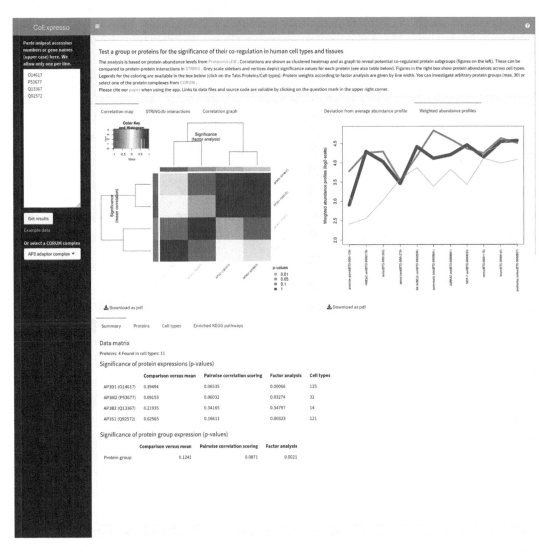

Fig. 8 CoExpresso statistically assesses and visualizes the behavior of any group of human proteins (here for the AP3 adapter complex in human cells)

2. Docker does not accept upper case letters in software names.

3. If you already run one of the applications, port **3838** will be occupied. To run an additional instance, change the first number to e.g. **3839** or higher.

4. You might need to activate JavaScript in the case they are blocked by a script blocker.

5. Here, upper and lower case letter matter, in contrast to usual rules for URL addresses.

6. You need to compile a custom library for VSClust as described in the instructions. This might become tricky on a Windows computer.

7. VSClust does not reorder the rows in the tables. Additional data columns can easily be re-added by copying the old columns from the original file into the output files.

8. The parameters are also described in the *Help* tab.

9. The input file format is restricted to comma-separated values (.csv). Files are required to contain only the numerical data that will be analyzed in addition to the following contents. In addition, each column can be described by a column name (only in the first row). Feature names for the rows (e.g., gene names, transcripts, peptides and proteins) have to be located in the first column and need to be unique as duplicated feature names are not accepted. An optional second column can be added upon the right setting of the respective parameter. Any other value needs to be numeric.

   When saving the table with programs like Excel, you might need to go through the following checklist:

   • Are the values really comma-separated? Open the file with a common text editor to see the "raw values."

   • Are there any artifacts left from prior calculations leading to non-numeric values (e.g., #DIV/0!)?

   • Are the row identifiers (first column) really unique?

   • Are there additional values left in the table that lead to additional, mostly empty columns?

10. You can analyze any quantitative set of features in VSClust. In the case of modified peptides (e.g., phosphorylated peptides), the second column describes the respective proteins. This information is then used in the functional analysis (last tab *Enriched terms (DAVID)*).

11. In the case of an input file that already contains estimated standard deviations, all columns but the last ones are considered different conditions while the last column contains the standard deviations.

12. VSClust can adapt the estimation of the variance to study designs with paired setup (e.g., samples taken from the same patients over different times).

13. The order of the columns defines experimental conditions and their replicates. We allow two different orders and require an identical number of replicates per condition. In the case your data consists of varying replicate numbers, you can include the required number of empty columns in the conditions with less replicate samples thus resembling the largest group.

   Layout of columns: Let's denote conditions by numbers (1–4) and replicates by letters (A–C). Then, the columns can be arranged as A1, A2, A3, A4, B1, B2, B3, B4, C1, C2, C3, C4 (Option *Replicates are grouped*) and ungrouped replicate columns as A1, B1, C1,..., A4, B4, C4,...

14. Most data clustering methods search for groups of proteins with similar expression profiles, independently on the amplitude of the changes. Hence, the values of each protein are scaled to have mean 0 and standard deviation 1. These values are often called z-values.

15. Biological data often comprises a hierarchical structure and mostly there is not only one "best" cluster number.

16. The clustering method assigns so-called membership values to each protein, describing the percentage the protein belongs to each cluster. Our implementation of both clustering methods discards proteins that cannot be assigned to the nearest cluster with more than 50%.

17. Validation indices are values calculated from the entire clustering result and help finding cluster numbers in the data set. VSClust relies on the two complementary validation indices *Minimum centroid distance* and *Xie-Beni index*. The former denotes a suitable cluster number by a large decay when increasing the number of clusters while the latter shows them by a minimum. These validation indices do not necessarily agree as they might capture the data structure differently.

18. The variance-based (also called variance-sensitive here) method uses the variance of the protein quantification to adjust the individual fuzzifier parameter of the method. Standard fcm-clustering uses one common fuzzifier for the entire data set.

19. Fuzzy c-means and VSClust are based on an optimization algorithm that is seeded from multiple randomly chosen initial states. Therefore, the results can be slightly different when running the method again.

20. You can run another data set on VSClust by pressing the button *Trigger server to reset file input* in the *File input* tab or just by reloading the web page in the browser.

21. This feature of VSClust depends on access to the DAVID server and therefore might not work.

22. Given the rather small number of 5 conditions, we do not expect large differences between both algorithms. This changes for larger numbers of conditions, where VSClust provides better resolution of the underlying patterns.

23. As the focus of this tutorial is on VSClust, we only shortly describe how to use the ComplexBrowser program. For a more extensive tutorial on ComplexBrowser, click on the question mark on the top right corner and select *Tutorial*.

24. Quality control is crucial in any study to check for unwanted bias, bad samples and batch effects. Therefore ComplexBrowser offers a set of visualizations to gather a general picture of data quality.

25. The two databases CORUM and Complex Portal rely on different approaches for the manual curation of protein complexes and therefore the number of identical complexes overlaps only slightly.

26. ComplexBrowser is the first software to calculate quantitative changes of protein complexes in proteomics data. It uses factor analysis for weighted averaging that was shown to perform well both in microarray data and to summarize peptides to proteins. This analysis also provides a noise estimation (column *Noise* in table) which is used to filter for protein groups that do not depict a common behavior.

27. ComplexBrowser summarizes the most changing protein complexes in the lower right panel. You can selectively look for most changing proteins by sorting the columns of the complex fold changes (*FC C2/C1*, …)

28. This is an experimental feature, and it might not work when using the application in Docker or when having a slow internet connection. In this case, we recommend calling CoExpresso from http://computproteomics.bmb.sdu.dk/Apps/ComplexBrowser or by pasting the protein accession names into the CoExpresso app.

29. CoExpresso tests for significance of co-regulation by comparing the observed profiles to a large number of randomized sets. Given the often low protein coverage of the ProteomicsDB data, we needed to employ a more sophisticated approach for the randomization. We use different approaches for the similarity within the proteins (correlation to the mean, pair-wise correlation, and factor analysis) which give slightly different results due to their different way of describing common behavior. For details, please read the paper [9].

Acknowledgments

We thank ON Jensen and A Rogowska-Wrzesinska for comments on the manuscript.

References

1. Bezdek JC (1981) Pattern recognition with fuzzy objective function algorithms. Plenum Press, New York

2. Schwämmle V, Jensen ON (2010) A simple and fast method to determine the parameters for fuzzy c-means cluster analysis. Bioinformatics 26:2841–2848

3. Smyth GK (2004) Linear models and empirical bayes methods for assessing differential expression in microarray experiments. Stat Appl Genet Mol Biol 3:1–25

4. Schwämmle V, Jensen ON (2018) VSClust: feature-based variance-sensitive clustering of omics data. Bioinformatics 34:2965–2972

5. Edwards AVG, Edwards GJ, Schwämmle V et al (2014) Spatial and temporal effects in protein post-translational modification distributions in

the developing mouse brain. J Proteome Res
13:260–267

6. Aebersold R, Mann M (2016) Mass-
spectrometric exploration of proteome structure and function. Nature 537:347–355

7. Gonçalves E, Fragoulis A, Garcia-Alonso L et al
(2017) Widespread post-transcriptional atten-
uation of genomic copy-number variation in
cancer. Cell Syst 5:386–398.e4

8. Michalak W, Tsiamis V, Schwämmle V et al
(2019) ComplexBrowser: a tool for identification and quantification of protein complexes in
large-scale proteomics datasets. Mol Cell Proteomics 18:2324–2334

9. Chalabi MH, Tsiamis V, Käll L et al (2019)
CoExpresso: assess the quantitative behavior
of protein complexes in human cells. BMC
Bioinformatics 20:17

10. Johansson HJ, Socciarelli F, Vacanti NM et al
(2019) Breast cancer quantitative proteome
and proteogenomic landscape. Nat Commun
10:1–14

11. Ruepp A, Brauner B, Dunger-Kaltenbach I et al
(2008) CORUM: the comprehensive resource
of mammalian protein complexes. Nucleic
Acids Res 36:D646–D650

12. Meldal BHM, Bye-A-Jee H, Gajdoš L et al
(2019) Complex Portal 2018: extended content and enhanced visualization tools for mac-
romolecular complexes. Nucleic Acids Res 47:
D550–D558

13. Zhang B, Pirmoradian M, Zubarev R et al
(2017) Covariation of peptide abundances
accurately reflects protein concentration differences. Mol Cell Proteomics 16:936–948

14. Hochreiter S, Clevert D-A, Obermayer K
(2006) A new summarization method for Affy-
metrix probe level data. Bioinformatics
22:943–949

15. Wilhelm M, Schlegl J, Hahne H et al (2014)
Mass-spectrometry-based draft of the human
proteome. Nature 509:582–587

Chapter 31

Automated Workflow for Peptide-Level Quantitation from DIA/SWATH-MS Data

Shubham Gupta and Hannes Röst

Abstract

Data-independent acquisition (DIA) is a powerful method to acquire spectra from all ionized precursors of a sample. Considering the complexity of the highly multiplexed spectral data, sophisticated workflows have been developed to obtain peptides quantification. Here we describe an open-source and easy-to-use work-flow to obtain a quantitative matrix from multiple DIA runs. This workflow requires as prior information an "assay library," which contains the MS coordinates of peptides. It consists of OpenSWATH, pyProphet, and DIAlignR software. For the ease of installation and to isolate operating system-related dependency, docker-based containerization is utilized in this workflow.

Key words OpenSWATH, pyProphet, DIAlignR, Data-independent acquisition, DIA, Retention time alignment, SWATH-MS

1   Introduction

Liquid chromatography coupled to tandem mass-spectrometer (LC-MS/MS) is widely used to analyze the proteome of biological samples. Currently, multiple methods using LC-MS/MS are available to practitioners which include targeted proteomics, shotgun proteomics, and data-independent acquisition (DIA) [1]. Over the last decade, DIA has gained traction for high-throughput and reproducible analysis [1–3]. Compared to traditional shotgun proteomics, in DIA/SWATH-MS peptides are isolated using a larger m/z window and are co-fragmented, which results in rich multiplexed MS2 spectra from multiple precursors. Experimental guidelines to generate DIA data are explained in Chaps. 16, 22–24. Since DIA spectra are convoluted with fragment-ions from many precursors, innovative strategies have been developed to extract signals for each peptide of interest. These approaches are divided into two categories: library based (peptide-centric) and non-library based (spectrum-centric). In general, library-based methods tend to be more sensitive and produce accurate quantitative results, especially

Katrin Marcus et al. (eds.), *Quantitative Methods in Proteomics*, Methods in Molecular Biology, vol. 2228,
https://doi.org/10.1007/978-1-0716-1024-4_31, © Springer Science+Business Media, LLC, part of Springer Nature 2021

if the assay library is prepared from the same sample [4]. In this chapter, we will focus on the library-based workflow.

Efficient ways to obtain a high-quality library and guidelines are detailed in the article by Schubert et al. [5] and in Chap. 22. Recent studies suggest that libraries can also be efficiently predicted using computational means alone [6, 7]. A library consists of a collection of MS coordinates uniquely describing the peptides of interest. This includes elution time of peptides, their charge state, fragment-ions that are most representative (unique and detectable), and their relative intensities. Each precursor is called an "assay," and this assay library is the key for a successful DIA data analysis. In addition, the assay library must include decoy-peptides for statistical scoring [8, 9]. Note the DIA data contains spectra from all ionized precursors which make it suitable for re-mining with a new library; however proper diligence is needed [10].

Compared to automated workflow for traditional mass-spectrometer data in which spectra are matched against a database, a DIA workflow focuses on identifying the correct chromatographic peak and its coelution profile; similar to targeted proteomics analysis [11]. Firstly, raw spectra files are converted from the vendor-specific format to a standard format such as mzML. For format-conversion, we are using the MSConvert tool [12].

Feature extraction is performed by first aligning library retention time (RT) to spiked-in standards, for example, the iRT peptides, followed by fetching MS2 extracted-ion chromatograms (XICs) using library coordinates, as shown in Fig. 1. OpenSWATH, supported by OpenMS, is an open-source software which performs this task [11]. OpenSWATH picks multiple peaks from XICs of a precursor; therefore, a statistical method is needed to assign a probability to each peak for being the correct one; in addition, chromatograms without any signal also need to be correctly labelled as such. pyProphet uses semi-supervised learning to calculate a discriminant score from which the p-value of each peak is estimated [8]. It also calculates the false discovery rate (FDR) or q-value (Fig. 1) to correct for multiple testing [13].

OpenSWATH is run independently on each run, which can cause inconsistency in quantification due to altered peak-boundaries, selecting different peaks across runs; and the strict error-rate control by pyProphet may also lead to missed peaks in certain runs [14]. Alignment of MS2 chromatograms across multiple runs establishes consistency and correspondence between the features. For this purpose, a recently developed software DIAlignR is employed: DIAlignR provides highly accurate retention-time alignment across heterogenous SWATH runs [15]. It uses q-values from pyProphet to pick a reference run for each peptide and aligns its XICs to the chromatograms of other runs (Fig. 2).

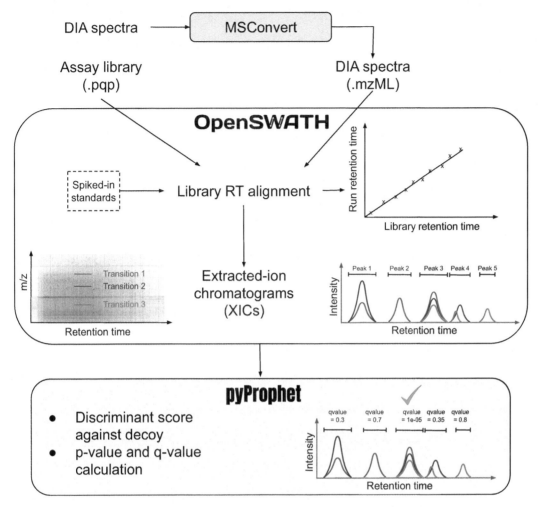

Fig. 1 Peak-identification workflow for a peptide precursor with three transitions (red, blue, and green) is illustrated. The first step is data conversion to the standard mzML format, subsequently OpenSWATH is used for RT alignment of library to the DIA run. OpenSWATH extracts MS2 chromatograms using library coordinates and identifies potential peaks. Successively, pyProphet identifies correct peaks using statistical scoring and FDR calculation

2   Materials

To explain the data analysis workflow an example dataset of *Streptococcus pyogenes* is used [11, 16]. The source data and intermediate result files can be downloaded from this link: http://www.peptideatlas.org/PASS/PASS01508. Although required applications (OpenMS, pyProphet) are publicly available, we suggest using Docker-based images for a convenient installation of the applications. For this tutorial, at least 50 GB storage space is required to store spectra and intermediate chromatogram files. The following material is needed to get started:

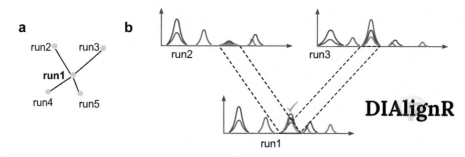

Fig. 2 (**a**) For retention-time alignment, DIAlignR picks a reference run for each precursor. (**b**) The XICs of the reference run are aligned to the corresponding XICs of other runs using a nonlinear approach. Thus, peak-boundaries from the reference run are mapped to the analysis run, which establishes consistency in quantification. Here, an alignment of XICs of run2 and run3 to a reference run1 is illustrated

1. Example dataset: The example dataset <u>PASS01508</u> has all the intermediary files for this tutorial. Download this repository and name it "PASS01508." Docker scripts are available at the root level. The "/lib" folder holds assay-library, alignment library, and SWATH acquisition scheme. The "/data" directory has MSConvert output mzML files and raw files inside the "raw" folder. The "/results" folder has OpenSwath output files in the "OpenSwathOutput" folder. It also has an "osw" and an "mzml" directory with features and chromatogram files required for alignment by DIAlignR.

2. SWATH spectra files: In the example dataset, *S. pyogenes* (strain SF370) was grown in 0% and 10% human plasma in biological duplicates. Four technical replicates for each sample were analyzed on an Eksignet nanoLC (with two-hour linear gradient) coupled to an AB SCIEX TripleTOF 5600 in SWATH-MS mode. Thus, the dataset has a total of 16 runs. The raw files are available at the /data/raw/ directory in <u>PASS01508</u>. The instrument produces two files per run, .wiff and .wiff.scan files, which should always be stored together. Besides the AB SciEx TripleTOF, other instruments can be used to acquire spectra in data independent acquisition mode (*see* **Note 1**).

3. Assay library: In a library-based workflow, the raw file is mined against an assay library which contains precursor sequences, their retention times, their transitions m/z values, and their intensities. For statistical scoring of peptide identifications, these libraries must contain suitable decoy transition groups [8, 9]. The assay library can be built from one or more shotgun runs [5] or downloaded from the SWATHAtlas database <u>http://www.swathatlas.org</u>. The SpyogenesAssayLibrary_decoy.pqp library used in this workflow is available in the /lib folder at <u>PASS01508</u>. The conversion to .pqp format is explained at the end of this chapter (*see* **Note 2**). To open the

library, either use online SQLite viewer (https://inloop.github.io/sqlite-viewer/) or install a local viewer such as the SQLite DB browser (https://sqlitebrowser.org/dl/).

4. Peptide assays for alignment: For targeted extraction, the library retention time needs to be aligned to each DIA run using linear or nonlinear methods [17, 18]. Generally, a set of spiked-in standard iRT peptides are used for this, however, in the absence of spike-in standards, the library can be aligned using high-intensity endogenous peptides without sacrificing quantitative accuracy [18]. In the example dataset, we have picked 20 endogenous peptides for linear alignment. This assay library is available in the /lib directory at PASS01508.

5. Docker software: Docker isolates dependencies between external software libraries and operating systems (Linux, Mac, Windows), providing reproducible computational results. Since Docker images are self-contained, they are lightweight and can be easily ported.

First, install the Docker engine on your machine. For Linux systems, follow guidelines at https://docs.docker.com/install, for Windows 10 Home install Docker Toolbox from https://docs.docker.com/toolbox/toolbox_install_windows, and for Windows 10 Pro or Mac install Docker Desktop from https://docs.docker.com/docker-for-windows.

On a Linux setup, use the following commands to get the OpenMS and pyProphet image:

```
sudo docker pull hroest/openms-executables-nightly:latest
sudo docker pull pyprophet/pyprophet:2.1.3
```

For Windows 10 Home:

Click on the start menu and type "Docker Quickstart Terminal." In this terminal window enter previous commands without *sudo* (*see* **Note 3**):

```
docker pull hroest/openms-executables-nightly:latest
docker pull pyprophet/pyprophet:2.1.3
```

6. Proteowizard: To convert raw files from vendor-specific format to open standardized mzML format, the msconvert tool from Proteowizard software suite is used. This software is available at http://proteowizard.sourceforge.net/download.html. For the software version and getting msconvert for a Linux system, *see* **Note 4**.

7. R and Rstudio: Alignment tool in this tutorial requires R (version > 4.0); therefore, install the latest version of R from https://cran.rstudio.com/. Rstudio integrates many tools to visualize data, to document scripts and to work efficiently with R. It can be downloaded from https://rstudio.com/products/rstudio/download/.

8. DIAlignR: This is an R-package which uses raw MS2 chromatograms for alignment of features across multiple runs. It can be installed from Bioconductor in R (R version > 4.0). To get this package, open RStudio and in the "Console" type these commands:

```
> if (!requireNamespace("BiocManager", quietly = TRUE))
      install.packages("BiocManager")
> BiocManager::install("DIAlignR")
```

3   Methods

Before proceeding to an automated analysis, it is crucial to understand DIA data, quality of chromatography and fragment-ion spectra. The Skyline software provides an open-source interactive solution for data visualization. We recommend using the Skyline-based workflow described previously by Röst et al. (2017) to assess data quality and manually inspect few precursors and their transitions [14].

Step 1: File Conversion to mzML: MSConvert, provided with ProteoWizard software suite, is used to convert vendor-specific raw spectra files into an open format such as mzML or mzXML. We will use mzML as it is a standardized format.

1. On a Windows setup, click on the *Start Menu* at the lower left corner (Windows icon). Type "MSConvert" in the *search* field and click on *MSConvert*.

2. Select *List of Files* then click on the *Browse* button and navigate to the folder where the .wiff and .wiff.scan files are present. Select the files and click on the *Add* button. Choose an *Output Directory* using the second *Browse* button. In the *Options* field select following parameters: Output format: mzML; Binary encoding precision: 64-bit; Write Index: yes; Use zlib compression: yes; TPP compatibility: yes; Package in gzip: yes; Use numpress linear compression: yes. Other boxes can be left unchecked (Fig. 3).

   (Optional) To perform centroiding, in the *Filters* menu select *Peak Picking* from drop down; Algorithm: Vendor; MS levels: 1–2, click on *Add* and then click *Start*. OpenSWATH

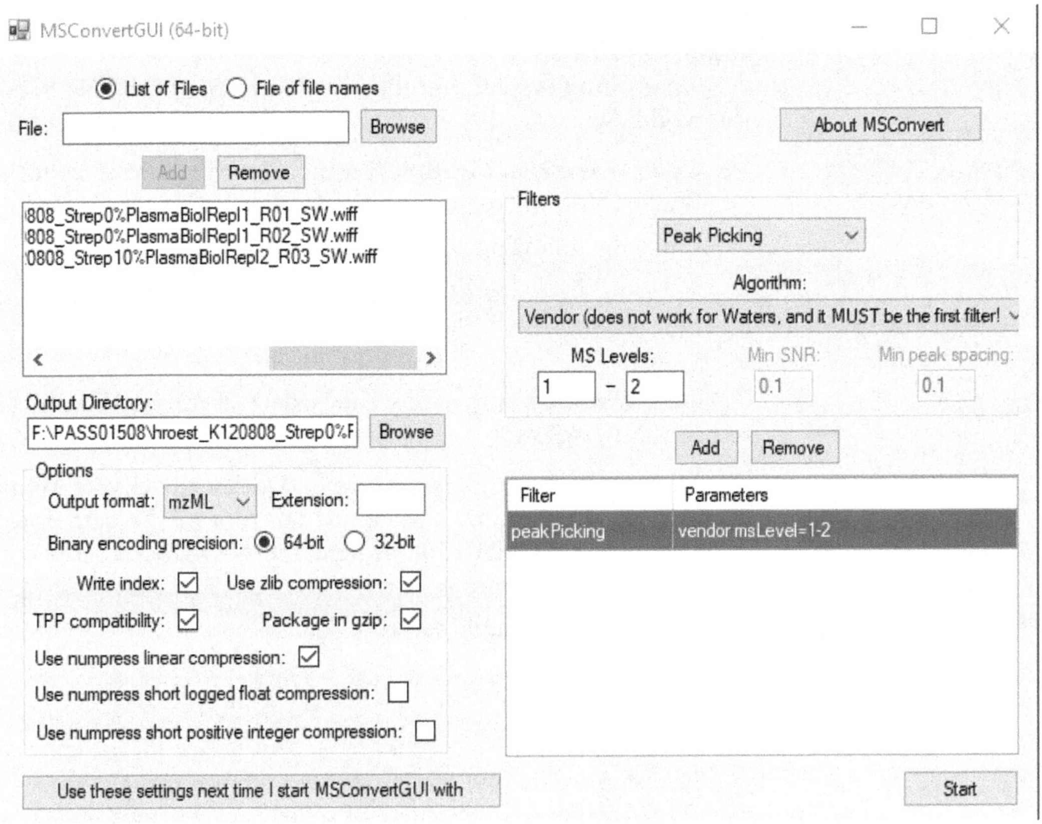

Fig. 3 Converting DIA spectra data to mzML format and centroiding spectra using MSConvert GUI

produces best results on profile data and should be used for any real-world analysis; however, to reduce file size and subsequent execution time, we chose to use centroiding in this workflow (*see* **Note 5**).

Step 2: OpenSWATH Workflow: *OpenSwathWorkflow* is the main command which efficiently integrates multiple tools from Open-Swath software. A basic command looks like this:

```
OpenSwathWorkflow -in filename.mzML.gz  \
                  -tr assayLibrary.pqp  \
                  -tr_irt iRTlib.TraML  \
                  -out_osw filename.osw \
                  -out_chrom filename.chrom.mzML
```

Where -in requires a spectra file, -tr requires an assay library, -tr_irt expects an iRT peptides library, -out_osw expects an output filename that shall contain picked peaks, and -out_chrom requires the filename that will contain MS2 chromatograms.

Firstly, move the MSConvert output files to the "data" directory which is located at the root of PASS01508. Make sure the assay library and the iRT library are present in the "lib" directory. For automated analysis:

On a **Linux system**, use the "oswDockerScript.sh" file to run this workflow:

1. Open "Terminal." In the "Terminal" navigate to the root of PASS01508, where the "oswDockerScript.sh" script is present.

2. Enter the following command at the Terminal:

```
sudo docker run -v `pwd`:/data --user $(id -u):$(id -g) hroest/openms-executables-nightly /bin/bash data/oswDockerScript.sh
```

On a **Windows system**, use the "oswDockerScriptWin.sh" file to run this workflow:

1. Click on the start menu and type "Docker Quickstart Terminal." In the terminal navigate to the root of the data downloaded from PASS01508, where the "oswDockerScriptWin.sh" script is present.

2. Enter this command (*see* **Note 3**):

```
docker run -v `pwd`:/data hroest/openms-executables-nightly /bin/bash data/oswDockerScriptWin.sh
```

This step yields two files (.osw and .chrom.mzML) for each run. The osw file is a collection of scored peak-groups in SQLite format. This contains potential peaks for each peptide and their scores. The chrom.mzML file contains all XICs of all transitions. For comparison, our results are stored at "results/OpenSwathOutput."

Besides the flags mentioned above, a few other important flags to control the execution of *OpenSwathWorkflow* are explained in Table 1. These could be added into the docker script used above. The explanation of flags can easily be obtained by following command:

```
sudo docker run -it hroest/openms-executables-nightly /bin/bash -c
"OpenSwathWorkflow --helphelp"
```

A web-portal http://www.openswath.org/en/latest/docs/openswath.html also goes into additional detail about Open-SWATH. Users are suggested to sign up for the mailing list General OpenMS discussion <open-ms-general@lists.sourceforge.net> for OpenSWATH-related inquiries.

Table 1
Few important flags in *OpenSwathWorkflow* command

Parameter	Explanation
--helphelp	Displays all flags and their explanations
Isolation window specific	
-swath_windows_file	A text file specifying the SWATH window out of which the fragment-ion traces will be extracted (*see* example file)
-sort_swath_maps	Sorts table specified in *swath_window_file*
-min_upper_edge_dist	The overlap (in m/z) between neighboring SWATH windows used for data acquisition (must not be used with *swath_window_file*). Usually set to 1.0 for regular SWATH-MS
Speed and memory optimization	
-batchSize	The number of chromatograms that are loaded into memory and analyzed at once. The smaller this number, the less memory required
-readOptions	Sets the memory strategy: by default, OpenSWATH loads all data into memory while "cache" will create a set of cached files on which OpenSWATH will work and not load any data into memory (more memory efficient). A good compromise between these two options is "cacheWorkingInMemory" which uses the cached files but loads the current Swath map into memory
-tempDirectory	Temporary directory to store cached files. (Must be used if *readOptions* are using a caching strategy)
-threads	Defines number of threads to parallelize extraction of XICs and their scoring
Extracted-ion-chromatogram (XIC)	
-RTNormalization:alignmentMethod	Selects method to align Library RT to sample RT space: "Linear," "interpolated," "lowess," and "b_spline." Recommended value with iRT peptides is 'linear' and with endogenous peptides select "lowess."
-rt_extraction_window	Size of the RT window in seconds to extract XICs from that peaks should be picked and scored. The RT window is centered around the aligned library RT. Default value is 600 s, which means extracting +/− 300 s around the expected RT. This value should be changed depending on the quality of library RT alignment
-extra_rt_extraction_window	Adds additional time to the rt_extraction_window only for XIC extraction. Peaks will not be picked and scored from this extra XIC. It is useful for visual inspection of chromatograms and for alignment tools such as DIAlignR

(continued)

**Table 1
(continued)**

Parameter	Explanation
-mz_extraction_window	Size of m/z extraction window (in Thomson) centered around library *m/z*. This value needs to be adjusted according to the instrument resolution. Its unit can be changed with *mz_extraction_window_unit*
-mz_extraction_window_unit	Defines the unit of *mz_extraction_window* and can be set to ppm
Peak smoothing	
-scoring:TransitionGroupPicker: PeakPickerMRM:peak_width	Peak-width in seconds required to define a non-spurious peak (adjust to your chromatography)
-scoring:TransitionGroupPicker: PeakPickerMRM:sgolay_frame_length	Number of points required for convolution by Savitzky-Golay smoothing. Increasing this parameter leads to stronger smoothing
-scoring:TransitionGroupPicker: PeakPickerMRM: sgolay_polynomial_order	Degree of polynomial fitted by Savitzky-Golay smoothing. Decreasing this parameter leads to stronger smoothing
-scoring:TransitionGroupPicker: PeakPickerMRM:use_gauss	If set as "true," Gaussian filter is used for smoothing
-scoring:TransitionGroupPicker: PeakPickerMRM:gauss_width	Gaussian width in seconds, estimated peak size in seconds
Debug mode	
-debug	Set as 1 to print all debug output
Quantification and scoring	
-TransitionGroupPicker: background_subtraction	Set as "exact" to remove background from peak signal. This option may increase the accuracy of the quantification
-use_ms1_traces	Extract the precursor ion trace(s) and use them for scoring

Step 3: False Discovery Rate (FDR) by pyProphet: After running OpenSWATH, a *q*-value for each feature across all runs is calculated using pyProphet. It requires the osw files and the corresponding assay library.

The following steps are performed by pyProphet:

1. Merges all osw files.

2. Trains a classifier on merged.osw. The classifier can be either "LDA" (Linear Discriminant Analysis) or "XGBoost." To see all flag use the help command:

```
pyprophet score --help
```

3. Employs (type I) error rate control at peptide level. The FDR-context can be "run-specific", "experiment-wide" and "global". To get an explanation of all options use help

```
pyprophet peptide --help
```

Basic commands for estimating *q*-value are given below:

```
pyprophet merge --template=assayLibrary.pqp --out=merged.osw *.osw
pyprophet score --in=merged.osw --classifier=XGBoost --level=ms2
pyprophet peptide --in=merged.osw --context=experiment-wide
```

For ease of use, we have included all the commands in a docker script. Before proceeding, make sure osw files from the previous steps are present in the "results" folder.

On a **Linux system**, open the "Terminal." In the "Terminal" navigate to the root of PASS01508, where the "pyprophetDockerScript.sh" script is present, then, execute following command:

```
sudo docker run -v `pwd`:/data --user $(id -u):$(id -g)
pyprophet/pyprophet:2.1.3 /bin/bash data/pyprophetDockerScript.sh
```

On a **Windows system**, click on the start menu and type "Docker Quickstart Terminal." In the terminal navigate to the root of PASS01508, where the "pyprophetDockerScriptWin.sh" script is present. Then, enter the following command:

```
sudo docker run -v `pwd`:/data pyprophet/pyprophet:2.1.3 /bin/bash
data/pyprophetDockerScriptWin.sh
```

pyProphet outputs a "merged.osw" file that has an FDR value for each analyte; in addition, it also outputs pdf files which contains summary figures on calculated discriminant score (*d*-score), *p*-values, and *q*-values. These files must always be consulted so that common errors, such as having an assay library without decoys, library with improper decoys, and library from another organism can be identified. In a successful analysis, pyProphet will have a clear separation between positives (true targets) and negatives (false targets and decoys), as shown in Fig. 4. The decoy distribution is used to estimate the *q*-value. To directly get a quantitative matrix from this output, *see* **Note 6**.

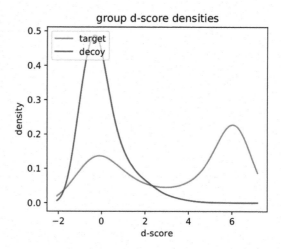

Fig. 4 Density-plot from calculated discriminant score (*d*-score). Generally, a bimodal distribution is expected for target peptides, whereas decoy peptides are expected to have a unimodal distribution. Based on these distributions, pyProphet computes *q*-values which are used to select a set of results with a well-defined FDR

Step 4: Multi-Run Alignments Using DIAlignR: DIAlignR has a set of functions for fetching XICs, selecting reference run for each analyte, performing chromatogram alignment and picking features based on the alignment. A single function *alignTargetedRuns* integrates all the functions and gives an intensity table. Prepare the data by moving the .chrom.mzml files to the "results/mzml" folder and the merged.osw file to the "results/osw" folder. To perform alignment:

1. Open Rstudio.
2. In "Console," use *setwd* command to navigate to the "PASS01508/results"directory.

```
> setwd("path/to/PASS01508/results")
```

3. To perform alignment and compute the intensity table, use the following command.

```
> library(DIAlignR)
> alignTargetedRuns(dataPath = ".", outFile = "strep_intensityTbl.csv")
```

alignTargetedRuns has many options to control the alignment algorithm and peak picking. A list of these options can be obtained using ?alignTargetedRuns in RStudio. Several of these options are explained in Table 2.

Table 2
Few options in *alignTargetedRuns* to control alignment of chromatograms

Options	Explanation
analytes	Specify analytes ("PEPTIDE_Charge") to be aligned. If unspecified, all analytes found in osw files are aligned
runs	Names of runs to be aligned. If unspecified all runs mentioned in RUN table of osw files are aligned
alignType	Defines the alignment strategy (default: 'hybrid'). There are three options- global: only aligns features from osw, this does not use XICs. local: only uses XICs for alignment. hybrid: uses XICs for alignment that is constrained using features from osw
simMeasure	Defines the type of metric should be used to construct pointwise similarity between two XIC-groups. XICs of a transition-group have fragment-ion intensities for each time-point. The metric defines the similarity across the time-points of two runs. Few common metrics are dotProduct, cosineAngle, dotProductMasked (default)
SgolayFiltOrd	Polynomial order of the Savitzky-Golay filter. It is suggested to be the same as *sgolay_polynomial_order* parameter of OpenSwathWorkflow
SgolayFiltLen	Length of the Savitzky-Golay filter. It is suggested to be the same as the sgolay_frame_length parameter of OpenSwathWorkflow
samplingTime	Cycle time used for DIA data. This is the time-difference between two neighboring points of MS2 chromatograms
analyteFDR	Maximum q-value (between 0 and 1) for peaks to be eligible for quantification
maxFdrLoess	Maximum q-value (between 0 and 1) for peaks to be eligible for global fit

DIAlignR also allows the user to plot aligned chromatograms. Use *getAlignObjs* and *plotAlignedAnalytes* functions to see aligned chromatograms.

```
> runs <- c("hroest_K120809_Strep0%PlasmaBiolRepl2_R04_SW",
"hroest_K120809_Strep10%PlasmaBiolRepl2_R04_SW")
> AlignObj <- getAlignObjs(analytes = 4618L, runs = runs, dataPath = ".")
> plotAlignedAnalytes(AlignObj, annotatePeak = TRUE)
```

Any issues related to DIAlignR can be reported at https://github.com/shubham1637/DIAlignR/issues.

4   Notes

1. SWATH-MS was originally developed for a TOF based mass-analyzer using SCIEX instrumentation. However, SWATH-like data-independent acquisition (DIA) data can be obtained from other types of instruments. One of the major considerations is the acquisition speed, which needs to allow sufficient sampling

during elution of an analyte from the LC column. Open-SWATH can analyze data from multiple vendors, including Waters, Thermo Fisher and Bruker. OpenSWATH, Skyline and other software have been used to analyze DIA data from Thermo Q Exactive instruments [19–22].

2. While OpenSWATH supports the .TraML format, to reduce the downstream file size, the SQLite based .pqp format is recommended. To convert a library from TraML to pqp format, open "Terminal" (in Linux) or "Docker Quickstart Terminal" (in Windows), as explained in the Methods section. Navigate to the location where the .TraML library is present, then run the following command:

```
sudo docker run -v `pwd`:/data --user $(id -u):$(id -g) hroest/openms-executables-nightly /bin/bash -c "cd data; TargetedFileConverter -in lib.TraML -out lib.pqp"
```

3. To execute docker commands for OpenSwath and pyProphet on a Windows 10 Home setup, the following precautions must be taken:

(a) To run docker containers in Windows and Mac systems, Linux commands can be used without having leading *sudo*.

(b) Make sure there are no spaces in the file path.

(c) Make sure PASS01508 data is in the C: drive (shared drive is not enabled for docker by default).

4. In this tutorial MSConvert is used from ProteoWizard release: 3.0.11252 (2017-8-22). For Linux systems, the docker image can be downloaded using the following command:

```
sudo docker pull chambm/pwiz-skyline-i-agree-to-the-vendor-licenses:latest
```

To convert wiff files to mzML, use commands from "msconvertOnLinux.sh" file available at PASS01508.

5. OpenSWATH produces optimal results on profile data. To obtain profile data, do not specify anything in the *Filters* menu, and remove any filter from the lower-right box, if present (Fig. 3). Centroiding reduces the number of peaks in the data and sometimes may remove low-intensity peaks, the XICs generated by OpenSWATH become less smooth and peak detection becomes less sensitive. If centroiding is performed, it is crucial to compare the results of OpenSWATH to those obtained on profile data on the same dataset to obtain an

accurate estimation of how centroiding affects the identification rate [14].

6. Alignment is a necessary step for consistent analysis across multiple runs. Nonetheless, this is a time-consuming step and under some circumstances is not required. A (un-aligned) peptide data-matrix can be directly obtained from **merged.osw** file, using the following command:

```
sudo docker run -v `pwd`:/data --user $(id -u):$(id -g)
pyprophet/pyprophet:2.1.3 /bin/bash -c "cd data; pyprophet export --in
Strep.merged.osw --out Strep.csv"
```

References

1. Aebersold R, Mann M (2016) Mass-spectrometric exploration of proteome structure and function. Nature 537:347–355. https://doi.org/10.1038/nature19949

2. Gillet LC, Navarro P, Tate S, Röst H, Selevsek N, Reiter L, Bonner R, Aebersold R (2012) Targeted data extraction of the MS/MS spectra generated by data-independent acquisition: a new concept for consistent and accurate proteome analysis. Mol Cell Proteomics 11:11. https://doi.org/10.1074/mcp.O111.016717

3. Collins BC, Hunter CL, Liu Y et al (2017) Multi-laboratory assessment of reproducibility, qualitative and quantitative performance of SWATH-mass spectrometry. Nat Commun 8:29. https://doi.org/10.1038/s41467-017-00249-5

4. Navarro P, Kuharev J, Gillet L et al (2016) A multicenter study benchmarks software tools for label-free proteome quantification. Nat Biotechnol 34:1130–1136. https://doi.org/10.1038/nbt.3685

5. Schubert O, Gillet L, Collins B et al (2015) Building high-quality assay libraries for targeted analysis of SWATH MS data. Nat Protoc 10:426–441. https://doi.org/10.1038/nprot.2015.015

6. Tiwary S, Levy R, Gutenbrunner P et al (2019) High-quality MS/MS spectrum prediction for data-dependent and data-independent acquisition data analysis. Nat Methods 16:519–525. https://doi.org/10.1038/s41592-019-0427-6

7. Gessulat S, Schmidt T, Zolg DP et al (2019) Prosit: proteome-wide prediction of peptide tandem mass spectra by deep learning. Nat Methods 16:509–518. https://doi.org/10.1038/s41592-019-0426-7

8. Rosenberger G, Bludau I, Schmitt U (2017) Statistical control of peptide and protein error rates in large-scale targeted data-independent acquisition analyses. Nat Methods 14:921–927. https://doi.org/10.1038/nmeth.4398

9. Elias J, Gygi S (2007) Target-decoy search strategy for increased confidence in large-scale protein identifications by mass spectrometry. Nat Methods 4:207–214

10. Guo T, Kouvonen P, Koh C et al (2015) Rapid mass spectrometric conversion of tissue biopsy samples into permanent quantitative digital proteome maps. Nat Med 21:407–413. https://doi.org/10.1038/nm.3807

11. Röst H, Rosenberger G, Navarro P et al (2014) OpenSWATH enables automated, targeted analysis of data-independent acquisition MS data. Nat Biotechnol 32:219–223. https://doi.org/10.1038/nbt.2841

12. Kessner D, Chambers M et al (2008) Proteo-Wizard: open source software for rapid proteomics tools development. Bioinformatics 24:2534

13. Storey JD, Tibshirani R (2003) Statistical significance for genome-wide studies. Proc Natl Acad Sci U S A 100:9440–9445

14. Röst HL, Aebersold R, Schubert OT (2017) Automated SWATH data analysis using targeted extraction of ion chromatograms. In: Comai L, Katz J, Mallick P (eds) Proteomics. Methods in molecular biology, vol 1550. Humana Press, New York

15. Gupta S, Ahadi S, Zhou W, Röst H (2019) DIAlignR provides precise retention time alignment across distant runs in DIA and targeted proteomics. Mol Cell Proteomics 18:806–817

16. Röst H, Liu Y, D'Agostino G et al (2016) TRIC: an automated alignment strategy for reproducible protein quantification in targeted proteomics. Nat Methods 13:777–783. https://doi.org/10.1038/nmeth.3954

17. Escher C, Reiter L, MacLean B et al (2012) Using iRT, a normalized retention time for more targeted measurement of peptides. Proteomics 12:1111–1121

18. Bruderer R, Bernhardt OM, Gandhi T, Reiter L (2016) High-precision iRT prediction in the targeted analysis of data-independent acquisition and its impact on identification and quantitation. Proteomics 16:2246–2256

19. Malmström L, Bakochi A, Svensson G et al (2015) Quantitative proteogenomics of human pathogens using DIA-MS. Proteomics 129:98–107. https://doi.org/10.1016/j.jprot.2015.09.012

20. Bruderer R, Bernhardt OM, Gandhi T et al (2015) Extending the limits of quantitative proteome profiling with data-independent acquisition and application to acetaminophen treated 3D liver microtissues. Mol Cell Proteomics 14(5):1400–1410. https://doi.org/10.1074/mcp.M114.044305

21. Egertson JD, Kuehn A, Merrihew GE et al (2013) Multiplexed MS/MS for improved data-independent acquisition. Nat Methods 10:744–746. https://doi.org/10.1038/nmeth.2528

22. Meier F (2019) et al, Parallel accumulation—serial fragmentation combined with data independent acquisition (diaPASEF): bottom-up proteomics with near optimal ion usage. bioRxiv. https://doi.org/10.1101/656207

INDEX

Katrin Marcus et al. (eds.), *Quantitative Methods in Proteomics*, Methods in Molecular Biology, vol. 2228,
https://doi.org/10.1007/978-1-0716-1024-4, © Springer Science+Business Media, LLC, part of Springer Nature 2021

CW00741362

Nine Hawker Fury IIs of Hawkinge-based 25 Squadron, freshly delivered from Brooklands in October 1936. All carry the unit's parallel black bands and most have a triangle on the fin in the colours of their flight and the squadron badge, a hawk alighting on a glove hand. Third and sixth from the back are flight leaders' aircraft. In the foreground is the CO's machine, K7270: this was further decorated with its fin and tailplanes painted black.

C⊙NTENTS

Edited by: **Ken Ellis**
With many thanks to: **Chris Gilson**
and **Steve Beebee** at *FlyPast*; **Ajay**
Srivastava at the RAF Museum
Contributing writers: **Steve Appleby**,
Daniel Ford, **Jonathan Garroway**, **Josh**
Lyman, **Bob Uppendaun**, **M L Wynche**
Group Editor: **Nigel Price**

Archive images: All **Key Collection**
unless noted
Artwork: **Andy Hay** and in fond memory
of **Pete West**

Art Editor: **Mike Carr**
Chief Designer: **Steve Donovan**

Production Editor: **Sue Blunt**
Deputy Production Editor: **Carol Randall**
Production Manager: **Janet Watkins**

Advertisement Manager: **Alison Sanders**
Advertising Production: **Debi McGowan**
Group Advertisement Manager:
Brodie Baxter

Marketing Executive: **Shaun Binnington**
Marketing Manager: **Martin Steele**

Commercial Director: **Ann Saundry**
Managing Director and Publisher:
Adrian Cox
Executive Chairman: **Richard Cox**

Contacts
Key Publishing Ltd, PO Box 100, Stamford,
Lincs, PE9 1XQ
Tel: 01780 755131
Email: flypast@keypublishing.com
www.keypublishing.com

Distribution: **Seymour Distribution Ltd,**
2 Poultry Avenue, London EC1A 9PT
Tel: **020 74294000**
Printed by: **Warners (Midland) plc,**
The Maltings, Bourne, Lincs, PE10 9PH

The entire contents of *RAF Centenary*
Celebration: Fighters is **copyright**
© **2017**. No part may be reproduced
in any form, or stored on any form
of retrieval system, without the prior
permission of the publisher

Front Cover:
Lovingly restored by
Hawker Restorations,
Hurricane I P2902
(G-ROBT) was test
flown on June 19,
2017, having been
salvaged as a hulk
from a beach near
Dunkirk in 1989.
Flown by Plt Off
Kenneth 'Mac'
McGlashan of 245
Squadron, it force-
landed amid the
retreating British
forces on May 31,
1940. Mac managed
to return to Britain on
the paddle steamer
Golden Eagle; his
Hurricane took nearly
another 50 years to
make it back.
DARREN HARBAR

Inset: *Tornado*
F.3s, top to bottom:
229 Operational
Conversion Unit,
29 Squadron, and
5 Squadron.
BRITISH AEROSPACE

This page:
Three generations
of RAF interceptors
joined for a special
formation on July
4, 1992 to mark the
disbandment of
the 'Firebirds', 56
Squadron on the
McDonnell Phantom
FGR.2 (middle) at
Wattisham. The unit
had flown Phantoms
since March 1976.
The 'number plate'
of 56 Squadron
was transferred to
the Tornado F.3s
of 229 Operational
Conversion Unit
(background), the
'new' 56 flying F.3s
until April 2008.
The squadron took
delivery of its first
Lightnings (F.1As)
in January 1961,
going on to F.3s and
retiring its last F.6s
in June 1976. For
the disbandment
formation in 1992,
British Aerospace
provided F.6 XP693
from Warton, where
it was being used
for Tornado air
defence variant radar
development trials.
BRITISH AEROSPACE

PER ARDUA

AD ASTRA

1918 TO 2018

All Fools' Day 1918 – April 1 – was an inauspicious date on which to start a new and independent element of Britain's armed forces. Yet the inauguration of the Royal Air Force became one of the most famous milestones in military aviation history.

For a while, the fledgling 'junior' service had to endure jibes about the 'Royal April Fool'. Yet, just 22 years later, it was the spearhead that prevented German ambitions to invade the United Kingdom.

Throughout its ten decades the RAF has been frequently tested, but seldom found lacking. Its Latin motto, 'Per Ardua ad Astra', best translated as 'Through Adversity to the Stars', well sums up its incredible achievements and heritage.

Royal ascent was granted to an Act of Parliament establishing the RAF and the Air Council on November 29, 1917. Prior to that there had been two air arms, with over-lapping operations, the Royal Flying Corps (RFC) and the Royal Naval Air Service (RNAS), each competing for resources and personnel. The RFC, established on March 13, 1912, had a Military Wing and a Naval Wing but from July 1914 the naval element became self-contained as the RNAS.

The Air Force Constitution Act provided for amalgamation of the RFC and the RNAS and set the date for this to come into full force as from April 1, 1918. At that point, the RAF was the largest air arm in the world: 188 frontline squadrons, 22,647 aircraft and 291,170 officers and men.

As well as on the Western Front, stretching across Europe, the new force was in action defending the home skies of Britain, in the Middle East, Africa and on the high seas. The armistice of November 11, 1918 brought the horrific conflict to an end and a dramatically reduced RAF faced a world changed politically, strategically and socially.

Countless words and images will be published during the centenary celebrations, and deciding how to present a *FlyPast* 'special' that paid tribute in an original manner took some pondering. The team settled on telling the story of the RAF through its fighters, its bombers and *two* publications. (Details of the magazine devoted to the bombers appear on page 2.)

WHAT *IS* A FIGHTER?

Eighty-seven years separate the Sopwith Snipe of 1918 from the Eurofighter Typhoon, which first entered RAF squadron service in 2005. A Bentley BR.2 230hp (171kW) nine-cylinder rotary propelled a fully loaded – 2,020lb (916kg) – Snipe at a maximum of 121mph (194km/h). Two

Eurojet EJ200 afterburning turbofans generate 20,250lb (90kN) of thrust to take all 46,297lb (21,000kg) of a Eurofighter Typhoon F.2 to twice the speed of sound. Both are single-seat fighters, otherwise they have so little in common.

What *is* a fighter? That's not as daft a question as at first may seem. In those ten decades the nature of the role has evolved. In the Snipe's day, its job was to climb to meet approaching enemy aircraft and engage them at close quarters. From 1931 the Hawker Demon two-seater's purpose was to patrol, lying in wait for incoming 'hostiles'.

There is an oft-quoted, partially tongue-in-cheek, query about the English Electric Lightning: "When does it go 'fuel critical'?" Answer: when it starts! It was conceived as the ultimate British point-defence fighter, trading endurance for dramatic acceleration and exceptional climb to engage high-flying 'bogies'. Likewise the Supermarine Spitfire and the Typhoon: they are *interceptors*.

Yet the McDonnell Phantom, and particularly the Panavia Tornado F.3, reprised the Demon's task. Their mission profile was that of a patroller, airborne and awaiting 'trade'.

In its current FGR.4 guise, Typhoon is a highly capable 'swing-role' aircraft, its near 20,000lb war load putting it in the 'bomber' category. In times

gone by, the twin-jet could have been referred to as a *fighter-bomber*; a name that was championed during World War Two by its namesake, the Napier Sabre-engined Hawker Typhoon. Even the Snipe could be fitted to carry bombs – up to four 20-pounders.

The RAF has sought to make its 'fighters' increasingly versatile, with fighter-bombers, close air support, anti-tank and tactical and strategic photo-reconnaissance abilities. The forthcoming Lockheed Martin Lightning II is intended to encapsulate all of this. The RAF describes it as a "multi-role supersonic stealth aircraft that will provide the UK with a hugely capable and flexible weapons and sensor platform."

Initially known as the Joint Strike *Fighter*, the Lightning II completely blurs the definition. As it is destined to replace the veteran Panavia Tornado GR.4 in RAF service, it's been consigned to our sister volume as a 'bomb-truck'.

TURKEYS AND TRIUMPHS

Inside this publication the reader will find every frontline fighter type that the RAF has operated in its 100 years. Already we can 'hear' the cries of anguish: "Only two pages on the Hurricane?" With limited space, to make sure that *all* the RAF's fighters get a mention, the most well-known have been slimmed down.

In the pages that follow are a few turkeys, but most of the RAF's fighters have been triumphs. The emphasis is less about derring-do, seasoned *FlyPast* readers will already be well-versed in the valour and exploits of RAF personnel.

Here, the intention is to be more concerned with where the subject fits in the RAF's centenary and its heritage, or even in world aviation history. The hope is that the reader will find more than a few "I never knew that" moments as we pay tribute to the finest air force in the world. ◉

Above left
First introduced to the Western Front in April 1917, the Bristol F.2b Fighter had an incredible career with the RAF. From 1919 its role had switched to army co-operation, and policing the empire. This example, built by Armstrong Whitworth, served with 5 Squadron from early 1920 until it was written off in the North Western Frontier of India in August 1922. 'Brisfits', as they were known, carried on in this role until 1932. PETE WEST

Left
The state-of-the-art fighter for the RAF in its first year was the Royal Aircraft Factory SE.5A. The hexagons were the badge of 85 Squadron, which used the type on the Western Front from August 1917 until it returned to Britain in February 1919 to disband. KEC

"In its current FGR.4 guise, Typhoon is a highly capable 'swing-role' aircraft, its near 20,000lb war load putting it in the 'bomber' category. Even the Snipe could be fitted to carry bombs – up to four 20-pounders."

SOPWITH
SNIPE
1918 TO 1928

D.VIIs – the most formidable of Germany's fighters. In full view of the trenches, he fought tenaciously, downing three of the enemy and driving off the others.

With wounds to both of his legs and an arm, Barker managed to force land E8102 behind the Allied lines. With his score brought to 50, an Allied record, he was awarded the Victoria Cross. On November 30, 1918 King George V presented the VC to Major W G Barker DSO* MC** who also

During World War One 'Sopwith' became a generic name for a British fighter. When the Royal Air Force came into being on April 1, 1918 several types that had originated from the drawing boards of the Kingston-on-Thames organisation were still in service, albeit in small numbers. These were the 1½ Strutter, Baby, Camel, Dolphin, Cuckoo, Snipe, Salamander and Dragon. Most had retired by the Armistice of November 11, 1918 but the Snipe was destined for greater longevity.

Among the many passions of Thomas Octave Murdoch Sopwith was aviation and he was awarded Aviators' Certificate No.31 on November 22, 1910. With Fred Sigrist, previously the engineer on 'Tommy's' motor yachts, the pair established Sopwith Aviation and Engineering in June 1912. Within four years the company was a giant with a vast factory and sub-contractors also building its designs.

As related in the section on the Hawker Woodcock, all of this came tumbling down in 1920, but was quickly reborn under the Hawker banner. Tommy Sopwith went on to head the massive Hawker Siddeley combine.

VICTORIA CROSS

The Snipe was developed as a private venture, taking into account the operational experiences of the Camel. The first examples reached the Western Front in September 1918 and the following month hit the headlines. Flying from Beugnatre, south of Arras, France, Canadian Major William George Barker took Snipe E8102 of 201 Squadron on a morning patrol. At 08:25 hours he bagged his 47th aerial victory, a German Rumpler.

This 'kill' took Barker's eye off the ball and five minutes later he was fighting for his life as he was 'bounced' by a 'circus' of no less than 15 Fokker

received the French Croix de Guerre and the Italian Medaglio d'Argento. The battered fuselage of E8102 was salvaged and on June 24, 1921 was transferred to Canada and is held by the Canadian War Museum.

CIVIL WARS

With its exceptional flying characteristics and its Bentley BR rotary engine readily available, the Snipe offered an opportunity to stock home and colonial units with a reliable and relatively new type.

By June 1919 about a dozen Snipes were shipped to Archangel in the Russian Arctic northeast of St Petersburg to aid the so-called 'White

SOPWITH SNIPE

Type:	Single-seat day fighter
First flight:	Late 1917; entered operational service in September 1918
Powerplant:	One 230hp (171kW) Bentley BR2 rotary
Dimensions:	Span 30ft 1in (9.16m), Length 19ft 9in (6.01m)
Weights:	Empty 1,312lb (595kg), All-up 2,020lb (916kg)
Max speed:	121mph (194km/h) at 10,000ft (3,048m)
Armament:	Two 0.303in machine guns mounted on the upper forward fuselage, firing through propeller arc
Replaced:	Sopwith Camel from September 1918
Taken on charge:	Circa 1,550; perhaps as many as 700 never entered service
Replaced by:	Variously, Gloster Grebe, Gloster Gamecock, Armstrong Whitworth Siskin and Hawker Woodcock 1924 to 1926. Type withdrawn from training in 1928

"...about a dozen Snipes were shipped to Archangel in the Russian Arctic northeast of St Petersburg to aid the so-called 'White Army' fighting against Bolshevik forces..."

Army' fighting against Bolshevik forces in the vicious civil war that broke out following the overthrow of the Tsarist regime in 1917. At least one Snipe is believed to have been captured by the 'Reds' and used against its previous operators.

Closer to home, Snipes were deployed to Ireland where the RAF was allowed to arm its aircraft in March 1921 as the Irish struggle for independence reached its peak. Operating from Oranmore, near Galway, on the west coast and Baldonnel, near Dublin, the RAF deployed Bristol F.2b Fighters and Handley Page O/400s, as well as Snipes.

Prior to the truce of July 11, 1921 a pair of Snipes were involving in foiling a Sinn Fein ambush of loyalist Auxiliaries, leaving five rebels dead and generating the headline 'Snipes for the Snipers' in a newspaper.

TWINS

On January 12, 1904 the Atcherley family celebrated in York the birth of twins. Both were destined for stellar careers with the RAF, ending up as AVM David Francis William Atcherley CB CBE DSO DFC and Air Marshal Sir Richard Llewellyn Roger 'Batchy' Atcherley KBE CB AFC*. The latter was a 'Sniper'.

Plt Off Richard joined 29 Squadron at Duxford on Snipes in July 1924. Transferring to 23 Squadron at Henlow, still on Snipes, in October 1925 Batchy set the ground crews an unusual task. A cylindrical fuel tank was fitted to the inside of the starboard undercarriage leg and 'plumbed' to the engine.

It is believed that Richard wanted to expand his aerobatic prowess to flying upside down and he was determined to set a record for sustained inverted flight. History fails to record if this venture was successful. By October 1928 Fg Off Batchy had other fish to fry; he became a part of the High Speed Flight ready to compete in the following year's Schneider Trophy at Calshot. (For more on Richard Atcherley, see the section on the Gloster Grebe.

Brother David chose the army and studied at Sandhurst from 1922. In the spring of 1927 the pull of the RAF was too strong and he transferred. He was taught to fly at 5 Flying Training School, Sealand, and *may* well have also sampled the Snipe. The school was one of the last to give up the venerable Sopwiths, including some two-seat conversions, the last examples retiring in 1928. ◎

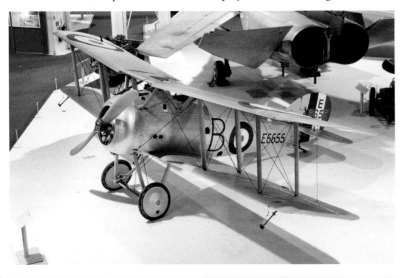

Above left
Built in the spring of 1918, Snipe E8050 in typical Royal Flying Corps colours. This machine did not see operational service with an RAF unit.
© ANDY HAY
www.flyingart.co.uk

Below left
Acquired by the RAF Museum on August 28, 2012, Hendon's Snipe is finished in the colours of 1 Squadron when the unit was based at Hinaidi, Iraq, in 1926. It is derived from a large collection of original parts, including a Humber-built Bentley BR2 acquired from Leicester Museum in 1972. The reconstruction was carried out by The Vintage Aviator Ltd in Wellington, New Zealand.
RAF MUSEUM
www.rafmuseum.org

GLOSTER GREBE AND

1924 TO 1933

B etween them, the Gloster Grebe and the Armstrong Whitworth Siskin heralded an important era for the RAF; they were its 'new generation' fighters, not inherited from the Royal Flying Corps. Even today, there is debate about which was first to enter service – with no hint of fudge, *both* were entitled to the accolade!

Having last flown Bristol F.2b Fighters in Palestine, 111 Squadron re-formed on October 1, 1923 at Duxford. At that stage its establishment was just one flight – half a dozen Grebe IIs – its personnel tasked with carrying out service trials of the new fighter.

From April 1924, the squadron also took on a flight of Snipes – a retrograde step as the type was 24mph (38km/h) slower than the Grebes, a sizeable performance difference by early 1920s' standards. Two months later 'Treble One' began to adopt the Siskin III as its sole equipment.

Siskins were delivered in full strength to 41 Squadron at Northolt in May 1924, with 25 Squadron at Hawkinge adopting an all-Grebe fleet the following October. So the Grebe was first to be issued to a unit, but that was only partial, while the Siskin was first *and* complete! The type was the beginning of a long line of Gloster fighters, stretching to the Javelin delta twin-jet of 1951.

Gloster had secured the services of experienced designer Henry Philip Folland, who had created the Royal Aircraft Establishment SE.5a, the best British fighter of the Great War. He was responsible for Gloster types through to the Gladiator before establishing his own enterprise, Folland Aircraft, in 1937.

The last frontline Grebes served with 25 Squadron, until July 1929. As well as the single-seater, Gloster developed the dual-control Mk.IIIDC which continued in use to 1931.

KING'S CUP

Plt Off Richard Llewellyn Roger 'Batchy' Atcherley (already mentioned in the section on the Sopwith Snipe) flew with 23 Squadron from Henlow, which converted to Gamecocks in April 1926, and gave a spirited demonstration of solo aerobatics in one at the annual Hendon display in July 1926.

During the 1920s and 1930s, RAF pilots and aircraft regularly entered civilian competitions, the most prestigious being the annual King's Cup races, established by George V in 1922. The July 1928 event centred on Hendon, and Batchy, by then a flying officer and instructing with the Central Flying School (CFS) at Wittering, entered in Grebe IIIDC J7520. The following month he was seconded to the RAF High Speed Flight.

The 1929 King's Cup was a gruelling 590-mile (949km) course from Heston in stages to Blackpool and back over two days, July 5-6. Entering under the pseudonym 'R Llewellyn', Batchy was again at the helm of J7520. Behind him, navigating, was Flt Lt George Hedley Stainforth. Having averaged 150mph, as the pair entered the Heston circuit at the end of the contest it was clear from the reception that they had won the coveted trophy.

Also taking part in the 1929 race was Flt Lt Edward Hedley 'Mouse' Fielden – personal pilot to Edward, Prince of Wales – flying a CFS Grebe IIIDC, J7519. In January 1936, he became the first commanding officer of the King's Flight when the prince, as Edward VIII, briefly took the throne. Mouse ended his career as AVM Sir Edward Fielden GCVO CB DFC AFC.

A month after clinching the King's Cup, Batchy was in a very different race, for the Schneider Trophy, thundering around the Solent at an average of 325mph in Supermarine

GAMECOCK

Above
The badge of 43 Squadron, inspired by the unit's Gloster Gamecocks.

Above left
A pair of Hawkinge-based Grebe IIs of 25 Squadron. Unit markings were parallel black bars.

"Gloster had secured the services of experienced designer Henry Philip Folland, who had created the Royal Aircraft Establishment SE.5a, the best British fighter of the Great War."

S.6 floatplane N248 as part of the High Speed Flight. He was not placed, but the RAF won the event for the second time: another win in 1931 would see the trophy awarded in perpetuity to Britain.

Flt Lt John Boothman took the honours on September 13, 1931 at Calshot. Part of the winning team was Flt Lt George Stainforth – Batchy's navigator in the 1929 King's Cup. Piloting Supermarine S.6B S1595 on the 29th, George became the first man

to fly at more than 400mph, clocking 407.5mph – a far cry from the sedate 150mph of the Grebe.

FIGHTING COCKS

Henry Folland produced the Gamecock, an improved version of the Grebe, from February 1925. As well as the more powerful Bristol Jupiter VII radial engine, the more rotund Gamecock had its machine guns mounted within the fuselage, either side of the cockpit. The Grebe

carried its guns Sopwith-style on the upper decking in front of the pilot.

The Gamecock was the last RAF fighter of wooden construction. The final examples were retired in July 1931 as 23 Squadron at Kenley took on Bristol Bulldogs.

From the spring of 1926, under Sqn Ldr A F Brook, 43 Squadron at Henlow was the first unit to equip with Gamecocks, forsaking Snipes. By the time it moved to Tangmere that December, its personnel were very impressed with the new fighter.

An unofficial squadron badge emerged, featuring the bird from which the Gamecock took its name. In July 1936, by which time 43 was flying Hawker Furies, the badge was officially approved, by King Edward VIII. As well as the pugnacious bird, it featured the Latin motto 'Gloria Finis' – glory is the end – taken from the family crest of former CO Sqn Ldr Brook.

Adopting Siskins from June 1928, the squadron retained the gamecock link throughout its existence and was proud to be known as the 'Fighting Cocks'. The last aircraft to carry 43's black and white chequers, and the gamecock, were Panavia Tornado F.3s – the unit disbanding at Leuchars on July 13, 2009. ◉

GLOSTER GREBE II

Type:	Single-seat day fighter
First flight:	1923; entered full RAF service October 1924 (Gamecock first flown February 1925; entered service March 1926)
Powerplant:	One 400hp (298kW) Armstrong Siddeley Jaguar IV radial
Dimensions:	Span 29ft 4in (8.93m), Length 20ft 3in (6.17m)
Weights:	Empty 1,720lb (780kW), All-up 2,614lb (1,185kg)
Max speed:	152mph (244km/h) at 10,000ft (3,048m)
Armament:	Two 0.303in machine guns mounted on the upper forward fuselage, firing through propeller arc
Replaced:	Sopwith Snipe from 1924 (Gamecock: Sopwith Snipe from 1926, Hawker Woodcock from 1928)
Taken on charge:	130, including two-seat Mk.IIIDCs (Gamecock: 90)
Replaced by:	Armstrong Whitworth Siskin III from 1928 (Gamecock: Bristol Bulldog from 1929)

ARMSTRONG WHITWORTH
SIKIN
1924 TO 1933

Modern-day warplanes can have their development programmes measured in decades, so the Siskin's five-year progress from prototype to service entry looks quite prompt. In that time, it morphed radically both in looks and construction, and its manufacturer changed its name.

The original Siskin, the SR.2, was ordered and designed in 1917, but it was two years before the first example appeared. The biplane was developed by the Siddeley Deasy Motor Car Company, which was acquired by the giant Armstrong Whitworth conglomerate in May 1919. The following year the aeronautical ventures of Siddeley Deasy were transformed into the Coventry-based Sir W G Armstrong Whitworth Aircraft Company – Armstrong Whitworth Aircraft (AWA) for short.

The SR.2 was a wooden-framed single-seat fighter of classic layout, but requirements were changing. By 1921 the Air Ministry was looking for all-metal aircraft and a redesign brought about the interim Siskin II in 1922, which featured a metal framed fuselage with wooden wings.

The Mk.III first appeared in May 1923, AWA referring to it as the 'All-Steel Siskin'. It was the first all-metal fighter to enter RAF service.

The variant differed considerably from its predecessors, adopting a sesquiplane format. The lower wing was of much shorter span and chord; the latter obviated the usual individual twin inter-plane struts, giving rise to the distinctive 'vee'-shape. In October 1925 the definitive Mk.IIIA took to the air, with a deeper fuselage and refined tail 'feathers'.

SHARE AND SHARE ALIKE

Service trials of Siskin IIIs were conducted by 41 Squadron at Northolt from February 1924 – the unit going completely operational that May. The Mk.IIIA was introduced into the RAF by 111

ARMSTRONG WHITWORTH SISKIN IIIA	
Type:	Single-seat day fighter
First flight:	1919; Mk.III May 7, 1923; entered RAF service in May 1924
Powerplant:	One 460hp (343kW) Armstrong Siddeley Jaguar IV radial
Dimensions:	Span 33ft 2in (10.1m), Length 25ft 4in (7.71m)
Weights:	Empty 2,061lb (934kg), All-up 3,012lb (1,366kg)
Max speed:	142mph (228km/h) at 15,000ft (4,572m)
Armament:	Two 0.303in machine guns mounted on either side of the fuselage, firing through propeller arc
Replaced:	Sopwith Snipe from 1924, Gloster Grebe and Gamecock from 1928
Taken on charge:	452
Replaced by:	Bristol Bulldog from 1929, Hawker Fury I from 1931

"Nine aircraft, in three flights of three, were to perform an aerobatic sequence, with each of the trios tied together... For most of the routine, the Siskins' wingtips could be as little as 6ft apart."

Above
Wearing the red stripe of Northolt-based 41 Squadron, Siskin III J7163, 1927. KEC

Above left
The prototype Siddeley SR.2 Siskin, C4541, in the spring of 1919.

Squadron at Duxford from September 1926.

The Siskin proved such a success that the Air Ministry eventually ordered 343 Mk.IIIAs – an incredible amount for the peacetime RAF. Throughout the 1920s and 1930s, and again in the 1950s, the powers that be were determined to keep as many design houses as possible busy with work. In the inter-war period this led to orders in tiny batches, preventing any economies of scale.

With similar thinking, the Air Ministry forced AWA to sub-contract Siskin production to competitors, with royalty payments. Blackburn at Brough, Bristol at Filton, Gloster at Hucclecote and Vickers at Brooklands all took part in its manufacture, with the parent company assembling just 87.

A dual-control variant, the Siskin IIIDC, was also built from new, and redundant fighter airframes were converted to two-seaters. The last of

these trainers was withdrawn in 1933.

After the Siskin, AWA enjoyed further success with the Atlas army co-operation biplane from 1927 to 1933. The company then became a bomber builder, initially with the twin-engined Whitley from 1938.

SHOW-STOPPER

Despite its angular appearance, the Siskin was a superb aerobatic mount. During the 1927 Hendon Pageant, Fg Off Allen Henry Wheeler of 111 Squadron made a name for himself with a solo display, including a roll executed in a near vertical climb.

Wheeler was a great friend of Richard Ormonde Shuttleworth, the founder of the famous collection that still carries his name, at first lending a hand to find and fly rare machines.

In August 1940 Richard was killed in a Fairey Battle and the Shuttleworth family asked Wheeler to manage the collection – which he would do, in one form or another, into the 1970s. His prowess on Siskins helped form a service career during which Air Cdre A H Wheeler

CBE OBE become commandant of the Aeroplane and Armament Experimental Establishment, Boscombe Down, in 1952.

With nine victories to his credit in 1917 and 1918, mostly flying Royal Aircraft Factory SE.5as with 24 Sqn, Sqn Ldr Cyril Nelson Lowe MC DFC was a much respected and popular choice to take command of 43 Squadron in 1927. He guided the 'Fighting Cocks' through the change over from Gloster Gamecocks to Siskin IIIAs at Tangmere in June 1928.

Lowe was always keen to demonstrate the skills of his pilots and, in his last year with the unit, he devised a show-stopper for the 1930 RAF display at Hendon. Nine aircraft, in three flights of three, were to perform an aerobatic sequence, with each of the trios tied together – a world first.

Cotton ropes, a reasonably generous 60ft (18.2m) long, were attached to the undercarriage. For most of the routine, the Siskins' wingtips could be as little as 6ft apart, but the rope attachment points were breakable to provide a failsafe device, should it be needed.

The display ended with a deliberate severing of the ties. The Siskins dived towards the crowd, pulling up in a manoeuvre called the 'Prince of Wales Feathers' when the three tied flights became nine individuals, each trailing their tethers as they turned to land in front of an ecstatic crowd. ◉

Above
The prototype two-seater, Mk.IIIDC J7000, first flew on October 31, 1924.

Left
A line-up of Siskin IIIs of 29 Squadron, with J8060, which served the unit 1931 to 1932, in the foreground. The unit's red 'triple X' and parallel line markings were repeated between the roundels on the upper wing. Based at North Weald, J8060 carried a camera gun on the top wing and the unit's badge, an eagle preying on a buzzard, on the fin.

HAWKER
WOODCOCK
1925 TO 1928

Woodcocks were issued to 3 Squadron at Upavon in May 1925. The only other fighter unit to fly the Woodcock was 17 Squadron – with parallel black and white 'zig-zag' heraldry – at Hawkinge, from February 1926.

After several incidents of wing failure, Mk.II J7513 was subjected to tests at the Royal Aircraft Establishment, Farnborough, in the spring of 1928. In August all Woodcocks were grounded.

A late casualty of World War One was the Sopwith Aviation and Engineering company. The liquidators were called in on September 11, 1920 and a great 'name' in British aviation was extinguished. Its founder, Thomas Octave Murdoch Sopwith, then 32, was determined to fight back.

A new company was formed on November 15, 1920 with the aims of manufacturing motorcycles, general engineering and, maybe, aircraft. This enterprise was named H G Hawker Engineering, in honour of Sopwith's incredibly talented designer-pilot-adventurer, Australian Harry George Hawker. (In 1933 the company was renamed Hawker Aircraft.)

Harry did not have long to savour the compliment Sopwith had paid him. On July 21, 1921 he was testing the one-off Nieuport Goshawk G-EASK at Hendon prior to racing it. At about 2,500ft (762m) the Goshawk made a violent turn to port, began a steep dive, started to burn and crashed.

Hawker was thrown clear, dying minutes later. The autopsy carried out on the 32-year-old discovered that he had advanced tubercular degeneration of the spine. This was highly likely to have curtailed his valiant life before very long.

HAWKER'S FIRST

The first of an illustrious line of fighters carrying the Hawker name did not have the distinctive looks of a Sopwith and was destined to serve with just two squadrons. The Woodcock was the first machine to the ordered by the RAF as a night-fighter from the very beginning.

Agility was not a requirement; stability was what was needed in the days when flying in the dark was full of risk. The all-wood construction Woodcock was not aerobatic, but it offered excellent visibility for its pilot and was a good gun platform. Unusually mounted outside of the fuselage, either side of the cockpit, the two Vickers machine guns could be reached by the pilot should a breech require clearing.

Designed by Bertram Thomson, the prototype Woodcock appeared in March 1923. Thomson's spell with Hawker was fleeting. Five months later the definitive Mk.II materialised, crafted by the gifted George Carter, but he could do little for the type's portly looks. (Carter later joined Gloster, beginning design of the Meteor jet in 1940.)

With a green stripe down the fuselage and across the top wing,

FRENCH CONNECTION

Fg Off Sidney Albert Thorn, known to his colleagues as 'Bill', flew with 17 Squadron at Hawkinge. (He went on to join Avro as a test pilot in 1934.)

Bill was given a most unusual task in June 1927 as the RAF lent one of its Woodcocks to Charles Augustus Lindbergh. The previous month, Lindbergh had become the first person to fly the Atlantic solo – earning him the nickname 'The Lone Eagle' – in the Ryan NYP monoplane *Spirit of St Louis*.

While visiting Britain, the American adventurer needed to return to Paris quickly prior to taking a ship back to the USA. Lindbergh was talked through Woodcock II J8295 at Hawkinge and set off to the French capital on June 2, 1927. Bill was detailed to make his way by surface transport to Le Bourget and bring the Woodcock back. ◉

HAWKER WOODCOCK II	
Type:	Single-seat day/night fighter
First flight:	March 1923; entered RAF service in May 1925
Powerplant:	One 420hp (313kW) Bristol Jupiter IV radial
Dimensions:	Span 32ft 6in, (9.9m), Length 26ft 2in (7.97m)
Weights:	Empty 2,014lb (913kg), All-up 2,979lb (1,351kg)
Max speed:	115mph (185km/h) at 10,000ft (3,048m)
Armament:	Two 0.303in machine guns mounted on either side of the fuselage, firing through propeller arc
Replaced:	Sopwith Snipe from 1925
Taken on charge:	63
Replaced by:	Gloster Gamecock from 1928

BRISTOL
BULLDOG

1929 TO 1937

A moment of recklessness: a very low, slow roll ended when the port wing hit the ground. The biplane cartwheeled and came to rest in a pile of twisted metal. Its 21-year-old pilot was pulled out, but looked to be a 'goner'. Surgery saved his life but not his legs; they had to be amputated.

Although he was determined to remain in the RAF, the unfortunate flyer was invalided out of the service in May 1933. Plt Off Douglas Robert Stewart Bader was destined to return to the RAF in November 1939. He was lauded as the 'Legless Ace', ending the war as a group captain and was knighted in 1976.

Douglas Bader was probably the most famous of many well-known names that piloted the RAF's state-of-the-art fighter of the early 1930s, the Bulldog. On that fateful day of December 14, 1931, Bader was flying Mk.IIA K1676 which had been issued to 23 Squadron at Kenley the previous April. Part of a trio of Bulldogs that had visited Woodley, near Reading; on departure, Bader turned back to the aerodrome for impromptu, and nearly fatal, aerobatics.

CHASING ORDERS

The exceptional two-seat Bristol F.2b Fighter entered service with the Royal Flying Corps in 1917 and became the backbone of the post-war RAF.

BRISTOL BULLDOG IIA

Type:	Single-seat day/night fighter
First flight:	May 17, 1927; entered RAF service in May 1929
Powerplant:	One 490hp (365kW) Bristol Jupiter VIIF radial
Dimensions:	Span 33ft 11in (10.33m), Length 25ft 2in (7.67m)
Weights:	Empty 2,412lb (1,094kg), All-up 3,530lb (1,601kg)
Max speed:	174mph (280km/h) at 10,000ft (3,048m)
Armament:	Two 0.303in machine guns mounted on either side of the fuselage, firing through propeller arc
Replaced:	Armstrong Whitworth Siskin and Gloster Gamecock from 1929
Taken on charge:	381, including 69 Bulldog TM advanced trainers
Replaced by:	Hawker Demon from 1933, Gloster Gauntlet from 1935, Gloster Gladiator from 1937

Bristol enjoyed lucrative contracts keeping the type in production, or reworking earlier examples, until 1927. The company churned out a string of hopeful prototypes from 1919 to 1925 but, try as it might, it could not break back into the RAF fighter market with a new design.

By 1926 the Air Ministry was changing its mind regarding fighters, seeking a higher performance interceptor, replacing the standing patrols as epitomised by the Armstrong Whitworth Siskins and Gloster Gamecocks of the day.

Against stiff competition, the Bulldog was triumphant in its Mk.II form. Along with smaller numbers of Hawker Furies, the two types formed the basis of British air defence in the first half of the 1930s. Having waited all that time to achieve a mainline fighter order from the RAF, the Bulldog was the last of the line for Bristol: the manufacturer's future lay with the ground-breaking Blenheim.

GLORIOUS HERALDRY

Bulldog squadrons performed en masse at the annual RAF displays at Hendon. Duxford-based 19 Squadron was the first to employ coloured smoke during its routines, in 1933.

As well as their formation aerobatics, Bulldogs are best remembered for their gloriously vivid squadron markings. Between the roundels on the upper wings and along the fuselage was a colourful band, often in the form of chequers: blue and white in the case of 19 Squadron. Propeller bosses and wheel centres carried a colour denoting which element of the squadron the aircraft belonged to: for example, red for 'A' Flight.

Bulldog IIA K2159 was delivered from the production line at Filton to 19 Squadron in September 1931. Two months previously, Sqn Ldr Clifford

Sanderson DFC (later Air Marshal Sir Clifford) became the unit's CO and he adopted K2159 as his personal aircraft.

To facilitate others forming up on his lead, the CO's aircraft was even more flamboyant. Bulldog K2159 carried additional blue and white chequers on the fin and elevators and streamers from the rudder and lower wings. Within the chequers on K2159's fin was 19's winged dolphin badge and the squadron leader's Royal Blue/Air Force Blue/Red 'stripes' were carried above the serial on the rudder.

BACK FROM OBLIVION

The RAF Museum's Bulldog has a Douglas Bader link. It was built as a demonstrator and civilian registered as G-ABBB in June 1930. Bristol offered the fighter to the Science Museum in February 1939. During the summer of 1955 the Bulldog was given the spurious serial 'K2496' for use as an 'extra' in Lewis Gilbert's film *Reach for the Sky*, based on Paul Brickhill's book on Bader, with Kenneth More playing the legless fighter pilot.

Following approaches from Bristol, the Bulldog was returned to its birthplace, Filton, for restoration to flying condition. On June 23, 1961 test pilot Godfrey Auty took G-ABBB up for a test flight and it was painted as 'K2227' in the red and white chequers of 56 Squadron.

Bristol handed the Bulldog to the Old Warden-based Shuttleworth

Collection on September 12, 1961 but it continued to be kept at Filton. In front of a huge crowd at Farnborough on September 13, 1964 Bristol deputy test pilot Ian Williamson displayed the biplane. At the top of a loop the Jupiter engine cut and the Bulldog plummeted to earth; miraculously Williamson suffered only minor injuries.

Parts were stored at Old Warden, while a large cache of bits was squirreled away by the RAF Museum at Henlow and later Cardington. In the 1980s, as Shuttleworth reviewed its stores, more of the Bulldog was transferred to RAF care. By 1992 it was clear that there was the basis of a static restoration. Tim Moore's Skysport Engineering brought 'K2227' back from oblivion and it was formally unveiled at Hendon on the last day of March 1999. ◉

Above
Looking backwards from his forward cockpit, the pilot of a Supermarine Southampton caught a Bulldog pulling away after an 'attack' on the flying-boat.

Left
Bulldog IIA K2159, the personal mount of Sqn Ldr Clifford Sanderson DFC, the CO of Duxford-based 19 Squadron. The device on the upper port wing is a camera gun. KEC

HAWKER FURY

1931 TO 1940

S ydney Camm changed the thinking of the Air Ministry and the prospects for his employer, Hawker, when the Hart two-seat day bomber was revealed in June 1928. Coupling exceptional aerodynamics and the groundbreaking supercharged Rolls-Royce Kestrel vee-format, 12-cylinder engine, the Hart had a top speed of 184mph (296km/h), a good 28mph faster than the RAF's premier fighter of the day, the Armstrong Whitworth Siskin. It could also show a clean pair of heels to the Siskin's replacement, the Bristol Bulldog.

The design concept was ripe for development and a single-seat fighter was quickly on the drawing boards in Camm's office at Kingston upon Thames. The result was the Hornet and it began flight testing at Brooklands in 1929. Further refinement led to the Fury, the first examples entering service with 43 Squadron at Tangmere in April 1931.

With the Fury, the RAF had its first real interceptor. The more sedate Bulldog had been labelled as such, but the greater speed, excellent climb and superb agility of the Hawker biplane placed it in a league of its own.

The idea of an interceptor was to dispense with expensive standing patrols, waiting for 'trade' to come their way. Furies could be left waiting for the 'scramble' call, quickly climbing to height to engage their quarry. Only with the advent of radar and co-ordinated fighter control techniques could this be achieved efficiently and that had to await the next generation of fighters – the Hurricane and Spitfire.

The Fury was the first RAF fighter to exceed 200mph in level flight and it was the darling of RAF events across Britain. Its cockpit layout was among the first to have provision specified for a seat-type parachute, which had been made a mandatory requirement from 1927.

REPLACING PERFECTION

The Hawker design office did not rest on its laurels with the Fury. Camm argued that the best replacement for a Fury was a more improved Mk.II and work started in 1934. Rolls-Royce had continued to 'tweak' the Kestrel and the new machine featured the 640hp (477kW) Mk.VI. Every aspect of the Fury was refined for the Mk.II, the most obvious outward features being the dapper spats on the mainwheels and a tailwheel replacing the skid.

Top speed on the Fury II was 223mph, an increase of 16mph on the Mk.I. The Fury I's climb to 10,000ft (3,048m) was achieved in an impressive four-and-a-half minutes, but the new model could get there 50 seconds faster. These small advances helped to keep the Fury 'cutting edge' while

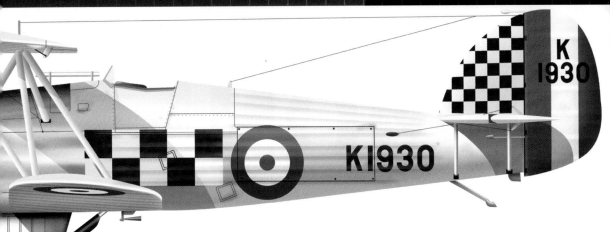

Left
Fury I K1930 was delivered to 43 Squadron, the 'Fighting Cocks', at Tangmere in April 1931. By June 1932 it was the personal aircraft of the CO, Sqn Ldr W E Bryant. It served with 43 until October 1935.
PETE WEST

the next generation was contemplated. (Turn to page 3 for Fury IIs in their element.)

The first Fury IIs joined the RAF with 25 Squadron at Hawkinge in October 1936. The last frontline examples were retired from 41 and 43 Squadrons in the first weeks of 1939. As Hurricanes and then Spitfires came on line, Furies were increasingly relegated to training schools where they provided a potent stepping stone for the rapidly swelling ranks of fighter

"Every aspect of the Fury was refined for the Mk.II, the most obvious outward features being the dapper spats on the mainwheels and a tailwheel replacing the skid."

pilots. Fury IIs were still in use with second line units in late 1940.

While the ink was drying on the Fury II drawings, in 1934 Camm and his team were scheming the ultimate Fury evolution. Tiring of unrealistic Air Ministry specifications, the Hawker board had prepared the way for a private venture.

A Fury II fuselage with an enclosed canopy was married to a monoplane wing of relatively deep section. At first a fixed, spatted undercarriage was proposed, but this was soon abandoned in favour of a wide track, retractable arrangement. Four machine guns, two in each wing, were chosen but they gave way to a breathtaking four on either side.

Rolls-Royce was also busy throwing its own resources into the pot, with the PV.12 engine, the nascent Merlin. Hopes were high that just as the Fury had broken the 200mph 'barrier', the Fury Monoplane as the new design was referred to at Kingston, would go beyond 300mph. All of this was but a short step away from the Hurricane. ⊙

Above
Having become lost on a sortie out of Netheravon on January 6, 1936 the pilot of 6 Flying Training School Fury K3737 attempted a forced landing at Childbury, Berkshire. As the inscription on the photo says: "The crosswinds got me!" The biplane was a write off.

Above left
Fury I K2065 of the Tangmere-based 1 Squadron, circa 1933. The unit markings were red stripes; note the small code 'J' above the main undercarriage. Behind is a Boulton Paul Sidestrand III of 101 Squadron.

HAWKER FURY

Type:	Single-seat interceptor
First flight:	March 25, 1931; entered RAF service April 1931
Powerplant:	One 525hp (391kW) Rolls-Royce Kestrel IIS V12
Dimensions:	Span 30ft 0in (9.14m), Length 26ft 8⅜in (8.14m)
Weights:	Empty 2,623lb (1,189kg), All-up 3,490lb (1,583kg)
Max speed:	207mph (333km/h) at 14,000ft (4,267m)
Armament:	Two 0.303in machine guns mounted on the upper forward fuselage, firing through propeller arc
Replaced:	Armstrong Whitworth Siskin from 1931
Taken on charge:	225 (Mks I and II) Including sub-contract by General Aircraft.
Replaced by:	Hawker Hurricane from 1938

HAWKER
DEMON
1931 TO 1939

Above right
Demons of 604 Squadron outside the hangars that now form the main display halls of the RAF Museum. The fin of K5406 in the foreground carries a version of the unit badge, a red shield winged in yellow with three curved swords.

Right
Five Demons of 64 Squadron, Martlesham Heath, November 1936. In the foreground is K4520, the CO's aircraft, with a red fin with a 'fighter spear' containing the unit's scarab beetle badge. The squadron's heraldry was red and blue interlocking zig-zags. KEC

Having created the Hart two-seat day bomber, which was faster than contemporary fighters, the Hawker company knew it had the basis of a family of warplanes. Exploiting the design's speed, it began a two-pronged programme – the single-seat Fury plus a much more straightforward two-seater fighter, reinventing a concept that was last employed by the Bristol F.2b in 1917 and 1918.

From chief designer Sydney Camm's drawing office, for the RAF the Hart also spawned the Demon two-seat fighter; the Audax army co-operation version and its replacement, the Hector; the Hardy 'colonial' general-purpose type; and the aircraft that succeeded the Hart, the Hind

The Fleet Air Arm benefited with the Nimrod single-seat fleet fighter and the Osprey reconnaissance two-seater, and there was a plethora of export versions with all manner of engine options.

This success transformed Hawker into the industry leader: indeed, from the late 1920s, the RAF was often referred to as the 'Hawker Air Force'. Such was the status of the company that its chief, Thomas Sopwith, masterminded the acquisition of Armstrong Whitworth, Avro and Gloster in 1935 to form the Hawker Siddeley Group.

AUXILIARIES

Transforming the Hart into a two-seat fighter was a relatively easy task. The prototype, converted from Hart J9933, appeared in late 1930. The first half-dozen production aircraft, known as Hart Fighters, were issued to 23 Squadron at Kenley in July 1931. After further refining, the new type was named Demon in July 1932.

The Air Ministry was very taken with the Demon, seeing it as a way to cut the cost of the interceptor fleet, enabling the Bristol Bulldog units to give up 'standing watch' patrols. Having pioneered the Hart fighter, 23 Squadron was the first

RAF unit to go fully operational with Demons, from April 1933, by which time it was resident at Biggin Hill.

The Demon was also considered ideal for bolstering the Auxiliary Air Force fighter force. At Hendon, 604 (County of Middlesex) Squadron was the first to re-equip, taking aircraft direct from the assembly line at Brooklands in June 1935. Also receiving brand new Demons up to late 1937 were the following 'Auxiliaries': 600 (City of London), 601 (County of London), 607 (County of Durham) and 608 (County of York).

Demons left frontline service in early 1939, handing over to Bristol Blenheim fighters, but the venerable biplanes carried on in secondary duties until the mid-1940s.

BOMBER DESTROYERS

By the summer of 1936 Hawker was gearing up to produce Hurricanes in huge numbers and needed to relocate manufacture of its adaptable biplanes. At Pendeford, Wolverhampton, Boulton Paul took on building Demons from August 1936.

The gunners in Demons had suffered considerably since the type had been introduced. Their

HAWKER DEMON

Type:	Two-seat day fighter
First flight:	1930; entered service April 1933
Powerplant:	One 525hp (391kW) Rolls-Royce Kestrel IIS V12
Dimensions:	Span 37ft 3in (11.35m), Length 29ft 7in (9.01m)
Weights:	Empty 3,067lb (1,391kg), All-up 4,464lb (2,024kg)
Max speed:	182mph (292km/h) at 16,000ft 4,876m)
Armament:	Two 0.303in machine guns mounted on either side of the fuselage, firing through propeller arc; one machine gun in rear position (in Fraser-Nash turret from 1936)
Replaced:	Bristol Bulldog from 1933
Taken on charge:	234, including sub-contract to Boulton Paul
Replaced by:	Gloster Gladiator and Bristol Blenheim I from 1938

colleagues in the rear positions of Harts, Hinds, Audaxes and Hardys experienced relatively gentle flight profiles, but Demons needed to be agile and gunners were exposed to the elements and g-forces.

To try to alleviate this, the Fraser-Nash company came up with a relatively lightweight and compact hydraulically activated gun turret. The gunner was still out in the open, but behind him were four sections of what was called a 'lobster back' which unfurled into the slipstream, affording some shelter.

The first Turret Demons were delivered in October 1936.

The idea of the two-seat fighter had worked well during the Great War, although Bristol F.2bs often required single-seater scouts to provide support. In the early 1930s, the Air Ministry envisaged that Demons could confront enemy bombers by flying alongside or through formations, with the rear gunners dishing out punishment.

When this 'bomber destroyer' notion was conceived, the opposition had been expected to arrive in lumbering biplanes or monoplanes without the benefit of escorts. Yet in Germany, well-armed, reasonably manoeuvrable bombers were becoming a reality, as were long-range fighters.

This did not stop the generation of purpose-built turret British fighters, resulting in the lacklustre Blackburn Roc for the Fleet Air Arm in 1938. The disastrous Boulton Paul Defiant was meanwhile devised for the RAF, designed to fight in a manner that had already been consigned to the history books. ◉

Above
A Fraser-Nash 'lobster back' turret fitted to a Demon, January 1935.

Left
Demons of Hendon-based 604 (County of Middlesex) Squadron, Auxiliary Air Force, July 1936. The unit colours were red and yellow triangles.

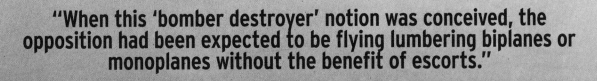

"When this 'bomber destroyer' notion was conceived, the opposition had been expected to be flying lumbering biplanes or monoplanes without the benefit of escorts."

GLOSTER
GAUNTLET AN

1935 TO 1945

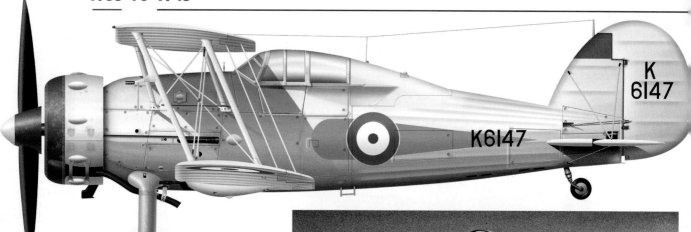

Above
Gladiator I K6147 of 3 Squadron, based at Kenley, 1937. This aircraft served the unit from March 1937 to April 1938. It was last recorded with 1415 Flight at Habbaniya, Iraq, in late 1943.
PETE WEST

Right
Gladiator Is of 73 Squadron at height in 1937. The squadron colours in the 'diamonds' were blue with yellow in the middle. In the background is K7985 as flown by Fg Off 'Cobber' Kain. KEC

After his success with the Grebe and the Gamecock, Gloster designer Henry Folland was determined to regain the company's pre-eminence in RAF fighters. When it came to replacing the Bristol Bulldog, Folland's answer was the Gauntlet of 1933, destined to be the last RAF open cockpit fighter.

Despite being powered by a draggy Bristol Mercury radial, the new fighter had the edge on the sleek Hawker Fury, by about 30mph (48km/h) at 15,000ft (4,572m).

Service trials began with the Bulldog-equipped 19 Squadron at Duxford in February 1935, the unit completely adopting the new type that May.

In July 1935 Gloster became part of the Hawker Siddeley Group and the company had access to Hawker's construction techniques and volume production expertise. This gave rise to a major revision of the Gauntlet: the Mk.II had 'dumbbell'-shaped spars and revised fuselage structure.

Gauntlet IIs were last in operational service with North Weald-based 17 Squadron in June 1939. In the Middle East, 6 Squadron withdrew its last examples in April 1940.

Gauntlets formed part of 430 Flight, based at Khartoum, Sudan, from early August 1940, to counter Italian forces. Flt Lt A B Mitchell in K5355 shot down a Regia Aeronautica Caproni Ca 133 tri-motor bomber-transport on September 7, 1940. The unit wound down its operations in the first weeks of 1941.

LAST OF THE BREED

The Gauntlet II provided the basis for Gloster's answer to Specification F7/30 which sought an interceptor to meet the RAF's needs for the second half of the 1930s. The private venture, SS.37, which was refined into the Gladiator, won out against a host of competitors, including monoplane prototypes from Bristol and Supermarine.

The bulk of the hopefuls had opted for the Air Ministry's favoured Rolls-Royce Goshawk V12. Featuring evaporative, or 'steam', cooling, the engine proved to be a road to nowhere and only about 20 Goshawks were completed.

When the Gladiator entered service the writing was already on the wall for its kind; it was the last RAF frontline biplane. Nevertheless, it provided a vital stepping stone to the next generation. Hawker and Supermarine seized the initiative with the private venture Hurricane and Spitfire multi-gun monoplane interceptors, cutting through the Air Ministry's inertia.

The Gladiator gave the pilot the luxury of a canopy; four machine guns – the extra ones being mounted in the lower wings; single-bay inter-plane struts in place of twin-bay; plus cantilever main undercarriage and split flaps on upper and lower wings. The result was the pinnacle of biplane achievement and well greeted by the squadrons.

D GLADIATOR

Left
A formation of Gauntlet IIs of 213 Squadron shortly after re-forming at Northolt, March 1937. When this sortie was flown station personnel had yet to apply the unit badge in the fighter 'spearhead' on the fins. Squadron colours were two black bands and a yellow one. The third aircraft from the left is K7806, the CO's machine, denoted by the black fin. KEC

First to receive the new type was the re-formed 72 Squadron, at Tangmere in February 1937. Gladiators were still in use during the Battle of Britain (see below) and the last of the breed was withdrawn from meteorological data gathering duties as late as January 1945.

INTO ACTION

When Britain went to war on September 3, 1939 the RAF had two Auxiliary Air Force squadrons equipped with Gladiators: 607 (County of Durham) at Usworth and 615 (County of Surrey) at Kenley. Eight days after redeploying to nearby Acklington, on October 17, 1939 Flt Lt John Sample of 607 downed a

Dornier Do 18 flying-boat over the North Sea – the Gladiator had been 'blooded'.

Both these Auxiliaries went to France as part of the Advanced Air Striking Force the following month. They returned to 'Blighty' in mid-May, with no combat losses, as the Battle of France raged.

Having previously flown Westland Lysanders, 16 Squadron partially re-equipped with Gladiator IIs in March 1940. With the retreat to Dunkirk in full swing, the unit deployed to Lympne in Kent on May 19. On the last day of the month Flt Lt George Shepley DFC was killed in combat over the French town. (His brother, Plt Off Douglas Shepley

of 152 Squadron, died in Spitfire I K9999 while opposing Junkers Ju 88s near the Isle of Wight on August 12, 1940.)

Re-formed at Roborough on August 1, 1940 with Gladiators, 247 Squadron was charged with defence of the maritime approaches to Plymouth until February 1941. The biplanes flew defensive patrols, mostly at night, and as such put the Gladiator into the Battle of Britain history books.

NORWEGIAN TRAGEDY

Germany began the invasion of Norway on April 8, 1940 and the Gladiators of 263 Squadron, which had re-formed in October 1939, were hastily embarked in HMS *Glorious* and settled upon the frozen surface of Lake Lesjaskog, southwest of Trondheim, Norway, as a

Above
An unusual angle showing the cockpit of a Gladiator and the access panels to the two 0.303in machine guns and their 400-round ammunition belts in the lower wing.

GLOSTER GLADIATOR I

Type:	Single-seat fighter
First flight:	September 1934; entered service February 1937 (Gauntlet first flew 1933, entered service May 1935)
Powerplant:	One 840hp (626kW) Bristol Mercury IX radial
Dimensions:	Span 32ft 3in (9.82m), Length 27ft 5in (8.35m)
Weights:	Empty 3,450lb (1,564kg), All-up 4,750lb (2,154kg)
Max speed:	253mph (407km/h) at 15,000ft (4,572m)
Armament:	Two machine guns, mounted on either side of the fuselage, firing through propeller arc; one machine gun under each lower wing
Replaced:	Bristol Bulldog and Gloster Gauntlet from 1937, Hawker Demon from 1938 (Gauntlet replaced Bulldog from 1935)
Taken on charge:	490 (Gauntlet: 128)
Replaced by:	Bristol Blenheim If and Hawker Hurricane from 1938, Supermarine Spitfire from 1939 (Gauntlet: replaced by Gladiator from 1937 and Hawker Hurricane and Spitfire 1938)

> "Gladiators of 263 Squadron settled upon the frozen surface of Lake Lesjaskog as a temporary – and tenuous – base on the 24th. The inevitable happened the following day when the Luftwaffe bombed the ice..."

Above
Line-up of Kenley-based Gauntlet IIs of 46 Squadron, 1937. In the foreground, K7796 carries a squadron leader's pennant under the cockpit and a red fin tip denoting 'A' Flight. The fuselage red arrow markings derive from the unit's three-arrow head badge, which is carried on the fin.

**Below,
left and right**
The RAF Museum at Hendon has two Gladiators on display: the remains of Mk.II N5628 of 263 Squadron, salvaged from the waters of Lake Lesjaskog, Norway, in 1968; and Mk.II K8042 in 87 Squadron colours.
RAF MUSEUM
www.rafmuseum.org

temporary – and tenuous – base on the 24th.

The inevitable happened the following day when the Luftwaffe bombed the ice and at least a dozen Gladiators were wrecked or went to the bottom. (The RAF Museum has the salvaged remains of one of them.)

The following day, Plt Off M A Craig-Adams of 263 Squadron found his Gladiator II N5633 spluttering with engine trouble. The topography was such that a forced landing was impossible and he took to his parachute. During the morning of May 22, by then operating out of Bardufoss, Craig-Adams was airborne again. He engaged a Heinkel He 111 but came off the worst, and was killed when N5698 crashed near Salangen.

While evacuating on *Glorious*, on June 8 the aircraft carrier was sunk. Among the losses were Hurricanes and personnel of 46 Squadron and ten pilots and seven Gladiators from the exhausted 263 Squadron.

FAITH, HOPE AND CHARITY

The most famous Gladiators of all were 'co-opted' into RAF service, having previously been on charge with the Fleet Air Arm.

Meanwhile, despite its vitally strategic position, Malta was devoid of air defence in April 1940. Having discovered Sea Gladiators in crates at Kalafrana, the island's Air Officer Commanding, Air Cdre F H 'Sammy' Maynard, decided they could be put to use.

Six pilots were gathered together, and as the first biplane was tested at Hal Far on April 19, 1940 the Fighter Flight Malta was born. Classically, ten days later the little unit had to be stood down as the Royal Navy reclaimed its fighters.

The 'turf war' was resolved on May 2, which was just as well as on June 10 Italy entered the war. The following day ten Savoia-Marchetti SM.79 Sparviero tri-motor bombers headed towards Malta. Piloting

N5520, Fg Off J L Waters dispatched one of the enemy, vindicating the creation of the impromptu unit. On the 16th, the opposition was such that Italian bombers were accompanied by fighters.

By the 22nd, the Fighter Flight was down to three operational Gladiators. The crews referred to this trio as 'Faith', 'Hope' and 'Charity' – and this stuck, being widely used in the press on the island and in Britain. Later the aircraft were retrospectively allocated these names with Waters' N5520 becoming 'Faith'. (The composite airframe that forms the present-day exhibit in Valletta is considered to be N5520.)

On June 28 four Hurricanes arrived, with more promised. The Fighter Flight was renamed 261 Squadron on August 1, by which time only one Gladiator was functioning. The exploits of a small number of pilots and their obsolete biplanes had become the stuff of legend. ◉

AVIATION SPECIALS

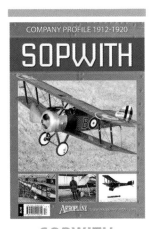

SOPWITH
Looks at all the types produced by this prolific manufacturer.

£6.99 inc **FREE** P&P*

UK WARBIRD DIRECTORY
A definitive guide to the historic ex-military aircraft flying in British skies today.

£7.99 inc **FREE** P&P*

1918: AN ILLUSTRATED HISTORY
This is the story of the Great War's final year.

£6.99 inc **FREE** P&P*

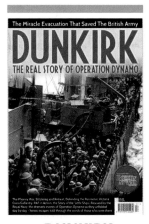

DUNKIRK
The story of the great evacuation is told, day-by-day.

£6.99 inc **FREE** P&P*

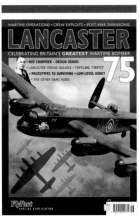

LANCASTER 75
Pays tribute to all who built, maintained and flew Lancasters, past and present.

£5.99 inc **FREE** P&P*

RAF BATTLE OF BRITAIN MEMORIAL FLIGHT
Spectacularly celebrates the Flight's activities and achievements.

£5.99 inc **FREE** P&P*

VALOUR IN THE AIR
A comprehensive salute to some of the greatest pilots ever take to the skies.

£6.99 inc **FREE** P&P*

BRITISH PHANTOMS
It is hard to believe that 25 years have passed since the RAF retired its 'Phinal Phantom', in October 1992.

£5.99 inc **FREE** P&P*

AVIATION SPECIALS

ESSENTIAL reading from the teams behind your **FAVOURITE** magazines

HOW TO ORDER

 OR

PHONE
UK: 01780 480404
ROW: (+44)1780 480404

FREE Aviation Specials App

Simply download to purchase digital versions of your favourite aviation specials in one handy place! Once you have the app, you will be able to download new, out of print or archive specials for less than the cover price!

IN APP ISSUES **£3.99**

HAWKER
HURRICANE

1937 TO 1945

How can the Hurricane be adequately described in about a thousand words and a handful of photographs? Well, just one word might suffice: saviour.

There can be no better way to pay tribute to this thoroughbred than to quote Owen Thetford from his exceptional book *Aircraft of the Royal Air Force since 1918*: "During [the Battle of Britain], Hurricane pilots shot down more enemy aircraft than all other defences, air and ground, combined."

As explained in the section on the Fury, Hawker chief designer Sydney Camm conceived the Hurricane as an extension of the construction philosophy introduced with the spectacularly successful Hart biplane. Consequently, when 'George' Bulman took the prototype, K5083, for its maiden flight on November 6, 1935 it was a low-risk programme and the board had already cleared the way to build a thousand examples.

So it was that on July 1, 1940 – nine days before the Battle of Britain began – Fighter Command's Order of Battle, across 11, 12 and 13 Groups, comprised 29 Hurricane, 19 Spitfire, 8 Blenheim and 2 Defiant squadrons. A month later Hawker had delivered at total 2,309 Hurricanes, nearly double that of the Spitfire.

Since December 1937, when 111 Squadron took delivery of the first operational examples at Northolt, a large cadre of pilots had got to know the Hurricane very well. During the Battle of Britain the Spitfire attracted the accolades, but it was the Hurricane that took the brunt and saved the day.

'MISSING' NUMBERS

Unlike the Spitfire, the Hurricane lacked development potential, but Camm went on to create entirely new designs – the Typhoon and Tempest – in a step-by-step approach. Hurricane IIc PZ865, still airworthy with the Battle of Britain Memorial Flight, was produced in July 1944, the last of 12,711 examples.

First appearing in September 1940, the Mk.II was produced in the greatest numbers and was the most versatile. The Mk.IIa packed eight machine guns, the Mk.IIb a dozen, the Mk.IIc four 20mm cannon and the Mk.IId 'tank-buster' a pair of 40mm cannon.

Underwing pylons brought about the 'Hurri-Bomber', toting initially two 250lb (113kg) and later a pair of 500-pounders. During 1943 the Mk.IV fighter-bomber was introduced with the capability of carrying eight 60lb rocket projectiles, four under each wing.

Canadian Car and Foundry built Hurricanes under licence at Fort William, Ontario. These machines were designated Mks X, XI and XII.

There was a Hurricane III, but it remained a proposal only. It would have been powered by Rolls-Royce Merlins built under licence by Packard in the USA. Two Mk.Vs entered flight testing in 1943, both conversions of Mk.IVs, with four-bladed propellers.

ONE OF 'THE FEW'

There is a temptation to highlight the exploits of one of the famous 'names' that flew Hurricanes during

the Battle of Britain, but 'unknown' pilots are just as deserving of our thoughts. In Kent, the incredible Shoreham Aircraft Museum has been laying memorials to pilots who died in the immediate area during those momentous days of 1940.

The latest hero to be commemorated was Sgt Jack Hammerton, of 615 Squadron, Northolt. Operational from August 1940, he was killed in former 1 Squadron RCAF Hurricane I 323 (previously L1886), which came down at Noah's Ark, south of Kemsing on November 6 – the anniversary of the prototype's first flight. The simple, but impressive memorial was dedicated in a service conducted on September 29, 2017.

In October the museum's magazine, *Friends of the Few*, recorded Jack's combat report for October 29, 1940 when he got to fire his guns at an enemy: "I got separated from my section after enemy aircraft had been sighted. I saw [it] some considerable height above me, so followed and reached a height of about 20,000ft when [it] dived... I followed down to about 7,000ft. I went in and delivered a short burst...

"I saw that the machine was twin engined and apparently a Ju 88. Again, I followed it out to sea and got in a long burst from astern and last saw it losing height with clouds of smoke coming from it."

On that day Kampfgeschwader 1's (KG 1) III Gruppe lost Ju 88A-3 2210 when it crash-landed at Bapaume in France and KG 4 lost Ju 88A-1 5014 in unknown circumstances. As Jack was the only RAF pilot to record an attack on a Ju 88 on the 29th, it seems highly likely that one of these fell to his guns.

The journal then turns to November 6, when Jack failed to return, he was: "chasing some enemy

aircraft and was last seen leaving the formation at about 25,000ft. ...on November 13, as recorded by the squadron diarist, news was received

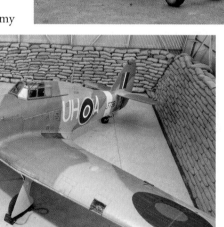

that Jack had sadly been discovered dead by his burnt-out Hurricane..."

The accident card for his Hurricane notes that control was lost due to lack of oxygen. Jack was laid to rest in his home town of Slough. The Battle of Britain ended officially on the last day of October 1940 and two days before that Jack had courageously engaged the enemy. He had probably been victorious but could not have known that.

Just 25 years old, Jack Hammerton's time in combat had been short and his story largely unknown. Thanks to the determination of the Shoreham Aircraft Museum one of 'The Few' has been graciously honoured. ⊙

Left
Hurricane I V6799 of 501 Squadron, based at Kenley, September 1940. PETE WEST

Above
Based at Warmwell from November 1941, Canadian 402 Squadron was used for publicity shots of the 'Hurri-Bomber' role: BE417 'K-for-King' (illustrated) has an exhaust glare shield fitted.

Left
Displayed at the RAF Museum Cosford, Mk.IIc LF738 wears the colours of 1682 Bomber Defence Training Flight. It served with this unit, based at Enstone, from April to August 1944. RAF MUSEUM
www.rafmuseum.org

HAWKER HURRICANE IIC

Type:	Single-seat fighter / fighter-bomber
First flight:	February 21, 1935; entered service December 1937
Powerplant:	One 1,280hp (954kW) Rolls-Royce Merlin XX V12
Dimensions:	Span 40ft 0in (12.19m), Length 32ft 2in (9.8m)
Weights:	Empty 5,800lb (2,630kg), All-up 7,800lb (3,538kg)
Max speed:	339mph (545km/h) at 22,000ft (6,705m)
Armament:	Four 20mm cannon, two in each wing; up to 1,000lb (453kg) of bombs
Replaced:	Mk.I initially: Gloster Gladiator from 1938
Taken on charge:	12,711 -all variants. Production also undertaken by Gloster, Austin Motors and Canadian Car and Foundry
Replaced by:	Hawker Typhoon intended replacement; widely supplanted by Supermarine Spitfire and US-supplied types

BRISTOL BLENHEIM

1938 TO 1944

Right
Blenheim IVFs of
254 Squadron,
circa 1941.

Captain Frank Barnwell, chief designer for Bristol, came up with a very clean-looking twin-engined eight-seater with retractable undercarriage that might appeal to businesses and the rich – a 1930s Learjet. Lord Rothermere, patron of the *Daily Mail* and a great supporter of British aviation, got wind of the Type 142 project and ordered one, which he declared would carry the name *Britain First*.

When it was evaluated in 1935, *Britain First* ruffled feathers. It had a top speed 54mph (87km/h) *faster* than the state-of-the-art Gloster Gladiator biplane fighter that had just been ordered for the RAF. Here was a twin with obvious potential to become a 'heavy' fighter or a bomber.

Point made, his lordship presented *Britain First* to the nation. In August 1935 an appreciative Air Ministry ordered 150 military-configured Type 142Ms off the Filton drawing boards. The incredible Blenheim had been conceived.

With a 'pallet' of four machine guns under the centre section, the Mk.If (F for fighter) was first issued to 25 Squadron at Hawkinge in December 1938. By then the production lines had standardised on the long-nosed Mk.IV. All Blenheim fighters were converted from bomber airframes: about 200 Mk.Ifs and 125 Mk.IVfs.

Blenheims were quickly adapted to the night-fighter role with Fighter Command and provided convoy protection with Coastal Command. Experience with the type enabled Barnwell to create the Beaufort, which became operational in the spring of 1940, taking over the maritime role, and in turn led to the incredible Beaufighter, which excelled with both Fighter and Coastal Command.

FATHER AND SON
Frank Barnwell did not bask in this success. On August 2, 1938 he died while piloting a single-seat ultra-light of his own design: the Barnwell BSW Mk.1 G-AFID, powered by a Scott two-cylinder. He was making the

BRISTOL BLENHEIM If

Type:	Two/three seat twin-engined day/night fighter
First flight:	June 25, 1936; entered service December 1938
Powerplant:	Two 840hp (626kW) Bristol Mercury VIII radials
Dimensions:	Span 56ft 4in (17.16m), Length 39ft 9in (12.11m)
Weights:	Empty 8,025lb (3,640kg), All-up 11,975lb (5,431kg)
Max speed:	230mph (370km/h) at 18,000ft (5,486m)
Armament:	One 0.303in machine gun in port wing, another in dorsal turret, four in under-fuselage fairing
Replaced:	Did not directly replace any type, units converting from Gloster Gladiators etc from 1937
Taken on charge:	1,134 (*all* Mk.Is), 3,296 (*all* Mk.IVs). Production sub-contracted to Avro and Rootes Securities. (Also built in Canada by Fairchild as the Bolingbroke.)
Replaced by:	Fighter Mk.Is gave way to Bristol Beaufighter from 1940

little machine's second flight when it plunged to the ground. The world had been robbed of a great talent.

Inspired by his father, John Barnwell joined the RAF and became a proud Blenheim pilot with 29 Squadron. On the night of June 18/19, 1940, he and Sgt K L Long were flying in Mk.If L6636 from Martlesham Heath. They were observed chasing an enemy bomber and there was a vigorous exchange of fire. One of the aircraft fell into the North Sea.

Barnwell and Long had been awarded a victory by the squadron.

As hours turned into days it was clear it was the Blenheim that had fallen and not the enemy. After 16 days, the body of Plt Off John Barnwell washed up on the Suffolk coast.

SEEING IN THE DARK
Among its many claims to fame, the Blenheim became the first type in the world to be fitted with airborne interception (AI) radar and use it in anger. As such it pioneered the way for Allied night-fighters, especially its successor in that role, the Beaufighter.

On the night of July 22/23, 1940, Fg Off Glyn 'Jumbo' Ashfield was piloting Mk.If L6836 of the Fighter Interception Unit (FIU) out of Tangmere. Plt Off G Morris was acting as observer, wireless operator and gunner. Based at Ford on the south coast, the FIU was tasked with making AI operational.

Facing the tail and intently monitoring the bulky AI Mk.III set was Sgt R Leyland. He watched as two blips gave him the height and bearing of a target and locked on to a Dornier Do 17Z, guiding 'Jumbo' to a point in the blackness where the Blenheim's four guns sent the bomber spiralling down into the English Channel.

Personnel at FIU continued to refine the system, standardising on the more practical AI Mk.IV. At the same time detachments from radar-equipped Blenheim squadrons came to Ford to be briefed and apply the new skills.

CHANGING TACTICS
The world's only flying Blenheim is operated by the Aircraft Restoration Company at Duxford. Based on a Canadian-built Bolingbroke IVT, G-BPIV is fitted with a genuine Mk.IF cockpit, from L5739.

Built by Avro, L6739 was issued to 23 Squadron at Wittering in early

1939 and struck off charge on the last day of 1940. To mark this legacy, the magnificent restoration flies in 23 Squadron's colours.

Converting in December 1938 from Hawker Turret Demons, 23 worked up on the new twins and changed base to adjacent Collyweston in May 1940. The unit's first engagement, on the night of June 18/19 – the evening John Barnwell was killed – was a costly affair: three aircrew and two Blenheims were lost.

Commanding 23's 'B' Flight, Flt Lt Myles Duke-Woolley managed to shoot down a Heinkel He 111 that evening – the unit's first 'kill'. The Luftwaffe bomber crashed off the Norfolk coast, its crew taken prisoner. Duke-Woolley's Mk.IF, *L-for-London*, had taken a lot of return fire from the Heinkel and declared beyond repair at Collyweston.

Myles was not impressed with night-fighting and requested a transfer to day fighters. He left the unit on September 12 to join the Hurricane-equipped 253 Squadron at Kenley.

No 23 Squadron moved on September 12 to Ford, the home of FIU, but on December 21 changed tactics. With the secret radar removed and the gun turret reinstated, the squadron embarked on night intruder

operations against Luftwaffe airfields in Normandy.

The aim was to catch enemy aircraft landing or taking off and engage them. If no such 'trade' was available, installations could be bombed. An additional advantage was that the presence of a Blenheim in the area would force the airfields' flare paths to be extinguished, disrupting the operations and possibly causing aircraft to divert.

From early January to mid-March 1941 the Blenheim IFs of 23 Squadron accounted for a pair of He 111s and a Focke-Wulf Fw 200 Condor over French territory, along with half-a-dozen 'probables'. But again, night operations came at a high price: 12 aircrew were killed, two taken prisoner and five Blenheims were destroyed.

One of them was piloted by the very determined Sgt Vic Skillen who attempted to *bomb* an He 111 on approach to Amiens-Glisy. Not content with this, he saw another Heinkel and turned to confront it, only to collide with his quarry. All aboard *X-for-X-ray* were killed, along with five on the Luftwaffe bomber. In April, 23 Squadron began converting to Douglas Havocs, ready for a new era. ◉

SUPERMARINE
SPITFIRE MERLIN

1938 TO 1951

K9794

Millions of words, tens of thousands of articles and thousands of books have been expended on the story of the Spitfire. There is so much to cover: 20,351 Spitfires of a bewildering number of variants and 2,408 navalised Seafires were produced, not to mention the post-war hopefuls, the Spiteful and the Seafang.

The exploits of its pilots from the Battle of Britain onwards are the stuff of legend, defending Britain, in Europe, North Africa, the Middle and Far East. Then there are exports; the vast majority by intent; the four operated by the Luftwaffe by chance. As well as Australia, Canada, Rhodesia and South Africa, more than 30 nations operated Spitfires or Seafires including the Soviet Union and the United States.

And the story continues to the present day. The Battle of Britain Memorial Flight operates a Mk.II, a Mk.V, a Mk.IX, a Mk.XVI and a pair of PR.XIXs, so the Spitfire is *still* in RAF service. Figures vary, but getting on for 200 Spitfires and Seafires survive, some as battered hulks, others as pristine museum pieces. Of that figure, around 60 are airworthy, with more being reconstructed to swell the numbers. Such are the capabilities of the industry supporting 'warbird' owners, it is possible to create a Spitfire airframe from little more than a salvaged maker's plate.

PRECIOUS PROTOTYPE
John 'Mutt' Summers test flew the hand-built Supermarine Type 300, K5054 – the prototype Spitfire – at Eastleigh, Southampton, on March 5, 1936. Watching was the pale-

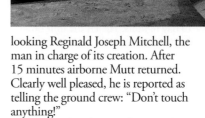

looking Reginald Joseph Mitchell, the man in charge of its creation. After 15 minutes airborne Mutt returned. Clearly well pleased, he is reported as telling the ground crew: "Don't touch anything!"

After initial evaluation the Aeroplane and Armament Experimental Establishment (A&AEE) at Martlesham Heath issued a glowing report on June 8/9, 1936: "In general, the handling of this aeroplane is such that it can be flown without risk by the average fully trained fighter pilot." This was exactly what the RAF was looking for – a super-plane that did not need to be piloted by supermen. The assessment was just as well: the pressure of world affairs was such that an order for 310 Mk.Is had been placed five days previously.

Type 300 K5054 was destined to be the only Spitfire that its designer saw in the air. Three years before that maiden flight, he had been diagnosed with a cancer. R J Mitchell CBE died on June 11, 1937, aged 42. Up to the last he had been poring over layouts to put cannons into the Spitfire's wings.

NIGHTMARE WING
Despite being part of the giant conglomerate Vickers-Armstrongs, Supermarine was a small concern, finding building Walrus amphibians challenging. Mass manufacturing a state-of-the-art fighter was a daunting prospect.

Problems emerged at an alarming rate. A string of sub-contractors was engaged to help with the huge capacity expansion required. Co-ordinating ten different companies, all reliant on one another to keep the flow going, was a never-ending ordeal.

That beautiful elliptical wing was a nightmare to build and a specification change to the leading edge did not help the timetable. Production of fuselages was out-pacing the wings and Supermarine was taking a lot of the flak from the Air Ministry.

Alternatives needed to be considered: perhaps the Itchen and Woolston factories in Southampton were better suited to other, less complex, types? The prospect that Supermarine could switch to building the 'rival' Hurricane was real.

"That beautiful elliptical wing was a nightmare to build and a specification change to the leading edge did not help the timetable. Production of fuselages was out-pacing the wings..."

The ministers held their nerve and a second contract for 200 Mk.Is was placed in March 1938. It was not until May 14, 1938 that the second Spitfire, Mk.I K9787, first took to the air. Thus, K5054 was an exceptionally precious airframe, for a worrying 27 months it was the only development airframe.

FIRST OF THE MANY

An eager A&AEE took delivery of K9787 on July 27, 1938 and there was not a moment to be lost. Eight days later, K9789 and K9790 touched down at Duxford to join 19 Squadron – the first of a vast number of RAF Spitfire units. Production was still piecemeal, only one more

was handed over to 19 in August (K9792) and another (K9795) in September. By that time more than 200 Hurricanes were operational.

Ten followed in October, one in November and on December 19, the unit had received 20 of the new fighter. The first loss came on September 20, 1938 when K9792 suffered an undercarriage malfunction while landing and ended up on its back. It had just 41 hours, 35 minutes 'on the clock'.

The dip in deliveries to 19 Squadron during November 1938 was because its neighbour unit, 66 Squadron, was also working up on the type. Its first example, K9803, arrived at Duxford on October 23, with more to follow.

Meanwhile, talks had begun with a giant of the manufacturing process, motor vehicle mogul William Morris – Lord Nuffield. The idea of 'shadow factories' – well-proven industries building aircraft to supplement the original constructor – had been commonplace in World War One. Construction of the vast Castle Bromwich Aircraft Factory, near Birmingham, began on July 12, 1938. The trickle of Spitfires was soon to become a flood.

In November 1944 the last major Rolls-Royce Merlin-powered Spitfire, the Mk.XVI, entered service. This version enjoyed a long career, the last examples being withdrawn in 1951.

The design of the Spitfire was such that it could be constantly evolved. In April 1942 a converted Mk.IV took to the air as the pioneer of the Spitfire's most radical metamorphosis. As page 56 explains, superb though the Merlin was its successor, the Griffon, was to take the Spitfire to new heights. ⊙

SUPERMARINE SPITFIRE Va

Type:	Single-seat day interceptor
First flight:	March 5, 1936; Mk.I entered service August 1938
Powerplant:	One 1,440hp (1,074kW) Rolls-Royce Merlin 45 V12
Dimensions:	Span 36ft 10in (11.22m), Length 29ft 11in (9.11m)
Weights:	Empty 4,981lb (2,259kg), All-up 6,417lb (2,910kg)
Max speed:	374mph (601km/h) at 13,000ft (3,962m)
Armament:	Eight 0.303in machine guns, four in each wing
Replaced:	Mk.Is took over from Gloster Gauntlets and Gladiators from August 1938
Taken on charge:	18,293 of all Merlin variants. Sub-contracts to Westland. Built at various Supermarine- and Vickers-run factories and at Castle Bromwich.
Replaced by:	Continued development, the Spitfire effectively replaced itself for much of its service life. Ultimately, Gloster Meteor and de Havilland Vampire directly took over the day-fighter role

Above left
Built as a Mk.I in July 1940, R6923 was converted to a Mk.Vb and was issued to 92 'East India' Squadron at Digby in November 1941. Coming off the worst in a tussle with Messerschmitt Bf 109s off Dover on June 21, 1941, Sgt G W Aston successfully baled out of 'S-for-Sugar'.

Left
As might be expected, the RAF Museum is custodian of several Spitfires. At Hendon, Mk.Vb BL614 is displayed in the colours of 222 'Natal' Squadron worn when flown by Sgt J W McDonald on two sorties of Operation 'Jubilee', the ill-fated raid on the French port of Dieppe, on August 19, 1942.
RAF MUSEUM
www.rafmuseum.org

BOULTON PAUL
DEFIANT
1939 TO 1945

Above
A flight of Defiants of 264 Squadron, taken between August 24 and 28, 1940. N1535 'PS-A' was shot down after combat with a Junkers Ju 88 at about 12:40 hours on the 24th: its crew, Sqn Ldr P A Hunter and gunner Plt Off F H King were killed. In the foreground, L7026 'PS-V' was brought down over Manston by Messerschmitt Bf 109s at 08:55 on the 28th: Plt Offs P L Kenner and C E Johnson perished.
PETER GREEN COLLECTION

Turret-equipped Hawker Demons entered service with the RAF in October 1936 and these 'bomber destroyers' struck a chord with the Air Ministry. As turret technology improved, so it seemed that a faster monoplane in the same vein would be a formidable extension of the armoury.

Specification F9/35 sought such a machine and Boulton Paul flew its prototype Defiant, K8310, on August 11, 1937. The Wolverhampton-based company had built Demons under sub-contract and via its Overstrand turret-armed bomber was at the forefront of power-operated guns.

The Fleet Air Arm was also seduced by the idea and the first Blackburn Roc, a development of the Skua dive-bomber, appeared in December 1938. Series production of Rocs was handed over to Boulton Paul. Rocs served with the Royal Navy from February 1940 but they were withdrawn from frontline units in August 1941.

While a fusillade from four close-coupled 0.303in Browning machine guns into the side or the belly of a bomber would be lethal, the turret fighter was neutered if its quarry

was escorted by fast, manoeuvrable fighters.

The Defiant and the Roc had more in common than their turrets. Both lacked any fixed, forward-firing guns. Once the enemy grasped this, frontal attacks – not even needing to be head-on – were very effective.

CLOAK OF DARKNESS
The first frontline unit to operate the Defiant was 264 Squadron at Martlesham Heath. Entering combat in the skies over Dunkirk in May

1940 at first Defiants basked in the enemy's unfamiliarity with the type. Reports of 'Hurricanes' that could fire backwards were greeted with incredulity by Luftwaffe intelligence officers as they debriefed very confused aircrew.

Soon this advantage vaporised, as word was distributed that the Defiant could be confronted relatively easily. At the start of the Battle of Britain, two squadrons were operational, 141 and 264. During a convoy patrol on July 19, Messerschmitt Bf 109s

BOULTON PAUL DEFIANT I

Type:	Two-seat turret day-/night-fighter
First flight:	August 11, 1937; entered service December 1939
Powerplant:	One 1,030hp (768kW) Rolls-Royce Merlin III V12
Dimensions:	Span 39ft 4in (11.98m), Length 35ft 4in (10.76m)
Weights:	Empty 6,078lb (2,756kg), All-up 8,350lb (3,787kg)
Max speed:	303mph (487km/h) at 16,000ft (4,876m)
Armament:	Four 0.303in machine guns in power-operated turret
Replaced:	New type, units converting from December 1939
Taken on charge:	1,060 of all versions
Replaced by:	Withdrawn as a fighter in September 1942; units adopting Bristol Beaufighter and de Havilland Mosquito

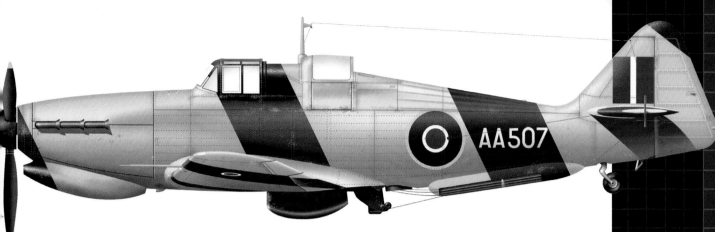

cut into 141's Defiants and *seven* were shot down or written off: ten aircrew were killed.

In the space of four days, August 24 to 28, the Defiants of 264 took a mauling: 14 men died and *ten* of the type were lost. It came as no surprise when the following month flights of the turret fighters were switched to the cloak of darkness. At first nocturnal Defiants relied on the eyesight of their crews, but during 1941 they were slowly equipped with radar.

Amid this carnage, 264 had been fighting back. At midday on August 26, the unit was taking punishment – that day three Defiants fell. But it was not all one-sided; Plt Off Frederick Desmond Hughes and his gunner Sgt Fred Gash dispatched a pair of Dornier Do 17s.

This combination of pilot and gunner made the transition to night flying and between mid-October and April 1941; the pair notched up three more victories – all Heinkel He 111s – to become 'aces'. Hughes, referred to as 'Hawk Eyes' in the press, was awarded a DFC and Gash a DFM for their efforts.

Hughes went on to Bristol Beaufighters and by January 1944 was a wing commander piloting de Havilland Mosquito XIIIs with 604 Squadron out of Scorton. On the night of 13th/14th 1945 he shot down a Junkers Ju 188 near Rotterdam to bring his total to an impressive 18½ victories – all but the

inaugural pair of Dorniers dealt with in the dead of night. Fred Hughes stayed with the RAF after the war and had a glittering career, retiring as an air vice-marshal in 1974.

ELECTRONIC SPOOFS

Defiants were withdrawn as night-fighters in July 1942 when 151 and 264 Squadrons handed over their Mk.IIs for Mosquito IIs. The brief domain of the turret fighter had ended.

The Boulton Paul two-seater was adapted as a platform for the top-secret 'Moonshine' transmitter. This could make German radar operators think that there were many more aircraft on their screens than was the case. Alongside this was 'Mandrel', a countermeasures device that

overwhelmed enemy radar.

In May 1942 Defiant Flight was formed at Northolt to try out the new 'toys'. On August 6, nine Moonshine-equipped aircraft helped to 'spoof' a USAAF raid by Boing B-17 Flying Fortresses. While circling Portland, the Defiants created the illusion of a massed attack on Cherbourg, where the balloon barrage was raised and a large force of fighters were scrambled. The B-17s headed for their intended target of Rouen, where they were greeted by little opposition.

'Moonshine' was utilised during the abortive amphibious assault on Dieppe of August 19. Three days later, 515 caused chaos by diverting more than 100 Luftwaffe interceptors away from Allied bombers hitting Rotterdam.

During October 1942 the work of Defiant Flight was acknowledged when it was renamed 515 Squadron. Mandrel was first deployed in early December. The unit continued with Defiants until December 1943, by which time Beaufighter IIs were taking over the risky work.

The Defiant achieved an honourable career in support roles, including gunnery training, air-sea rescue and target towing. The final examples left service in April 1945. ◎

BREWSTER
BUFFALO
1940 TO 1942

Needs must... With offices in New York, members of the British Purchasing Commission had deep pockets backed by the nation's dwindling gold reserves. The delegates were offering cash to the American aircraft industry in a scramble to bolster the capabilities of the RAF as Europe plunged from 'Phoney War' into all-out conflict.

By ordering straight off the production line in the early weeks of 1940 the hope was that the hardware would begin to arrive in the spring of the following year. Fingers were crossed that what they had bought would be a war-winner.

The team needed to find manufacturers with spare capacity, or the ability to increase build rates. The problem was that, as the USA was beginning to expand its own armed forces, any organisation that didn't have a bulging order book was probably in that state for a reason.

Just a short distance from the commission's offices in New York was the Brewster Aeronautical Corporation at Long Island. The company's first design, the SBA-1 two-seat monoplane naval scout bomber, attracted an order for 30 from the US Navy in 1938. Brewster encountered problems putting it into production and the batch was completed by the Naval Aircraft Factory at Philadelphia, Pennsylvania, but even then it was 1941 before the type entered service, already obsolete.

BREWSTER BUFFALO

Type:	Single-seat fighter
First flight:	December 1937; RAF operational trials in September 1940; entered service with 67 Squadron in March 1941
Powerplant:	One 1,100hp (820kW) Wright Cyclone GR-1820
Dimensions:	Span 35ft 0in (10.66m), Length 26th 0in (7.92m)
Weights:	Empty 4,479lb (2,031kg), All-up 6,840lb (3,102kg)
Max speed:	292mph (469km/h) at 20,000ft (6,096m)
Armament:	Two 0.303in machine guns in the nose, firing through the propeller arc; one 0.303in in each wing
Replaced:	Direct acquisition from USA, plus stocks intended for Belgium, 1940
Taken on charge:	198, including 28 diverted from Belgian Air Force order
Replaced by:	Withdrawn at fall of Singapore February 1942

In December 1937 Brewster flew the prototype F2A, destined to be the first monoplane fighter to enter service with the US Navy. Manufacture took time to get rolling: it was June 1940 before US Navy unit VF-3 began to shake down the new fighter. Operational evaluation included embarking on the carrier the USS *Saratoga*.

ORDERING A TURKEY
By the time F2As were touching down on the deck of the *Saratoga*, the RAF had placed an order for 170 F2A-2 models, named Buffalo, France had fallen to Germany's blitzkrieg and Britain was bracing itself for invasion. A batch of 28 F2A-3s destined for Belgium was

diverted and the first of these was assembled at Burtonwood, straight off the Liverpool docks, in August.

The pilots of VF-3 were not at all pleased with their F2A-1s. The undercarriage proved to be so weak that deck landings were banned until Brewster could sort a 'fix' and the armour plating was also insufficient. It was back to the drawing boards for revisions.

Inevitably, the mods to the landing gear and strengthened armour put the weight *up* and the performance *below* requirements. A more powerful Wright Cyclone radial would have to be fitted. After 507 F2As had been delivered, the US Navy come to its senses and ordered the Grumman F4F Wildcat; the first

entering service in December 1940.

Despite the problems at Long Island, the first Buffalo Is arrived in Britain in February 1940 with the order completed in August. To hasten the introduction to service, 71 Squadron at Church Fenton took some for service trials alongside its Hawker Hurricanes. At Boscombe Down, the Aeroplane and Armament Experimental Establishment (A&AEE) was also busy technically evaluating the American fighter.

Boscombe pilots were far from impressed. Re-arming the machine guns in the wing was overly complex. When the nose-mounted guns were fired at low altitude, an oily film was deposited on the windscreen. Carbon monoxide levels inside the cockpit were dangerously high. This alarming trait could be cured, but the other flaws would require a redesign.

By November 1940 the pilots at 71 Squadron had rejected the Buffalo outright. At A&AEE, the type was always running hot – oil temperatures were excessive, even in Britain's relatively cold climes.

What was to be done with the troublesome fighters? A decision had already been made. Batches of crated Buffalos were already on the high seas destined to defend the hot and humid skies of Singapore.

UNEQUAL STRUGGLE

At Kallang, 67 and 243 Squadrons re-formed in March 1941 and worked up on the Brewsters. More examples were shipped to the Far East: 453 Squadron Royal Australian Air Force at Sembawang and 488 Squadron Royal New Zealand Air Force, also at Kallang, took deliveries in August and October 1941, respectively.

Last to take the type was 146 Squadron in India in April 1942, replacing Curtiss Mohawks. The unit put up with the poor performing Brewsters for a matter of weeks, taking on Hurricanes with great relief.

Pearl Harbor on December 7, 1941 changed the face of World War Two. That night Japanese forces began the invasion of Malaya, working down the peninsula. For eight days from February 7, 1942 Singapore was ferociously defended from land, sea and air.

It was a valiant but inevitably unequal struggle.

The Buffalos were no match for the Japanese Nakajima Ki-43 *Oscars* or Mitsubishi A6M *Zekes*. Their guns frequently jammed and the engines lacked power. Despite this, in the chaos of the air battles, the RAF, Australia and New Zealand units accounted for an impressive number of enemy aircraft, including at least two *Zekes*. Nevertheless, by

the middle of February, the Buffalo units defending Singapore had been obliterated.

HISTORY REPEATED

Meanwhile, Brewster had created the SB2A Buccaneer dive-bomber. The US Navy took about 220, but they were destined for secondary duties. In its initial shopping frenzy, the British Purchasing Commission ordered 450, and then another 300, calling the type Bermuda.

The first examples arrived in September 1942 and they were greeted at Boscombe Down with the same derision as the Buffalo.

Of the 226 shipped to the UK none saw service; all going straight into store. Some were palmed off on the Australians and the Canadians. Upwards of 350 were cancelled in early 1943.

Brewster had opened a brand new factory at Johnsville, Pennsylvania, to cope with expected Buccaneer/Bermuda orders. To fill the void, the US Navy issued contracts to build 735 of the superb Vought F4U-1 Corsair, under the designation F3A.

History repeated itself: poor delivery schedules and quality control issues dogged the programme. Another contract for 773 Corsairs was torn up and in July 1944 the Brewster Aeronautical Corporation closed its doors. ◉

Above
Buffalo I W8144 of 67 Squadron based at Kallang, March 1941. This machine was written off in a landing accident at its base on September 19, 1941.
STEVE NICHOLLS © 2017

Below
Buffalos of 243 Squadron on a sortie out of Kallang, Singapore, in mid-1941. The censor has obliterated the unit's 'WP-' codes, on those that carry them. A Blenheim IV is tagging along in the background. KEC

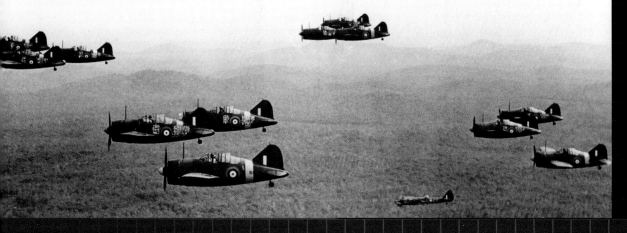

BRISTOL
BEAUFIGH

1940 TO 1960

There are times when the luxury of a blank sheet of paper on a designer's drawing board is not an option. Time constraints require a more practical solution. Some of the best aircraft of World War Two were adaptations of another type – such was the Beaufighter.

Bristol's formidably-armed fighter was the third of a dynasty of twin-engined warriors that owed its lineage to newspaper mogul Lord Rothermere. As shown on page 28, the Type 142 corporate transport quickly evolved into the Blenheim, a stalwart of the early years of World War Two.

A major re-think of the Blenheim format met Specification 10/36 for an anti-shipping aircraft capable of carrying two torpedoes. The result was the Beaufort, which first flew on October 15, 1938. The Beaufort is beyond the remit of this publication – its story can be found in the companion volume on bombers – details elsewhere.

'SPORTS MODEL'

Bristol designer Frank Barnwell was killed in a light aircraft of his own creation in August 1938. His assistant, Leslie George 'LG' Frise, took over the Filton design office.

Barnwell and Frise had been examining the possibilities of a four-cannon fighter, as had Westland's designer, W E W Petter – see page 38. Concentrated fire from 20mm guns clustered in the forward fuselage would provide unprecedented 'punch'.

Fighter Command needed a two-seat 'heavy' fighter for day and night interceptions and Coastal Command

BRISTOL BEAUFIGHTER VI	
Type:	Two-seat, twin-engined night-fighter
First flight:	July 17, 1939; entered service September 1940
Powerplant:	Two 1,670hp (1,245kW) Bristol Hercules IV radials
Dimensions:	Span 57ft 10in (17.62m), Length 41ft 8in (12.69m)
Weights:	Empty 14,600lb (6,622kg), All-up 21,600lb (9,797kg)
Max speed:	333mph (535km/h) at 16,000ft (4,876m)
Armament:	Four nose-mounted 20mm cannon and six 0.303in machine guns, three in each wing
Replaced:	Bristol Blenheim IVs as night-fighters and coastal fighters from 1940
Taken on charge:	5,584 of all variants. Sub-contracts to Fairey, Rootes Securities and the Department of Aircraft Production, Australia
Replaced by:	Bristol Brigand from 1950

was seeking a nimble and hard-hitting torpedo/strike aircraft. After the Beaufort's debut, 'LG' came up with an inspired transformation of its airframe. Retaining the wings, centre section, tail unit and undercarriage, he created a new slim-line fuselage with a pilot and observer in tandem. Gone were the troublesome Bristol Taurus radials; in their place was a pair of beefy Hercules.

This gelled as the Type 156 which was given a name that reflected the thinking behind it: the Beaufort-Fighter, or Beaufighter. A more advanced version was called the 'Sports Model' by the Filton design staff.

Just over six months after the initial layouts had been presented, Cyril Uwins took the prototype, R2052, into the air for the first time on July 17, 1939 and an order for 300 was signed. Without weaponry and operational fittings, R2052

had a maximum speed of 335mph (539km/h) at 17,000ft (5,180m) – the Hurricane I was capable of 316mph.

FIGHTER AND COASTAL

From July 10, 1940 the Battle of Britain was raging and the need to replace the Blenheim in the night-fighter role had become urgent. Trials of the Beaufighter with airborne interception radar revealed a formidable potential that could be deployed with rapidity. The RAF quickly realised that no further development was needed; the new Bristol was already a 'Sports Model'.

Sub-contracting went into full swing, the 100th 'Beau' was ready for service on December 7, 1940 and in November 1942 the 1,000th appeared – an exceptional achievement.

The first Beaufighter unit was 25 Squadron at Debden in September 1940. The inaugural Coastal Command squadron was 252 at

TER

Chivenor three months later.

Another powerplant was deemed prudent as 'insurance' and the Rolls-Royce Griffon was put forward but the Merlin XX was substituted. The first Beaufighter II (as the Merlin version was designated) flew in July 1940. The Mk.II became a dedicated night-fighter and 600 Squadron at Colerne was the first to adopt it in April 1941.

Beaufighters replaced Beauforts in the coastal strike role from May 1942. At North Coates, 254 Squadron started training with torpedoes in August 1942 and its first 'ops' were made off the Norwegian coast before the year was out. The 'TorBeau' had been born.

The ultimate strike version, with radar, a torpedo and fuel for a long patrol was the Mk.X. It was a heavily-laden beast on take-off – as much as

25,400lb (11,521kg), that's 400lb *more* than *two* fully-loaded Blenheim Is! The first 'Tens' entered service with 248 Squadron at Predannack in June 1943.

DAY TRIPPERS

At Thorney Island on June 12, 1942, Beaufighter IC T4800 was ready for its 236 Squadron crew, Flt Lt A K Gatward RCAF and observer Sgt G Fern. They flew a low-level sortie, initially across Normandy. Two special pieces of 'ordnance' were on board – weighted French Tricolore flags.

In the lead-up to Bastille Day, they made a spectacular arrival in Paris flying down the Champs-Élysées, dropping a flag close to Napoleon's Arc de Triomphe. As *C-for-Charlie* careered on along the avenue, the Kriegsmarine Headquarters in the Place de la Concorde filled Gatward's

gunsight. Four 20mm cannon raked the building – Coastal Command had left an indelible 'calling card'! *Charlie* streaked overhead and the other Tricolore fluttered down. Gatward and Fern received a DFC and a DFM respectively for their audacious exploit.

LONG CAREER

Beaufighters were in action beyond the Japanese surrender of August 15, 1945. In Burma, the TF.Xs of 27 Squadron at Mingaladon shadowed Allied troops as they rounded up straggling Japanese soldiers. A burst from the 20mms often helped to quell resistance. To further this, the unit dropped over 250,000 leaflets explaining that World War Two was a thing of the past.

'Peace' in the theatre did not last long. In Malaya communist dissidents were becoming more organised. With typically British understatement, what became an 18-year conflict was known nonchalantly as 'The Emergency' and in military-speak as Operation 'Firedog'.

A detachment of 84 Squadron Beaufighter TF.10s – RAF designations changed from Roman numerals to Arabic in 1948 – was sent to Kuala Lumpur in July 1948 and supplemented by 45 Squadron the following year. It fell to the latter unit to carry out the last Beaufighter strike, on a settlement in Johore, on February 7, 1950.

Production came to a halt in Britain during September 1945, with TF.X SR919 taking the accolade as the last of the breed. That machine and others were converted into the final variant, the target-towing TT.10 from May 1948. This workhorse was destined for an extraordinarily long life.

Taking off from Seletar, Singapore, on May 12, 1960 in TT.10 RD761 Fg Off H Marshall made the RAF's final Beaufighter sortie. Eight days later, RD761 was struck off charge, stripped of spares and its forlorn carcass left on the scrap dump. An ignominious end to a formidable aircraft. ◉

Above
Aircrew of 600 Squadron at Predannack in the spring of 1942; with a Mk.VI behind. The unit was busy countering the Luftwaffe's so-called 'Baedeker' raids – striking at tourist towns, including Bath and Exeter.

Below
A TF.X of 455 Squadron RAAF unleashing a salvo of eight rocket projectiles in mid-1944.

Below left
Having scrapped the last Beaufighter in 1960, the RAF Museum had to turn to the Portuguese Air Force for its TF.X RD253. It was officially presented in July 1965 and went on display at Hendon on March 15, 1971.
RAF MUSEUM
www.rafmuseum.org

WESTLAND
WHIRLWIND
1940 TO 1943

In the early hours of June 21, 1943 five Whirlwind Is of 137 Squadron lifted off from Manston. Led by the commanding officer, Sqn Ldr J B Wray DFC, the twin-engined fighters were on their way to the Luftwaffe airfield at Poix-en-Picardie, southwest of Amiens. Equipped as 'Whirly-bombers' with a 250-pounder under each wing, four of the flight bombed and strafed their objective and returned to base safely.

F/Sgt John Barclay, in P6993, failed to find Poix, but spotted a locomotive and brought it to a satisfying halt amid billows of steam. On the return leg, one of his Rolls-Royce Peregrine V12s lost power and Barclay had to compensate for the asymmetric power with a boot full of opposite rudder.

By the time he reached Manston, the one good engine was running on fumes and he executed a wheels-up landing. Barclay was unhurt, but P6993 was a write-off. That turned out to be 137 squadron's final operational sortie in Whirlwinds; the unit was busy working up on Hawker Hurricane IVs.

At Warmwell 263 Squadron, the first unit to have operated the Whirlwind, from July 1940, became the last to fly it. As well as 137 and 263 only one other squadron, 25 at Martlesham Heath flew Whirlwinds for operational trials starting in June 1940.

Ramrod 109, a flight of eight of 263's Whirlwinds, left Warmwell on November 26, 1943. They were bound for Cherbourg, a target the pilots knew well. Leading, in P6983, was the unit's commanding officer since June, Sqn Ldr Ernest Baker DSO DFC*.

They dive-bombed the port from 14,000ft (4,267m). There was an extensive flak barrage and all aircraft came back with impact scars; P7046 limped home on just one of its Peregrines. *Ramrod 109* was the last frontline sortie by Whirlwinds. The unit was getting to grips with its new Hawker Typhoon Ibs.

The service life of the Whirlwind had amounted to just 43 months. The finales for both 137 and 263 Squadrons were fighter-bomber sorties, with the Whirlwind's four guns as a secondary weapon. This was a pity, because those 20mm cannon were what the twin-engined fighter was all about.

CONCENTRATED FIREPOWER

Specification F37/35 of February 1936 was seeking a fighter fitted with four 20mm cannon, a long range and an all-around view for its pilot. At Yeovil, Westland's gifted but headstrong 27-year-old designer, William Edward 'Teddy' Willoughby Petter, set about creating a radical twin-engined beauty which solicited an order for 340 straight off the drawing board. The prototype, L6844, had its maiden flight on October 11, 1940.

The twin-engined layout allowed the four Hispano 20mm guns to be clustered in the nose, to provide formidable concentrated firepower. The prototype briefly tested a 37mm cannon in the nose during the summer of 1939 but the results from the quartet of 'Hissos' were impressive enough.

The 20mm cannon could spew 650 shells a minute at a muzzle velocity of 2,800ft per second. Each was fed by a drum holding 60 shells, so pilots had to be sparing with the trigger. Even so, roughly a one-second burst from the four guns hurled 40 shells, each weighing 9oz (255g), a devastating torrent of 22lb 8oz (10.2kg) of hot metal at the target.

GIFTED BUT DOOMED

Development and manufacturing problems meant that the first production deliveries did not take place until June 1940. Despite its potential, by then the Whirlwind was doomed. Pilots found the Whirlwind underpowered but a delight to fly and a potent ground-attack platform. Its high landing speed precluded the twin from many RAF airfields. For ground crews the Peregrine, the ultimate development of the Kestrel of Hawker Fury fame, proved to be very troublesome.

The Whirlwind was the only frontline type to have adopted the Peregrine, making the powerplant an 'orphan'. The Air Ministry

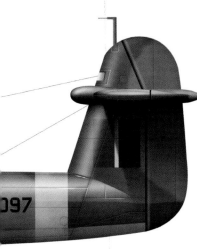

WESTLAND WHIRLWIND

Type:	Single-seat, twin-engined fighter / fighter-bomber
First flight:	October 11, 1938; entered operational service December 1940
Powerplant:	Two 885hp (660kW) Rolls-Royce Peregrine I V12
Dimensions:	Span 45ft 0in (13.71m), Length 32ft 9in (9.98m)
Weights:	Empty 8,310lb (3,769kg), All-up 10,379lb (4,707kg)
Max speed:	360mph (579km/h) at 15,000ft (4,572m)
Armament:	Four nose-mounted 20mm cannon
Replaced:	New type, units converting from December 1940
Taken on charge:	112 - 340 originally ordered
Replaced by:	Withdrawn by December 1943, replaced by Hawker Hurricane IVs and Typhoons

If the Peregrine was demanding, the Hawker fighter's 24-cylinder H-format Napier Sabre was a nightmare!

AERODYNAMIC PURITY

Not one Whirlwind has survived, which is a great shame as it was a most attractive looking aircraft. Petter had previously designed the army co-operation Lysander and was destined to conceive

Left
Armourers loading a 60-round drum into the nose of a Whirlwind; the entire nose nacelle detached to allow ease of maintenance to the 20mm guns. The small fairing offset to starboard, just above the propeller blade, housed the camera gun.

"...a one-second burst from the four guns hurled 40 shells, each weighing 9oz, a devastating torrent of 22lb 8oz of hot metal at the target."

the incredible English Electric Canberra jet bomber, initiate the format of the Lightning supersonic interceptor and create the Folland Gnat lightweight fighter/advanced trainer.

In his book *Adventure with Fate* (Airlife, 1984) Westland test pilot Harald Penrose highlighted the yawning gap between the practicality of the pilot and the designer's adherence to aerodynamic purity: "...I had emphatically disagreed with Petter's perilous decision to run the exhaust pipes *through* the petrol tanks in the wing to avoid parasitic resistance.

"I stressed that this was a *fighting* machine and one bullet through the tank and exhaust would set the whole thing on fire. [Petter] coldly told me: 'You pilots have to accept a *few* risks you know.'" ◎

Above
Believed to be taken at Drem, P6979 'HE-Q' of 263 Squadron in late 1940. This machine served throughout with 263, from November 1940. It was shot down near Cherbourg on October 24, 1943; Flt Lt P T R Mercer was killed. KEC

Left
Bombing up 137 Squadron's P6997 with a pair of 250-pounders, mid-1943. It served with the unit, mostly from Manston, from January to June 1943 when the unit re-equipped with Hurricane IVs.

deemed that Westland would be better to build Supermarine Spitfires and Seafires, and Rolls-Royce to concentrate on the Merlin. The plug was pulled and the last of 112 Whirlwinds rolled out of the Yeovil factory in December 1941.

That left two squadrons, 137 and 263, flying a dwindling number of aircraft with spares for their troublesome engines becoming ever rarer. For the ground crews of 263 Squadron, there was no respite as the unit took up the Typhoon.

BELL
AIRACOBRA
1941 TO 1942

Tasked with building a very heavily armed single-seat fighter, Bob Woods, designer for the Bell Aircraft Corporation, came up with a radical solution – the XP-39 Airacobra, which first flew in June 1938. A pair of machine guns were installed mid-way along each wing to keep the wingspan short, the aerofoil slim, and allow space for the undercarriage and fuel tanks.

Woods placed a pair of machine guns in the upper nose, geared to fire through the propeller arc. A 37mm cannon was mounted below these, its muzzle firing through the propeller hub. This was similar to the Messerschmitt Bf 109D of 1937. To achieve this in the Luftwaffe fighter, the cannon was shoe-horned underneath the inverted 'vee' 12-cylinder Daimler-Benz DB 601A engine.

At the Bell factory in Buffalo, New York, Woods had other ideas. He banished the Allison V-1710 upright V12 to *behind* the cockpit, driving the propeller through an extension shaft. The nose was not totally devoted to firepower, it also housed part of the type's other major innovation – the nosewheel of the tricycle undercarriage. Access for the pilot was via a car-type door on either side of the cockpit.

On October 8, 1939 France ordered 200 Airacobras, roughly the equivalent of the United States

Army Air Force (USAAF) P-39C, but opted for a 20mm, not 37mm, cannon. With the writing on the wall for the fall of France, the British Purchasing Commission took over the French Airacobra order on April 13, 1940 and upped the contract to 675 units. Initially the new type was given the name Caribou, but this changed to Airacobra by the time deliveries began in July 1941.

Trials by the Air Fighting Development Unit (AFDU) at Duxford included comparisons with a Spitfire V and a captured Bf 109E. These showed the Airacobra matched them in both speed and agility below 15,000ft (4,572m) but, because of its single-stage supercharger, above 20,000ft it was condemned as "utterly useless".

FIRST AND ONLY

On August 6, 1941 a pair of Airacobras landed at Matlask to be handed over to the resident 601 Squadron, which was flying Hawker Hurricane IIBs. The commanding officer, Sqn Ldr Edward John 'Jumbo' Gracie DFC, had been informed his unit was going to fast-track the Airacobra into frontline service. To take advantage of AFDU's increasing knowledge of the type, 601 moved to Duxford on August 16.

It was not long before Gracie's pilots were agreeing with their AFDU colleagues. The Airacobra was not suited to engaging fighters at medium height – 20,000ft – let alone the 30,000ft-plus that Supermarine Spitfire Vs could achieve.

BELL AIRACOBRA

Type:	Single-seat fighter
First flight:	June 4, 1938; entered service August 1941
Powerplant:	One 1,150hp (857kW) Allison V-1710 V12, mounted in mid-fuselage, behind the pilot, driving the propeller via an extension shaft
Dimensions:	Span 34ft 0in (10.36m), Length 34ft 2in (10.41m)
Weights:	Empty 5,360lb (2,431kg), All-up 7,380lb (3,347kg)
Max speed:	358mph (576km/h) at 15,000ft (4,572m)
Armament:	One 20mm cannon firing through the propeller hub, two machine guns firing through the propeller arc, four machine guns - two in each wing
Replaced:	Direct acquisition from USA, plus others intended for France, 1941
Taken on charge:	Around 55 accepted out of an original order for 675
Replaced by:	Withdrawn from combat March 1942

Ground crew were as disgruntled as the pilots: the V-1710 took a lot of learning and arming the seven guns was a complex and time-consuming process.

Gracie realised that the 20mm cannon potentially made the Airacobra effective against surface targets. For initial 'ops', it was decided to use it against shipping and 601 deployed to Manston on October 6, 1941.

In the dying light of the 9th of that month, Flt Lt Jaroslav Himr and Sgt Briggs set off on a low-level sortie to the French coast at Dunkirk and, as 601's operations record book (ORB) described:

"severely hurt the feelings of a trawler!"

In his feature *Rejected and Accepted,* in the June 2016 *FlyPast,* Andy Thomas described the final Airacobra operations. On the morning of the 10th: "Plt Off Jiri Manak lifted off in AH595 and attacked several barges in the canals behind Dunkirk in a 45-minute sortie. Having flown down from Duxford in AH583 the CO found nothing of interest. Weather then precluded any further ops.

"Early on the 11th Himr carried out a weather check over the Channel, returning with news that things looked suitable for

operations. Accordingly at 08:00 hours Himr, Manak and Fg Off Chivers took off to escort eight 615 Squadron Hurricanes on a shipping reconnaissance off Boulogne."
The ORB recorded: "Proceedings commenced with a before breakfast sweep up the 'single man's side' of the coast to Ostend ... Nothing of interest was seen except flak."

"That report stands as a requiem for the Airacobra's RAF operational service. At 14:35 Gracie led Manak, Briggs and Himr back to Duxford where they landed 40 minutes later. So ended the first, and as it transpired, final detachment for 601's American fighters, which had flown just eight sorties."

AXED AND DIVERTED

The Air Ministry had already terminated the order with Bell and was busy trying to divert those that were in the process of delivery or on the production line at Buffalo. On the afternoon that 601 Squadron's Manston detachment returned to Duxford, a delegation of Soviet officers were visiting to inspect the Airacobra.

Britain had offered to transfer to Russia examples that had been earmarked for the RAF and 212 were handed over. Soviet tactics on the Eastern Front were at medium and low level, and the former RAF Airacobras, and large numbers of US-supplied P-39s with 37mm cannon, became potent ground-attack weapons.

The USAAF took on 179 Airacobra Is which were designated P-400s to emphasise their different level of armament and equipment. Another 54 were lost at sea when the vessels bringing them across the Atlantic to Britain fell foul of U-boats.

In March 1941 the first and only RAF Squadron to fly the mid-engined, tricycle undercarriage fighter began to convert to Spitfire Vs. Although the RAF shunned the type, production of the Airacobra totalled 9,558, mostly for the USAAF and the Soviet Union. ◉

Above
Airacobra I AH601 'Skylark XII' as flown by Sqn Ldr E J Gracie, OC of 601 Squadron, Duxford. Note the unit's winged sword badge below the windscreen and the squadron leader's pennant on the door. Ferried to Duxford on October 10, 1941, it was written off in a forced landing at its base on December 12, 1941. PETE WEST

Left
A rare colour image of Sqn Ldr Gracie's AH601 being re-armed, with the nose access panel removed. The ground crew to the left are handling belts of 20mm shells.
P J HULTON VIA
ANDY THOMAS

CURTISS
MOHAWK, TOMAH

1941 TO 1945

Above
Tomahawk IIB AN413 of 112 Squadron, based at Sidi Barrani, Egypt, autumn 1941. Note the crudely applied kangaroo nose-art and the name 'Mam'. This machine was damaged by Bf 109s over Sidi Rezegh on November 20, 1941 and was written off in a forced landing. © ANDY HAY-www.flyingart.co.uk

After a gallant, but futile, fight in Greece trying to hold back the German invasion of April 1941 with its Gloster Gladiator biplanes, 112 Squadron re-established itself at Fayid, Egypt, in July. Its new equipment was the Tomahawk IIb and the unit was destined to remain loyal to the Curtiss fighter family until 1944.

Tomahawks gave way to Kittyhawk Is in December 1941 and Mk.IIIs in the autumn of 1942. These machines crossed the Mediterranean in July 1943 to Malta and onwards to Sicily and Italy. In June 1944, the squadron traded in the Kittyhawk IVs it had been flying for two months, in favour of North American Mustangs.

As the pilots got used to their new mounts, the groundcrew began to apply 112 Squadron's hallmark – the 'shark mouth' – on the nose (page 54). Since 1941 the unit had often been referred to as the 'Shark Squadron' in the press.

Under the nose of the Tomahawk was a gaping air scoop that cooled the Allison V-1710-33 engine. That 'mouth' was too tempting for personnel of 112 Squadron determined to make their aircraft distinctive. Either side of the nose, eyes were painted ahead of the exhaust stubs. Around the air intake a fearsome mouth with fangs, inspired by the vicious carnivore of the sea, was applied.

The much-enlarged intake of the Kittyhawk lent itself to more extensive decoration and larger

CURTISS KITTYHAWK III	
Type:	Single-seat fighter-bomber
First flight:	XP-40: October 14, 1938; Kittyhawk entered service April 1942. (Mohawk - Curtiss Model 75 - May 15, 1935)
Powerplant:	One 1,600hp (1,193kW) Allison V-170-81 V12
	Mohawk: one 1,200hp (895kW) Wright Cyclone GR-1820 radial
Dimensions:	Span 37ft 4in (11.36m), Length 31ft 2in (9.49m)
Weights:	Empty 6,400lb (2,903kg), All-up 8,500lb (3,855kg)
Max speed:	362mph (582km/h) at 5,000ft (1,524m)
Armament:	Six 0.50in machine guns in the wings; up to 1,000lb (453kg) of bombs under the centre section and under wing
Replaced:	Direct acquisition from USA, plus stocks of Mohawks and Tomahawks intended for France in 1940. Kittyhawks replaced Tomahawks from April 1942
Taken on charge:	Mohawk: about 236. Tomahawk: 885. Kittyhawk: 1,758
Replaced by:	Mostly North American Mustangs from 1944

teeth. Completely unofficial, these markings stayed with the unit for the rest of the war. Further colour was provided by the red propeller spinners that were a standard recognition feature for Allied fighters in the Mediterranean theatre.

By the end of 1941, in Burma, the Curtiss Hawks of the American Volunteer Group, the 'Flying Tigers', were adorned with 'shark mouths'. Other United States Army Air Force P-40 units took up the idea.

It is generally accepted that 112 Squadron was the first Allied unit to paint 'shark mouths' on its aircraft. But 112's 'trademark' was not original, that honour fell to the Luftwaffe, at least two years before.

Early in 1940 the Messerschmitt Bf 110s of Zerstörergeschwader (ZG) 26 had both cowlings of their Daimler-Benz DB 601B engines embellished with 'shark mouths'. Sister Bf 110 unit ZG 76, based in Sicily in 1941, also carried the motif and it is quite likely that it stirred the imaginations of the artists at 112 Squadron.

FIRST 'KILL'

France placed a massive order for Curtiss Hawk 75As in 1938 in the hope of bolstering its defences. Of the 1,000 commissioned, most were based upon the United States Army Air Corps (USAAC) P-36A, powered by the Pratt & Whitney R-1830 radial.

WK. KITTYHAWK

When war with Germany broke out on September 3, 1940, the Armée de l'Air had close to 300 Hawk 75A-1s. Five days later its Hawks clashed with Bf 109Es and two of the Luftwaffe fighters were shot down. These were the first air-to-air combat losses that Germany had suffered in the conflict.

With the fall of France, Britain took on some of the Hawk 75 order, while the rest were cancelled. This was not a great blow to the manufacturer, the Curtiss-Wright Corporation at Buffalo, upstate New York. Unlike its neighbour, Brewster (see page 34), Curtiss was a long-term master of fighters and large-scale production and was busy gearing up on the phenomenal P-40 Warhawk.

French refugee Hawks and diverted examples with the R-1830s were of little use to the RAF. These Hawk 75A-1s, -2s and -3s were designated Mohawk I, II and III respectively. Pilots evaluating them had to master French throttles, which worked in the opposite sense to American and British ones. (The French pulled back to increase power.)

After evaluation at the Aeroplane and Armament Experimental Establishment at Boscombe Down, the Wright Cyclone GR-1820 radial-powered Hawk 75A-4, became the Mohawk IV. Boscombe's pilots found the handling of the Mohawk to be delightful, but declared its performance unsuitable for the fast-climbing, hard-turning, high-flying combat of the European theatre.

The Air Ministry sent some Mohawks to South Africa and others to the Far East. The first operational unit was 5 Squadron at Dum Dum in India from December 1941. These Mohawks, along with those 146 and 155 Squadrons, saw extensive action against the Japanese during the Burma campaigns. The last Mohawks were retired by 155 at Tulihal in India in January 1944, giving way to Supermarine Spitfire VIIIs.

DESERT 'HAWKS

The Hawk 75 provided Curtiss with the basis for a far more advanced fighter, the P-40 Warhawk. The airframe was cleaned up and the draggy Pratt & Whitney or Wright radial gave way to a 1,040hp (775kW) Allison V-1710-33 ◎➤

"Under the nose of the Tomahawk was a gaping air scoop that cooled the engine. That 'mouth' was too tempting for personnel of 112 Squadron determined to make their aircraft distinctive."

Above
Taken from a Westland Lysander, Tomahawk IIb AK533 of 2 Squadron, South African Air Force. It appears to have been overhauled and resprayed with the original paint layer around the serial number having been masked off. KEC

Above right
Tomahawk IIBs AK185 and AK276 of Croydon-based 414 Squadron Royal Canadian Air Force, September 1941.

Far right
The RAF Museum's Kittyhawk IV wears the colours of 112 Squadron's FX760, which flew with the unit in Italy in June 1944. The 'question mark' was occasionally used when a unit's complement of aircraft exceeded the number of letters of the alphabet. The exhibit is based on the remains of former 80 Squadron Royal Australian Air Force A29-556, salvaged from Papua New Guinea in 1974. RAF MUSEUM
www.rafmuseum.org

Above right
Kittyhawk I AK998 served with 450 Squadron Royal Australian Air Force. It failed to return from a sortie out of Gambut, Egypt, on May 29, 1942.

P-40D Warhawk, with a 1,600hp Allison and true fighter-bomber capabilities. The first Kittyhawk I had its maiden flight on May 22, 1941 with deliveries beginning on August 27. The first 20 carried four wing guns, the remainder had three in each wing and were designated Mk.IA, the equivalent of the P-40E. The final Kittyhawk was the Mk.IV, a version of the P-40N.

V12. The prototype XP-40 had its maiden flight in October 1938 and Curtiss never looked back; in November 1944 the 13,738th and last example, a P-40N, rolled out of the factory.

The RAF was introduced to the P-40 through France. As part of

that country's ordering frenzy, 140 Model H81A-1s – equivalent of the USAAC's P-40A – were requested. These became Tomahawk Is with the RAF and the British signed up for 110 Mk.IIs and 635 Mk.IIIs.

Although deficient above medium altitude, Tomahawks proved ideal for tactical reconnaissance and 26 Squadron at Gatwick became the first of 24 British-based units in February 1941.

Most of the Tomahawks were destined for North Africa and, as related above, 112 Squadron was the first to use the type in combat in the theatre, starting in mid-1941. The natural replacement for the Tomahawk was the next member of the family, the Kittyhawk.

Keeping to the 'Hawk' theme, the RAF honoured Orville and Wilbur Wright, by naming its latest Curtiss fighter after the venue of the world's first sustained powered flight on December 17, 1903 at Kittyhawk, near Dayton, Ohio. The company founded by the brothers had merged with Curtiss in August 1929.

The Kittyhawk was based on the

With the RAF Kittyhawks were devoted to the Desert Air Force and followed the combat to Sicily and into Italy. The 'Sharks', 112 Squadron, along with 250 Squadron introduced it to operations in April 1942. With the ever-changing nature of the war in North Africa, the Kittyhawk units worked a 'cab rank' system, responding on demand to the needs of the army with close and precise support.

The last of the breed was retired by 250 Squadron, at Lavariano in Italy, in August 1945. As with the 'Sharks', 250 converted to another US fighter success story, the Mustang.

While Brewster Buffalos and Bell Airacobras were disappointing, thanks to the French legacy, the RAF chose well with its Curtiss fighters. The Mohawk-Tomahawk-Kittyhawk series might not have been the most agile or the fastest, but thanks to the industrial might of Curtiss, they could be churned out in great numbers and proved dependable and efficient warplanes. ◎

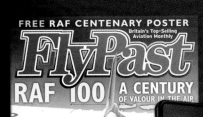

DE HAVILLAND
MOSQUITO
1941 TO 1963

Right
Built in 1946, Mosquito PR.34 RG177 ended its days with 81 Squadron (illustrated) at Seletar, Singapore. While landing at Seletar on March 17, 1955, RG177 swung and to bring it to a halt, Fg Off A J 'Colly' Knox brought up the undercarriage. This wrote off RG177, but saved Knox and potentially others in the way of the careering aircraft.
PETER GREEN COLLECTION

Right
Operating in the bomber support role on the night of December 3/4, 1943 Mk.II DD739 of 456 Squadron Royal Australian Air Force failed to return to its base at Fairwood Common. Plt Off J L May and Fg Off L R Parnell were both killed.

and as vitally needed as it had been before and that, provided we did not permit orthodoxy especially in the shape of officialdom to stifle us, we could do better still the second time. Our scheme was to discard every item of equipment that was not essential, design for a two-man crew and *no* rear armament, relying on high speed for defence."

Sir Geoffrey "...estimated that a year could be saved in production due to the simplicity of wood construction as compared to metal". He concluded:

When the Panavia consortium was established in 1969, it announced the swing-wing Multi-Role Combat Aircraft (MRCA) programme. This took on the name Tornado in 1974 and the air defence version is featured on page 94. Among the press corps at the launch were some who well remembered the *original* MRCA, conceived in great secrecy in 1940.

The de Havilland Mosquito was a remarkable aircraft that mastered the roles of fighter-bomber, night-fighter, intruder, bomber and photo-reconnaissance. Its frontline RAF career extended all the way to 1955, while the target-tug versions, bomber-based TT.35s, were honourably retired in May 1963.

The Mosquito is all the more incredible because it was not created in response to a need from the Air Ministry. It was the product of industry working out what was really required and stumping up the costs to bring the concept to reality.

The concept was to create an aircraft so fast that it did not need defensive armament, and could be built quickly and relatively simply. The inspiration was the DH.4 high-speed bomber of World War One.

In his autobiography *Sky Fever* (Hamish Hamilton, 1961) Sir Geoffrey de Havilland CBE described the rationale: "We were confident that this formula would be as novel

DE HAVILLAND MOSQUITO FB.VI

Type:	Two-seat, twin-engined fighter-bomber
First flight:	Prototype November 25, 1940; photo-recce Mk.I entered service September 1941
Powerplant:	Two 1,230hp (917kW) Rolls-Royce Merlin 23 V12
Dimensions:	Span 54ft 2in (16.5m), Length 40ft 6in (12.34m)
Weights:	Empty 14,300lb (6,486kg), All-up 22,300lb (10,115kg)
Max speed:	380mph (611km/h) at 13,000ft (3,962m)
Armament:	Four 20mm cannon and four 0.303in machine guns, all in the nose; up to 2,000lb (907kg) of bombs
Replaced:	Douglas Boston from 1943
Taken on charge:	6,439 of all variants in Britain, 7,781 including overseas production. Sub-contracts by Airspeed, Percival, Standard Motors. Also by de Havilland Aircraft in Australia and Canada
Replaced by:	Night-fighter version retired in 1951, Gloster Meteor NF.11s and de Havilland Vampire NF.10s taking up the role. Photo-recce Mosquitos began giving way to English Electric Canberra PR.3s in 1953

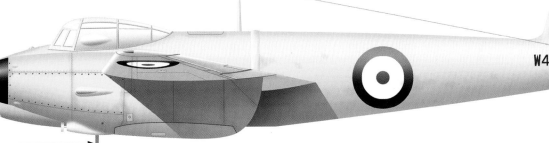

W4050

"All this gave us a wonderful opportunity to make an outstanding war aeroplane in almost record time."

Sir Geoffrey's 30-year-old son, Geoffrey, piloted the prototype on its maiden flight at Hatfield on November 25, 1940. It had been designed and built in just 11 months at Salisbury Hall, London Colney, a couple of miles to the southwest and now the home of the de Havilland Aircraft Museum.

Three versions were offered initially: bomber, night-fighter and photo-recce. From these stemmed 39 British variants, including folding-wing Mosquitos for the Fleet Air Arm.

PHOTOGRAPHIC EVIDENCE

The Mosquito first entered service with the Photographic Development Unit in September 1941 at Benson.

Photo-recce versions roamed as far as the Balkans from British bases and served in the Mediterranean and the Far East.

Perhaps the most famous exploit of the reconnaissance Mosquitos was a sortie over Peenemünde on Germany's Baltic coast on October 3, 1943 by a PR.IX of Leuchars-based 540 Squadron. On November 13, a WAAF photo-interpreter, Fg Off Constance 'Babs' Babington Smith was poring over the images and spotted a little aircraft on Karlshagen airfield, part of the Peenemünde complex.

She had discovered the V-1 flying-bomb and very soon was to link it with the mysterious 'ski ramps' to be found nearby. Another 540 Squadron sortie was staged on November 28 to

confirm her observations and the campaign against the 'doodle-bugs' began.

NOCTURNAL INTRUDERS

First unit to receive Mk.II night-fighters was 157 Squadron at Castle Camps, in January 1942 followed by 151 at Wittering in April. There was a rivalry between the two Cambridgeshire bases as to which would get off the mark first.

Commanding Officer (CO) of 151 was New Zealander Wg Cdr Irving 'Black' Smith, who had flown Hawker Hurricanes and Boulton Paul Defiants with 151. On June 24 he and Flt Lt Shepherd were airborne in W4095 and they inaugurated the nocturnal Mosquito in fine style by dispatching a Heinkel He 111 and a Dornier Do 217 ten minutes apart. These were Smith's only 'kills' piloting Mosquitos: his total came to eight, one

"The gun was 12ft 5in long and with its auto-feed mechanism was 38in high. It weighed 1,800lb... Far from being horrified at the thought of installing such a beast in his masterpiece, Mosquito designer Ronald Bishop set to adapting the Mk.VI as a priority."

Above
Mosquito II W4087 of 157 Squadron, Castle Camps, January 1942. PETE WEST

Right
The crew of a Castle Camps-based Mosquito VI of 605 Squadron boarding for an intruder operation, autumn 1943. KEC

Below
Four Mosquito II night-fighters in the mist at Hatfield in the first days of January 1942. The image had escaped the censor's brush, as 'arrow head' radar aerials are visible on the noses. In the left foreground in W4090, which joined 151 Squadron at Wittering that April. Middle and right are W4092 and W4088, both of which were issued to 157 Squadron at Castle Camps that February. BRITISH AEROSPACE-KEC

in a Defiant, the remainder with Hurricanes.

It was August before 157 Squadron chalked up a Mosquito victory and it was the CO sharing the honours. On the 22nd, Wg Cdr Gordon Slade and Plt Off Truscott took down a Do 217.

Slade had already been introduced to the Mosquito. Serving with the Aeroplane and Armament Experimental Establishment, he took the prototype Mosquito, W4050, up for trials on February 24, 1941. While taxiing back in over rough ground at Boscombe Down, the tailwheel jammed and the fuselage fractured. The machine was fitted with the fuselage intended for the first photo-recce prototype, W4051, and re-flew at Boscombe on March 14.

At Ford on the south coast, 23 Squadron was eagerly awaiting its first Mosquito, to replace a mixed fleet of intruder Douglas Bostons and Havocs. The tactics of an intruder was to take the fight to the Luftwaffe's bases, catching tired and off-guard crews as they reached the apparent sanctuary of their turf.

Stripped of its radar, Mk.II DD670 *S-for-Sugar* arrived at Ford on July 6, 1942. The following night it was in action, crewed by the CO, Wg Cdr Bertie 'Sammy' Hoare DFC*, and W/O J F Potter. Over Chartres, France, the pair could not believe

their luck: ahead was a Do 217 whose crew was so relaxed they had put their navigation lights on. They shot the unsuspecting enemy down. Hoare ended up with nine 'kills', six on Mosquitos. *S-for-Sugar* led a charmed life; it was retired from RAF service in January 1946.

BITE OF THE TSETSE
One of the most outstanding aspects of the Mosquito is its link to a London factory that specialised in making machinery for the cigarette industry. Desmond Molins turned the

skills of his workforce to developing automatic loading systems for guns. In 1940 the company devised belt feeds for the 20mm Hispano cannon.

The British army hoped to mount its ubiquitous 6lb (2.72kg) anti-tank artillery piece within a fast-moving armoured vehicle. From February 1942 the Molins Machine Company devised a compact, lightweight system to feed shells into the breech. With the advent of the German Tiger tank, the army quickly lost interest as the 6-pounder's shells just bounced off the new target.

The Air Ministry had been looking at the use of 40mm cannon in fighters and it was suggested that the work done by Molins on the 6-pounder, a 57mm weapon, might fit the bill. The gun was 12ft 5in (3.6m) long and with its auto-feed mechanism was 38in (96.5cm) high. It weighed 1,800lb (816kg) and the force of its recoil was 8,000lb. Far from being horrified at the thought of installing such a beast in his masterpiece, Mosquito designer Ronald Bishop set to adapting the Mk.VI as a priority.

Coastal Command was enthusiastic as a gun that could fire a 6lb shell

trawler on November 4. Three days later the unit engaged its first U-boat, caught on the surface.

Usually, sorties by 248 Squadron would comprise a pair of Mk.XVIIIs escorted by up to eight Mk.VIs. U-boats, freighters and tankers were not the only targets on the receiving end of the Mosquito's 6-pounder.

Sqn Ldr Tony Phillips of 248 described an engagement of March 10, 1944: "I immediately observed a submarine proceeding on the surface with an escorting minesweeper. Whilst making my approach an escorting Ju 88 appeared in the

The epic round trip, to and from Singapore, lasted from December 16, 1943 to July 14, 1944 where the personnel and drawings disembarked. Twelve days later the US Navy sank I-29, taking its German technological hardware to the bottom with it.

In February 1945 the Molins-equipped version was relinquished by 248 Squadron, but between April and May 1945 the North Coates-based 254 Squadron flew Mk.XVIIIs. Beaufighters and Mosquitos with underwing racks for rocket projectiles proved to be easier to use and far

with a muzzle velocity of 2,600ft per second would be a lethal ship hunter. The Airborne 6-Pounder Class M Gun, mostly referred to as the Molins gun, had 23 rounds, one in the breech and 22 in the 'hopper'. Shell cases were not ejected; they were retained within the fuselage.

A written-off fighter-bomber Mk.VI fuselage was used for static firing trials that began on April 29, 1943. A former Boscombe Down Mk.VI, HJ723, became the flying prototype in June 1943. Trials showed that the feed mechanism needed beefing up to withstand the forces generated in combat manoeuvres. Initially, the project was known as Tsetse – like the Mosquito a tiny fly with a nasty bite - and the new variant was designated Mk.XVIII.

SUB-HUNTING

A detachment of 618 Squadron came down from Skitten in Scotland to join the resident 248 Squadron at Predannack in Cornwall to hasten operational trials of the new weapon. The latter flew its first sortie on October 24, 1943. A crew from 618 perished while attacking an armed

sights. I pressed the gun button and the Junkers disintegrated. I then attacked the submarine which was seen to be hit."

The Junkers was a Ju 88C-6 based at Cazaux, France. The submarine is believed to have been Commander Kinashi Takakazu's I-29 of the Imperial Japanese Navy on a 'Yanagi' supply mission. Having cracked the German 'Enigma' code, the team at Bletchley Park had warned of I-29's approach and a major operation was launched to eliminate the Japanese courier. The submarine survived the efforts of the Mosquitos of 248, Bristol Beaufighters and Consolidated Liberators and docked in the impregnable U-boat pens at Lorient the following day.

The Japanese submarine brought armament technicians to liaise with their German counterparts and is reputed to have had gold bullion on board. On the return trip, I-29 is thought to have been carrying a Walter HWK 509 rocket motor as fitted to the Messerschmitt Me 163 Komet, a Junkers Jumo 004B from the Me 262 jet fighter programme and German specialists with blueprints.

more effective and only 27 Mosquito VIs were converted to Mk.XVIIIs.

FINAL CURTAIN

In November 1950 the last of 6,439 British-built Mosquitos, NF.38 VX916, was rolled out of the factory at Hawarden, near Chester. It saw no operational flying with the RAF and was handed over to the Yugoslav Air Force in December 1951.

The sister publication *RAF Centenary Celebration - Bombers* covers the last use of military Mosquitos in Britain - target-tug TT.35s at Exeter in May 1963.

The last frontline sortie was flown by Fg Off A J 'Colly' Knox in 81 Squadron PR.34A RG314 out of Seletar, Singapore, on December 15, 1955. He was detailed to take pictures of a suspected terrorist base in the Malayan jungle. The previous March, quick thinking by Knox had saved his and perhaps other lives when PR.34 RG177 swung violently on landing; he selected undercarriage 'up' and the Mosquito slid safely to a halt. The jet age had dawned for 81 Squadron, Gloster Meteor PR.10s were its new equipment. ◎

Above
Molins gun-equipped Mk.XVIII NT225 of 248 Squadron, based at Portreath. It was painted with black and white 'Invasion Stripes' as an easy identification feature in the crowded skies over Normandy during D-Day, June 1944,

Above left
A co-pilot's view of a formation of Mk.VIs of 4 Squadron, based at Gütersloh, Germany, in 1947. The aircraft bottom left, RS678, was delivered new to the unit from the Airspeed factory at Christchurch in 1946. The unit moved to Wahn in late 1947 and while landing there on June 17, 1948 RS678 suffered brake failure, swung badly and the undercarriage collapsed. It was a write-off, but its crew was unhurt.

DOUGLAS
HAVOC

1941 TO 1943

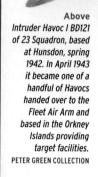

Ed Heinemann, chief engineer for Douglas, designed a string of successful combat aircraft, from the Dauntless dive-bomber of 1938 through to the A-4 Skyhawk delta wing fighter of 1954. Working at the El Segundo, California, plant, he developed a twin-engined light bomber, the Model 7, in 1938. It was hoped to secure a contract from the United States Army Air Corps.

In February 1939 the French ordered 100 machines based on the Type 7, but demanded many modifications that altered the airframe considerably. The result was the DB-7, and the first example took to the air in August 1939, powered by two 1,200hp (895kW) Pratt & Whitney Twin Wasp radials. The more powerful DB-7A, with a pair of Wright R-2600 Double Cyclones of 1,600hp followed.

With the fall of France in June 1940, undelivered DB-7s and -7As were taken over by the RAF, as Boston Is and IIs, respectively. Some of these wore the colours of the Belgian Air Force, having been diverted by the French in an attempt to bolster its neighbour. Additionally, the DB-7B was ordered direct and to British specifications. The first of these was the Boston III and these saw extensive service, as detailed in the companion volume on bombers see elsewhere for details.

Boston Is and IIs were not suitable as light bombers, but Fighter Command was desperate to supplement its night-fighter force. Accordingly, the Havoc programme came about with the majority of modifications being carried out at Burtonwood.

The original glazed nose was removed and a new-build structure holding the Airborne Interception (AI) Mk.IV radar with its distinctive 'arrowhead' aerials and eight machine guns was fitted. Designated Havoc I, the first of these stop-gap fighters was issued to 85 Squadron at Debden in April 1941.

Martin-Baker at Denham devised a similar conversion of Boston Is to Havoc IIs. The main difference was an elongated nose carrying the AI Mk.IV and no less than *a dozen* machine guns. Havoc IIs supplemented and then replaced the Mk.Is with 85 Squadron.

With night flying modifications, though retaining the glazed nose section and upper rear gun position, Havocs were also used for intruder sorties into France. Known as 'Moonfighters', 23 Squadron at Middle Wallop was the first to accept this variant, in March 1941.

PANDORA'S MINES

Before the night-fighter Havocs entered service, a bizarre version was the first to join a frontline squadron, in December 1940. Under Operation 'Mutton', 420 Flight was formed at Christchurch on September 25, 1940 with Handley Page Harrow bomber-transports. Four days later the unit moved to Middle Wallop where tactics were devised for a weapon called the Long Aerial Mine (LAM).

Developed from 1938, LAM was designed to disrupt enemy bomber formations. The idea was to fly ahead and above of an inbound raid and drop up to 120 mines at regular intervals. Floating down by parachute, each LAM had a 1lb (0.45kg) explosive charge and trailed a weighted cable 2,000ft (609m) long. An enemy bomber snagging the cable produced a shock load that released

From September 1941 a number of Turbinlite Havoc units were deployed to airfields already hosting Hawker Hurricanes. The theory was that the Havoc would guide the Hurricane to a 'bogey' using its radar and, at the last moment, illuminate the enemy with the searchlight to help the fighter make a 'kill'.

At Hibaldstow, 1459 Flight had

the parachute, deployed a small drogue 'chute that, in theory, brought the mine into contact with the wing or fuselage.

At 16,000ft, a Harrow could manage about 160mph (257km/h) while a formation of Heinkel He 111s would be travelling at 240mph. Something faster was needed and about 20 Havocs were adapted, under the codename 'Pandora', with the LAMs clustered in the bomb bay. The first Pandora Havoc was issued to 420 Flight in October 1940, and on December 7 the unit was renamed as 93 Squadron.

Operational trials were abandoned in November 1941 as impractical. Getting a Havoc to the right place at the right time was difficult enough, but assessing wind speed and the direction of the release track was nigh on impossible. Stray LAMs could not be dropped over Britain – bomb disposal teams had enough on their plate!

FLYING SEARCHLIGHT

Although radar could bring a night-fighter close to its target, the pilot still had to use his eyes for the 'kill'. Air Ministry 'boffin' Wg Cdr William Helmore approached the General Electric Company (GEC) with his idea for an airborne searchlight. The result was a 2,700 million candela 'Turbinlite', and former light aircraft designer Leslie 'Baron' Baynes managed to shoehorn both the searchlight and the radar into the nose of a Havoc – but there was no room for guns.

a fleet of Havoc Is, IIs and Boston IIIs and it began to work with the Hurricane IIs of 253 Squadron. The only recorded Turbinlite-Hurricane combination victory took place from Hibaldstow on April 30, 1942. Piloting Havoc II AH484 Flt Lt Winn guided Flt Lt Yapp of 253 Squadron to a successful shoot down of an He 111.

In September 1942 the searchlight units were re-organised: 1453 and 1459 Flights became 532 and 538 Squadrons. Self-contained, with a mixture of Douglas twins and Hurricanes, they no longer relied on co-located fighters. The system was ponderous and was overtaken by improvements in airborne radar and fighter control. The Turbinlite units disbanded in January 1943.

The work put in by Helmore and GEC to create an airborne searchlight had a successful spin-off. Inspired by the Turbinlites, Sqn Ldr Humphrey Leigh devised a searchlight housed in a retractable 'dustbin' in the lower rear fuselage of Coastal Command Vickers Wellingtons and Warwicks. First employed in mid-1942 Leigh Lights were used to illuminate surfaced U-boats at night to great effect. ◎

DOUGLAS HAVOC I

Type:	Two-seat, twin-engined night-fighter
First flight:	August 17, 1939 (DB-7); Havoc I entered service March 1941
Powerplant:	Two 1,200hp (895kW) Pratt & Whitney Twin Wasp radials
Dimensions:	Span 61ft 4in (18.69m), Length 46ft 11in (14.3m)
Weights:	All-up 20,329lb (9,221kg)
Max speed:	295mph (474km/h) at 13,000ft (3,962m)
Armament:	Eight nose-mounted 0.303in machine guns
Replaced:	Direct acquisition from USA
Taken on charge:	About 253 conversions
Replaced by:	Withdrawn from operations October 1942

HAWKER
TYPHOON

1941 TO 1945

DN406 PR◉F

Even before the Supermarine Spitfire had entered squadron service, the Air Ministry began the process of replacing it and the Hawker Hurricane, issuing Specification F18/37 in March 1938. Twelve machine guns or six 20mm cannon were the armament options, and the preferred engines the Rolls-Royce Vulture or Napier Sabre, both 24-cylinder monsters.

Hawker won the development contract and designer Sydney Camm decided to build prototypes with both powerplants: the Tornado having the Vulture and Typhoon the Sabre. The Tornado was the first into the air, on October 6, 1939, but this programme was brought to a halt in mid-1941. The Vulture was troublesome, and Rolls-Royce needed to concentrate on mass production of Merlins.

The prototype Typhoon had its maiden flight on February 24, 1940 and became the RAF's first fighter to exceed 400mph (643km/h). The Mk.Ia featured the dozen machine guns; the Mk.Ib carried four 20mm and became the standard model. Half-way through production, the cumbersome framed cockpit and car-like access door were replaced by a Perspex teardrop-shaped canopy offering greatly enhanced visibility.

Structural problems and development of the Sabre engine delayed entry into service. A series of fatal accidents in which the tail section separated looked set to cancel the entire programme. Hawker developed a 'fix' and the Typhoon went on to become a famed close support aircraft. The thick wing profile allowed it to become a fighter-

bomber, in particular accommodating rails underwing for eight 60lb (27kg) rocket projectiles. These, coupled with the 20mm cannon, made the Typhoon a formidable presence over the battlefield.

DOWN IN THE DRINK

The first Typhoons were delivered to 56 Squadron at Duxford in September 1941. Development problems caused an agonising eight-month delay until May 30, 1942 when the unit was declared operational.

In January 1943 New Zealander Thomas Henry Vicent Pheloung – known to his colleagues as 'THV' – took command of 56, which by then was at Matlask. Pheloung was in DN374 on a patrol off the Norfolk coast on March 15 when anti-aircraft fire from a warship forced him to bale out. He was picked up by a Supermarine Walrus amphibian and

returned to his unit.

In late April, 56 Squadron was tasked with showing off the Typhoon to the press, including a formation flight for the benefit of photography maestro Charles E Brown. Sadly, THV was not to see Brown's work in print. On June 20, he was leading a flight of 56 Squadron in EK174 *C-for-Charlie* near St Dizier, France. A victim of flak once again, THV failed to return.

It was late in the year before clearance came from the Air Ministry for the Typhoon images to be published.

WAKE-UP CALL

A fine testament to the character of Typhoon pilots was displayed on the morning of New Year's Day 1945 when the Luftwaffe unleashed Operation 'Bodenplatte' (Base Plate). This was a massive attack designed to catch enemy aircraft on the ground

Left
In April 1943 the
Typhoons of 56
Squadron were
introduced to the
press with Sqn Ldr
Pheloung leading
the unit for a photo-
shoot. Third from
the camera is DN317
'C-for-Charlie', which
was destroyed by fire
on landing at its base
on April 23, 1943. The
stripe on the upper
wings was yellow, a
recognition aid used
from September 1942
to mid-1943.

Left
In April 1943 the
Typhoons of 56
Squadron were
introduced to the
press with Sqn Ldr
Pheloung leading
the unit for a photo-
shoot. Third from
the camera is DN317
'C-for-Charlie', which
was destroyed by fire
on landing at its base
on April 23, 1943. The
stripe on the upper
wings was yellow, a
recognition aid used
from September 1942
to mid-1943.

Left
At the time of writing,
the RAF Museum's
Typhoon MN235 was
on 'sabbatical' at
the Canada Aviation
and Space Museum
at Rockcliffe, Ottawa.
This machine was
shipped to the USA
in March 1944 for
evaluation and in
1949 was taken on
the inventory of the
Smithsonian Institute.
In exchange for a
Hurricane, MN235
was returned to
the UK in January
1968: it is illustrated
after assembly at
Shawbury. The only
intact example of a
Typhoon in the world
went on display at
Hendon in November
1992. ROY BONSER-KEC

at airfields in Belgium and the
Netherlands.

Led by Major Heinz Bär,
Jagdgeschwader 3 was to hit B78,
Eindhoven, the base of the 2nd
Tactical Air Force's 124 and 143
Wings of Typhoons along with other
units. A mixed force of 70 Focke-Wulf
Fw 190s and Messerschmitt Bf 109s
descended on B78 at 09:25 hours.

Just before he screamed across the
airfield at very low level Bär, flying
Fw 190 *Red 13*, had shot down two
Typhoons – his 204th and 205th
victories. Behind, Flt Lt H P Gibbons
of 168 Squadron had Bär's wingman
in his sights. He'd taken off moments
before in MN486 *D-for-Dog* and was
probably hoping upon hope the guns
were loaded. They were, and he was
seen to blow the tail off the 'Butcher
bird' his was pursuing. But Bf 109s
were queuing up to shoot Gibbons

"To his left, Fw 190s were rolling in at tree-top height. He turned his aircraft to face the raiders, hit the brakes and throttle to raise the tail so that he could momentarily shoot at them."

down; he was killed defending
his base.

Waiting for colleagues to get
airborne ahead of him, Plt Off

Andy Lord of 438 Squadron Royal
Canadian Air Force (RCAF) watched
in horror as both wings of his MN607
G-for-George erupted in flame. With
his aircraft disabled, Lord was so mad
that he unstrapped, climbed down
from the cockpit and retaliated with
the only weapon he had – his pistol.
Quickly, common sense prevailed, and
he took cover in an old bomb crater.

On the other side of the airfield, the
Typhoons of 440 Squadron RCAF
were taxiing out to a runway; they
were also badly shot up. Within the
line, Plt Off Dick Watson was another
pilot who let his rage take form.

To his left, Fw 190s were rolling in at
tree-top height. He turned his aircraft
to face the raiders, hit the brakes and
throttle to raise the tail so that he
could momentarily shoot at them.
Watson claimed one as damaged,
but it was not awarded to him. Pity,
it would have been an extraordinary
triumph. ◉

HAWKER TYPHOON IB

Type:	Single-seat fighter / fighter-bomber
First flight:	February 24, 1940; issued to 56 Squadron in September 1941, but it was May 1942 before operations began
Powerplant:	One 2,180hp (1,626kW) Napier Sabre IIA 24-cylinder H-format
Dimensions:	Span 41ft 7in (12.67m), Length 31ft 11in (9.72m)
Weights:	Empty 8,800lb (3,991kg), All-up 13,250lb (6,010kg)
Max speed:	412mph (663km/h) at 18,000ft (5,486m)
Armament:	Four 20mm cannon, two in each wing; two 1,000lb (453kg) bombs or a maximum of 16 rocket projectiles
Replaced:	Original specification to replace the Hawker Hurricane and Spitfire, introduced into service as individual type
Taken on charge:	3,333 all but the first 14 built by Gloster at Hucclecote
Replaced by:	Withdrawn from service 1946

NORTH AMERICAN
MUSTANG

1942 TO 1947

Top right
Mustang I AG367 'Z-for-Zebra' of 26 Squadron, based at Gatwick early 1942. The unit used Mk.Is from January 1942 up to the spring of 1944, returning to them from December 1944 to June 1945 when Rolls-Royce Griffon-engined Spitfire XIVs arrived.
PETE WEST

Below
Owned by the appropriately named Sharkmouth Ltd, Goodwood-based two-seater P-51D Mustang G-SHWN was painted in the colours of Mk.IV 'KH774' of 112 'Shark' Squadron in 2015. The original KH774 served with the unit in Italy on sorties over the Balkans in 1945. See page 42 for details of the original of the markings.
DARREN HARBAR

Arguably the finest interceptor and escort of the war, the Mustang was a gift from Britain to the United States. To be fair, this success story also required the vision and determination of North American Aviation (NAA) of Inglewood, California, and the Detroit, Michigan-based Packard Motor Car Company.

James Howard 'Dutch' Kindelberger, NAA's president, was a far-sighted entrepreneur. He was keen to expand the business, so took on a 'Grand Tour' of Europe over the winter of 1938-1939. He examined the air forces of Britain, France and Germany and as many manufacturers as would tolerate him in their factories.

Kindelberger concluded that the European market wanted high-speed, impressive climb, great agility and *plenty* of guns, ideally 20mm. These qualities were lacking in existing American products. Along with NAA chief engineer John Leland 'Lee' Atwood, the pair started to sketch out a future fighter that utilised the latest in aerodynamics – including laminar flow wings – and manufacturing techniques.

Working from an office in New York, the British Purchasing Commission (BPC) was on a 'shopping' expedition across American industry to boost the UK's armed forces. Since 1938, the RAF had been a contented NAA customer, having acquired the first of a large number of Harvard advanced trainers. In the early spring of 1940 the BPC, knowing NAA's quality and production capacity, asked if Inglewood could build additional Tomahawks (see page 42) under licence from Curtiss, which was having problems coping with demand.

Not keen to churn out an older generation fighter from a rival, Kindelberger pitched the NA-73X. This excited the BPC, but an agreement signed on May 29, 1940 – dangling an initial order for 320 units as an incentive – stipulated that construction of the prototype take no more than 120 days.

THREE DAYS EARLY

Atwood gathered a design team, headed by Raymond Rice and including Larry Waite, Edward Horkey and Edgar Schmued. German-born Schmued was later to conceive another NAA thoroughbred, the F-86 Sabre. Building the NA-73X took 117 days and it had its maiden flight on October 26, 1940. Though clearly a winner, it was handicapped by its Allison V-1710, the only liquid-cooled V12 format engine available in the USA.

The first Mustang I, AG345, flew at Inglewood on May 1, 1941 and the first examples were shipped to Britain in October 1941. Designed in five sub-assemblies, the Mustang was intended for rapid manufacture and the production line moved into full swing.

Performance above 15,000ft (4,572m) decayed rapidly and a Supermarine Spitfire could run rings around a Mustang I. However, the Aeroplane & Armament Experimental Establishment at Boscombe Down was impressed with the Mustang's speed and its four-hour endurance. It was ideal for tactical-reconnaissance, and 26 Squadron at Gatwick was the first to get the type in May 1942. The RAF received 662 Mustang Is and the improved Mk.IIs.

A MARRIAGE MADE IN DERBY

When the BPC was trying to persuade NAA to build Curtiss fighters, it had also visited Packard in Detroit to get it to produce Merlins to help Rolls-Royce meet the spiralling demand for that exceptional engine for the RAF and Fleet Air Arm. A deal was signed in June 1940 for 9,000 units.

an order for 1,200 Merlin-powered Mustangs on July 20.

At the Rolls-Royce test airfield at Hucknall, Mustang I AL975, which had arrived in the UK on April 30, 1942, first flew with the new powerplant that October. It clocked 433mph (697km/h) – a legend had been born.

Meanwhile, the United States Army Air Force (USAAF) was disinterested in the Mustang. Pearl Harbor – December 7, 1941 – changed all of that. The fourth and tenth British specification Mk.Is off the Inglewood line were evaluated as XP-51s in early 1942. The first P-51As were ordered in August for delivery beginning in March 1943.

In Britain, the idea of marrying the Mustang with the Merlin was blossoming. On June 15, 1942 an agreement was reached that Rolls-Royce at Derby would receive a trio of Mustangs and fit Merlin 65s to them. Assuming all would go well, the Air Ministry approved

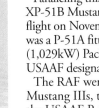

Paralleling this, the first of two XP-51B Mustangs had its first flight on November 30, 1942. This was a P-51A fitted with a 1,380hp (1,029kW) Packard V-1650-3, the USAAF designation of the Merlin 61.

The RAF went on to order 946 Mustang IIIs, the equivalent of the USAAF P-51B and 'C. The first unit to take the variant was 65 Squadron at Gravesend in December 1943. This was followed by 874 of the tear-drop canopied Mk.IVs, based on the P-51D. It was February 1947 before 217 Squadron at Nicosia on Cyprus retired the last operational RAF Mustangs.

By the middle of 1944 NAA plants at Inglewood and Dallas, Texas, reached peak production of 22 units a day – close to one every hour. A grand total of 15,367 Mustangs was built.

Let's leave the Mustang story with a question. If the BPC had not knocked on NAA's door in early 1940, NAA would almost certainly have built its fighter for the USAAF after Pearl Harbor. But would it ever have been powered by a Merlin? ◎

NORTH AMERICAN MUSTANG III

Type:	Single-seat fighter / fighter-bomber
First flight:	NA-73X - October 26, 1940; Mustang I entered RAF service May 1942
Powerplant:	One 1,680hp (1,253kW) Packard Rolls-Royce V-1650-7 Merlin V12
Dimensions:	Span 37ft 1in (11.3m), Length 32ft 3in (9.82m)
Weights:	Empty 7,000lb (3,175kg), All-up 9,200lb (4,173kg)
Max speed:	442mph (711km/h) at 25,000ft (7,620m)
Armament:	Four 0.50in machine guns, two in each wing; up to 1,000lb (453kg) of bombs
Replaced:	Direct acquisition from USA
Taken on charge:	2,482 of all versions
Replaced by:	Final examples replaced by Hawker Tempests 1947

Left
Newly delivered Mustang Is of 2 Squadron at Sawbridgeworth in the autumn of 1941. Second from the camera, AG623 'W-for-William' flew into a hill in Dorset on May 26, 1943, killing its pilot.

Left
American owner/ pilot Bob Tulius presented his 1944-built P-51D Mustang N51RT to the RAF Museum on November 11, 2003. It wears the colours of the USAAF's Debden-based 4th Fighter Group, 336th Fighter Squadron, and was known as 'The Duck'. KEN ELLIS

Left
Delivered to the Aeroplane and Armament Experimental Establishment at Boscombe Down in June 1944, Mk.IV KH589 was the first 'bubble' canopy Mustang to be received by the RAF. It was used initially for performance assessment, gunnery and cockpit trials until retirement in October 1946. Built at Inglewood, California, it carried traces of its USAAF serial number, 44-13332, for most of its time in Britain.

SUPERMARINE
SPITFIRE GRIFFON

1943 TO 1957

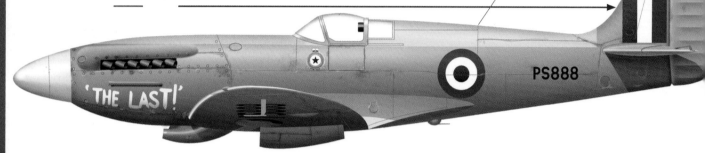

PS888

'THE LAST!'

With little ceremony, a Spitfire was rolled out of the factory at South Marston, near Swindon on February 24, 1948. This brought to an end nearly a decade of production, the first Spitfire I, K9787, having emerged in the summer of 1938 at Eastleigh.

The secret of the type's longevity was its ability to evolve, stage by stage. After the death of Reginald Joseph Mitchell in June 1937, Joseph 'Joe' Smith masterminded the Spitfire and Seafire family. (See page 30 for the Merlin Spitfire.)

Part of this transformation was a new engine developed from January 1939 – the Rolls-Royce Griffon. Based on Merlin thinking, although it was more powerful, it was kept compact. This allowed established Merlin-powered types to adopt the new engine with relative ease. The Griffon turned left-handed, opposite to its forebear: this took some getting used to as pilots converted to the later Spitfires.

The prototype Griffon version was the Mk.IV, DP845 having its maiden flight on November 27, 1941. To avoid confusion with the Mk.I-based photo-reconnaissance PR.IV, the Griffon Mk.IV was re-designated as the Mk.XX.

This is a good moment to explain that in 1944 the Air Ministry decreed that once a type reached Mk.XX, subsequent versions would adopt Arabic numbers: hence the Spitfire XX was followed by the Mk.21. In early 1948 Roman numerals were dropped altogether, eg Spitfire PR.XIXs became PR.19s that year.

The first Griffon Spitfire to enter

SUPERMARINE SPITFIRE XIV	
Type:	Single-seat fighter / fighter-reconnaissance
First flight:	Mk.IV - November 27, 1941; Mk.XII entered service February 1943
Powerplant:	One 2,050hp (1,529kW) Rolls-Royce Griffon 65 V12
Dimensions:	Span 36ft 10in (11.22m), Length 32ft 8in (9.95m)
Weights:	Empty 6,576lb (2,982kg), All-up 8,500lb (3,855kg)
Max speed:	448mph (720km/h) at 26,000ft (7,924m)
Armament:	Two 20mm cannon, one in each wing, four 0.303in machine guns, two in each wing; up to 1,000lb (453kg) of bombs
Replaced:	Continued development of the airframe, not specifically replacing a type
Taken on charge:	2,042 of all Griffon variants
Replaced by:	Through continued development the Spitfire effectively replaced itself for much of its service life. Ultimately, Gloster Meteor and de Havilland Vampire directly took over the day fighter role

RAF operational service was the Mk.XII with 41 Squadron at Llanbedr in February 1943. Other fighter versions were the high-altitude Mk.XIV, from January 1944, and the Mk.XVIII of 1946. The pressurised, unarmed, long-range photo-recce PR.XIX had its debut with 542 Squadron at Benson in June 1944.

The final three versions, Mks 21, 22 and 24, broke with the elliptical planform wing and adopted the

Left
*Personnel of 2
Squadron arranged
for a formal
photograph at
Wunstorf, West
Germany, in 1951.
To the left are
Spitfire FR.14s and
to the right PR.19s.
During that year the
unit converted to
Gloster Meteor FR.9s
and PR.10s. In the
background
are visiting DH
Mosquito B.35s.*

J H McElroy DFC. In one pass he shot down two of the RAF machines: Sayers was killed; McElhaw baled out.

Cooper saw another red-nosed Spitfire, clearly marked with Star of David roundels. He applied full throttle to out-climb the Mk.IX, but it was obvious that ground fire had damaged his FR.18. A burst of fire from the Israeli shattered Cooper's instrument panel and hit the engine; he took to his parachute.

Later RAF Hawker Tempest FB.6s from Deversoir, Egypt provided top cover for six FR.18s of 208 as they searched for the overdue aircraft. They were attacked by Israeli Spitfires, a

laminar flow technology that was at the core of the North American Mustang's success. At Wunstorf in Germany 80 Squadron inaugurated the Spitfire F.24 into the RAF in January 1948.

FROM ALL SIDES
With the British mandate in Palestine due to end, the date for withdrawal was fixed as May 22, 1948. Hard-

Left
*Built in December
1945, Spitfire F.21
LA331 joined Biggin
Hill-based 600 (City
of London) Squadron,
Royal Auxiliary Air
Force, in April 1497.
From mid-1950 the
unit's mixed fleet of
F.21s and F.22s gave
way to Gloster Meteor
F.4s.* KEC

line Zionists were demanding mass settlement in the area, while Arab nationalists had ambitions for the same territory. On May 14 the State of Israel was proclaimed and the situation slid into open warfare.

At the RAF base at Ramat David in Palestine, the peace was shattered at 06:00 hours on the 22nd. A Spitfire appeared overhead and dropped a bomb. Pilots of 208 Squadron rushed to their Spitfire FR.18s but the 'rogue' had a head start and got away.

Two other attacks were repulsed that day by the RAF. Believing the aircraft were Israeli, pilots were shocked to recognise the green, white and green markings of the Royal Egyptian Air

Force. Four REAF Spitfires were dispatched and RAF Regiment guns took out a fifth. The Egyptians made a public apology, claiming they had intended to attack the Israeli-held airfield at Megiddo.

Briefly 208 Squadron deployed to Nicosia on Cyprus, but in November 1948, the unit moved to Fayid in the Suez Canal Zone. During a tactical recce on January 7, 1949 Fg Offs G Cooper and T McElhaw and Pilot IIs F Close and R Sayers came under attack from Israeli ground forces. Close's Spitfire caught fire and he took to his parachute.

An Israeli Spitfire IX appeared, flown by Canadian war 'ace' Sqn Ldr

Tempest was shot down and its pilot killed. This episode was as disastrous as it was unexpected and generated worldwide comment. A ceasefire was called almost immediately and by the end of January hostilities between Israel and Egypt ended.

END OF THE LINE
The last offensive sorties by RAF Spitfires took place over the jungles of Malaya in May 1951. Based at Tengah, Singapore, 60 Squadron retired its FR.18s that month, converting to fighter-bomber de Havilland Vampires.

Singapore was also the venue for the final operational sortie by an RAF Spitfire. On April 1, 1954 the commanding officer of 81 Squadron, Sqn Ldr W P Swaby, took off from Seletar in PR.19 PS888 to bring the curtain down on an incredible era.

Spitfires with red, white and blue roundels were still flying with the Temperature and Humidity ('THUM') Flight under contract by Short Brothers and Harland at Woodvale. Spitfire PR.19 PS853 conducted the last weather-reconnaissance sortie on June 10, 1957. Former THUM Flight PR.19s PM631 and PS853 are operated by the Battle of Britain Memorial Flight from Coningsby, continuing the RAF's Spitfire heritage. ◎

Left
*Test flown at South
Marston on February
27, 1946, F.24 PK724
was sent to store
at 33 Maintenance
Unit at Lyneham
eight months later. It
was never issued to
service, becoming an
instructional airframe
in November 1955.
It joined the RAF
Museum's collection
in February 1970.*
RAF MUSEUM
www.rafmuseum.org

GLOSTER
METEOR
1944 TO 1982

Right
The NAAFI van provides a diversion for personnel of 124 Squadron at Molesworth in the summer of 1945 with three Meteor IIIs in the background.

Below
Meteor F.4 VW299 of 203 Advanced Flying School, Stradishall, 1950. In 1956 it was converted into a U.16 unmanned target drone. PETE WEST

When 'Gerry' Sayer took the Gloster E28/39 W4041 up from Cranwell's grass runway on May 15, 1941 he became Britain's first ever jet pilot. The event also began a dynasty that ended with the last Javelin delta-winged fighter in April 1960.

Three months before this moment of history – February 7, 1941 – the Air Ministry had issued a contract to Gloster for a dozen prototypes to meet Specification F9/40, a single-seat jet fighter. Gloster designer George Carter was already at work on what would become the RAF's first operational jet. Originally it was to be called Thunderbolt, but as the United States Army Air Force started to take deliveries of Republic P-47 Thunderbolts (see page 62) in March 1942, Meteor was chosen instead.

Available turbojets lacked the power to make the new machine single-engined. The twin-jet layout allowed for a concentration of firepower in the nose and like the E28/39 – a tricycle undercarriage. Mounting the engines in the wings allowed powerplant options to be installed far more easily than if they were buried in the fuselage.

Three pioneer turbojets were in the frame: the Rover-built Whittle W.2B (the basis for the Rolls-Royce Welland and the Derwent), the Halford H.1 (which became the de Havilland Goblin) and the Metropolitan-Vickers F.2 (leading to the Beryl and the Sapphire). First Meteor into the air was the H.1-powered

DG206 on March 5, 1943.

The first truly militarised version was the Mk.I with Wellands which first flew on January 12, 1944. As will be seen below, no time was wasted in getting what became the only Allied jet to go into combat in World War Two operational. Improvements resulted in the interim Mk.III and the Mk.IV, by which time the fighter had standardised on the Derwent.

Carter set to work in 1947 on the ultimate day fighter version which had the potential to expand into other roles, including ground-attack and reconnaissance. The result was the exceptionally clean-looking F.8 which first entered RAF service with 245 Squadron at Horsham St Faith in June 1950. This became the backbone of RAF air defence up to the mid-1950s and examples were still conducting target facilities sorties in 1982.

An elongated cockpit accommodating two pilots in tandem – instructor in the rear, pupil in front – was devised and the first T.7s were taken on charge in August 1949.

Prior to this, every pilot converting to the Meteor went solo every time. The trainer led the way to the next generation Meteor, see page 70.

JETS VERSUS ROCKETS

Two Meteor Is were delivered to Culmhead on July 12, 1944 and 616 Squadron, commanded by Sqn Ldr Andrew McDowell DFM*, became the first RAF jet unit. Nine days later 616 moved to Manston to take part in intercepting V-1 flying-bombs.

During the evening of August 4, Fg Off T D 'Dixie' Dean in EE216 got a 'doodlebug' in his sights, but his guns jammed. He flew alongside the V-1 and carefully toppled it with his wing tip; the gyros inside the flying-bomb failed and it crashed in the Kent countryside. Minutes later, Fg Off J K Rodger took out another V-1, this time using the cannon.

Dean shot down his third flying-bomb on the 10th. By the last day of August, the launch ramps across the Channel had been over-run and 516 Squadron had attained 13 V-1 'kills'.

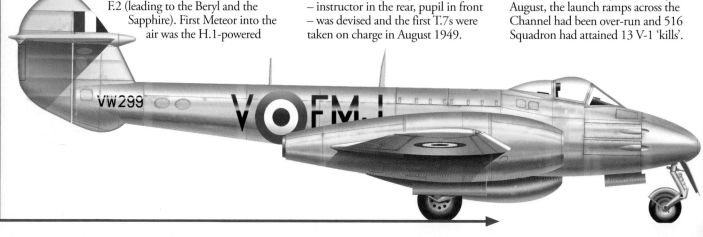

VW299   V·FM·I

The squadron moved to Colerne in January 1945, re-equipping with Mk.IIIs. Keen to maximise the propaganda value of the Meteor and the possibility of combat with Luftwaffe jets, a detachment was sent to Melsbroek, Belgium, in February. The entire unit settled into Gilze-Rijen in the Netherlands in April.

There was no jet-on-jet action but 616 became proficient in strafing ground targets. Moving with the Allied advance, the squadron disbanded at Lübeck in Germany in August 1945.

RECORD BREAKERS

With the war won, it was clear that the Meteor had phenomenal potential and it was decided to attempt establishing the world absolute airspeed record. This was a joint Gloster and RAF endeavour and two Meteor IIIs were upgraded to Mk.IV status: EE454 *Britannia* and EE455 *Forever Amber*.

Test pilot Eric Greenwood flew for Gloster and Gp Capt H J 'Willie' Wilson, who had served with 616 Squadron. It was Wilson who took the laurels, over Herne Bay on November 7, 1945. His average speed in EE454 being 606.25mph (975.63km/h).

On June 14, 1946 the RAF revived a unit that had made history in September 1931 by securing the Schneider Trophy in perpetuity for Britain. This was the High Speed Flight, then flying Supermarine S.6 floatplanes, but this time the pilots were issued with specially modified Meteor IVs, known as Star Meteors.

Commanding the special unit was Gp Capt E M 'Teddy' Donaldson

DSO AFC and among the pilots he had gathered were Sqn Ldr Neville Duke DSO DFC and Sqn Ldr Bill Waterton AFC – destined to become test pilots for Hawker and Gloster, respectively. On September 7, 1946 Donaldson piloted EE549 around a calibrated course off Littlehampton, to establish a new world record of 615.8mph (991km/h), Mach 0.81.

This was not to be the only world record clinched by aircraft flying from nearby Tangmere. Duke returned in September 1953 in the prototype Hunter, WB188 (see page 82). Flying the same 'race track' off the Sussex coast on September 7, 1953 Duke took WB188 to 727.6mph to clinch the record and on the 19th took the 100km closed-circuit record to 709.2mph for good measure.

In 1992 the RAF Museum paid the Tangmere Military Aviation Museum a huge compliment by presenting on loan the two aircraft that had flashed through the skies off Littlehampton: Meteor IV EE549 and Hunter Mk.3 WB188. ◎

GLOSTER METEOR F.4

Type:	Single-seat, twin-engined day interceptor
First flight:	March 5, 1943; Meteor I entered service July 1944
Powerplant:	Two 3,500lb st (15.56kN) Rolls-Royce Derwent 5 turbojets
Dimensions:	Span 43ft 0in (13.1m), Length 41ft 4in (12.59m)
Weights:	Empty 11,217lb (5,088kg), All-up 14,545lb (6,597kg)
Max speed:	550mph (885km/h) at 30,000ft (9,144m)
Armament:	Four nose-mounted 20mm cannon
Replaced:	Later Supermarine Spitfire variants and Gloster Meteor III from 1947
Taken on charge:	2,534 of all day fighter versions
Replaced by:	Gloster Meteor F.8 from 1950

HAWKER
TEMPEST
1944 TO 1955

Above
A trio of Tempest Vs of 501 Squadron, based at Bradwell Bay, September 1944.

Originally referred to as the Typhoon II, the Tempest was, essentially, a thin-winged version of its forebear. With an elegant, elliptical, laminar flow wing and a longer fuselage, the Tempest was 30mph (48km) faster than its older brother and could carry a similar warload.

Typically, designer Sydney Camm decided to explore powerplant options for his new fighter. The Mks I, V and VI stayed with the Typhoon's Napier Sabre, while the Mk.III was fitted with a Rolls-Royce Griffon. The aerodynamics of the Tempest were so clean that it could take a much more practical radial, the Bristol Centaurus, with little penalty in performance – this was the Mk.II.

Only the Mk.V saw combat in World War Two, as the fastest British-built fighter of the conflict and earned fame as the champion V-1 flying-bomb killer. Tempests brought down 638, about half of the 'doodlebugs' destined for England that were destroyed by the RAF.

The first Tempest unit was 486 Squadron Royal New Zealand Air Force, taking delivery at Tangmere in January 1944. The Mk.VI, a tropicalised Mk.V, served in the Middle East from December 1946 until 1950. A batch of 80 Mk.Vs was converted into target-tugs from 1950. These were the last of the RAF Tempests to fly, being retired from Pembrey in July 1955.

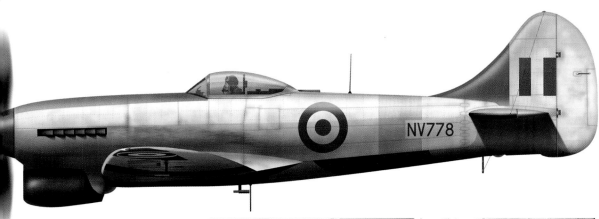

NV778

JET VICTORY

As the Allies moved ever closer to Germany, the Luftwaffe increasingly adopted 'tip-and-run' raids by single aircraft – such intruders were referred to as 'Rats'. By the autumn of 1944, Messerschmitt Me 262 twin-engined, swept-wing jets were among the Rats.

Based at Volkel, Netherlands, the pilots of 3 Squadron's Tempest Vs were on the alert for aircraft passing them at low level. Plt Off R W Cole spotted a Me 262 at very low level streaking towards Nijmegen. He dived, clocking 530mph (852km/h), and got on the enemy's tail. At 200 yards (182m) Cole opened fire and the jet streamed smoke from its port engine, rearing upwards before plunging to oblivion. Its pilot took to his parachute safely.

The airfield complex at Rheine, north of Münster, was known to be a major Me 262 base. Armed reconnaissance sorties of were frequently staged but these were risky ventures, as the area was surrounded by extensive flak batteries.

On November 26, 1944 Plt Off Cole, in JN822, was on such an operation when he fell victim to intense anti-aircraft fire. It was

Cole's turn to bale out, landing a short distance from the Me 262's lair. He spent the remainder of the war as a prisoner of war.

JUNGLE STRIKES

Intended for use in the Far East in the war against Japan, the Centaurus-engined Tempest II entered RAF service 19 months after the Mk.V – joining 183 Squadron at Chilbolton in August 1945. With its 18-cylinder, twin-row sleeve-valve Bristol Centaurus V radial rated at 2,520hp (1,879kW), the Tempest II was the most powerful RAF single-seat fighter.

The Tempest FB.2 was a good candidate for use in Malaya, the long-term conflict against communist terrorists. This was

referred to in typical British understatement as 'The Emergency' in diplomatic circles and Operation 'Firedog' by the RAF.

Its ability to carry underwing racks for a quartet of 60lb (27kg) rocket projectiles under each wing was not its most attractive attribute for the theatre. The FB.2 excelled because its radius of action with full weaponry was around 500 miles (804km) – superior to the incumbent Supermarine Spitfire FR.18s of 28 and 60 Squadrons.

Previously based in Germany, the FB.2s of 33 Squadron set off for the Far East on July 2, 1949 and arrived at Changi in Singapore on August 8. The unit's first major operation was in the Negri Sembilan region to the southeast of Kuala Lumpur on October 21.

This area required transits of up to 220 miles, but the Tempest's endurance allowed 33 Squadron to loiter over potential targets, hoping to flush out terrorists, and to assess damage after a strike.

By the spring of 1951, the Tempests of 33 Squadron were beginning to show their age and the ground crews were facing an ever-increasing struggle to keep even ten aircraft on line. Replacements began to arrive in April, in the form of de Havilland Hornet twins – see page 64. The last 'round engine' Tempests were withdrawn from the RAF when the final six veterans of 33 Squadron retired in June. ◉

HAWKER TEMPEST V

Type:	Single-seat interceptor / fighter-bomber
First flight:	September 2, 1942; entered service January 1944
Powerplant:	One 2,200hp (1,641kW) Napier Sabre IIB 24-cylinder H-format
Dimensions:	Span 41ft 0in (12.49m), Length 33ft 8in (10.25m)
Weights:	Empty 9,000lb (4,082kg), All-up 12,850lb (5,828kg)
Max speed:	442mph (711km/h) at 20,500ft (6,248m)
Armament:	Four 20mm cannon, two in each wing; provision for up to 2,000lb (907kg) of ordnance under wing
Replaced:	Partially supplanted Hawker Typhoons
Taken on charge:	1,394 of all versions. 50 Mk.IIs sub-contracted to Bristol
Replaced by:	Final examples gave way to de Havilland Vampire FB.5s in 1950

Above
Built in late 1944, Tempest V NV778 saw no service and was stored, returning to its birthplace, Langley, in March 1950. It was converted to a TT.5 target-tug and test flown by Neville Duke on November 9, 1950. It was issued to 233 Operational Conversion Unit, Pembrey – illustrated – on October 27, 1952 and served until July 12, 1955 when it made its last flight. After service as a 'gate guardian', it was acquired by the RAF Museum in August 1965. © ANDY HAY
www.flyingart.co.uk

Above left
The RAF Museum's Tempest II is displayed in the colours of 5 Squadron's PR536. This machine initially served with 5 Squadron at Peshawar, India, in 1947. It was transferred to the Royal Indian Air Force as HA457. Its battered forward fuselage was recovered to Britain in 1979 and used as the basis of a composite restoration. It was handed over to the RAF Museum on November 13, 1991. RAF MUSEUM
www.rafmuseum.org

REPUBLIC

THUNDERBOLT

1944 TO 1946

"A single-engined Dakota with no windows", was how a ground crew member of 30 Squadron described the Thunderbolt to the writer. The Americans called Republic's monster the 'Jug' – short for juggernaut.

The Thunderbolt certainly attracted a lot of comment from those who flew, maintained or supported the RAF's examples. Other than the handful brought to the Aeroplane and Armament Experimental Establishment (A&AEE) at Boscombe Down for evaluation or trials, the bulk of the 826 taken on charge were to be found in small numbers in Egypt and mostly with South East Asia Command (SEAC) in India and Burma.

A pilot who flew with 135 Squadron, the first unit to take on the Thunderbolt, talked about his preferred tear-drop canopied version – the Mk.II. "Taxiing it was an act of faith that there was nothing in front of you. You learned to memorise features to either side of the track out to the runway as a way of marking your progress. The view over the 'bonnet' was non-existent. If you chose to 'weave' to gain a look forward, you very soon found you were exhausted and you couldn't feel your feet [from working the rudder pedals]."

"The cockpit was vast... If you copped for small arms fire, or worse, that penetrated the cockpit, you could always undo your straps and run around, minimising the chances of any of the 'nasties' hitting you!"

The test pilots at A&AEE agreed regarding the view from the cockpit, which was determined as "disappointing". Other elements achieved "satisfactory": stability, cockpit layout, control harmonisation and manoeuvrability. The Thunderbolt was a smooth ride; trimmed off and at around 12,000ft (3,657m) it was vibration free.

ADJUSTING TO THE 'JUG'

The Air Ministry saw no requirement for the Thunderbolt in the European theatre, but it was ideal to replace the increasingly outclassed Hawker Hurricane IIs in SEAC. All the RAF Thunderbolts were the equivalent of the United States Army Air Force P-47D: the Mk.Is having the 'razorback' fuselage fairing and the 'birdcage' canopy, the Mk.IIs had the tear-drop shaped canopy.

As deliveries were direct to India, pilots had no prior experienced of the Thunderbolt. To overcome this, 1331 Conversion Unit at Risalpur, in present-day Pakistan, and 1670 (Thunderbolt) Conversion Unit was established at Yelahanka, Bangalore, in April and June 1944 respectively. The introduction to the new machine was a 'talk through' the cockpit by an instructor standing on the wing root of a stationary Thunderbolt, lectures and question and answer sessions; followed by a solo blast around the circuit.

Comparison with the Hurricane illustrates how radically different the new mount was. The Hawker had a top speed of around 339mph (545km/h), the Republic 427mph. The Hurricane had 1,280hp (954kW) at its disposal, the Thunderbolt boasted 2,300hp. The big US fighter

needed that power to move its all-up weight of 14,600lb (6,622kg) around the sky: that was well over twice the Hurricane's 6,600lb.

In July 1944 a more integrated approach was initiated at Fayid in Egypt when 73 Operational Training Unit began tuition on the Thunderbolt. Aircrew destined for SEAC would transit Egypt, making the location convenient; the weather conditions were much more stable, and the airspace was not in a war zone.

CAB-RANK SERVICE

At Chittagong, on the Burmese border, 135 Squadron was the first to take the Thunderbolt to the front line. In May 1944, it gave up its Hurricane IIs. Offensive sorties began in October and by December the unit was moving down the Arakan coast keeping pace with the advance against the Japanese, reaching Cox's Bazaar by April.

The Jug gained a good reputation as a reliable and stable fighter-bomber. With a standard bomb load of three 500-pounders, one under the centre section and one on a pylon under each wing, the Thunderbolt could interdict supply lines and hit armoured vehicles. The battery of eight '50-cal' machine guns allowed pilots to pick off targets of opportunity after delivering their bombs.

Operations in SEAC were staged on a 'cab-rank' basis. Fighters would respond to calls for ground support. The instigation of aircrew acting as forward air controllers, interpreting the needs of the ground forces through the realities of operations and assessing the battlefield in a first-hand manner, hugely improved the efficiency of the Thunderbolts' strikes.

Deploying back to Chakulia in India, 135 Squadron was re-numbered as 615 Squadron in June 1945, but the 'new' unit disbanded three months later. The last RAF Thunderbolts with an operational squadron flew from Kemajoran on Java, changing to Spitfire FR.18s in the final weeks of 1946. ◎

Above
Thunderbolt II KJ348 of 73 Operational Training Unit, Fayid, Egypt, 1944. PETE WEST

Left
The RAF Museum's Thunderbolt was built at Evansville, Indiana, and handed over to the USAAF as 45-49295 in June 1945. It remained in the USA and was supplied to the Yugoslav Air Force, as 13064, in May 1952. It was brought to the UK in 1985 and was acquired by the museum the following year. It wears the colours of 'KL216' operated by 30 Squadron, 1945. RAF MUSEUM
www.rafmuseum.org

REPUBLIC THUNDERBOLT II

Type:	Single-seat fighter-bomber
First flight:	May 6, 1941; entered service May 1944
Powerplant:	One 2,300hp (1,715kW) Pratt & Whitney Double Wasp
Dimensions:	Span 40ft 9in (12.42m), Length 36ft 2in (11.02m)
Weights:	Empty 10,000lb (4,536kg), All-up 14,600hp (6,622kg)
Max speed:	427mph (687km/h) at 25,000ft (7,620m)
Armament:	Eight 0.5in machine guns, four in each wing; up to 2,000lb (907kg) of bombs
Replaced:	Direct acquisition from USA; mostly supplanting Hawker Hurricane IIs
Taken on charge:	826 - all versions
Replaced by:	Hawker Tempest IIs from late 1946

DE HAVILLAND
HORNET

1946 TO 1955

Right
Rocket-equipped
Hornet F.3 PX314 of
41 Squadron, based
at Church Fenton,
1948.

Entering service in May 1946, the shapely Hornet was the fastest and the last of all of the RAF's piston-engined fighters. Like the Hawker Tempest II, it was intended for use in the campaign against Japan – its twin engines providing greater safety on long sorties over the Pacific.

As with the Mosquito before it, the Hornet began as a private venture, employing similar thinking to how the Beaufighter had been morphed out of the Beaufort by Bristol. The concept of a single-seat, twin-engined, cannon-armed fighter was not new to the RAF, the unfortunate Westland Whirlwind (page 38) which served in small numbers from 1940 to 1943 having achieved that distinction.

The Hornet bore only a familial likeness to the Mosquito; it was an entirely new design taking in all of the lessons learned since the 'Wooden Wonder' first flew in 1940. The fuselage employed the same production technique as the Mosquito – moulded ply. The wooden 'sandwich' was laid up over concrete moulds for each half of the fuselage; split vertically down the oval cross section – in the same manner as a plastic model kit.

Interior bracing, bulkheads and other fittings were added to each half before they were joined together

using a 'V-shaped' butt joint. This was reinforced by plywood strips in rebates inside and out.

The Hornet's laminar flow wing combined wood and metal. The internal structure was metal, the lower wing skin was aluminium and the top surface was plywood. The Hornet was the first aircraft in Britain – if not the world – to employ wood-to-metal bonding.

This was achieved thanks to the invention of Redux adhesives by the Aero Research company. Dr Norman de Bruyne had been experimenting

in structural adhesives and composite materials at his factory opposite Duxford airfield since the early 1930s. The brand name Redux was derived from REsearch DUXford. De Bruyne designed two light aircraft, the Snark and the Ladybird, to test out his theories. Now owned by the Hexcel Corporation, the Duxford factory continues to innovate in advanced composites for the aerospace and other industries.

Rolls-Royce developed Merlins with minimal frontal area to keep the Hornet as slim-line as possible. The

DE HAVILLAND HORNET F.3

Type:	Single-seat, twin-engined fighter / fighter-bomber
First flight:	July 28, 1944; entered service May 1946
Powerplant:	Two 2,030hp (1,514kW) Rolls-Royce Merlin 130/131 V12
Dimensions:	Span 45ft 0in (13.71m), Length 36ft 8in (11.17m)
Weights:	Empty 12,880lb (5,842kg), All-up 20,900lb (9,480kg)
Max speed:	472mph (759km/h) at 22,000ft (6,705m)
Armament:	Four nose-mounted 20mm cannon, up to 2,000lb (907kg) of bombs
Replaced:	North American Mustang IV and Supermarine Spitfire XVI from 1946; Bristol Brigand B.1, Hawker Tempest F.2 and Supermarine Spitfire F.24 from 1951
Taken on charge:	198 - all versions. Type also served with the Fleet Air Arm
Replaced by:	Gloster Meteor F.4 from 1950; de Havilland Vampire FB.9 and Venom NF.2 from 1955

PX232

to Arabic numerals in 1948) were used.

At Horsham St Faith – these days Norwich Airport – 64 Squadron was the first to operate the Hornet, taking delivery in May 1946. The co-located 65 Squadron received its examples the following month. The final British-based Hornets were retired in 1951.

It was with the Far East Air Force (FEAF) that Hornets saw most use, including action in strikes against terrorists in the jungles of Malaya. The twin-engined fighter could carry a 1,000lb (453kg) bomb under each wing, but 60lb rocket projectiles were the weapon of choice during Operation 'Firedog'.

engines were 'handed' the Merlin 130 in the starboard wing turning to the left, the Merlin 131 in the port wing turning to the right. This counter-rotation cancelled the swing created should an engine fail on take-off or landing.

LAST OF THE PISTONS

The prototype Hornet, RR915, was first flown at Hatfield on July 28, 1944. During trials, it reached 485mph (780km/h) and the fully equipped and heavier production F.3 was not much slower at 472mph.

By the time the Hornet was ready to be issued to RAF units, the Japanese had surrendered. Three fighter versions, Mks I, III and IV, and the photo-recce PR.II (changing

The first unit in the FEAF to accept the Hornet was 33 Squadron, replacing its Tempest FB.2s with the twins in March 1951 at Butterworth, Malaya. In May 1955 the surviving examples from 33, 45 and the Far East Air Force Training Squadron were ferried to the Maintenance Unit (Far East) at Seletar, Singapore. There they joined by the Hornets of 80 Squadron which had arrived from Kai Tak, Hong Kong, all to await the scrap merchant. The RAF's piston-engined frontline fighter era had ended. ◉

Above
Hornet F.1 PX232 of 65 Squadron, based at Horsham St Faith, in the spring of 1946. The unit moved to Linton-on-Ouse in August 1946 and retired the Hornets in favour of Gloster Meteor F.4s in 1951.
PETE WEST

Left
The ramp at Butterworth in the spring of 1954; in the foreground is Hornet F.3, behind Mosquito T.3 VP349, both serving with the Far East Air Force Training Squadron. KEC

Left
Based at Butterworth, a Hornet F.3 of 33 Squadron, circa 1954. It carries rails for two rockets – and it has launched two of these – and a 500-pounder bomb on a pylon, under each wing. KEC

Below left
Many visitors to the de Havilland Aircraft Museum at London Colney pay little attention to the lump of concrete outside the main display hall – until they read the placard. It is a mould to create a half fuselage of a Hornet; the Mosquito shared the same construction technique. KEN ELLIS

DE HAVILLAND
VAMPIRE
1946 TO 1970

Right
Vampire FB.5s WA442 and WA387 of 185 Squadron on a sortie out of Luqa, Malta, 1952. Both carry the unit badge, flanked by red-blue-red bars on the nose. 'Kilo', in the foreground, has a red-blue-red pennant behind the roundel on the tail boom. KEC

Below right
Vampire NF.10 WP252 of 25 Squadron in its intended environment at West Malling, circa 1952. To the right is a Meteor NF.11.

It wasn't an inspiring name – Spidercrab – thankfully, common sense prevailed and the much more dramatic Vampire was substituted for Britain's second jet fighter. Lacking the experience that pioneer Gloster had built up, de Havilland was determined to enter the marketplace. Just seven months after the prototype Meteor had its maiden flight, the first Vampire was test flown at Hatfield, on September 29, 1943.

De Havilland had been awarded the development contract partially as 'insurance' should the Gloster project come unstuck, but specifically to create a nimble single-engined fighter. For the prototype, chief designer Ronald Bishop initially had to manage with a meagre thrust of 2,700lb st (12kN) from the DH Goblin turbojet, with more promised.

The first-ever British jet, the Gloster E28/39 of 1941, had demonstrated that a considerable amount of thrust was lost in the jet pipe that vented at the tail. Mounting the turbojet away from the centre of gravity was fraught with complexities, so Bishop came up with the 'pod and boom' layout.

The pilot sat in a pressurised cockpit immediately in front of the Goblin and beneath him were four 20mm cannon. To keep the weight down, the Mosquito's moulded ply construction techniques were employed to create

the 'pod'; even in the jet age, wood still had a place.

Like the Mosquito, the Vampire was to become a major money-spinner for de Havilland. More than 3,000 were built, many for export with licence construction being undertaken in Australia, France, India, Italy and Switzerland. Development extended to two-seaters and the improved Venom – see page 74.

The first RAF unit to accept the Vampire I to service was 247 Squadron at Chilbolton and later Odiham, from April 1946, replacing

Hawker Tempest IIs.

The little jet spawned a whole series of 'firsts': introducing the Royal Auxiliary Air Force to the new age with 605 Squadron at Honiley in July 1948; the Middle East Air Force in 1949 and the Far East Air Force a year later. As will be seen below, the Vampire also conquered the Atlantic and was the first turbojet night-fighter.

Vampires were used extensively by the Second Tactical Air Force in West Germany, particularly the fighter-bomber versions. After the Mk.I, the Mk.III offered a much-needed

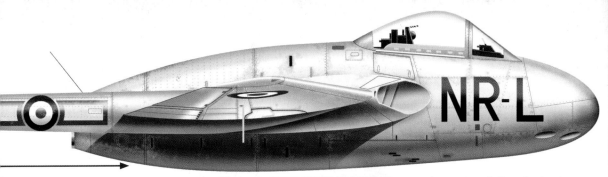

DE HAVILLAND VAMPIRE I

Type:	Single-seat interceptor / fighter-bomber
First flight:	September 29, 1943; entered service April 1946
Powerplant:	One 3,100lb st (13.78kN) de Havilland Goblin 2 turbojet
Dimensions:	Span 40ft 0in (12.19m), Length 30ft 9in (9.37m)
Weights:	Empty 6,372lb (2,890kg), All-up 12,390lb (5,620kg)
Max speed:	540mph (869km/h) at 20,000ft (6,096m)
Armament:	Four nose-mounted 20mm cannon
Replaced:	Hawker Tempest IIs from 1946, air defence Supermarine Spitfire units from 1947
Taken on charge:	1,057 – all single-seat variants. Production sub-contracted to English Electric and Fairey
Replaced by:	Fighter-bomber variants by de Havilland Venoms from 1952

"To keep the weight down, the Mosquito's moulded ply construction techniques were employed to create the 'pod'; even in the jet age, wood still had a place."

increase in range, from 730 miles (1,174km) to 1,145 miles.

For the attack role, the FB.5 featured a 2ft (0.6m) reduction in span (from 40ft) and a beefed-up undercarriage. The first example joined 16 Squadron at Gutersloh in Germany in December 1948. The tropicalised FB.9 followed.

Meteor F.8s replaced the frontline Vampire force by 1951 with the twin-bombers transferring to the operational training role until late 1953.

ATLANTIC CROSSING

Blue Leader Oxspring, *Blue 2* Wright, *Blue 3* Evans, *Red Leader* Woolley, *Red 2* Skinner and *Red 3* Wood departed Odiham on July 1 on a very special deployment in their Vampire F.3s. They returned on July 26 having made history: the six pilots had crossed the Atlantic twice – the first

ever by jet aircraft. In a classic piece of RAF folklore, it was a close-run thing...

Goodwill visits to Canada and the USA were increasing in frequency, but it was the domain of the 'heavies', Avro Lancasters, Lincolns and transports. Now that it was in the jet age, there would come a time when the RAF would need to move fighters long distances to flashpoints, so why not try it out on a visit to friends?

The Odiham wing of Vampire F.1s – 54, 72 and 247 – had gained a good reputation for its aerobatic demonstrations, in solos, trios and sixes. Its pilots had the skills to stage displays to North American audiences. With greater range, the F.3 was rolling off the production line. The stage was set and planning started.

Six brand new F.3s – VT863, VT864, VT868, VT869, VT871, VT873 – were delivered from English Electric at Samlesbury. They were fitted with a 100-gallon (454-litre) drop tank under each wing.

The commanding officer (CO) of 54 Squadron, wartime Supermarine Spitfire 'ace' from Mk.Is to Mk.XIVs, 29-year-old Sqn Ldr R W 'Bobby' Oxspring DFC** was 'Vampire Leader'. The five who flew with him on the initial sectors were: Hurricane 'ace' Flt Lt E W 'Ricky' Wright DFC, a flight leader on 247 Squadron; Pilot 1 S Evans; former wartime Spitfire pilot Flt Lt Frank G Woolley DFC*; Pilot 2 R J Skinner and Pilot 1 William C Wood. (Pilot 1 and 2 were ranks that were short-lived; Evans, Skinner and Wood had previously been warrant officers.)

Two others were designated to fly solo aerobatic displays in North America: Flt Lts N W 'Jeep' Heale and C I Colquhoun. Deputy ⊙➤

Above left
Vampire FB.5 VZ336 of 605 (County of Warwick) Squadron, Royal Auxiliary Air Force, based at Honiley, 1951-1952. The squadron 'flash', either side of the roundel, was a rectangle, with pale blue centre and red outline. PETE WEST

Above
With the jet age, 112 Squadron perpetuated the unit's famous 'shark mouth' – see page 42. *Vampire FB.5 WA345 in the squadron hangar at Jever, West Germany, 1952. Behind is the Meteor T.7 continuation trainer, still carrying the 'Fassberg flash' that 112 carried on its aircraft when based at the airfield of the same name, the year before.*

Left
Vampire FB.5s of 501 (County of Gloucester) Squadron, Royal Auxiliary Air Force, lined up at Filton, circa 1956. The aircraft in the foreground, VZ112, carries the stencilled warning, 'No Weight Here' under the tailplane and the unit's black and yellow triangles either side of the roundel on the tail boom. In the far background is an Avro Ashton. BOTH KEC

Above
A quartet of 60 Squadron Vampire FB.9s over Singapore on a sortie out of their base at Tengah, March 1953. The lead aircraft, WR202 'Lima' carries a dark blue band around the rear of the tail boom, denoting 'B' Flight. The three behind are painting with the unit's colours, a black rectangle with silver lightning flash, either side of the roundel on the tail boom.

Right
Personnel of 23 Squadron in front of the NF.10s at Coltishall in 1952. Most of the Vampires are carrying the unit badge on the nose; all feature its blue-red-blue-red squares either side of the roundel. In the background are Meteor NF.11s of 141 Squadron, which shared the base with 23.

44-85464, was former Republic P-47 Thunderbolt 'ace' Colonel David C Schilling. Fully intending to be the first jet pilots to make the Atlantic crossing, the Americans had to content themselves with conducting the initial eastbound journey. As a courtesy, the F-80s dropped in to Odiham on July 21 before completing the venture at Fürstenfeldbruck, Germany. 'Fox Able One' inaugurated regular USAF deployments to Europe that have continued ever since.

NOCTURNAL PIONEER

With its wealth of experience gained from its family of Mosquito night-fighters, de Havilland decided to develop a private venture side-by-side, two-seat, radar-equipped

"The world's first fully operational jet night-fighter unit, 25 Squadron at West Malling, started to take delivery of its Vampire NF.10s in July 1951."

Vampire Leader was Sqn Ldr R N H 'Buck' Courtney DFC*, CO of 74 Squadron and deployment commander was Wg Cdr D S Wilson-MacDonald DSO DFC.

The Vampires were to be escorted across by three DH Mosquitos and a trio of Avro York transports. In charge of the Mosquitos was former 'Dam Buster' Sqn Ldr H B M 'Mickey' Martin DSO* DFC** AFC.

Headwinds delayed the westward flight out of Stornoway on the Isle of Lewis as the mixed formation transited to Iceland and on to the airfield at Bluie West 1, Greenland. The next leg was the big one; 780 miles (1,255km) to Goose Bay, Labrador, arriving there on July 14 – they'd crossed the Atlantic.

What followed was an intense schedule of displays and hand-shaking, culminating in an appearance at the International Air Exposition at Idlewild, New York – now John F

Kennedy International Airport.

On the westbound trip, the Vampire team had met up at Goose Bay with pilots of 16 Lockheed F-80A Shooting Stars of the USAF 56th Fighter Wing taking part in exercise 'Fox Able One'. Leading, in the brightly coloured

Vampire for the export market. Featuring an entirely new 'pod' that was very reminiscent of the forward fuselage of the Mosquito, the prototype had its maiden flight on August 28, 1949.

The Egyptian Air Force placed an order for this new version two months

DE HAVILLAND VAMPIRE NF.10

Type:	Two-seat night-fighter
First flight:	August 28, 1949; entered service July 1951
Powerplant:	One 3,350lb st (14.9kN) de Havilland Goblin 3 turbojet
Dimensions:	Span 38ft 0in (11.58m), Length 34ft 7in (10.54m)
Weights:	Empty 6,984lb (3,167kg), All-up 13,100lb (5,942kg)
Max speed:	538mph (865km/h) at 20,000ft (6,096m)
Armament:	Four nose-mounted 20mm cannon
Replaced:	De Havilland Mosquito NF.36 from 1951
Taken on charge:	78
Replaced by:	De Havilland Venom NF.2 and Gloster Meteor NF.12 / NF.12 from 1953

version and installing ejector seats.

In the mid-1960s, the T.11 had been largely replaced by the Folland Gnat T.1 and was finally removed from the training syllabus in November 1967. The type continued in support roles and the last RAF examples with the Central Air Traffic Control School at Shawbury, operated by contractors Marshalls (Engineering) Ltd, were phased out at a ceremony on November 16, 1970.

Vampire T.11 XH304 was handed on to the Central Flying School at

later and production began, but the political situation in the Middle East was such that the Arab country was denied its night-fighters. The RAF realised this represented an opportunity to begin replacement of its Mosquito force ahead of the arrival of Meteor NF.11s – see page 70.

The world's first fully operational jet night-fighter unit, 25 Squadron at West Malling, started to take delivery of its Vampire NF.10s in July 1951. The nocturnal Vampires were only ever intended as a stop-gap and withdrawn from frontline service in 1954.

TRAINER FINALE
Until 1952, frontline Vampire units had two-seat Meteor T.7s detached to provide continuation training. Unlike the Gloster jet, where the trainer led to the development of the night-fighter, with the Vampire the NF.10 inspired the creation of the prolific T.11 trainer.

The prototype of the side-by-side two-seater first flew on November 15, 1950 and 538 were taken on by the RAF from March 1952. A rolling programme of modifications kept the T.11 fleet up to date, exchanging the cumbersome 'bird cage' canopy for a clear-vision

Little Rissington for operation by the 'Vintage Pair' heritage flight, along with Meteor T.7 WA669. Tragedy struck on May 25, 1986 when both aircraft collided at an airshow at Mildenhall; the crew of the Meteor were killed, XH304's pilots successfully ejected. ◎

GLOSTER
METEOR NIGH

1951 TO 1966

Right
NF.11s of 68 Squadron,
based at Wahn, low
over snow-clad
Germany in early 1953.
It was 1955 before
68 applied squadron
markings. KEC

Right centre
Marshalling Meteor
NF.14 WS725 of 25
Squadron at West
Malling, circa 1955.

LOYAL SERVANT

In July 1951 Tangmere-based 29 Squadron was the first unit to accept operational NF.11s, upgrading the NF.12 in early 1958.

By 1948 the Gloster design office was heavily involved in creating the Meteor T.7 trainer and the F.8 day-fighter – see page 58. It was also formulating its response to Specification F24/48 for an all-weather and night-fighter to replace the Mosquito force. The T.7 was to be the basis for this radar-carrying Meteor.

Something had to give, and in 1949 it was decided to hand over design responsibility and production to sister company Armstrong Whitworth (AW). The Hawker Siddeley Group had been formed in 1935, encompassing AW, Avro, Gloster and Hawker. Each division continued to trade under its own names, but work was frequently shared around.

The new fighter married the T.7 airframe to a longer nose housing the Airborne Interception Mk.10 radar, the tail of the F.8 and longer span wings – from 37ft 2in (11.34m) to 43ft. Previous fighter Meteors had all carried 20mm guns in the nose, but this was not possible in the night-fighter variant so two cannon were housed in each wing. The nocturnal Meteor programme was planned to be a progression of versions, with steady upgrades.

The radome profile was tried out on T.7 VW413 and this paved the way for AW's prototype NF.11, WA546, which was flown for the first time on May 31, 1950 from

GLOSTER METEOR NF.11

Type:	Two-seat, twin-engined night-fighter
First flight:	May 31, 1950 (night-fighter prototype); entered service July 1951
Powerplant:	Two 3,700lb st (16.45kN) Rolls-Royce Derwent 8 turbojets
Dimensions:	Span 43ft 0in (13.1m), Length 48ft 6in (14.78m)
Weights:	Empty 12,019lb (5,451kg), All-up 16,542lb (7,503kg)
Max speed:	541mph (870km/h) at 10,000ft (3,048m)
Armament:	Four nose-mounted 20mm cannon
Replaced:	De Havilland Mosquito NF.36s from mid-1951
Taken on charge:	556 of all versions; built by Armstrong Whitworth
Replaced by:	Gloster Javelin from 1957

Baginton, the present-day Coventry Airport. Most of the series was built at Bitteswell; the last rolling out in 1954.

On the production line, the NF.11 was followed by the NF.12, which employed American APS.21 radar. The NF.13 was essentially a NF.12 configured for operations in the Middle and Far East.

The final version was the NF.14, which did away with the sideways-opening, heavily framed canopy, substituting a clear-view, rear-sliding version. With a slightly longer fuselage, the NF.14 was the best looking of the night-fighter Meteors.

The RAF ordered a total of 556 of all four variants. As well as the RAF, Belgium, Denmark, Egypt, France, Israel and Syria took night-fighter Meteors.

At West Malling, 25 Squadron brought the NF.12 and the NF.14 into service in 1954. Several units flew a mixture of Mk.12s and Mk.14s at the same time.

The NF.13 flew with two units: 39 and 219 Squadrons, both originally based at Kabrit in Egypt. First with the variant was 39, from March 1953. Across the ramp, 219 Squadron traded in its NF.13s for Venom NF.2s – see page 74 – in 1955. Relocating to Luqa on Malta in January 1955, the Meteors of 39 Squadron ploughed on until the unit disbanded in mid-1958.

From 1956 the Gloster Javelin delta-winged all-weather/night-fighter began to replace the Meteors – see page 82. At Tengah, Singapore, 60 Squadron changed role in October 1959, exchanging

T-FIGHTER

Left
Meteor NF.14 WS733 served West Malling-based 25 Squadron from April 1954 to December 1956. The squadron flashes, either side of the roundel, were silver rectangles with black edging top and bottom.

Below left
The Tangmere-based Meteor NF.11s of 29 Squadron during the Queen's Coronation Review of the RAF at Odiham, July 1953.
KEC

Bottom left
The RAF Museum's Meteor NF.14, WS843, 'flies' inside the National Cold War Exhibition at Cosford in the colours of the Linton-on-Ouse-based 264 Squadron. This machine saw no frontline service, serving with 228 Operational Conversion Unit at Leeming and finally with 1 Air Navigation School at Stradishall. Retired in January 1966, it became part of the museum's collection on March 13, 1967. RAF MUSEUM
www.rafmuseum.org

its Venom FB.4s for Meteor NF.14s while it awaited Javelins. Conversion to the Javelin FAW.9 began in September 1961, and 60 Squadron had the honour of being the last unit to fly frontline Meteors.

Singapore was also the venue of another Meteor phase-out ceremony. Developed in 1956 for the Royal Navy, about 25 NF.11s were converted to TT.20 target-tugs. In November and December 1959 WD645 and WD678 were loaned to 1574 Flight at Seletar. These machines provided sterling service towing target sleeves for resident and visiting fighter units until they were struck off charge in August 1963.

Surplus Mk.14s were adapted as navigator trainers under the designation NF(T).14 and served with 1 Air Navigation School at Stradishall from 1960. These machines were the last of the night-fighter Meteors to fly with the RAF, giving way to the Hawker Siddeley Dominie T.1 – based on the HS.125 executive jet – by mid-1966.

With the retirement of 60 Squadron's NF.14s in 1961 the operational reign of the Meteor was brought to an end. Seventeen years previously, 616 Squadron had accepted its first pair of Mk.Is in July 1944. Lacking the charisma of the Spitfire, the Gloster twin jet was equally as crucial to RAF history. ◎

> **"Previous fighter Meteors had all carried 20mm guns in the nose, but as this was not possible in the night-fighter variant so two cannon were housed in each wing."**

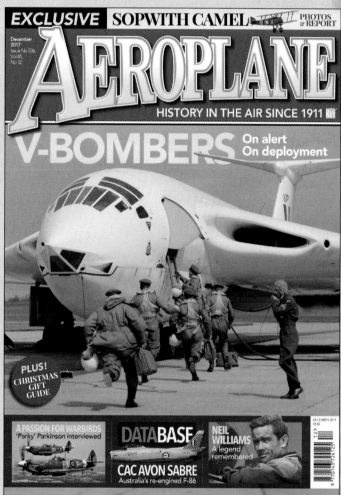

DE HAVILLAND
VENOM
1952 TO 1957

Il-28 *Beagle* bombers. One of the objectives for the Venoms would be to neutralise as many EAF aircraft as possible while they were still on the ground.

As dawn broke on November 1 Operation 'Musketeer' was unleashed. Waves of Venoms left Cyprus to attack Egyptian airfields to follow on from overnight bombing raids. In the June 2016 issue of *FlyPast*, Andy Thomas described the exploits of the RAF's Venoms in his feature *Venomous Action*. He takes up the story: "Eight FB.4s from 6 Squadron, led by the

While the Venom shared the same 'pod and boom' layout as its predecessor, the Vampire, it was an altogether different beast. Powered by the DH Ghost 103 turbojet, the Venom FB.1 had 50% more thrust, was nearly 50mph (80km/h) faster, had twice the range and a much greater weapon load than the Goblin-powered Vampire I that entered service in 1946.

Often described as a swept-wing aircraft, the Venom's leading edge was swept, but the trailing edge remained straight. It was the North American Sabre that introduced the RAF to swept-wings. Distinctive fixed wingtip fuel tanks accounted for the Venom's increased range. These allowed the wing pylons to carry up to 2,000lb (907kg) of bombs, or for long-range deployments, additional fuel tanks.

The prototype Venom had its maiden flight on September 2, 1949. Production examples were delivered as a priority to the Second Tactical Air Force in Germany, replacing fighter-bomber Vampires. The first unit to accept FB.1s was 11 Squadron at Wunstorf in November 1952.

With a redesigned tail and an ejection seat fitted as standard, the Venom FB.4 was introduced in 1955. Like the Mk.1, it was used in Germany, the Middle East and with the Far East Air Forces. The final single-seat Venoms, the FB.4s of 28 Squadron at Kai Tak, Hong Kong, were withdrawn in July 1962.

DE HAVILLAND VENOM FB.I

Type:	Single-seat fighter-bomber
First flight:	September 2, 1949; entered service November 1952
Powerplant:	One 4,850lb st (21.57kN) de Havilland Ghost 103 turbojet
Dimensions:	Span 41ft 8in (12.69m), Length 31ft 10in (9.7m)
Weights:	All-up 15,400lb (6,985kg)
Max speed:	597mph (960km/h) at 22,000ft (6,705m)
Armament:	Four nose-mounted 20mm cannon; up to 2,000lb (907kg) of bombs
Replaced:	Fighter-bomber de Havilland Vampires from 1952
Taken on charge:	525 - all single-seaters (Night-fighters: 744)
Replaced by:	Final phase-outs 1960-1962, by Hawker Hunters

MUSKETEERS

On July 26, 1956 Egyptian President Gamal Abdel Nasser nationalised the Suez Canal Company. This set off a crisis that turned into conflict, with Britain and France attempting to protect financial and strategic interests and with Israel motivated almost entirely by opportunism.

As relations decayed, large forces were mustered, among them three RAF Venom FB.4 units – 6, 8 and 249 Squadrons – arrived at Akrotiri, Cyprus. The jets were dispersed around the perimeter track under canvas, armed and placed on standby.

The Egyptian Air Force (EAF) had received large deliveries of modern Soviet hardware, including swept-wing Mikoyan-Gurevich MiG-15 *Fagots*, MiG-17 *Fresco* fighters and Ilyushin

CO Sqn Ldr Peter Ellis in WR404, were first off and headed for Kasfareet and Kabrit, arriving over them shortly after 6am. Ellis: 'We barrelled down with the first section, followed by the second section. We encountered only light flak.' Several MiGs were destroyed on the ground.

"Meanwhile a quartet of Venoms of 8 Squadron, headed by Sqn Ldr Joe Blyth in WR509, hit Abu Suier and Fayid. He recalled: 'We got a lot of MiGs – our four got 11. I got five of them, but then I was first in. I said no claim unless they went up in flame and smoke.'

"Targeting Kabrit, Sqn Ldr 'Jock' Maitland in WR499 led 249 Squadron, leaving six MiGs wrecked. One of these Venoms, WR504, was piloted by Fg Off Tony Gronert who

remembered: 'Our long dive worked out very well and we all had a good strike on aircraft parked neatly in lines on the dispersal.'

"Throughout the day, further waves of Venoms continued anti-airfield strikes with 249 'visiting' Fayid, Kasfareet and Kabrit. Along with his wing man, Fg Off Tom Lecky-Thomson destroyed six MiGs and a Vampire with cannon fire. Lecky-Thomson: 'I found aircraft camouflaged around the perimeter track and we had a field day.'

"During the first day the Venom squadrons flew over 100 sorties claiming 59 EAF aircraft, mainly MiGs, taken out on the ground; a dozen FB.4s suffered minor damage.

"The eradication of the EAF continued the following day with a further 44 aircraft destroyed during the morning. Later attacks were staged against the Egyptian army camp at Huckstep near Cairo where over 1,000 tanks and vehicles were spotted and attacked by Venoms and French Republic F-84F Thunderstreaks.

"Fg Off Dick Hadlow of 8 Squadron, pilot of WR405, noted: 'I still have vivid memories of bursts of flak ahead and above me. One shell was close enough to cause a large chip out of my windscreen.'

"Phase Two of Musketeer, on November 3 and 4, required strikes on Huckstep and targets of opportunity in preparation for the air and seaborne landings that began on the 5th. That morning the Venoms roamed over Egypt and several were damaged by flak.

"On the 6th Egyptian strongpoints were attacked, during which Maitland acted as 'Master Bomber', although the breaking weather proved troublesome. Pressure from the United States brought about a ceasefire on the 7th and withdrawal from the Canal Zone was not far behind."

INTO THE NIGHT

In the same manner that the single-seat Vampire grew into the two-seat NF.10 night-fighter, the Venom family was extended to include radar-carrying versions. Initiated as a private venture the prototype first flew on August 22, 1950. Development was protracted, and it was not until November 1953 that 23 Squadron at Coltishall took delivery of its first NF.2s.

Improved versions followed, the NF.2A dispensed with the 'bird cage'

canopy for a clear vision version and the electrics were enhanced. In June 1955 the definitive night-fighting Venom, the NF.3, was taken on charge by 141 Squadron at Coltishall.

At Stradishall, 89 Squadron relinquished the RAF's last frontline Venom NF.3s in November 1957. The service life had been brief, but changes in radar technology and aerodynamics had outstripped the small airframe's capacity to evolve. As page 82 explains, the RAF was about to receive its first purpose-built all-weather night-fighter, the mighty Javelin.

Above
Venom NF.2A WL873 of Waterbeach-based 253 Squadron; it served the unit from September 1955 to August 1957. It carries a squadron leader's pennant under the windscreen: the unit's boss was Wg Cdr Peter Finlayson. The unit's red and bright green opposing triangles with a silver dot in the middle, feature either side of the roundel on the tail boom.

Left
A quartet of Fassberg-based Venom FB.1s of 98 Squadron, late 1953. The aircraft are carrying the red 'Fassberg' flash on the nose. KEC

NORTH AMERICAN
SABRE
1953 TO 1956

There was considerable celebration at Leuchars in July 1954 when the 'Fighting Cocks', 43 Squadron, accepted the first Hawker Hunter F.1s into operational service. The swept-wing jets replaced the Gloster Meteor F.8s that the unit had been flying since 1950. In April the following year, 98 Squadron at Jever, in the Cold War frontline of West Germany, began replacing its de Havilland Venom FB.4s with Hunter F.4s – a giant leap in capability.

Development problems meant that the Hunter's service entry was slipping from the hoped-for 1952 or 1953 – see page 80. It was obvious Britain's air defence and its commitment to the North Atlantic Treaty Organisation would be outclassed if a conflict with the Soviet Union broke out. The Korean War had started in June 1950 and the advent of the swept-wing, transonic Mikoyan-Gurevich MiG-25 *Fagot* had caused shock waves across America and its allies.

On the other side of the 'Iron Curtain', the client states of the USSR had deployed MiG-15s and the more capable MiG-17 *Frescos* were gathering. During 1955 the MiG-19

Farmer, supersonic in level flight, entered service. Britain needed a stop-gap fighter and it turned to the MiG-15's adversary in the skies of Korea, the swept-wing, transonic, North American F-86 Sabre.

North American was committed to mass producing F-86s for the USAF, so the RAF's Sabres were manufactured by Sabre licensee Canadair at Cartierville, Montreal, Canada. Designated CL-13 by Canadair, the jets were the equivalent of the USAF's F-86E. To distinguish them from the machines destined for the Royal Canadian Air Force, they were known as Sabre Mk.2s and

Mk.4s and the RAF designated them F.2s and F.4 accordingly.

The first Sabre to join an RAF frontline unit was taken on charge by 3 Squadron in mid-May 1953. Part of the Second Tactical Air Force (2nd TAF) and based at Wildenrath, 3 Squadron took the RAF into the swept-wing era. In the UK, 66 Squadron at Linton-on-Ouse inaugurated the Sabre into Fighter Command in December 1953.

The Sabre interlude was brief; the 2nd TAF units – 71 Squadron at Brüggen and 234 at Geilenkirchen – retired the final examples in May 1956. The Hunter F.4, which had been declared a 'super-priority', had become available in large numbers.

ACROSS THE POND

To accelerate deliveries of the Canadian-built Sabres, Operation 'Becher's Brook' was established to 'convoy' the jets across the Atlantic. The origin of the codename dates back to 1839 and the first official Grand National steeplechase held at Aintree, Liverpool, when Captain Martin Becher fell from his horse, Conrad. With the rest of the field at full gallop, he sheltered from their hooves in the lee of the enormous fence that had toppled him. The captain later described this as "a formidable obstacle" and the fence was named in his honour. It seemed

XB958

an appropriate label for the regular, and urgent, ferrying of Sabres across 'The Pond'.

'Becher's Brook One' set off from St Hubert, Quebec, on December 9, 1952 with nine Sabres. The route took them to Goose Bay, Labrador; Bluie West 1, Narsarssuaq, Greenland; Keflavík, Iceland and Prestwick, Scotland, which was reached on December 19. In the case of the first sortie, the flight ended at Abingdon, but mostly they terminated at Kemble, the home of 5 Maintenance Unit (MU).

After all of the perils of the Atlantic, the risks involved at any stage were brought home when 31-year-old F/Sgt Alan Pugh died as XB534 plunged

into the ground a few miles from Prestwick. He became separated from two other Sabres also making an approach to land.

In March 1953 'Becher's Brook Two' took place, using the same routing: this time more than 30 Sabres participated. The last of the 'convoys' crossed the Atlantic in May 1954.

MUSEUM PIECE

This publication is highlighting the aircraft held by the RAF Museum and we'll take the opportunity to zoom in on its Sabre, which graces the National Cold War Exhibition at Cosford. First flown at Cartierville on April 16, 1943, XB812 was ferried across the

Atlantic as part of 'Becher's Brook Five', arriving at Kemble on June 3.

It was issued to 112 Squadron at Brüggen on January 29, 1954 and taken on charge by 93 Squadron at Jever on May 11. The Sabre was retired to 33 MU at Lyneham on June 1, 1955 and its brief RAF career ended. It was ferried to contractor Aviation Traders at Stansted three months later for refurbishing so that it could join another air arm under the US

Below
Framed by the structure of No.1 Hangar and a de Havilland Tiger Moth at Speke – Liverpool Airport – Sabre F.4s XB611 and XB542 in the care of contractor Airwork General Trading in mid-1956. Both were destined for the Yugoslav Air Force in 1957.

Mutual Defense Assistance Program.

The Sabre went to the Italian Air Force and it is known to have served from Pratica di Mare, south of Rome. By 1976 it was in use as an instructional airframe with a college in Rome where it was 'discovered' by The Fighter Collection (TFC) from Duxford which arranged for its acquisition and restoration for the RAF Museum. Returned to the colours XB812 wore with 93 Squadron, the Sabre was delivered to Hendon on January 31, 1994, moving to Cosford in January 2006.

In exchange for sourcing and restoring XB812, TFC received former 'gate guardian' Supermarine Spitfire V EP120. On September 12, 1995 EP120 took to the air again, civil registered as G-LFVB. The Mk.V is still based at Duxford and is a classic example of how the 'warbird' and museum communities can work together to achieve a 'win-win' situation. ✪

Above
The RAF's Museum's Sabre F.4 XB812 is dramatically exhibited within the National Cold War Exhibition at Cosford. It sports the colours of 93 Squadron's 'U-for-Uniform' that it wore in West Germany, 1954-1955.
RAF MUSEUM
www.rafmuseum.org

NORTH AMERICAN SABRE

Type:	Single-seat interceptor
First flight:	October 1, 1947; entered service May 1953
Powerplant:	One 5,200lb st (23.12kN) General Electric J47 turbojet
Dimensions:	Span 37ft 1in (11.3m), Length 37ft 6in (11.43m)
Weights:	Empty 10,000lb (4,536kg), All-up 16,500lb (7,484kg)
Max speed:	670mph (1,078km/h) at 35,000ft (10,668m)
Armament:	Six nose-mounted 0.5in machine guns
Replaced:	Interim equipment pending service entry of Hawker Hunter
Taken on charge:	460 – first three were F.2s, remainder F.4s
Replaced by:	Hawker Hunter F.4 from 1954

SUPERMARINE
SWIFT
1954 TO 1960

So much was expected of the first jet to enter operational RAF service from the hallowed Supermarine stable. Yet the Swift turned out to be part of the nightmare that brought about the hurried supply of stop-gap North American Sabres – see page 76.

The prototype Swift had its maiden flight on August 5, 1951. By that time the Air Ministry had ordered 150 as 'insurance' in case the Hawker Hunter – page 80 – turned out to be a turkey.

Understandably, there was considerable rivalry between the two companies. Operating from Tangmere, on September 7, 1953 Hawker test pilot Neville Duke took the modified Hunter prototype WB188 to 727.6mph (1,170.9km/h) to clinch the world speed record. Determined to steal Duke's laurels, Supermarine reacted quickly and in Libya on the 25th Mike Lithgow piloted Swift F.4 WK198 to 735.7mph.

Such was the nature of record-breaking in the 1950s that the ink was hardly dry on the Fédération Aéronautique Internationale's certificate when it was announced that just *two days* later a Douglas F4D-1 Skyray had seized the

record again for the USA, this time at 752.9mph. (Today, WB188 is displayed at Tangmere Military Aviation Museum, on loan from the RAF Museum and the fuselage of Lithgow's WK198 is at the Brooklands Museum, Weybridge.)

LOW DOWN PHOTO-RECCE

Swift F.1s entered service with 56 Squadron in February 1954. As the first British-built swept-wing jets in frontline service and pioneering powered-operated ailerons, the type generated a lot of interest but was soon to hit the headlines for the wrong reasons.

The initial F.1 and F.2 fighters were followed by a batch of F.3s which introduced reheat and eight further improved F.4s, of which WK198 basked in brief glory. A plan to build unarmed PR.6 photo-reconnaissance (PR) versions to supplant the Gloster Meteor PR.10 force was cancelled. A dozen F.7s served as guided missile test-beds.

With the permanent grounding of the F.1s and F.2s in 1955, the Swift was rejigged as the FR.5 tactical PR fighter to replace the Meteor FR.9. With a modified wing, incorporating a distinctive

'saw-tooth' leading edge, the FR.5 had a lengthened nose incorporating three cameras. Range was increased by a belly-mounted 220 imp gallon (1,000-litre) drop tank. With afterburning on its Rolls-Royce Avon 114 and flying at low and medium levels, the FR.5 became a potent asset with the Second Tactical Air Force in West Germany.

At Gütersloh, 2 Squadron introduced the photo-recce Swift to service, in March 1956. Three months later, 79 Squadron exchanged its Meteor FR.9s for Swifts at Wunstorf; it operated FR.5s until disbanding on New Year's Day 1961. In March 1961, 2 Squadron retired its hard-working Swifts; the last Supermarine frontline type to serve the RAF had bowed out.

FATEFUL 14 MONTHS

There can be no better way to describe the introduction of the Supermarine jet into service than to refer to *Swift Justice*, by Gp Captain Nigel Walpole (Astonbridge 2000). It is simply the best book ever written on the subject: Nigel flew FR.5s with 2 Squadron.

The first F.1, WK209, touched down at Waterbeach on February

Below
A trio of 2 Squadron Swifts FR.5 low down on a sortie out of Geilenkirchen, 1956-1957. The squadron 'bars', each side of the roundel were black with a white triangle superimposed. Leading is XD958 which served with 2 Squadron from June 1956 to March 1957 when it moved to 79 Squadron at Gütersloh. In making the transfer, XD958 gained the unique distinction of being the only Swift to serve with more than one frontline unit.

20, 1954 to be handed over to Sqn Ldr G J 'Twinkle' Storey's 56 Squadron. Nigel described the initial reaction to the new fighter: "Much may have been known already about its basic characteristics, but this did nothing to lessen the impact of its unique, bullish shape and the manner of its high-speed join, break and landing, and the audience was duly impressed. It may have been 56 Squadron's 'A' Flight Commander, Flt Lt 'Mac' McCaig, who christened it the 'Whistling Bullet'; sadly it would be called many other things in the next fateful 14 months."

As it worked up, 56 was visited by all sorts of 'brass', politicians and foreign dignitaries, but a string of incidents was soon to bring this to a close. To return to *Swift Justice*: "On May 7 Storey ejected safely from WK209, having entered a violent spin while practising stalls.

"Ominously, OC Flying Wing [Wg Cdr Mike Giddings] had some confusing aileron control problems in WK208 on May 3. He landed with some difficulty, but the fault could not be replicated on the ground and the aircraft was returned to service.

"Ten days later, Fg Off Neil Thornton, a young [he was 20] and well-liked first tour pilot on his second Swift sortie, had an aileron lock on take-off in the same aircraft [WK208]. It rolled out of control and hit the ground before Neil could complete his ejection and he was killed. This led to the grounding of all Swifts at Waterbeach pending modification to the control system selectors and warning indicators."

Flying resumed with four F.1s in August. On the 25th: "Fg Off John Hobbs had problems with WK213. After an uneventful training sortie, he re-joined the circuit to find that only the two mainwheels of his undercarriage would lock down. Re-selections and the use of emergency air not only failed to dislodge the nosewheel but rendered the main wheels unsafe, leaving John

with the choice of a wheels-up landing or ejection." He successful ejected.

Also in August, the first F.2s arrived and they: "proved a little more troublesome in the air and ten knots had to be added to

Left
Since 1995, the RAF Museum's Swift, FR.5 WK281, has been displayed on loan to the Tangmere Military Aviation Museum in Sussex. Built in late 1956 at South Marston, it joined 79 Squadron at Gütersloh in April 1959 and was still on charge when the unit disbanded on January 1, 1961. It joined the station collection at Colerne on March 14, 1967 and went on display at Hendon in 1989.
RAF MUSEUM
www.rafmuseum.org

the approach speed. Incidents abounded: Fg Off Al Martin suffered a flame-out on the approach to land caused by a crack in the casing of the engine's acceleration control unit, but managed to reach the runway undershoot."

In September Wg Cdr Giddings and Fg Off Martin suffered hydraulic failures – the latter two times. In early October Flt Lt 'Hoppy' Hoppitt had a similar incident, and the Swifts were again grounded. On returning to service, they were limited to 25,000ft (7,620ft), 550 knots and Mach 0.9 indicated. Serviceability remained poor and on March 15, 1955 the Air Staff pulled the plug on the Swift as a pure fighter. ◎

SUPERMARINE SWIFT FR.5

Type:	Single-seat day interceptor / fighter-reconnaissance
First flight:	August 5, 1951; F.1 entered service February 1954, FR.5 February 1956
Powerplant:	One 7,145lb st (31.78kN) Rolls-Royce Avon 114 turbojet
Dimensions:	Span 32ft 4in (9.85m), Length 42ft 3in (12.87m)
Weights:	Empty 13,435lb (6,094kg), All-up 21,673lb (9,830kg)
Max speed:	713mph (1,147km/h) at sea level
Armament:	Two nose-mounted 30mm cannon
Replaced:	Gloster Meteor F.8 from 1954 (F.1); Gloster Meteor FR.9 and PR.10 from 1956 (FR.5)
Taken on charge:	175, all versions
Replaced by:	Hawker Hunter from 1955 (fighter versions); Hawker Hunter FR.10 from 1961 (FR.5)

HAWKER
HUNTER
1954 TO 1994

WV269

Comparisons are dangerous things and there will be readers shocked at the suggestion that the Hunter was the Spitfire of its day. But there are so many similarities: a stunning looking machine, a long service history and an incredible number of roles, variants and exports. The two thoroughbreds also shared a dodgy start that threatened the programme's progress.

There was a lot riding on the Hunter; it was designed to be the spearhead of Britain's airspace and the guardian of the UK's interests worldwide for the bulk of the 1950s. Should the Hunter falter, the plan was that the Supermarine Swift would pick up the slack. The Air Ministry wished to hedge its bets still further, the Hawker jet was to have powerplant options; the preferred Rolls-Royce Avon and the back-up Armstrong Whitworth Sapphire. As shown on page 76, hasty acquisition of North American Sabres was needed in the light of the Swift's failure and the Hunter's delays.

Neville Duke first flew the Avon-powered

prototype Hunter, WB188, from Dunsfold, on July 20, 1951. It was followed ten months later by WB195, the F.1 development aircraft with provision for guns. The initial Sapphire example and prototype F.2, WB202, took to the skies in November 1952. Sapphire-powered Hunters were produced by Armstrong Whitworth at Bitteswell.

Much thought had been put into making the Hunter a formidable interceptor. The controls were all powered to make it very agile. Turn-arounds

were to be kept to a minimum by two 'firsts' for the RAF: a single-point refuelling system and a self-contained four-cannon pack that slotted in and out of a bay in the lower nose. The gun 'cassette' was a brilliant concept, but it was the main issue in the Hunter's tardy entry into full service.

HAWKER HUNTER F.6	
Type:	Single-seat day interceptor
First flight:	July 20, 1951; F.1 entered service July 1954
Powerplant:	One 10,000lb st (44.48kN) Rolls-Royce Avon 203 turbojet
Dimensions:	Span 33ft 8in (10.25m), Length 45ft 10½in (13.98m)
Weights:	Empty 14,400lb (6,531kg), All-up 17,750lb (8,051kg)
Max speed:	715mph (1,150km/h) at sea level
Armament:	Four nose-mounted 30mm cannon; up to 1,000lb (453kg)
Replaced:	De Havilland Venoms and North American Sabres from 1954 (F.1 and F.2)
Taken on charge:	1,073 - all versions; includes sub-contract to Armstrong Whitworth
Replaced by:	FGA.9 fighter-bombers gave way to Hawker Siddeley Harrier and McDonnell Phantom from 1969

ALL QUIET BEHIND

Much 'Western' and Soviet technology in the late 1940s and 1950s relied on, or was inspired by, captured German hardware, blueprints or 'boffins'. Such was the ADEN 30mm cannon that was derived from one developed for the Luftwaffe by the Mauser company. The British version was conceived by the Armament Development team at the Royal Small Arms Factory at ENfield, hence ADEN.

During the Hunter's early development, which was initiated in the summer of 1948, there was feverish debate about the need for guns at all. Unguided 'dumb' rockets were considered more efficient.

Left
Taken from a Gloster Meteor T.7, a quartet of Hunter F.4s of the North Weald-based 111 Squadron, 1955-1956. The unit flew Mk.4s from mid-1955 to November 1956 when it converted to F.6s which took on the famous 'Black Arrows' colours the following year as 'Treble-One' became the RAF aerobatic team.

Thankfully, the Hunter was allowed to swerve around this debate. The 'gun or no gun' argument has raised its head periodically all the way to the Eurofighter Typhoon – page 96 – which ironically packs a *Mauser* cannon.

ADENs were carried aloft for the first time on May 5, 1954; nearly two years after the first Hunter had flown. Royal Aircraft Establishment Bristol Beaufighter TF.10 RD388 returned from the initial firing trial with considerable damage, ominously caused by spent shell links hitting the airframe. Modified Gloster Meteor F.8 WK660 was dispatched to Canada in January 1955 for further tests.

Declared a 'super priority' programme by Churchill's Conservative government of the day, the first production F.1, WT555, flew in May 1953. It was christened *State Express* to reflect how much the industry was gearing up for mass production. Delays occurred when it was decided to install an air brake underneath the lower rear ⊙➤

Above
Built as an F.4, WV318 was converted into a two-seater T.7 in 1959. By the early 1960s it was serving with the Central Flying School, and illustrated visiting Leuchars in August 1965. It was one of the last-ever Hunters to fly for the RAF and was part of the ceremony at Lossiemouth in March 1994. Sold in early 1996, it became G-FFOX and flew in 111 Squadron colours.
ROY BONSER-KEC

attack to take over from obsolete DH Venoms. At Khormaksar in Aden, 8 Squadron accepted the first Mk.9s in January 1960. The F.6 was also adapted as the FR.10 for tactical reconnaissance.

From the mid-1960s, Hunters switched roles, becoming the workhorses of the operational conversion and tactical weapons units. The final examples, specially modified T.7As training Hawker Siddeley Buccaneer S.2 aircrews, were retired at Lossiemouth in March 1994. Thus the curtain came

Above
Blackpool-built F.4 WW658 of 98 Squadron, circa 1955. It has not had the 'Sabrinas' shell link collectors fitted under the nose. The squadron 'bars' were red with white zig-zags; these are repeated either side of the unit badge on the nose.

Right
Hunter XE556 was completed as an F.6, but was converted to FR.10 status for Jever-based 2 Squadron. Tac-recce Hunters replaced Supermarine Swift FR.5s with 2 Squadron from March 1961 until late 1970 when conversion to the McDonnell Phantom FGR.2 began. KEC

fuselage, but this was a simple 'fix' compared with the tribulations of the ADEN gun. During trials above 10,000ft (3,048m), pressing the gun button could result in it all going quiet behind the pilot as the Hunter's Avon 113 flamed-out. To compound the angst at Rolls-Royce and Hawker, this did not happen to Sapphire-powered examples.

At Leuchars, 43 Squadron accepted the first four F.1s in July 1954. Wartime DH Mosquito 'ace', Sqn Ldr Roy Lelong was to bring the Hunter up to readiness as soon as possible, while accepting some flying limitations as investigations on the gun issue continued. By October the 'Fighting Cocks' were declared operational, but with an interceptor that could not fire its guns safely.

After a lot of detective work, it was discovered that the gases emitted by the gun muzzles were still combustible as they flowed backwards, into the turbojet intakes. This surged the Avon, often to the state of flame-out – re-lights were possible, but the height loss could be nerve-racking. The answer lay in the Avon 115 and that sounded the death knell of the F.1.

The lessons learned from the Beaufighter in May 1954 resulted in a modification that ruined the Hunter's clean looks. While the shell cases of ADEN cannon flew clear, the links that formed the ammunition belt could cause considerable damage as they fell away.

Streamlined fairings under the forward fuselage collected the spent links. These bulbous additions became known as 'Sabrinas', after a well-endowed 1950s actress. The ADEN gun went on to be a great success; versions equipped the

Gloster Javelin, English Electric Lightning, HS Harrier and SEPECAT Jaguar.

FOUR DECADES

The Hunter F.4 was issued to 111 Squadron at North Weald in June 1955. This addressed the flame-out issue and improved the interceptor's range. In parallel, the Sapphire-powered F.5 was introduced.

These two variants paved the way for the ultimate Hunter pure fighter, the F.6, which was delivered to frontline units in October 1956. From the outside it was distinguished by a 'saw-tooth' on the leading edge of the wing, inside it was fitted with a much more powerful Avon 200 series. With the advent of the F.6, production was standardised and the Sapphire version was dropped.

A month after the first F.6 took off the T.7 two-seat advanced trainer also had its maiden flight: both types went into widespread squadron service.

With the Lightning due to come on strength in the early 1960s, withdrawn F.6s were converted into FGA.9s, optimised for ground

down on four decades of incredible service to the RAF and military aviation. Or had it?

From early in the type's long career Hunters were the darlings of the test organisations. Institutions such as the Royal Aircraft Establishment (RAF), the Empire Test Pilots School (ETPS) and the Aeroplane and Armament Experimental Establishment found its performance suitable for most trials, its airframe to be robust and easily modified along with ease of maintenance.

Although the test fleets are not technically RAF units, they are run completely on a military basis. They have always been staffed by personnel on secondment, or former aircrew and, of course, the aircraft proudly wear red-white-blue roundels.

Hunters in this guise extended the type's longevity still further. Built in 1956, Mk.6 XE601 was destined for a life devoted to trials work. Upgraded to FGA.9 status it made its last flight from Boscombe Down on July 15, 1999. It was disposed of in 2004, and exported to Canada in 2012.

"...specially modified T.7As training Hawker Siddeley Buccaneer S.2 aircrews were retired at Lossiemouth in March 1994. Thus the curtain came down on four decades of incredible service..."

After operational use that included 8 Squadron from Khormaksar in the late 1960s, T.7 XL612 joined the RAE in February 1970, moving on to ETPS at Boscombe Down five years later. This two-seater made the final flight of a British military operated Hunter on August 10,

2001 – designer Sydney Camm's incredible jet had spanned two centuries.

But the Hunter's qualities were still in demand. Established in 2000 and based at Scampton, civilian contractor Hawker Hunter Aviation flies a trio of former Swiss Air Force

F.58s and an ex-Fleet Air Arm T.8B on threat simulation and trials support missions for British air arms and the aerospace industry. These fly with military serials, camouflage and RAF 'low vis' roundels. That all adds up to 60-plus years of unbeaten loyalty. ◉

GLOSTER
JAVELIN

1956 TO 1967

Right
Navigator Flt Lt Pat O'Mahoney and pilot Flt Lt John Tritton in front of FAW.1 XA621 of 46 Squadron at Odiham, 1957. This unit introduced the Javelin into RAF service; XA621 served from April 1956 to March 1958.

Richard Walker became chief designer for Gloster in 1948 and his major assignment was the creation of the Javelin. Gone was the RAF designation prefix 'NF' for night-fighter, the Javelin was an 'FAW' – Fighter, All-Weather'. This was to demonstrate that it was for operation around the clock in all visibility conditions.

By the time the English Electric Lightning took over from the Javelin, it was taken for granted that the RAF's fighters could do their job no matter what the conditions. The prefix contracted back to the old-fashioned 'F-for-Fighter'.

The Javelin represented what later would be called a 'weapon system', a complicated mixture of aerodynamics, powered flying controls, sophisticated weaponry and radar.

GEORGE MEDAL

Gloster chief test pilot Bill Waterton carried out the first flight of the prototype, WD804, from Moreton Valence on November 26, 1951. During WD804's 99th flight on June 29, 1952 Waterton encountered elevator flutter but did not abandon it, bringing it back for a very risky forced-landing so that it could be examined for faults. He was awarded the George Medal for this heroic act.

Eight days after Waterton's horrific accident it was announced that the RAF was buying the Javelin in quantity under so-called 'super priority' status. Between 1953 and 1956 three Gloster test pilots lost their lives in accidents with Javelins.

On April 8, 1960 'Dickie' Martin was in command of FAW.8 XJ128. This was the 427th Javelin and the final maiden flight of the long line of Gloster fighters.

Upgrades to FAW.9 status occupied the company for a while, but in 1964 the Hucclecote factory was sold off and the famous name Gloster disappeared.

Right
Navigator Flt Lt Pat O'Mahoney and pilot Flt Lt John Tritton in front of FAW.1 XA621 of 46 Squadron at Odiham, 1957. This unit introduced the Javelin into RAF service; XA621 served from April 1956 to March 1958.

"In a 20th Century equivalent of 'gunboat diplomacy' 29's Javelin FAW.9s were to apply pressure while the diplomats got into gear."

ONE TO NINE

Development of the Javelin ran from the Mk.1 to Mk.9 and with the exception of the T.3 radar-less dual-control trainer, all featured radar or armament changes, upgraded engines, control and aerodynamic refinements. The Javelin brought about several 'firsts' in RAF service: the delta planform, air-to-air missiles and air-to-air refuelling capability via a fixed probe on modified FAW.9Rs.

First frontline unit with FAW.1s was 46 Squadron at Odiham in February 1956, while 23 Squadron became operational on FAW.9s at Coltishall in October 1960.

The plethora of variants has been regarded in some circles as indicative of a bad concept requiring constant 'fixes'. The design was complex to meet a challenging, and changing, specification and the Javelin went on to serve the RAF well.

The rolling series of improved versions was along the lines of American military aircraft programmes which were typified by the adage: "Wait till you see the B-model!" Each variant of the Javelin introduced new elements to Fighter Command incrementally, to the advantage of production and aircrew conversion. This step-by-step process was not new; the Hunter had gone through a similar evolution.

TROUBLE-SHOOTERS

After Rhodesia's unilateral declaration of independence (UDI) from the Commonwealth on November 11, 1965, Britain imposed economic sanctions. RAF Avro Shackletons were deployed to maintain a patrol of the waters off the port of Beira, Mozambique, to prevent sanction-busters.

out the ground crew and support equipment. After several stop-overs, the Javelins arrived at N'dola, on December 4. A detachment was set up at Lusaka, where crews sat it out on quick reaction alert duties.

Sanctions impacted heavily on the RAF presence: Zambia had previously got its fuel via Rhodesia. An airlift of fuel, for the Javelins, and to help the Zambian armed forces, was established by Lyneham-based 99 and 511 Squadron Bristol Britannias. Up to its cessation on the last day of October 1966, a staggering 3.5 million gallons had been flown in.

During the deployment, two FAW.9s were written off, thankfully without injury to the crews. On June 2, 1966 XH890 landed heavily at N'dola, burst a type and departed the runway. The Javelin fleet was close to phase-out and it was therefore uneconomic to repair. On the 27th XH847 was transiting from N'dola to Khormaksar, Aden, for routine maintenance. The starboard mainwheel locked on touch-down; it ran off the runway and was wrecked.

June 1966 was not a good month for the Javelin force. At Tengah, Singapore, 64 Squadron's XH709 lost control at 5,000ft and its crew successfully 'banged out'. The powered flying control unit actuating the tailplane had failed.

As was often the case with RAF assets, it fell to the Far East Air Force to be the last to operate the 'Flat Iron', as the Javelin was fondly nicknamed. At Tengah, 64 Squadron paid off the last examples in June 1967 as the fighter force became completely Lightning-shaped.

The last ever flight by a Javelin took place on January 24, 1975 when the Aeroplane and Armament Experimental Establishment's fabulously red-and-white painted FAW.9 XH897 was ferried from Boscombe Down to the Imperial War Museum at Duxford. ◉

Eight days after UDI, Wg Cdr K Burge, the commanding officer of 29 Squadron at Akrotiri on Cyprus, was told to put his crews on standby to deploy to Zambia, north of Rhodesia. In a 20th Century equivalent of 'gunboat diplomacy' 29's FAW.9s were to apply pressure, while the diplomats got into gear.

A trio of Akrotiri-based Handley Page Hastings of 70 Squadron flew

GLOSTER JAVELIN FAW.9

Type:	Two-seat all-weather fighter
First flight:	November 26, 1951; FAW.1 entered service in February 1956
Powerplant:	Two 11,000lb (48.92kN) Armstrong Siddeley Sapphire 205 turbojets
Dimensions:	Span 52ft 0in (15.84m), Length 56ft 9in (17.20m)
Weights:	Empty 38,100lb (17,282kg), All-up 43,165lb (19,579kg)
Max speed:	620mph (997km/h) at 40,000ft (12,192m)
Armament:	Two 30mm cannon, one in each wing; four de Havilland Firestreak air-to-air missiles under wing
Replaced:	Gloster Meteor night-fighters from 1956
Taken on charge:	427 - all versions, including sub-contract by Armstrong Whitworth
Replaced by:	English Electric Lightning from 1965

Above left
FAW.6 XA815 of Stradishall-based 89 Squadron with it barn door-like flaps down to keep it alongside the camera-ship. This aircraft served with 89 from October 1957 to September 1958 when it was re-numbered as 85 Squadron.

Left
As 72 Squadron completed its transition from Meteor NF.14s to Javelin FAW.4s at Leconfield, the opportunity was taken to celebrate the unit's Gloster heritage. In June 1959 a carefully choreographed formation was staged with Wg Cdr V G Owen-Jones DFC in FAW.4 XA755, Sqn Ldr I P W Hawkins in NF.14 WS724 and Gloster chief test pilot Geoff Worral in the company's Gladiator I 'K8032'. On November 25, 1960 Gloster presented the Gladiator to the Shuttleworth Collection at Old Warden, where it continues to fly.
GLOSTER-KEC

Above left
The RAF Museum's Javelin FAW.1 XA564 'flies' within the National Cold War Exhibition at Cosford. It was built in 1954 and did not serve operationally, becoming an instructional airframe in 1957; it joined the museum collection in September 1975.
RAF MUSEUM
www.rafmuseum.org

ENGLISH ELECTRIC
LIGHTNING
1970 TO 1993

971

Above right
Lightning T.4 XM971 was issued to 226 Operational Conversion Unit at Middleton St George in June 1963. The OCU moved to Coltishall in April 1964. During a flight over Norfolk on January 2, 1967 the radar 'bullet' in the nose collapsed and the engines ingested debris from it: both crew ejected successfully. PETE WEST

Right
Leuchars-based Lightning F.6s of 23 Squadron at play in 1968. 'M-for-Mike', XR754 was built in 1965 and served with 23 from February 1968 to November 1975.

"A manned fighter force, smaller than at present but adequate for this limited purpose, will be maintained and will progressively be equipped with air-to-air guided missiles. Fighter aircraft will in due course be replaced by a ground-to-air guided missile system." So wrote Duncan Sandys in his Defence White Paper of April 1957. We are still waiting for this prophesy to materialise.

A large number of projects fell by the wayside thanks to this Conservative government policy document which was presented to parliament on April 4, 1957. That very day, 200 miles (321km) from Westminster, Wg Cdr Roland Prosper Beamont climbed into English Electric (EE) P.1B XA847 at Warton and took it for a successful first flight. This was the true prototype of what was officially named Lightning in 1958.

Too far down the road to save any money by axing the programme, the Lightning survived the Sandys initiative.

Work on formulating what became the first single-seat British fighter to achieve supersonic speeds in level flight, began on the desk of W E W Petter in the summer of 1948. While working on the design of the Canberra, Petter came up with the extreme sweep, the ailerons taking the place of wing tips, and the positioning one engine above the other that characterised the P.1.

Petter left EE for Folland in October 1950. He may have *sketched* the Lightning, but F W 'Freddie' Page was the man who turned it into thundering reality.

Beamont had flown the first of two P.1As, WG760, on August 4, 1954.

"'Fighter aircraft will in due course be replaced by a ground-to-air guided missile system.' ...We are still waiting for this prophesy to materialise."

In more recent times, this would have been called a proof-of-concept aircraft because the P.1B was a considerable refinement. In a nod to the Hawker Hunter, the Lightning had a removable pack located in the bottom of the forward fuselage carrying a pair of Firestreak, or the later Red Top, air-to-air missiles.

The Lightning adopted a similar step-by-step evolution as the Gloster Javelin. The first F.1s were delivered to 74 Squadron at Coltishall in June 1960 – of which more anon. The interim F.2 appeared in December 1962. Having moved to Leuchars, 74 also inaugurated the most numerous Lightning variant, the F.3 into service in April 1964. Delivered from 1960 was the first two-seat trainer version, the T.4.
It was followed by the F.3-based T.5 in 1965.

The final RAF fighter version was the F.6 which was first issued to 5 Squadron at Binbrook in December 1965. The last RAF Lightning, F.6 XS938, had its maiden flight at Warton on June 30, 1967 and was delivered to Leuchars-based 23 Squadron on August 28. Production continued for Kuwait and Saudi Arabia.

TIGERS FIRST
Gleaming silver with black spines, a roaring tiger's head against a black fin – this was the colour scheme adopted

XM971

by 74 Squadron to spellbind crowds at the Farnborough airshow in September 1961. The audience was looking at the first Lightnings to enter frontline service and there were no less than nine of the arrow-like interceptors rolling in formation.

Known as the 'Tigers' since World War One, 74 joined the jet age at Colerne in June 1945, taking delivery of Gloster Meteor IIIs. It graduated to Mk.IVs in December 1947, followed by F.8s in October 1950. (The RAF adopted Arabic numerals for its designations in early 1948.) Hunter F.4s arrived in March 1957, introducing 74 to the swept-wing era. The Mk.4s were only interim; the first of the much more capable F.6s touched down eight months later.

In June 1959 the 'Tigers' moved to nearby Coltishall to prepare for the introduction of the Lightning, under Sqn Ldr John Howe, a year later. Complex to operate, demanding to maintain, but a quantum leap from previous fighters, the personnel of 74 Squadron had their work cut out honing tactics and ironing out technical glitches.

The Lightning was an ideal 'flag waver' and, as a new programme, likely to attract export orders. To add to its pioneering role, 74 was also tasked with making appearances at airshows at home and abroad. This led to the idea of an aerobatic team. That massed formation at Farnborough was the catalyst for 'The Tigers', by 1961 the RAF's principal display team.

In March 1964 the unit forsook Norfolk and display flying, heading north to Leuchars to convert to the Lightning F.3 and a period of intensive shakedown of the new variant. Two years later, 74 upgraded to the F.6 in readiness for another challenge.

In June 1967 the unit moved to Tengah, Singapore. Its 13 F.6s were supported by Handley Page Victor tankers along the route. With the draw-down of the RAF in the Far East, 74 Squadron was disbanded on August 25, 1971.

BOWING OUT

From May 1969 the McDonnell Phantom – page 88 – began the phase-out of the Lightning fleet. At Binbrook, 5 and 11 Squadrons were the last to relinquish the ultimate swept-wing fighter, in May 1988.

But Britain wasn't done with the Lightning. At the Lightning's birthplace, Warton, British Aerospace inheritors of EE's heritage had need of the type's performance. Four F.6s were used as 'targets' to help develop the Panavia Tornado fighter version's radar.

First to retire was XS928 in the spring of 1992; today it is proudly on static display within the Warton complex. On December 23, 1992 XP693 and XR773 were ferried to Exeter and in 1998 sailed for a new life in South Africa. The last ever flight by a Lightning in the UK took place on January 21, 1993 when Sqn Ldr Peter Orme AFC took XS904 to Bruntingthorpe to be cherished by the Lightning Preservation Group. ◎

ENGLISH ELECTRIC LIGHTNING F.3

Type:	Single-seat, all-weather interceptor
First flight:	August 4, 1954; F.1 entered service June 1960
Powerplant:	Two 11,250lb st (50.04kN) Rolls-Royce Avon 210 turbojets
Dimensions:	Span 34ft 10in (10.61m), Length 55ft 3in (16.84m)
Max speed:	1,390mph (2,236km/h) at 36,000ft (10,972m)
Armament:	Two nose-mounted 30mm cannon and two de Havilland Firestreak air-to-air missiles
Replaced:	Hawker Hunter F.6 from 1960, Gloster Javelin from 1965
Taken on charge:	254 – all versions
Replaced by:	McDonnell Phantom from 1969

I FEAR NO MAN

Above left
A pair of newly delivered 74 Squadron Lightning F.1s on a sortie out of Coltishall in late 1960. To the right is the uncoded XM165, to the left XM140 'M' has the unit's black and yellow 'Tiger stripes' painted on in the wrong order: XM165's are in the correct style.

Above
The badge of 74 Squadron with the head of a tiger and the motto 'I fear no man'.

Left
The RAF Museum displayed Lightning F.6 XS925 at Hendon. Built in 1967, it last served with 11 Squadron at Binbrook and flew for the final time on July 21, 1987. It moved to Hendon on April 28, 1988 and wears the pennant of Wg Cdr J C 'Jake' Jarron. Under the port wing root is the in-flight refuelling probe and it carries 260-gallon over-wing tanks. RAF MUSEUM www.rafmuseum.org

McDONNELL
PHANTOM
1969 TO 1992

A new era dawned on May 27, 1958 at St Louis, Missouri, when McDonnell XF4H-1 Phantom 142259 blasted off on its maiden flight. Ordered by the US Navy, it was also adopted by the US Air Force, entering service in December 1960 and January 1962, respectively. The XF4H-1 was the first of 5,195 built up to 1981, widely exported and still in service.

With the cancellation of the Hawker Siddeley P.1154 supersonic short/vertical take-off and landing strike fighter in February 1965, the RAF and the Royal Navy were in the market for something to take its place. The Phantom was the obvious choice and the Labour Government demanded at least 50% participation by the British aerospace industry.

So the 'Rolls-Royce Phantom' was born. The fitting of Spey turbofans involved considerable redesign of the rear fuselage and the air intakes, and the naval version had an extendable nose leg for take-off from carriers. As well as the Rolls-Royce involvement, the British Aircraft Corporation and Shorts became major sub-contractors.

Two versions were ordered, the interceptor with ground-attack capability FG.1 (USAF designation F-4K) and the ground-attack/tactical reconnaissance FGR.2 (F-4M). The first British machine was YF-4M XT852, which became airborne on February 17, 1967.

In May 1969, Coningsby-based 6 Squadron became the first operational RAF unit, with FGR.2s. Four months later, 43 Squadron inaugurated the FG.1 into frontline use at Leuchars. From 1974, the SEPECAT Jaguar – page 92 – began to take over the ground-attack role and the Phantom force became dedicated to air defence. It fell to 74 Squadron at Wattisham to retire the last Phantoms in October 1992, giving way to Panavia Tornado F.3s – page 94.

HOT-ROD
After the Falklands conflict, April to June 1982, the decision was made to base air defence Phantoms at Port Stanley: 29 Squadron deploying in October. This left a gap in the UK's defensive cover and 15 former US Navy F-4Js were purchased to equip 74 Squadron, which re-formed at Wattisham in October 1984.

Designated F-4J(UK) in RAF service, the 'Jay' was powered by 17,900lb (79.6kN) thrust General Electric J79 turbofans. This was a 'hot-rod' with a maximum speed of 1,584mph (2,549km/h) at 48,000ft (14,630m), so was around 300mph faster than the Spey-models. Postings at 74 Squadron became very much in demand! The stop-gaps fighters were withdrawn in early 1991.

ALCOCK AND BROWN
Phantom FGR.2 XV424 touched down at Aldergrove – now Belfast International Airport – on February 12, 1969 at the end of its transatlantic delivery. It was

destined to cross 'The Pond' twice more. Having been prepped for service, XV424 flew to Coningsby in April and was one of ten FGR.2s on strength with the RAF's first unit, 6 Squadron, when it stood up on May 6. In early 1972, XV424 left 6 Squadron.

Over the next six years, XV424 served, in turn, with the following units: 54 Squadron, 228 Operational Conversion Unit (OCU), 6, 41, 6, 29 and 111 Squadrons. It returned to Wattisham in December 1978, and joined 56 Squadron.

During the spring of 1979, squadron personnel decided to commemorate the 60th anniversary of the first direct transatlantic flight – 56 was uniquely qualified for this. Between June 14-15, 1919 Captain John Alcock DSC and Lt Arthur Whitten Brown had flown a distance of 1,890 miles (3,041km) in 15 hours 57 minutes in a Vickers Vimy. The crossing was made from St John's, Newfoundland, to Clifden, Ireland. (The Vimy is on display at

the Science Museum in London.)

A special colour scheme was devised by artist Wilf Hardy and applied to XV424, as the primary, and XV486, as the reserve. The Phantoms deployed to Goose Bay, Labrador, ready for the anniversary flight, which took place on June 21.

On board was a crew of three. The pilot was Sqn Ldr A J N 'Tony' Alcock MBE, a flight commander on 56 and nephew of Sir John Alcock (he and Arthur Brown had been

Above
Phantom FG.1 XV574 of 43 Squadron, Leuchars, 1972. This machine served with the unit from September 1969 to April 1977. © ANDY HAY www.flyingart.co.uk

Left
The Alcock and Brown commemorative Phantom FGR.2 XV424 on display at the September 1979 'Battle of Britain At Home' display at Coningsby. KEC

Left
On display at the RAF Museum Hendon since 1992, FGR.2 XV424 in the colours of 56 Squadron. RAF MUSEUM www.rafmuseum.org

"This was a 'hot-rod' with a maximum speed of 1,584mph at 48,000ft, so was around 300mph faster than the Spey-models. Postings at 74 Squadron became very much in demand!"

knighted for their 1919 exploit). Flt Lt Norman Browne, a Buccaneer navigator, had been temporarily seconded to 56 and was forgiven for the 'e' at the end of his surname.

And the third member? A toy black cat called Twinkletoes. This mascot had been carried on the 1919 adventure, so was regarded as knowing the way!

Alcock and Browne were somewhat faster than the Vimy, taking 5 hours 40 minutes for the crossing. In that time, XV424 topped up five times from Handley Page Victor K.2 tankers of 57 Squadron.

In November 1979, the special colour scheme was removed and XV424 – by then with 228 OCU – became the first RAF Phantom to wear the air defence grey colour scheme. Between 1980 and 1988, XV424 served mostly with the OCU, apart from brief spells with 29 Squadron at Coningsby (1980) and 92 Squadron at Wildenrath, West Germany (1985-1986).

Issued to 56 Squadron at Wattisham in March 1988, XV424 embarked on its last operational sortie on July 13, 1992 and, on the last day of the month, flew for the final time. It was taken by road to the RAF Museum at Hendon on November 12, 1992 and has been turning heads there ever since. ◎

McDONNELL PHANTOM FGR.2

Type:	Two-seat, all-weather fighter with strike and reconnaissance capability
First flight:	XF4H-1 May 27, 1958, YF-4M (FGR.2) February 17, 1967; entered service May 1969
Powerplant:	Two 12,250lb st (54.48kN) Rolls-Royce Spey 202/203 turbofans
Dimensions:	Span 38ft 4in (11.68m), Length 57ft 7in (17.55m)
Weights:	Empty 31,000lb (14,061kg), All-up 58,000lb (26,308kg)
Max speed:	1,386mph (2,230km/h) at 40,000ft (12,192m)
Armament:	One nose-mounted 20mm cannon; four Sky Flash or Sparrow air-to-air-missiles under the centre section and four pylon-mounted Sidewinder air-to-air missiles
Replaced:	English Electric Lightning from 1969
Taken on charge:	185 - all versions, including Royal Navy usage
Replaced by:	SEPECAT Jaguar from 1974 and Panavia Tornado F.3 from 1987

HAWKER SIDDELEY
HARRIER GR

1969 TO 1990

squadron personnel at Wittering thinking and they announced a new motto: 'Twice Vertical'.

As the Pegasus grew in power, so Harriers were upgraded to GR.1A status with Mk.102 engines and the definitive first-generation variant, the GR.3, was fitted with a Pegasus 103. It also carried a laser ranging and marked-target seeker in an elongated nose.

D ominating the heritage of the Harrier is its part in the Falklands conflict. This was certainly the type's 'finest hour', but during its development and squadron service there were many other exploits where superlatives could be heaped on the diminutive 'jump jet'.

The P.1127 prototype, XP831, was a private venture by Hawker utilising the exceptional Bristol Siddeley Pegasus vectored thrust turbofan. Nozzles located in tandem on the fuselage sides rotated from the aft position, providing thrust in a conventional manner, to downwards, enabling the aircraft to hover.

Flight testing started on October 21, 1960 with tethered hovers at Dunsfold. Test pilot 'Bill' Bedford carried out a careful, step-by-step approach. The P.1127 was roaded to the Royal Aircraft Establishment at Thurleigh and on March 13, 1961 Bedford carried out the type's inaugural sortie as a conventional jet. He flew it back to Dunsfold on the 25th.

As the hovers increased in complexity and piloting experience grew, on September 12, 1961 Bedford took XP831 from a vertical take-off to a conventional landing and from a 'normal' take-off to a vertical landing. The V/STOL fighter concept was proven!

(The prototype P.1127, part of the RAF Museum collection, is on display at the Science Museum in London.)

The next phase was the proof-of-

HAWKER SIDDELEY HARRIER GR.3	
Type:	Single-seat tactical fighter/ground attack with reconnaissance capability
First flight:	September 12, 1961 (P.1127, first full transition); GR.1 entered service July 1969
Powerplant:	One 21,500lb st (95.63kN) Bristol Siddeley Pegasus 103 vectored-thrust turbofan
Dimensions:	Span 25ft 3in (7.69m), Length 46ft 10in (14.27m)
Weights:	Empty 13,300lb (6,032kg), All-up 23,000lb (10,432kg)
Max speed:	730mph (1,174km/h) at sea level
Armament:	Two 30mm cannon under centre section; up to 5,000lb (2,268kg) of ordnance on centre-line and wing pylons
Replaced:	Hawker Hunter FGA.9s from 1969
Taken on charge:	138 - all versions GR.1 to T.4
Replaced by:	SEPECAT Jaguar from 1977, British Aerospace/McDonnell Douglas Harrier GR.5 from 1987

concept Kestrel FGA.1, which first flew on March 7, 1964. This was evaluated by RAF, West German and US Air Force pilots by a special unit at West Raynham.

Looking very much like its forebears, the Harrier GR.1 was a much more sophisticated aircraft, the product of nearly a decade of evolution. In July 1969 at Wittering, 1 Squadron became the world's first operational V/STOL strike fighter unit.

Formed at Farnborough on May 13, 1912 as a Royal Flying Corps unit, 1 Squadron's initial equipment were observation balloons taken over from the Royal Engineers. This started

From 1969 the first operational conversion trainer, the T.2, began testing and it was followed by the T.4, the two-seat equivalent of the GR.3.

MOMENTOUS DAYS
Four Harrier GR.3s of 1 Squadron landed on the deck of HMS *Hermes* as it steamed south, bound for the Falkland Islands, on May 18, 1982. They had taken off from the *SS Atlantic Conveyer* which had brought the jets from Ascension Island to a rendezvous with the aircraft carrier.

First on board was 1 Squadron's 'Boss', commanding officer Wg Cdr Peter Squire in XZ972, followed

L TO T.4

XZ129

by Flt Lt John Rochfort in XV789, Sqn Ldr Peter Harris in XZ988 and Sqn Ldr Jerry Pook in XZ997. (This last Harrier is part of the RAF Museum's collection and is displayed at Hendon.)

The story of XZ998's war is a dramatic example of those crowded and momentous days. The Harriers were soon in action; Pook, flying XZ998, and Flt Mark Hare set off at daybreak on the 21st on an armed reconnaissance of a suspected helicopter base near Mount Kent, west of Port Stanley. The pair found helicopters parked out and Hare used his 30mm ADEN cannon to good effect, destroying a Boeing Chinook and badly damaging a Sud Puma.

That afternoon Harris took XZ988 on an uneventful recce of the San Carlos area, on the western shore of East Falkland.

On the 23rd, Harper was flying XZ988 as part of a four-ship sent to render an airstrip at Dunnose Head, on West Falkland, unusable. The Harriers attacked with 1,000lb (453kg) bombs, but caused little damage.

Next day, four Harriers with cover from a pair of Fleet Air Arm Sea Harrier FRS.1s raided the runway at Stanley with parachute retarded 1,000-pounders. The 'Boss' was piloting XZ988 in the second wave, but the runway remainder operative.

A return visit was made around 15:00 hours on the 25th with Pook in XZ988; again the runway held firm. Quickly turned around, XZ988 was heading back for Stanley at 16:30, this time with Harris at the helm. Results remained insubstantial.

At 16:12 on May 27 Sqn Ldr Tony Iveson was at the controls as he XZ998 shot up and off the 'ski jump' bow of *Hermes*. Armed with cluster bomb units (CBUs), he and Hare, were off to strike artillery and troop concentrations close to Goose Green, in the middle of East Falkland. They were supporting British paratroop forces on the long 'yomp' to Stanley.

Having dropped their CBUs on enemy soldiers, the pair returned for a strafing run. Iveson's aircraft was hit by anti-aircraft fire and he

was forced to part company with the blazing XZ988. The Harrier plummeted into the ground and exploded, Iveson floated down to a soft landing.

Using his evasion training, Iveson was able to stay well clear of Argentine troops and was rescued by a Royal Marines Westland Gazelle AH.1 helicopter on the 30th. He was flown to *Hermes* on a Fleet Air Arm Westland Sea King on June 1.

SECOND GENERATION

Rethought, redesigned and enlarged by a joint McDonnell Douglas and British Aerospace team, the Harrier GR.5 first appeared in 1985 and the GR.3 fleet began to be phased out. The Mk.5's usage was that of strike fighter and later still as a precision bomber.

Due to form up at Marham in the summer of 2018, the RAF's first operational Lockheed Martin F-35B will be 617 Squadron which will take over the Panavia Tornado GR.4's, strike role. Accordingly, the 'big wing' Harriers and the 'Lightning II' can be found in the sister journal on RAF bombers. ◉

Above
Harrier GR.3 XZ129 of 3 Squadron, based at Gütersloh, West Germany, 1978.
PETE WEST

Below
A 1 Squadron GR.1 during off-airfield trials in 1970. Straw bales have been arranged to create a makeshift blast pen.

SEPECAT JAGUAR
1974 TO 2007

Collaboration with France was not limited to the Concorde supersonic airliner project, at Warton the British Aircraft Corporation was hard at work with Bréguet on the Jaguar attack and advanced trainer project. The organisation established to oversee the co-operative venture in 1966 basked in the name Société Européenne de Production de l'Avion d'Ecole de Combat et d'Appui Tactique, unsurprisingly the abbreviation SEPECAT rolled much better off British tongues.

At first the Jaguar was intended to replace the Folland Gnat T.1 and Hawker Hunter T.7 in the advanced trainer role, but this was changed when the incredible Hawker Siddeley Hawk was ordered in 1972. Instead the Jaguar was developed as a ground-attack platform, enabling the McDonnell Phantom force to concentrate on air defence.

Bréguet chief test pilot Bernard Witt flew the first prototype, E.01, from Istres, France, on September 8, 1968. 'Jimmy' Dell piloted the first British-assembled single-seater, XW560, on October 12, 1969 taking it supersonic during the 50-minute sortie.

Jaguars excelled in the strike role and their systems and armament were upgraded throughout their operational lives, ending up with the GR.3A variant. On August 30, 1971 the prototype T.2 two-seater with attack capability had its maiden flight. Converting from Phantom FGR.2s, 54 Squadron took the first operational Jaguar GR.1s into RAF service at Lossiemouth in March 1974, moving five months later to Coltishall. The last unit to fly the Jaguar was 6 Squadron, disbanding at Coningsby in 2007.

WESTBOUND CARRIAGEWAY
The Jaguar was designed to operate from 'off-airfield' strips, even from grass, and in 1975 British Aerospace found a way to vividly demonstrate this. The second production GR.1, XX109, had its maiden flight from Warton on November 16, 1972. After evaluation at the Aeroplane and Armament Experimental Establishment at Boscombe Dow it was back with the trials fleet at its birthplace by March 1975.

By that time construction of

"They were in the thick of it from the first day when a four-ship formation hit an Iraqi army encampment: the jets were later to attack elements of the Republican Guard."

the M55 motorway, running from Preston to Blackpool, was drawing to a close. Warton's Jaguar marketing department couldn't miss the opportunity to prove that such an 'improvised' airstrip was perfect for the new strike fighter.

On April 26, 1975 deputy chief test pilot Tim Ferguson landed XX109 on the westbound carriageway. Apart from his groundcrew, he was greeted by a phalanx of reporters and plenty of public, many of whom had trudged a long way across fields to witness the event.

In a 'bombing-up' demonstration, XX109 was fitted with a quartet of 1,000-pounders. With after-burners on, Tim rolled under the bridge carrying the Weeton to Wrea Green road and took off, returning to Warton. (This Jaguar is on show at the City of Norwich Aviation Museum at Norwich Airport.)

THE THICK OF IT

Jaguars were the first RAF frontline aircraft to deploy to the Gulf as part of Operation 'Granby' following the Iraqi invasion of Kuwait. The Coltishall Wing (6, 41 and 54

Squadrons) and personnel from 226 Operational Conversion Unit, Lossiemouth, formed 'JagDet' under Wg Cdr Bill Pixton at Thumrait, Oman, and Muharraq, Bahrain, from August 1990.

To provide self-protection, the Jaguars carried AIM-9L Sidewinder air-to-air missiles on over-wing pylons thus maintaining the maximum number of under-wing and centre-section stations for offensive weaponry. In 611 sorties and 920 flying hours, the Jaguars dropped 1,043 bombs and fired 9,600 rounds of 30mm cannon shells.

JagDet's role during First Gulf War was classic ground-attack or battlefield air interdiction in 1990s parlance. They were in the thick of it from the first day January 17, 1991 when a four-ship formation hit an Iraqi army encampment: the jets were later to attack elements of the Republican Guard.

Targets varied, on the 26th the Jaguars neutralised a Chinese-

supplied 'Silkworm' coastal defence missile site. Four days later a large landing craft was sunk. By March 13, the Jaguars were back at what had become the type's ancestral home, Coltishall.

The Jaguar force was to see more action. Working from Gioia del Colle in Italy, the jets took part in Operation 'Grapple' from 1992 and later 'Deliberate Force', both over the skies of Bosnia. Starting in 1993 Jaguars were back in Iraqi airspace, flying from Incirlik, Turkey. This was Operation 'Warden', later 'Northern Watch', patrolling norther Iraq to prevent strikes against Kurd settlements. This was not at all bad for an aircraft originally intended as an advanced trainer. ◎

Above
Based at Brüggen, West Germany, GR.1 XX746 flew with the resident 17 Squadron from August 1982 to June 1984. It is currently in use as an instructional airframe at Cosford. © ANDY HAY www.flyingart.co.uk

Below left
Within the 'Test Flight' exhibition at the RAF Museum, Cosford, is the much-modified Jaguar GR.1 XX765, Active Control Technology test bed. It was the world's first to fly with an all-digital quadruplex 'fly-by-wire' control system, on October 21, 1981 with Chris Yeo at the helm. KEN ELLIS

SEPECAT JAGUAR GR.IA

Type:	Single-seat strike fighter
First flight:	October 12, 1969 (British prototype S-06); entered service
Powerplant:	Two 7,300lb st (32,47kN) Rolls-Royce/Turboméca Ardour 104 turbofans
Dimensions:	Span 28ft 6in (8.68m), Length 55ft 2½in (16.82m)
Weights:	Empty 15,432lb (6,999kg), All-up 34,621lb (15,704kg)
Max speed:	990mph (1,593km/h) at 36,000ft (10,972m)
Armament:	Two nose-mounted 30mm cannon; provision for up to 10,500lb (4,762kg) of ordnance on centre-line pylon and two pylons on each wing
Replaced:	McDonnell Phantom from 1974, Hawker Siddeley Harrier from 1977
Taken on charge:	203 - all versions
Replaced by:	Panavia Tornado, phase-out completed 2007

PANAVIA
TORNADO ADV

1984 TO 2011

Right
A pair of 29 Squadron F.3s inside a hardened aircraft shelter at Coningsby in April 1987 – the month the unit became operation on the new type. KEC

Below right
A stack of F.3s over their Coningsby base, top to bottom: 229 Operational Conversion Unit, 29 Squadron and 5 Squadron. BRITISH AEROSPACE-GEOFF LEE

The prototype of the interim Tornado F.2 version was first flown on October 27, 1979 and a small number began to enter service with 229 Operational Conversion Unit (OCU) at Coningsby in November 1984. The Foxhunter radar was subject to delay and, for a while, the F.2s were not fitted with radar, carrying ballast instead. This 'modification' was nicknamed 'Blue Circle' to sound like a top-secret device, but was actually from the well-known brand of cement reputed to be in the nose!

One of the most important aircraft in recent RAF heritage is the Tornado, the Anglo-German-Italian swing-wing strike aircraft, now in the twilight of its long career. The British aircraft industry referred to this machine as the IDS, standing for Interdictor/Strike. This was to distinguish it from a version unique to the British part of the tri-national programme – the ADV, air defence variant.

(The Tornado IDS appears in the sister publication on RAF Bombers, details on page 2.)

The fighter Tornado was to replace the English Electric Lightning and McDonnell Phantom and was conceived as a stand-off interceptor, with the ability to loiter up to 350 miles (563km) away from its base to tackle incoming Soviet bombers long before they got close to British airspace. Capable of carrying long range air-to-air missiles, the ADV was to acquire and attack multiple targets using the advanced GEC-Marconi Foxhunter radar. A retractable in-flight refuelling probe was housed in the port side of the forward fuselage.

In the ADV role, the F.3 could receive datalink information on targets from patrolling Boeing

"...for a while, the F.2s were not fitted with radar, carrying ballast instead. This 'modification' was nicknamed 'Blue Circle' to sound like a top-secret device but was actually from the well-known brand of cement reputed to be in the nose!"

Sentry AEW.1 airborne early warning platforms. An anti-radar role was introduced so that F.3s can pass information on an opponent's radar site back to the Sentry or ground stations for onward relay to other aircraft or ground forces.

COUNTER-AIR

Fully operational F.3s joined 29 Squadron at Coningsby in April 1987. Unlike the Phantoms before, the Tornado interceptor fleet saw action as the final decade of the 20th century erupted into turmoil.

For Operation 'Granby', the RAF contribution to 'Desert Shield/ Storm', the Coningsby and Leuchars squadrons deployed throughout the conflict, with the first deployment arriving at Dhahran, Saudi Arabia, in August 1990. Along with Royal Saudi Air Force Tornado ADVs and McDonnell Douglas F-15 Eagles, the RAF F.3s mounted a round-the-clock combat air patrol over northern Saudi airspace.

There was only one encounter, on January 18, 1991 when a pair of Iraqi Dassault Mirage F1s turned and high-tailed it back to base. The F.3 detachment's last 'op' took place on March 8 and the Tornados were back in the UK on the 13th.

PANAVIA TORNADO F.3

Type:	Two-seat stand-off interceptor
First flight:	October 27, 1979 (F.2), F.3 entered service April 1987
Powerplant:	Two 16,250lb st (72.28kN) Turbo-Union RB.199-34R Mk.104 turbofans
Dimensions:	Span (extended) 45ft 7in (13.89m), Length 59ft 3in (18.06m)
Weights:	Empty 31,500lb (14,288kg), All-up 50,700lb (22,997kg)
Max speed:	1,480mph (2,381km/h) at 40,000ft (12,192m)
Armament:	One nose-mounted 27mm cannon; four Skyflash or AMRAAM air-to-air missiles under the centre section and four Sidewinder air-to-air missiles on two under wing pylons
Replaced:	McDonnell Phantom from 1987
Taken on charge:	162 - F.2 and F.3
Replaced by:	Eurofighter Typhoon from 2003

The force was kept busy working on Operation 'Bolton', later 'Southern Watch', monitoring airspace over southern Iraq, 1998 to 2003.

The F.3s were back in the Middle East for the Second Gulf War – Operation 'Telic' – March to May 2003 and its aftermath, British forces finally pulling out in May 2011. During this time, Tornado F.3s of 43 and 111 Squadrons were involved in counter-air operations, to deny the airspace to non-friendly incursions.

All Tornado F.3 operations ended on March 31, 2011 with the Eurofighter Typhoon force taking up the role. ◎

EUROFIGHTER
TYPHOON

FROM 2003

Top right
Two-seat Typhoon T.3 ZJ815 was delivered to Coningsby on November 29, 2007 for use with 17 Squadron. Its job complete, by late 2016 it had joined the Reduce To Produce programme at Coningsby for older Typhoon airframes.
PETE WEST

Right
In April 2017, Typhoon FGR.4s of 3 Squadron from Coningsby deployed to Mihail Kogalniceanu air base near Constanta in Romania. The operation was set up and managed by 135 Expeditionary Air Wing. The detachment was completed in September and, during that time, the RAF personnel, their hosts and other NATO allies worked to secure airspace in the Black Sea region.
CROWN COPYRIGHT -SGT NEIL BRYDEN

Alongside the banks of the River Ribble, a prototype fighter screamed into the air from the British Aerospace airfield at Warton. That machine was DA.2 ZH588; the first British Eurofighter 2000, later named Typhoon, the date was April 6, 1994.

Today, Typhoons continue to blast off from the same runway for the RAF and export customers and look set to do so for years to come. With total 'fly-by-wire' technology controlling all the flying surfaces, including the distinctive canard foreplane, some functions voice-actuated or by the hands-on-stick-and-throttle system, the pilot has command of an exceptionally advanced aircraft.

Piloting DA.2 that day in 1994 was Chris Yeo. He and everyone else around him could not be aware that it was to be the *last* maiden flight of any significant from-new *manned* British aircraft programme to the present day. No such event was envisaged within at least the next decade, if at all.

The four-nation Eurofighter consortium – Germany, Italy, Spain and the UK – was established in June 1986, to create and produce a highly agile, twin-engined advanced air superiority fighter. Each nation would have an assembly line for its own Typhoons while work for export examples to date have flown from Warton, but the order placed by Kuwait in 2016 will be met from the Italian line at Caselle, Turin.

The first aircraft, DA.1, flew on March 27, 1994 from Manching, Germany. This and the British-assembled DA.2 were powered by Turbo Union RB.199-34R-104Es turbofans, similar to those in the Panavia Tornado. The first example with the intended Eurojet EJ200 was DA.4 ZH590, which was flown from Warton in March 14, 1997. (DA.2 is displayed at the RAF Museum, Hendon, DA.4 at the Imperial War Museum, Duxford.)

Initial versions for the RAF were the interim T.1 two-seat advanced trainer and the F.2, which was optimised as an

EUROFIGHTER TYPHOON F.2

Type:	Single-seat air superiority and strike aircraft
First flight:	March 14, 1997 (EJ200-powered prototype); entered full squadron service July 2005
Powerplant:	Two 13,490lb st (90kN) Eurojet EJ200 turbofans
Dimensions:	Span, 35ft 11in (10.95m), Length 52ft 4in (15.96m)
Weights:	Empty, 24,240lb (10,995kg), All-up 46,297lb (21,000kg)
Max speed:	Mach 1.8 in level flight, service ceiling 55,000ft (16,764m)
Armament:	13 store stations, four AIM-120 AMRAAM air-to-air missiles under the centre section. 27mm Mauser cannon in starboard forward fuselage. Provision for Paveway bombs and a range of stand-off missiles
Replaced:	Panavia Tornado F.3 from 2003
Taken on charge:	160 ordered
Replaced by:	In service

interceptor. All of these were upgraded to T.3 (from T.1) and FGR.4 (from F.2) status. During negotiations relating to the British order, there was debate about the merits of fitting a cannon, a 27mm Mauser, but this was retained. The Mk.4 is referred to as 'swing-role' by the RAF, having the ability to adopt air superiority or precise strike duties.

WARPLANE

First unit to receive Typhoons was the evaluation unit 17 Squadron – see below. At Coningsby, 11 Squadron was formally activated on the type on July 1, 2005. The Typhoon took over responsibility

for British quick reaction alert (QRA) on June 29, 2007.

The turbulent politics of the 21st century meant the RAF's latest warplane was soon in action. Typhoons were deployed to Gioia del Colle, Italy, on March 20, 2010 to take part in Operation 'Ellamy', enforcing a no-fly zone over Libya. They carried out the type's first frontline patrol the following day.

As the Libyan situation worsened, on April 12, 2011 Typhoons dropped 1,000lb (453kg) Paveway II precision-guided bombs on an armoured vehicle parking area. Flying from Akrotiri, Cyprus, a detachment of Typhoons turned bomber yet again from

ZJ815

December 3, 2015. This time the objectives were in Syria or northwest Iraq, the targets were elements of the so-called Islamic State.

TWICE FIRST

Deliveries of the first T.1s and Tranche 1 F.2s to the RAF first Typhoon unit, 17 Squadron, in 2003 were remarkably cost-effective. While detailed development and testing were carried out, 17 was temporarily based at BAE Systems Warton, so that manufacturer and operator could maximise co-operation by sharing the same ramp.

Re-locating by May 19, 2005 to Coningsby, 17 Squadron was officially re-formed as the Typhoon Operational Evaluation Unit. With this experience behind its personnel, the unit was destined to be the first to operate the next cutting-edge aircraft to join the RAF, the Lockheed

Martin F-35B Lightning II. (Details of the F-35B appear in our sister journal on RAF bombers.)

Established in this role from April 12, 2013 at Edwards Air Force Base, California, USA, 17 became a joint RAF-Royal Navy Test and Evaluation Squadron. Collaborating with the US Air Force, Marine Corps and Navy, 17's task is to develop and hone how the fifth-generation combat air platform will be used when it enters full service with 617 Squadron at Marham during the summer of 2018. The Operational Conversion Unit, 207 Squadron is scheduled to be activated at the Norfolk airfield on July 1, 2019. 🎯

The second frontline Eurofighter Typhoon unit was
11 Squadron, which re-formed at Coningsby on March
29, 2007 with F.2 air defence versions. (An example
is illustrated visiting Abu Dhabi in 2011.) Typhoons
took over responsibility for United Kingdom quick
reaction alert interception on June 29, 2007. Today,
11 Squadron operates the FGR.4 'swing-role'
Typhoon and the unit spearheads development
of the type's air-to-surface capability.
COPYRIGHT EUROFIGHTER-K TOKUNAGA